VOLUME ONE HUNDRED AND TWELVE

ADVANCES IN PROTEIN CHEMISTRY AND STRUCTURAL BIOLOGY

Therapeutic Proteins and Peptides

Academic Press is an imprint of Elsevier
50 Hampshire Street, 5th Floor, Cambridge, MA 02139, United States
525 B Street, Suite 1800, San Diego, CA 92101-4495, United States
The Boulevard, Langford Lane, Kidlington, Oxford OX5 1GB, United Kingdom
125 London Wall, London, EC2Y 5AS, United Kingdom

First edition 2018

ISBN: 978-0-12-814340-7
ISSN: 1876-1623

For information on all Academic Press publications
visit our website at https://www.elsevier.com/books-and-journals

Publisher: Zoe Kruze
Acquisition Editor: Ashlie M. Jackman
Editorial Project Manager: Alina Cleju
Production Project Manager: James Selvam
Cover Designer: Victoria Pearson

Typeset by SPi Global, India

VOLUME ONE HUNDRED AND TWELVE

ADVANCES IN PROTEIN CHEMISTRY AND STRUCTURAL BIOLOGY

Therapeutic Proteins and Peptides

Edited by

ROSSEN DONEV

MicroPharm Ltd, United Kingdom

ELSEVIER

ACADEMIC PRESS

An imprint of Elsevier

CONTENTS

CONTRIBUTORS

Sandra Aguiar
Centro de Investigação Interdisciplinar em Sanidade Animal (CIISA), Faculdade de Medicina Veterinária, Universidade de Lisboa, Avenida da Universidade Técnica, Lisboa, Portugal

Nandika Bandara
University of Alberta, Edmonton, AB, Canada

Bruno Baron
Institut Pasteur, Plate-Forme de Biophysique de Macromolécules et de leurs Interactions, Paris Cedex 15, France

Sherry Bhalla
Center for Computational Biology, Indraprastha Institute of Information Technology, New Delhi, India

Elizabete S. Cândido
Centro de Análises Proteômicas e Bioquímicas, Programa de Pós-Graduação em Ciências Genômicas e Biotecnologia, Universidade Católica de Brasília, Brasília; S-inova Biotech, Programa de Pós-Graduação em Biotecnologia, Universidade Católica Dom Bosco, Campo Grande, Brazil

Marlon H. Cardoso
Programa de Pós-Graduação em Patologia Molecular, Faculdade de Medicina, Universidade de Brasília; Centro de Análises Proteômicas e Bioquímicas, Programa de Pós-Graduação em Ciências Genômicas e Biotecnologia, Universidade Católica de Brasília, Brasília; S-inova Biotech, Programa de Pós-Graduação em Biotecnologia, Universidade Católica Dom Bosco, Campo Grande, Brazil

Régis Cohen
Service d'Endocrinologie, Centre Hospitalier de Saint-Denis, Saint-Denis, France

Yves-Marie Coïc
Institut Pasteur, Unité de Chimie des Biomolécules, UMR 3523, Paris Cedex 15, France

Frederico A. da Silva
Centro de Investigação Interdisciplinar em Sanidade Animal (CIISA), Faculdade de Medicina Veterinária, Universidade de Lisboa, Avenida da Universidade Técnica, Lisboa, Portugal

Gayan H. De Zoysa
School of Chemical Sciences, The University of Auckland, Auckland, New Zealand

Joana Dias
Centro de Investigação Interdisciplinar em Sanidade Animal (CIISA), Faculdade de Medicina Veterinária, Universidade de Lisboa, Avenida da Universidade Técnica, Lisboa, Portugal

Ilkcan Ercan
International Biomedicine and Genome Institute, Dokuz Eylul University, Balcova, Izmir, Turkey

Octávio L. Franco
Programa de Pós-Graduação em Patologia Molecular, Faculdade de Medicina, Universidade de Brasília; Centro de Análises Proteômicas e Bioquímicas, Programa de Pós-Graduação em Ciências Genômicas e Biotecnologia, Universidade Católica de Brasília, Brasília; S-inova Biotech, Programa de Pós-Graduação em Biotecnologia, Universidade Católica Dom Bosco, Campo Grande, Brazil

Kursad Genc
Institute of Health Science, Dokuz Eylul University, Balcova, Izmir, Turkey

Sermin Genc
International Biomedicine and Genome Institute; Institute of Health Science, Dokuz Eylul University, Balcova, Izmir, Turkey

Mahmoud Ghomi
Laboratoire Matrice Extracellulaire et Dynamique Cellulaire (MEDyC), UMR 7369, Université de Reims, Faculté des Sciences, Reims Cedex 2; Groupe de Biophysique Moléculaire, Sorbonne Paris Cité, Université Paris 13, UFR Santé-Médecine-Biologie Humaine, Bobigny, France

Hugh D. Glossop
School of Chemical Sciences, The University of Auckland, Auckland, New Zealand

Pedro M.P. Gois
iMed.ULisboa—Research Institute for Medicines, Faculty of Pharmacy, Universidade de Lisboa, Lisbon, Portugal

Joao Goncalves
iMed.ULisboa—Research Institute for Medicines, Faculty of Pharmacy, Universidade de Lisboa, Lisbon, Portugal

Antonella Grigoletto
University of Padua, Padua, Italy

Yang He
Shanghai Institute of Materia Medica, Chinese Academy of Sciences, Shanghai; University of Chinese Academy of Sciences, Beijing, China

Belén Hernández
Laboratoire Matrice Extracellulaire et Dynamique Cellulaire (MEDyC), UMR 7369, Université de Reims, Faculté des Sciences, Reims Cedex 2; Groupe de Biophysique Moléculaire, Sorbonne Paris Cité, Université Paris 13, UFR Santé-Médecine-Biologie Humaine, Bobigny, France

Ryan E. Holcomb
Legacy BioDesign LLC, Johnstown; Colorado State University, Fort Collins, CO, United States

Yongzhuo Huang
Shanghai Institute of Materia Medica, Chinese Academy of Sciences, Shanghai; University of Chinese Academy of Sciences, Beijing, China

Ezgi Karaca
International Biomedicine and Genome Institute, Dokuz Eylul University, Balcova, Izmir, Turkey

Sergei G. Kruglik
Laboratoire Jean-Perrin, Sorbonne Université, UPMC Univ. Paris 06, CNRS UMR 8237, Paris, France

Rajesh Kumar
Center for Computational Biology, Indraprastha Institute of Information Technology, New Delhi; Bioinformatics Centre, CSIR-Institute of Microbial Technology, Chandigarh, India

Vinod Kumar
Center for Computational Biology, Indraprastha Institute of Information Technology, New Delhi; Bioinformatics Centre, CSIR-Institute of Microbial Technology, Chandigarh, India

Feng Li
Harrison School of Pharmacy, Auburn University, Auburn, AL, United states

Tiansheng Li
HTL Biosolutions, Newbury Park, CA, United States

Jun Liu
Coherus Biosciences, Redwood City, CA, United States

Eduardo López-Tobar
Instituto de Estructura de la Materia, IEM-CSIC, Serrano 121, Madrid, Spain

Mark C. Manning
Legacy BioDesign LLC, Johnstown; Colorado State University, Fort Collins, CO, United States

Ana M. Manuel
iMed.ULisboa—Research Institute for Medicines, Faculty of Pharmacy, Universidade de Lisboa, Lisbon, Portugal

Cláudia Martins
i3S—Instituto de Investigação e Inovação em Saúde; INEB—Instituto Nacional de Engenharia Biomédica, Universidade do Porto, Porto, Portugal

Katia Maso
University of Padua, Padua, Italy

Chalamaiah Meram
University of Alberta, Edmonton, AB, Canada

Elias Moreira
CESPU—Instituto de Investigação e Formação Avançada em Ciências e Tecnologias da Saúde, Gandra, Portugal

Rui P. Moura
CESPU—Instituto de Investigação e Formação Avançada em Ciências e Tecnologias da Saúde, Gandra, Portugal

Karen G.N. Oshiro
S-inova Biotech, Programa de Pós-Graduação em Biotecnologia, Universidade Católica Dom Bosco, Campo Grande, Brazil

Gianfranco Pasut
University of Padua; Veneto Institute of Oncology IOV—IRCCS, Padua, Italy

Esperanza Pearl
School of Chemical Sciences, The University of Auckland, Auckland, New Zealand

Fernando Plüger
Groupe de Biophysique Moléculaire, Sorbonne Paris Cité, Université Paris 13, UFR Santé-Médecine-Biologie Humaine, Bobigny, France

Gajendra P.S. Raghava
Center for Computational Biology, Indraprastha Institute of Information Technology, New Delhi; Bioinformatics Centre, CSIR-Institute of Microbial Technology, Chandigarh, India

Samilla B. Rezende
S-inova Biotech, Programa de Pós-Graduação em Biotecnologia, Universidade Católica Dom Bosco, Campo Grande, Brazil

Roberto Russo
iMed.ULisboa—Research Institute for Medicines, Faculty of Pharmacy, Universidade de Lisboa, Lisbon, Portugal

Santiago Sanchez-Cortes
Instituto de Estructura de la Materia, IEM-CSIC, Serrano 121, Madrid, Spain

Bruno Sarmento
i3S—Instituto de Investigação e Inovação em Saúde; INEB—Instituto Nacional de Engenharia Biomédica, Universidade do Porto, Porto; CESPU—Instituto de Investigação e Formação Avançada em Ciências e Tecnologias da Saúde, Gandra, Portugal

Vijayalekshmi Sarojini
School of Chemical Sciences, The University of Auckland, Auckland, New Zealand

Nan Shang
University of Alberta, Edmonton, AB, Canada

Flávia Sousa
i3S—Instituto de Investigação e Inovação em Saúde; INEB—Instituto Nacional de Engenharia Biomédica; ICBAS—Instituto Ciências Biomédicas Abel Salazar, Universidade do Porto, Porto; CESPU—Instituto de Investigação e Formação Avançada em Ciências e Tecnologias da Saúde, Gandra, Portugal

Kemal Ugur Tufekci
International Biomedicine and Genome Institute, Dokuz Eylul University, Balcova, Izmir, Turkey

Salman Sadullah Usmani
Center for Computational Biology, Indraprastha Institute of Information Technology, New Delhi; Bioinformatics Centre, CSIR-Institute of Microbial Technology, Chandigarh, India

Jianping Wu
University of Alberta, Edmonton, AB, Canada

PREFACE

Protein-based therapeutics have proved to be highly successful in clinical practice. Due to their highly recognized potential, there is vast current research on proteins and peptides that can be employed as therapeutics for treatment of diverse diseases and disorders. A review article on therapeutic proteins has classified them into the following groups based on their pharmacological activity: (1) replacing a protein that is deficient or abnormal; (2) augmenting an existing pathway; (3) providing a novel function or activity; (4) interfering with a molecule or organism; and (5) delivering other compounds or proteins (Dimitrov, 2012).

This thematic volume of the *Advances in Protein Chemistry and Structural Biology (APCSB)* dedicated to therapeutic proteins and peptides focuses on several major topics in this area. It begins with rational design of stable liquid formulations of therapeutic proteins (Chapter 1). In the next five chapters (Chapters 2–6) the authors discuss in great details all current strategies for delivery of therapeutic peptides and proteins which include the use of nanoparticles from different origins and with various designs, chimeric small antibody fragments, small cell-penetrating peptide-based techniques, etc. The usage of each of these strategies is exemplified in different diseases and/or disorders.

Chapter 7 reviews the in silico tools and databases used for designing peptide-based vaccine and drugs. Next, Chapter 8 is focused on the function of proteins and peptides in elderly health as well as their effects on preventing aging-related diseases. These include sarcopenia, skeleton disorders, gastrointestinal disorders, cardiovascular disease, and Alzheimer's disease. A number of mimetic peptide derivatives of erythropoietin and their application for treatment of neuroinflammation and neurodegeneration have been discussed in depth in Chapter 9. The final two chapters of this thematic volume of the *APCSB*, Chapters 10 and 11, focus on the design and applications of antimicrobial peptides. Considering that the indiscriminate use of antibiotics has become a great concern, mainly because it has led to multidrug resistance, these two chapters discuss the application of antimicrobial peptides as a replacement treatment.

We believe that the articles in this volume would inspire new ideas in researchers and specialists from a wide number of fields such as protein chemists, pharmacologists, medical doctors, cell biologists, biomedical researchers, immunologists, structural biologists, computational biochemists, and others working in the exciting field of protein and peptide therapeutics. This volume is also intended as a scholarly material for use by students with interest in the above topics.

DR. ROSSEN DONEV
MicroPharm Ltd and Biomed Consult
United Kingdom

REFERENCE

Dimitrov, D. S. (2012). Therapeutic proteins. *Methods in Molecular Biology*, *899*, 1–26. https://doi.org/10.1007/978-1-61779-921-1_1.

CHAPTER ONE

Rational Design of Liquid Formulations of Proteins

Mark C. Manning*,†,1, Jun Liu‡, Tiansheng Li§, Ryan E. Holcomb*,†
*Legacy BioDesign LLC, Johnstown, CO, United States
†Colorado State University, Fort Collins, CO, United States
‡Coherus Biosciences, Redwood City, CA, United States
§HTL Biosolutions, Newbury Park, CA, United States
1Corresponding author: e-mail address: Manning@LegacyBioDesign.com

Contents

Abstract

Twenty years ago, a number of eminent pharmaceutical scientists collaborated on an article describing a rational approach to developing stable lyophilized protein formulations (Carpenter, Pikal, Chang, & Randolph, 1997). Since that time, no corresponding document for rational development of liquid formulations of proteins has appeared. Certainly, many of the principles underpinning rational protein formulation have been

Advances in Protein Chemistry and Structural Biology, Volume 112
ISSN 1876-1623
https://doi.org/10.1016/bs.apcsb.2018.01.005

known for some time, but no overarching scheme has ever been described in the literature. Now the time has come to provide a framework for the rational design of protein formulations as aqueous solutions. The objective of this review is to lay out four concepts that will guide one to obtaining a stable liquid protein formulation. Additionally, the aim will be to identify factors that are intrinsic to the stabilization of any protein, not just a particular class of proteins, such as monoclonal antibodies (Uchiyama, 2014; Wang, Singh, Zeng, King, & Nema, 2007) and to provide guidelines aiming to effect stabilization. Noting that all approaches to stabilization face validation that must be performed empirically, it is hoped that the rational strategies described here will help the formulation scientist in their daily tasks and inspire continued advancement of the science involved in protein formulation.

Twenty years ago, a number of eminent pharmaceutical scientists collaborated on an article describing a rational approach to developing stable lyophilized protein formulations (Carpenter, Pikal, Chang, & Randolph, 1997). Since that time, no corresponding document for rational development of liquid formulations of proteins has appeared. The general aim here is to identify factors that are intrinsic to the stabilization of any protein, not just a particular class of proteins, such as monoclonal antibodies (Uchiyama, 2014; Wang, Singh, Zeng, King, & Nema, 2007) and to provide four concepts as guidelines for stabilization of proteins as aqueous solutions. In addition, the formulation scientist often is faced with challenges beyond simply stabilization. There may be a need to optimize the solubility of the protein or reduce the viscosity of the resulting solution. While these are important, they are not the primary focus of this review. However, the discussion of colloidal stability will certainly provide clarity as to the approaches one must take to achieve greater solubility or lower viscosity.

In general, all proteins experience four types of instability. These are chemical instability, conformational instability, colloidal instability, and interfacial instability. If one can devise strategies to mitigate each one of these degradation pathways, one should obtain a liquid product that is stable for an extended period of time, likely for at least 2 years if stored at 2–8°C.

The formulation scientist can control four aspects of a drug product (DP) formulation. These include (i) the selection of the dosage form (liquid, dried product, frozen solution, etc.), (ii) the choice of excipients and their concentrations, (iii) the process by which the formulation is manufactured, and (iv) identifying the specific primary and secondary packaging needed for the DP. In this review the focus will be on excipient selection (including the choice of the optimal pH) for liquid formulations of proteins, although aspects of the process (i.e., handling) and packaging will be discussed as well. The interaction of excipients with proteins is the key to understanding the proper strategy and approaches toward stabilization of biotherapeutics (Kamerzell, Esfandiary, Joshi, Middaugh, & Volkin, 2011).

1. CHEMICAL INSTABILITY

All proteins and peptides in aqueous formulations are susceptible to numerous chemical degradation processes. This is of particular concern for liquid formulations, especially compared to lyophilized and frozen formulations, because of both the motion of the protein and facile movement of potential degradants through solution, which can be substantial even at 2–8°C. The former (protein motion) leads to exposure of susceptible sites on the protein and the latter (degradant diffusion) allows for potential degradants to reach those sites effectively. These dynamics can be impacted by any additional species in the solution, aside from the protein and water. Therefore, excipients can have an impact on any chemical degradation process (albeit to varying degrees).

While a great deal is now known about degradation routes, sequence sensitivities, and the role of local structure (not so much dynamics) on chemical instability, the picture is far from complete. A comprehensive set of accurate tools for stability prediction is unavailable, but some useful algorithms have been developed. In addition, we have some helpful rules of thumb for both sequence and structural liabilities. For example, the more solvent exposure of a sequence known to be susceptible to degradation, the greater the liability for reactions like glycation (Gadgil et al., 2007), crosslinking with formaldehyde (Toews, Rogalski, & Kast, 2010), or Met oxidation (Chu, Yin, Brooks, et al., 2004; Duenas, Keck, DeVos, Jones, & Cleland, 2001; Griffiths & Cooney, 2002; Pan et al., 2009; Teh et al., 1987).

Given the challenges of even just identifying the relative reactivity of certain residues, predicting the effects of excipients is even more challenging. While excipients might have a global effect on proteins (e.g., on the overall conformation), what is usually more important for chemical degradation processes is the local effect of excipients at a particular reactive site. These local effects are generally not known, except for what may be gleaned from H/D exchange experiments or molecular simulations.

Despite these challenges, general and specific formulation strategies can be employed to improve the chemical stability of a given protein product once these instabilities are identified. The following section reviews the most common types of chemical instabilities encountered with aqueous protein formulations as well as the interplay of physical instability with chemical degradation processes. In addition, mitigation of chemical instabilities by minimizing liabilities during the discovery phase and applying appropriate formulation strategies during the development phase is discussed.

1.1 Overview of Common Chemical Instabilities

While the details of the mechanisms of the various chemical degradation routes are outside the scope of this chapter, the formulation scientist should be aware of them from references in the literature (Manning, Chou, Murphy, Payne, & Katayama, 2010; Manning, Patel, & Borchardt, 1989; Topp et al., 2010). Summaries of the key aspects of the most common chemical degradation processes are provided below.

1.1.1 Oxidation

Oxidation proceeds through one of three general mechanisms: metal-catalyzed oxidation (MCO), light exposure, or free radical generation by processes other than those mediated by light or metals (Chakravarti & Chakravarti, 2007; Hawkins & Davies, 2001; Torosantucci, Schöneich, & Jiskoot, 2014). Each leads to the generation of a variety of reactive oxygen species (ROS) that can cause damage at a wide range of amino acid residues. In fact, a more recent review on protein oxidation has identified oxidative instability at 14 of the 20 naturally occurring amino acids (Trnková, Dršata, & Boušová, 2015). In general, Met and Trp residues are most susceptible to oxidation, with Met capable of being oxidized by even molecular oxygen (Chu & Trout, 2004; Torosantucci et al., 2014; Trnková et al., 2015). Typically, oxidation reactions are not very sensitive to pH, unlike the hydrolytic reactions listed below (Chu, Yin, Brooks, et al., 2004). Like deamidation, the rates can be affected by both local and global conformational differences (Pan et al., 2006).

1.1.2 Deamidation

Deamidation refers to the hydrolysis of the amide side chain of Asn and Gln residues to form their corresponding carboxylic acid derivatives (Aswad, Paranandi, & Schurter, 2000; Wakankar & Borchardt, 2006). Although deamidation can occur via acid-catalyzed direct nucleophilic attack of water on the amide, the more common mechanism is intramolecular cyclization involving nucleophilic attack by the succeeding amide nitrogen to form a succinimide (also called cyclic imide) intermediate (Aswad et al., 2000; Li, Borchardt, Topp, Vander Velde, & Schowen, 2003; Manning et al., 1989). The formation of the succinimide is the rate-limiting step, and subsequent hydrolysis of the succinimide forms either aspartate (Asp) or isoaspartate (isoAsp) (Li et al., 2003). Unlike oxidation, deamidation is quite sensitive to pH (Patel & Borchardt, 1990; Topp et al., 2010). The most reactive sites tend to be Asn followed by amino acids with little steric bulk to interfere with the cyclization (i.e., Gly), or ones possessing neighboring amino acid that can hydrogen bond to the Asn carboxyl oxygen (e.g., Asn, Asp, Ser, His), thereby rendering it more reactive (Topp et al., 2010).

1.1.3 Hydrolysis

Also called proteolysis, these terms refer to nonenzymatic cleavage of an amide bond within the polypeptide chain leading to chain scission (Pan, Ricci, & Trout, 2011; Smith & Hansen, 1998). This reaction is both acid- and base-catalyzed (Smith & Hansen, 1998). Moreover, it occurs much faster at Asp residues (Inglis, 1983; Smith & Hansen, 1998), where an internal cyclization process facilitates chain scission. Of all the amide linkages, it appears that Asp-Pro is the most reactive in terms of hydrolytic degradation (Eng et al., 1997; Kameoka, Ueda, & Imoto, 2011). Hinge region hydrolysis in monoclonal antibodies (mAbs) is also quite common (Cordoba, Shyong, Breen, & Harris, 2005; Kroon, Baldwin-Ferro, & Lalan, 1992; Rustandi & Wang, 2011; Vlasak & Ionescu, 2011), with the reaction proceeding via many different mechanisms, including radical-centered pathways (Yan & Boyd, 2011; Yates et al., 2010) and metal-catalyzed mechanisms (Ouellette et al., 2009; Rustandi & Wang, 2011; Smith et al., 1996). Hinge region hydrolysis is sensitive to pH (Rustandi & Wang, 2011) and can be catalyzed by buffers (Salinas, Sathish, Shah, Carpenter, & Randolph, 2010).

1.1.4 Other Notable Chemical Degradation Processes

Many other chemical reactions, in addition to those listed above, have been observed for polypeptides. Condensation reactions are one such example and involve modification of the protein via covalent reactions with excipients. Glycation is one type of condensation reaction where a reducing sugar can react with Lys residues on a protein to form a variety of different end products (also known as the Maillard reaction or nonenzymatic browning) (Ledl & Schleicher, 1990). Chemical instabilities like Asp isomerization are not discussed in this review, but many of the same principles which are discussed here for slowing deamidation also apply to Asp isomerization (Wakankar & Borchardt, 2006).

1.2 Connections of Chemical Instability With Physical Instabilities

Both conformational and interfacial instabilities can lead to enhanced susceptibility to chemical degradation. They can lead to chemical instabilities by increasing the solvent exposure of susceptible groups, by increasing the solvent accessible area of those groups, and by increasing the local dynamical fluctuations so that those fluctuations lead to a higher transient exposure. Furthermore, interfacial instability can lead to more direct contact with degradant species, such as oxygen present at interfaces or silicone oil.

It has been reported that deamidation can increase the propensity for aggregation (Dunkelberger et al., 2012; Harms et al., 2004; Nilsson, Driscoll, & Raleigh, 2002), while oxidation of various types can do the same (Gaudiano et al., 2005; Koudelka, Dehle, Musgrave, Hoffmann, & Carver, 2012; Luo et al., 2011; Schöneich, 2000). On the other hand, there are number of studies showing that oxidation slows aggregation (Mulinacci, Poirier, Capelle, Gurny, & Arvinte, 2011; Palmblad, Westlind-Danielsson, & Bergquist, 2002; Sancataldo et al., 2014). Oxidation can also impact conformational stability (Sancataldo et al., 2014; Torosantucci et al., 2014). Likewise, physical instability can impact chemical reactivity (Abaturov, Burstein, & Nosova, 2008; Luo et al., 2011; Zakrzewska, Krowarsch, Wiedlocha, Olsnes, & Otlewski, 2005). The point here is that chemical instability can have profound effects on the physical stability of a protein and vice versa. So, although chemical and physical instabilities may be discussed separately in this review, be aware that they cannot be completely dissociated when developing a formulation and stabilization strategy.

1.3 Identification of Chemical Instability Liabilities in Discovery

A good approach to protein stabilization starts with the drug discovery phase. Potential liabilities can be removed or selected out at this point in time, whereas once the molecule goes forward to product development, it is usually too late to change the sequence. The challenge, of course, is limited resources of time and materials. Much, however, can be accomplished via computational methods, even though the available tools are far from comprehensive. Structures can be obtained via homology modeling or even better via X-ray diffraction or NMR (if the protein is small enough). High-throughput crystallization can be done (Broecker et al., 2016; Liang, Zhu, & Fang, 2016), and synchrotron sources are readily accessible (Pham et al., 2017). At any rate, given a structure, the first thing that a formulation scientist should do is simply look at the surface for exposed residues (e.g., residues within the CDRs of antibodies), as these will be most prone to oxidation and glycation. Even better, performing molecular dynamics simulation should give an accurate view of these liabilities.

Methods exist to predict the proclivity of a protein to undergo oxidation, deamidation, and hydrolysis (Aledo, Canton, & Veredas, 2017; Chennamsetty et al., 2015; Jia & Sun, 2017; Lauer et al., 2016; Lorenzo, Alonso, & Sanchez, 2015; Sydow et al., 2014). While these methods are far from 100% accurate, they have been validated on several systems and should form the starting point for the internal development of correlations between structure/sequences and propensity to chemical degradation. The identification of liabilities can lead to removal by mutation or candidate selection. The challenge, of course, is false positives or negatives, but continued application and development of these methods should lead to substantial improvement of predictive tools. Ideally, before the final molecule is sent on for development, experimental screening would be performed for liabilities. This activity is now known as developability assessment and is commonly performed for biopharmaceutical drug candidates (Lorenz et al., 2014). Often, aspects of stability are part of key criteria for moving a molecule forward (Obrezanova et al., 2015).

1.4 Formulation Strategies for Addressing Chemical Instability

Once the protein molecule has been chosen and the relative propensity to chemical degradation assessed, it is left to the formulation scientist to achieve the desired stability. While every protein presents its own particular challenges, the following ideas and principles can be applied to achieve the maximal chemical stability in a rational fashion.

1.4.1 Deamidation

By far, the most important factor for controlling deamidation is pH (Manning et al., 1989; Topp et al., 2010) (see Table 1). Note that the mechanism of deamidation changes as a function of pH, leading to a V-shaped rate profile, which itself depends on sequence and conformation. The result is that acidic conditions (generally pH 3–6) yield a lower intrinsic rate of deamidation (Manning et al., 1989; Topp et al., 2010). Of course, there

Table 1 General Principles for Reducing Deamidation in Peptides and Proteins

1. Employ an acidic pH, typically between pH 3 and pH 6
2. Use a minimal buffer concentration to avoid buffer catalysis
3. Control polypeptide chain flexibility
4. Reduce the storage temperature
5. Focus on the most sensitive sites (e.g., Asn-Gly)
6. Explore excipients that have been reported to slow deamidation specifically

is a trade-off, since reducing the pH can affect other degradation processes. Moreover, pH can affect both the global and local protein conformation and dynamics. In addition to pH, it is worth mentioning that certain buffers can catalyze deamidation reactions (Patel & Borchardt, 1990), leading to the trade-off of buffering capacity vs potential increase in deamidation rates. Thus, the amount of buffer should be limited in order to minimize degradation rates (Manning et al., 2010). Also, as in general with all chemical reactions, the rate of degradation follows Arrhenius behavior, so lower temperatures will slow the deamidation process (Robinson & Rudd, 1974).

It has been shown that conformational stabilizers (i.e., excluded solutes or osmolytes) can slow deamidation by both minimizing exposure and sterically hindering the reaction at susceptible groups (Miroliaei & Nemat-Gorgani, 2001; Singh & Singh, 2003; Stratton et al., 2001; Yin et al., 2017). Since the flexibility of a polypeptide chain plays a role in deamidation (Kosky, Dharmavaram, Ratnaswamy, & Manning, 2009), susceptible sites within well-defined regions of higher order structure generally exhibit slower deamidation rates, presumably by limiting access to the conformations needed for facile intramolecular cyclization (Box & Smith, 2000; Kossiakoff, 1990).

Salts and other electrolytes can have an effect on deamidation through both electrostatic screening and protein conformation. Generally, there is a trade-off for charge screening, which also depends on the charge of the protein. For highly charged proteins, some charge screening is generally favorable (Laber et al., 2017), but too much can destabilize the protein. For proteins of lower charge, less screening is typically warranted. It has been shown that addition of amines to a formulation can slow the rate of deamidation (Shiraki, Tomita, & Inoue, 2016; Tomita & Shiraki, 2011). There have been reports of sugars (Singh & Singh, 2003) and polyols

(Yin et al., 2017) being effective at slowing deamidation as well. Ligand binding, which not only stabilizes the protein conformationally (see below) but can also rigidify the protein, has been shown to slow deamidation (Shojaee, Sabbaghian, & Ebrahim-Habibi, 2016). Finally, increasing solution viscosity has also been shown to impede deamidation (Li, Hageman, & Topp, 2002).

1.4.2 Oxidation

There are a number of approaches one can take to reduce oxidative damage of proteins (see Table 2). One of the first approaches in addressing oxidation issues could be to include antioxidants (Yin et al., 2004). These are typical free radical scavengers, such as ascorbic acid. The use of polyols as free radical scavengers has also been reported (Gebicki et al., 2010; Ji, Zhang, Cheng, & Wang, 2009; Lecour et al., 1998). Chelators of redox-active metals that generate free radicals, such as EDTA, can be effective at mitigating MCO (Ji et al., 2009; Zhou et al., 2010). Be aware that a free radical scavenger, such as ascorbate, can actually increase the ability of metals, like iron, to generate ROS (Bridgewater & Vachet, 2005; Jensen, Kolvenbach, Roy, & Schöneich, 2000).

In addition, sacrificial agents, such as free methionine, can reduce the concentration of free radicals (Ji et al., 2009), or even more importantly, reduce the concentration of direct oxidants, like peroxides (Wang, Ignatius, & Thakkar, 2014). However, these additives can also change the local dynamics and increase solvent exposure of a reactive group, leading to greater instability.

The best approach, therefore, is to reduce primary exposure to reactive oxidizing species (Chu, Yin, Brooks, et al., 2004; Chu, Yin, Wang, &

Table 2 General Principles for Reducing Oxidation in Peptides and Proteins

1. Reduce degree of solvent exposure
2. Include sacrificial additives or free radical scavengers
3. Eliminate the presence of redox-active metals
4. Use chelating agents
5. Protect from light
6. Reduce the storage temperature
7. Eliminate or reduce oxygen in the headspace

Trout, 2004c; Griffiths & Cooney, 2002; Pan et al., 2009). This can be done by sourcing excipients that contain a minimum amount of the contaminants that lead to oxidation (Saineni, Chandrasekharan, & Nassar, 2014; Wang et al., 2014; Waterman et al., 2002). These contaminants include peroxides (such as are often in polysorbates) (Waterman et al., 2002; Wu, Levons, Narang, Raghavan, & Rao, 2011), metals from excipients (Sadineni, Chandrasekharan, & Nassar, 2014; Wang et al., 2014), and metals which can be introduced from equipment during the manufacturing process (Sadineni et al., 2014). Conformational stabilizers can also reduce exposure to ROS, leading to enhanced stability (DePaz et al., 2000; Li, Patapoff, Nguyen, & Borchardt, 1996). Finally, pH can affect the rate of oxidation processes, as can salt, buffer types, and buffer concentrations. However, pH does not directly affect Met oxidation, although it may indirectly affect it through conformational rearrangement (Chu, Yin, Wang, & Trout, 2004b; Pan et al., 2006).

Direct exposure to oxygen via the head space can be detrimental, as can exposure to light that can lead to singlet oxygen. Thus, headspace can be filled with nitrogen, which has been shown to slow oxidation (Malek-Sabet, Masaumian, Zeinali, Khalilzadeh, & Mousaabadi, 2013; Qi et al., 2009). Also, care can be taken both during processing (Østdal, Weisbjerg, Skibsted, & Nielsen, 2008; Sreedhara et al., 2016) and during storage (Qi et al., 2009) to reduce exposure to light. Most therapeutic proteins are quite sensitive to photooxidation (Kerwin & Remmele, 2007). This type of oxidation can be reduced by proper packaging such as the use of amber vials, wrap-around labels, or placing the product in boxes. Note that there have been some reports of a particular container causing an increase in oxidation (Nguyen, 1994; Yokota, Saito, Masuoka, Kaniwa, & Shibanuma, 2000), so packaging plays an important role in modulating oxidation.

Reducing the storage temperature should also slow down oxidation, but it is important to keep in mind that the solubility of gases in water increases with decreasing temperature. This could lead to an increase in oxygen levels at 2–8°C (Fransson, Florin-Robertsson, Axelsson, & Nyhlén, 1996), which could accelerate oxidation, even overcoming the reduced reaction rate related to Arrhenius behavior.

1.4.3 Hydrolysis

The rates of nonenzymatic hydrolysis processes are known to be strong functions of pH. There tends to be an optimal pH, above and below which the rates of hydrolysis are greater, depending on the properties of a specific protein (Li, Fort, Kessler, & Wang, 2009). As a result, there is a trade-off in choosing a pH value to reduce deamidation and other degradation processes and choosing a pH value to reduce hydrolysis. As hydrolysis generally occurs at highly solvent exposed sites, such as hinge regions in multidomain proteins or regions without secondary or tertiary structure, minimizing those regions (and susceptible sequences) during design is a must. In general, reducing solvent exposure should reduce rates of hydrolysis. It should be noted that conformational stabilizers might not always have an effect in reducing solvent exposure in highly exposed regions. However, they can be tried, together with adjusting salt content and buffer type.

1.4.4 Summary

Various mechanisms have been reported for chemical degradation of peptides and proteins. Of these, deamidation, oxidation, and proteolysis are the most common. A number of bona fide approaches are known by which one can slow these processes significantly, as described above. For hydrolytic reactions, they focus on identifying the optimal pH, minimizing buffer concentrations, storing at lower temperatures, and minimizing the flexibility of the peptide backbone. For oxidation (as well as for glycation), reducing the solvent exposure of reactive side chains is critical. While pH has a much smaller effect, the proper choice of excipients can be quite effective at reducing the extent of oxidative damage.

2. CONFORMATIONAL INSTABILITY

Protein conformations in solution are dynamic and subject to change upon exposure to external stresses. A three-dimensional folded protein may have multiple thermodynamically favorable conformations that depend on the formulation conditions. Both chemical and physical degradation can potentially alter a proteins conformation in solution. Physical degradation of proteins includes changes in secondary, tertiary, and quaternary structures, as well as undesirable adsorption to surfaces and aggregation (Chi et al., 2003; Manning et al., 2010). Aggregation can occur when some external factor changes the conformation of a protein, thereby inducing an aggregation event. Such factors include the temperature and pH of a protein solution, ionic strength, formulation excipients, change in pH from formulation to physiological pH, freeze–thaw, agitation, and air/water interfaces. For proteins where a conformational change is the rate-limiting step, the kinetics are first order, as aggregation is a unimolecular process (Morris, Watzky, & Finke, 2009).

While there are other general mechanisms of aggregation, this one is most directly connected to conformational stability (Philo & Arakawa, 2009). Many therapeutic proteins, including mAbs, exhibit first-order aggregation kinetics under certain conditions (Hartmann, Saptharishi, Yang, Mitra, & Soman, 2004; Zidar, Susteric, Ravnik, & Kuzman, 2017). This section will examine various aspects of conformational stability, with a particular emphasis on how one selects excipients to increase conformational stability.

2.1 Determination of Conformational Stability

Protein conformational stability under different formulation conditions can be characterized by a variety of biophysical methods. Potential methods that can be used to assess the conformational stability of a protein during formulation development are listed in Table 3.

2.2 Impact of Conformational Stability on Aggregation

Protein aggregation, in most cases, is irreversible and generally arises from interactions between proteins in their nonnative conformations (e.g., unfolded or partially unfolded conformations). During formulation development, irreversible aggregation is of most concern as these nonnative, associated species may present the greatest risk with respect to immunogenicity (Rosenberg, 2006; Rosenberg, Verthelyi, & Cherney, 2012). Thus, considerable effort is directed at developing the right combination of analytical methods and formulation strategies to monitor aggregation and prevent aggregation from occurring (Carpenter, Kendrick, Chang, Manning, & Randolph, 1999; Carpenter et al., 2009). Since the major cause of irreversible aggregation is formation of unfolded or partially unfolded proteins, an evaluation of and stabilization against unfolding, as well as enhancing the stability of the native conformation is generally needed to achieve a robust formulation (Freire, Schön, Hutchins, & Brown, 2013). Conformational stability can be measured by a number of biophysical techniques, as mentioned previously (see Table 3). In cases where the actual free energy of unfolding (ΔG_u) of a protein is unavailable, the melting transition (T_m) may be utilized as a suitable surrogate or indicator of protein conformational stability. Melting transitions are usually measured using DSC. Spectroscopic techniques such as far-UV CD (Joshi, Shivach, Yadav, & Rathore, 2014) and FTIR (Barth, 2007; Cerdá-Costa, De la Arada, Avilés, Arrondo, & Villegas, 2009; Manning, 2005; Tidy, Lam,

Table 3 Methods Used to Determine the Conformational Stability of Proteins

Methods	**Structure Attributes**
Far-UV CD spectroscopy	Secondary structure
FTIR spectroscopy	Secondary structure
Raman spectroscopy	Secondary and tertiary structure
Near-UV CD spectroscopy	Tertiary structure
UV–vis absorbance spectroscopy	Tertiary structure and aggregation (A_{360})
Fluorescence spectroscopy	Tertiary structure
ANS binding	Surface hydrophobicity (tertiary structure)
Differential scanning calorimetry (DSC)	Thermal stability (T_m)
Differential scanning CD (DSCD)	Thermal stability (T_m)
Differential scanning fluorescence (DSF)	Thermal stability (T_m)
Differential scanning light scattering (DSLS)	Thermal stability (T_{agg})
Dynamic light scattering (DLS)	Aggregation
AUC	Aggregation
HIAC	Particle counts
MFI or FlowCAM	Particle counts
HDX-MS or HDX-FTIR	Excipient or solvent binding with protein

Fimognari, Mamo, & Hackett, 2017) can also be used to determine thermal transitions of protein secondary structure. Fluorescent dye-binding experiments have been employed to probe protein conformational stability as well (Groenning, 2010; Haidekker & Theodorakis, 2005; Kayser, Chennamsetty, Voynov, Helk, & Trout, 2011; Taddei et al., 2001).

In general, protein aggregation rates tend to accelerate at temperatures close to their T_m, which is due to the accumulation of unfolded or partially unfolded structures near T_m (Chakroun, Hilton, Ahmad, Platt, & Dalby, 2016). As a result, the conformational stability of a protein often correlates with aggregation rates (Poshusta, Katoh, Gertz, Dispenzieri, & Ramirez-Alvarado, 2013), provided the degree of reversibility is comparable across the different formulations (Remmele, Nightlinger, Srinivasan, & Gombotz, 1998). Formulation screening studies conducted at elevated temperatures, however, may not necessarily represent the mechanism of aggregation found at a normal or intended storage temperature (such as 2–8°C). Although protein aggregation kinetics at temperatures below T_m may follow Arrhenius behavior, aggregation rates at elevated temperatures cannot always be extrapolated to 2–8°C (Chakroun et al., 2016). In spite of this limitation, formulation screening studies at elevated temperatures can add great value in understanding conformational stability of proteins, as well as providing guidelines on how to address temperature excursions of drug products. In order to investigate the effect of conformational stability on aggregation rate, some groups have employed a comprehensive approach utilizing protein engineering to design more conformationally stable variants, as with human γD-crystallin (Costanzo et al., 2014; Sahin et al., 2011). In this particular study, a large number of these protein variants were prepared, and the resultant measurement of their aggregation rates indicated that improved conformational stability can mitigate aggregation.

For proteins containing multidomain structures, the effect of conformational stability on aggregation rate may be more complex than for proteins containing only a single domain. This arises from the fact that proteins with multidomain structures, such as mAbs, typically have non-Arrhenius aggregation kinetics (Drenski, Brader, Alston, & Reed, 2013; Kayser, Chennamsetty, Voynov, Helk, Forrer, et al., 2011; Nicolai, Britten, & Schmitt, 2011). In many cases, unfolding of the least conformationally stable domain leads to aggregation. In the case of mAbs, this is usually the CH_2 domain, where its stability correlates with aggregation propensity (Andersen, Manno, Rischel, Thorolfsson, & Martorana, 2010; Kameoka, Ueda, & Imoto, 2011; Mueller et al., 2013). On the other hand, colloidal stability often plays an even more important role in controlling mAb aggregation (Banks et al., 2012; Kalonia et al., 2016; Li, Ogunnaike, & Roberts, 2010; Yageta, Lauer, Trout, & Honda, 2015). This was shown in one recent study where the k_D (i.e., colloidal stability), and not a measure of conformational stability, was the best indicator of the propensity of a mAb to aggregate at high concentrations (Zidar et al., 2017). This result shows that the ability of conformational screening to accurately predict the rank ordering of formulations can be variable (Goldberg et al., 2017; Youssef & Winter, 2013; Zidar et al., 2017).

Using a fluorescent dye-binding technique, it was discovered that at high incubation temperatures the unfolding of a mAb governs the aggregation kinetics (Kayser, Chennamsetty, Voynov, Helk, & Trout, 2011). This suggests that conformational stability may have an inverse correlation with aggregation kinetics. In fact, an earlier study with human muscle acylphosphatase (AcP) yielded a similar result (Taddei et al., 2001), where it was found that mutations stabilizing the native α-helical conformation appeared to slow down the aggregation rate, but mutations destabilizing the native α-helix increased the rate of aggregation. A similar approach has been taken for the Fc portion of antibodies (Chen et al., 2016). For native state aggregation, there are numerous reports indicating that electrostatic interactions govern aggregation rates (Chakroun et al., 2016), and that mutations suppressing native state electrostatic interactions may slow down aggregation (Austerberry et al., 2017).

One final aspect of the relationship between partially folded states and aggregation must be emphasized. The degree of structural rearrangement in the nonnative, aggregation competent state may be quite subtle. The overall tertiary structure may be similar, if not identical to the native state. The hydrodynamic radius may only be slightly expanded in the aggregation prone state (Chiti & Dobson, 2009, 2017). The similarity to the native state is so great that this aggregation competent state is often referred to as N*. So, anything that the formulations scientist can do to shift the equilibrium of the native state ensemble from the more expanded, aggregation competent species toward more compact states will be effective at increasing the storage stability (Freire et al., 2013). Approaches to accomplish this conformational stabilization are discussed below.

2.3 Screening Formulations Based on Conformational Stability/Stabilization

Since the major cause of irreversible aggregation is an increase in the amount of unfolded or partially unfolded species (i.e., N*) in solution, screening formulation excipients or conditions which inhibit unfolding is one of the primary approaches that can be taken for a formulation development program. Formulation excipients which influence the conformational stability of a protein can be screened using a number of biophysical techniques as mentioned previously (see Table 3), with DSC being one of the more commonly used techniques. Although protein unfolding or partial unfolding cannot be completely captured by measuring the melting transition (T_m), this metric can be a reasonably good indicator of a proteins conformational stability. The T_m values measured by DSC are commonly used for screening formulation excipients, and it has become an increasingly popular tool as relatively high-throughput experimental design can be achieved.

There are number of noteworthy examples describing the use of excipients to increase conformational stability (Baynes & Trout, 2003; Kamal, Ahmad, & Rao, 2011; Nicoud, Cohrs, Arosio, Norrant, & Morbidelli, 2015; Timasheff, 1998). One example for a simple globular protein involves G-CSF, where both conformational stability and colloidal stability were simultaneously evaluated (Chi et al., 2003; Raso et al., 2005). For this protein, it has been reported that the native state exists in equilibrium with a conformationally altered state which is prone to aggregation. Stabilization of the conformationally altered state appeared to reduce aggregation. In another report, the effect of polyols on the conformational stability of lysozyme was investigated by measuring T_m with DSC (Singh & Singh, 2003). The results of this study demonstrate the utility of using T_m for formulation screening.

Arakawa et al. have provided a comprehensive review on the effects of additives on protein aggregation (Arakawa, Ejima, et al., 2007; Arakawa, Tsumoto, Kita, Chang, & Ejima, 2007; Hamada, Arakawa, & Shiraki, 2009). The effect of L-Arg-HCl on protein solubility and stability was particularly emphasized in their reviews, and the impact of L-Arg-HCl has been widely reported by a number of other research groups (Kamerzell et al., 2011), including its ability to modulate aggregation (Baynes, Wang, & Trout, 2005). The mechanism of stabilization of L-Arg-HCl for reducing protein aggregation appears to be due to its capacity to increase the solubility of unfolded species (Baynes et al., 2005; Reddy, Lilie, Rudolph, & Lange, 2005). The interaction of proteins with Arg and the molecular basis for stabilization by this particular amino acid has been examined in detail (Blobel et al., 2011; Shukla & Trout, 2011).

Another more recent report describing the use of excipients to stabilize proteins discusses the use of an L-Arg-HCl and L-Glu combination to enhance the conformational stability of mAbs (Kheddo et al., 2014). In this study, the temperatures for the onset of aggregation (T_{agg}) and the unfolding transition (T_m) were both measured using static light scattering (SLS) and intrinsic fluorescence (Kheddo et al., 2014). It was discovered that the combination of L-Arg-HCl and L-Glu increased the values of T_{agg} for a set of three mAbs. However, this excipient combination increased the T_m for only one of the three mAbs. The use of T_{agg} appears to be helpful during initial formulation screening studies to gain insight into conformational stability; however, it remains to be seen whether the value of T_{agg} can be used as a guide for long-term storage stability. In another study examining the prediction of long-term storage stability with thermal unfolding measurements, it was found that increasing the thermal unfolding temperature of the Fab transition of a mAb appeared to be a better formulation strategy than increasing either the onset (T_{onset}) or midpoint (T_m) value of the first unfolding transition (Brader et al., 2015).

2.4 Formulation Strategies for Improving Conformational Stability

There are two general formulation approaches for increasing the conformational stability of a protein. The most general is the addition of an excluded solute (also referred to as cosolvents or osmolytes in some literature references). This mechanism is the most generalized approach toward increasing conformational stability by proper selection of excipients, as it can work for any protein where the overall surface area increases in the unfolded state compared to the native state (which is almost always true) (Holthauzen, Auton, Sinev, & Rosgen, 2011; Street, Bolen, & Rose, 2006; Timasheff, 1998). Of course, the degree to which this works in practice must be tested. As shown in Fig. 1, the exclusion of the solute leads to an increase in the chemical potential of the native state (N), as negative binding is thermodynamically unfavorable. However, the effect is even greater for the denatured (D) state, as its overall surface area is larger. Thus, the total effect is to increase the energy gap between the native (N) and denatured (D) states, which corresponds to an increase in the free energy of unfolding (ΔG_u). As most common excipients (sugars, polyols, polymers, most amino acids, some salts)

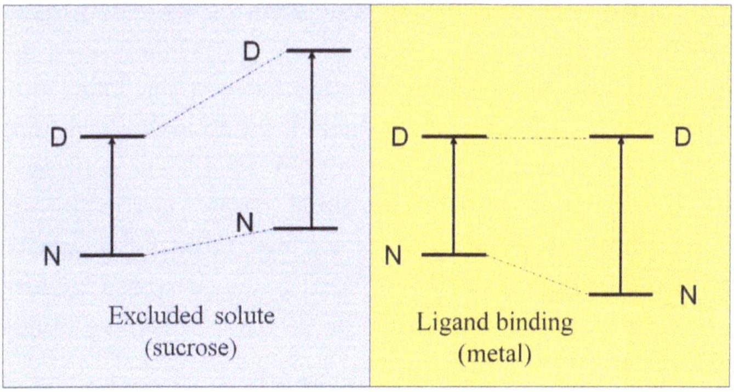

Fig. 1 Generalized approach for increasing the conformational stability of a protein, either by the excluded solute mechanism (*left panel*) or by ligand binding to the native state (*right panel*).

are preferentially excluded from the protein surface, this mechanism applies for most any liquid protein formulation. The effects of these excluded solutes are quite weak, so the more that can be added to the formulation, the greater the impact on stability. Also, the effect tends to be roughly linear with concentration (Chalikian, 2016; Poddar, Ansari, Singh, Movahedi, & Ahmad, 2010; Tiwari & Bhat, 2006), with larger effects for proteins with low intrinsic stability (Kamal et al., 2011). It should also be noted that the effects are additive. In other words, there is so synergistic effect of adding two or more excluded solutes (Holthauzen & Bolen, 2007).

As stated above, many common excipients are excluded from the surface of the protein, leading to a net increase in conformational stability. This includes virtually all sugars and polyols, as well as nearly any amino acid (arginine is the primary exception). In addition, polymers are almost always excluded (although these effects are often referred to in terms of macromolecular crowding). With polymer excipients, one can imagine using PEG, dextrans, and even other proteins to stabilize a target protein molecule (Chebotareva, Kurganov, & Livanova, 2004; Ellis & Minton, 2006). Still, most examples in the literature tend to focus on the use of sugars and/or polyols. It should be noted that polyols can improve the stability of proteins via the excluded solute mechanism, but they may also decrease solubility or increase viscosity as well (Auton, Rosgen, Sinev, Holthauzen, & Bolen, 2011; Kumar, Singh, & Kalonia, 2012). Larger polyols tend to be better stabilizers than smaller polyols (Bhat & Timasheff, 1992; Poddar et al., 2010). Finally, the effectiveness of a given polyol has been reported to be related to their hydrophobicity (Abbas, Sharma, Patapoff, & Kalonia, 2011; Kumar, Dixit, Zhou, & Fraunhofer, 2011). Of course, the relative effectiveness of a given conformational stabilizer must be determined experimentally for a given protein.

As the excluded solute interactions are fairly weak, using more of an excluded solute will be beneficial, as the effect is roughly linear in excipient concentration (Chalikian, 2016; Tiwari & Bhat, 2006). In many formulations, the overall tonicity is constrained to be isotonic with bodily fluids, thereby limiting the amount of excluded solute that can be added. However, if the drug product is intended to be delivered by intravenous (IV) infusion, then the tonicity limits may be loosened, allowing a hypertonic product to be developed.

It should be noted that despite decades of developing protein formulations using the excluded solute effect as a guiding principle, there have been reports, many appearing recently, that have called into question this mechanistic framework (Bruździak, Panuszko, & Stangret, 2013; Panuszko, Bruzdziak, Kaczkowska, & Stangret, 2016), suggesting that another mechanism may be operable to achieve increased conformational stability (Bruzdziak, Panuszko, Jourdan, & Stangret, 2016). In part, it is because the degree of preferential exclusion does not correlate with the increase in conformational stability. In some cases, excipients, like trehalose, were found not to be excluded, yet confer increased stability (Barnett et al., 2016).

The other general mechanism for conformational stabilization of a protein involves ligand binding. In this process, the ligand binds to the native state, reducing its chemical potential, as the binding is thermodynamically favorable (Cimmperman et al., 2008). If there is no appreciable interaction with the denatured state, the net free energy of unfolding, ΔG_u, is increased and the conformational stability has been enhanced (Fig. 1). Numerous examples of this approach toward protein stabilization can be found in the literature. For example, fatty acids analogs, such as sodium octanoate, are often added to human serum albumin (HSA) products. This excipient binds directly to the protein in a high-affinity site, providing a significant increase in conformational stability (Kawai et al., 2017). Ligand binding can be seen for a wide range of excipients, including cofactors (Curnow & Booth, 2010; Gruia et al., 2017), buffers (Zbacnik et al., 2017), metals (Kumari, Rosenkranz, Fitter, & Kayastha, 2011; Niu et al., 2010), anions (Muzammil, Kumar, & Tayyab, 2000; Raibekas et al., 2005; Ramos & Baldwin, 2002), polyanions (Derrick et al., 2007), and polycations (Karamitros & Labrou, 2017). Even surfactants can bind to proteins, providing an increase in conformational stability by this mechanism (Bam et al., 1998).

2.4.1 Summary

Loss of native structure, even if to a minor extent, can render a protein more aggregation prone. Therefore, addition of stabilizing excipients can improve conformational stability, thereby increasing the long-term storage stability for many proteins. Stabilization by excipients appears to be due either to preferential exclusion or through a ligand-binding mechanism (see Fig. 1). Stabilization by preferential exclusion, which includes the effects of polyols, sugars, polymers, and most amino acids, is general in nature and will work for stabilization of most any protein. The effect is relatively weak, so higher concentrations will provide greater stabilization. Overall, the increase in stabilization by this mechanism appears to be roughly linear with the concentration of the excipient. On the other hand, ligand binding can be equally effective, even with low concentrations of the excipient. Even weak binding can increase conformational stability by a kcal/mol or more. However, not all proteins will exhibit ligand binding, meaning that screening for ligand binding may prove futile. Still, a wide range of excipients (metals, anions, polymers, surfactants, buffers) have been shown to stabilize proteins via the ligand-binding mechanism.

As important as conformational stability can be to achieving long-term storage of a therapeutic protein, it is not the only aspect of stability that can affect real-time stability. Colloidal stability (described below) can be equally important or even more so for controlling physical stability of proteins. As a result, screening of formulations solely based on changes in conformational stability can be risky, especially if one attempts to compare formulations that are quite different in terms of reversibility upon thermal stress.

3. COLLOIDAL STABILITY

Over the past two decades, the importance of colloidal behavior in the stabilization of proteins has become more widely appreciated. Colloidal stability refers to the fact that protein molecules can behave as independent particles due to their size (Neal, Asthagiri, & Lenhoff, 1998), and as a result can have net attractive or repulsive interactions with each other. These colloidal protein–protein interactions (PPIs) modulate important physical properties such as viscosity, solubility (Cheng, Bianco, Sandler, & Lenhoff, 2008; Franco & Pessôa Filho, 2013; Neal et al., 1998), crystallization (Tessier, Vandrey, et al., 2002; Wilson & DeLucas, 2014), phase behavior (Ahamed et al., 2007; Pellicane, Costa, & Caccamo, 2004; Rakel, Galm, & Bauer, 2015), partitioning into two-phase systems (Kress & Brandenbusch, 2015), and aggregation (Chi et al., 2003; Ho, Middelberg, Ramage, & Kocher, 2003; Hofmann, Winzer, Weber, & Gieseler, 2016; Krishnan et al., 2002; Le Brun, Friess, Bassarab, & Garidel, 2010; Olsen, Andersen, Randolph, Carpenter, & Westh, 2009; Quigley & Williams, 2015; Zhang & Liu, 2003). Colloidal behavior has also been shown to be important in the clearance of antibodies in vivo (Kelly et al., 2015). Thus, efforts to understand the colloidal behavior of a given protein molecule should be undertaken during the formulation development process to aid in designing an effective final formulation. The following section reviews the techniques generally used to measure colloidal stability of protein solutions. In addition, the effects of colloidal behavior on solubility, viscosity, and phase separation are discussed. Finally, the impacts of excipients on modulating colloidal stability are presented.

3.1 Methods for Measuring Colloidal Stability

The second osmotic virial coefficient (A_{22} value) is a physical parameter that is used to directly measure the colloidal stability of a protein solution (Hofmann, Winzer, Weber, & Gieseler, 2014). Traditionally, SLS has been employed to determine A_{22} from dilute solutions of proteins (Dieterle, Blaschke, & Hasse, 2013; Hofmann et al., 2014; Minton, 2016), although many other methods have been used to obtain similar colloidal stability metrics (Geng, Cheung, Narasimhan, Shameem, & Tessier, 2014). Although one can obtain an accurate measure of colloidal stability from A_{22} SLS measurements, the technique itself is material- and labor intensive (Fernández & Minton, 2008). SLS measurements require that multiple protein concentrations be evaluated, as the A_{22} value comes from the slope of a Zimm plot which graphically represents the light scattered from the protein as a function of scattering angle and protein concentration (Nguyen, Wiencek, & Kirsch, 2003). Cleanliness is important when making these measurements, and care must be taken to ensure samples are rigorously free of dust particles. Moreover, the effects of the excipients must also be considered to obtain accurate measurements (Dieterle et al., 2013; Wills & Winzor, 2017). Finally, as the concentration of the protein increases, higher order terms, such as the third osmotic virial coefficient (A_{33} value), can become important and need to be taken into consideration. There has been some discussion about the real and apparent values of A_{22} from SLS data (Siderius, Krekelberg, Roberts, & Shen, 2012), while the use of SLS for higher protein concentrations has been debated as well (Blanco, Sahin, & Roberts, 2011).

As a consequence, other techniques have evolved to obtain virial coefficients describing the colloidal interaction of proteins. Other methods have also been described that allow one to determine B_{22} values, which is the designation for the second virial coefficient obtained from methods other than static light scattering. These include SANS (Liu, Fratini, Baglioni, Chen, & Chen, 2005; Receveur, Durand, Desmadril, & Calmettes, 1998), SEC (Bajaj, Sharma, & Kalonia, 2004; Bloustine, Berejnov, & Fraden, 2003; Petsev, Thomas, Yau, & Vekilov, 2000; Printz, Kalonia, & Friess, 2012; Winzor, 2016), HIC (To & Lenhoff, 2007), SAXS (Kaieda, Lund, Plivelic, & Halle, 2014; Mosbaek, Konarev, Svergun, Rischel, &

Vestergaard, 2012), AUC (Schaink & Smit, 2007), osmotic pressure (Schaink & Smit, 2007), NMR (Liu, Gong, Dong, & Tang, 2016), X-ray scattering (Inouye, Houde, Temel, & Makowski, 2016), and self-interaction nanoparticle spectroscopy (Geng, Wittekind, Vigil, & Tessier, 2016; Jayaraman et al., 2014; Liu et al., 2014; Sule et al., 2011).

Most often, B_{22} values are measured using self-interaction chromatography (SIC), which was first described in 1996 (Patro & Przbycien, 1996) and then elaborated by many other groups (Ahamed, Ottens, van Dedem, & van der Wielen, 2005; Deshpande et al., 2009; Hedberg, Heng, Williams, & Liddell, 2015; Hedberg, Heng, Williams, & Liddell, 2016; Martin & Lenhoff, 2011; Quigley, Heng, Liddell, & Williams, 2013; Tessier, Lenhoff, & Sandler, 2002; Tessier, Vandrey, et al., 2002; Valente, Payne, Manning, Wilson, & Henry, 2005; Winzor, Scott, & Wills, 2007). Together, these articles provide a primer for using SIC to determine B_{22} values. In many cases, the values obtained by SIC correlate well with those determined from SLS (Johnson, Parupudi, Wilson, & DeLucas, 2009; Le Brun, Friess, Bassarab, Muhlau, & Garidel, 2010; Quigley & Williams, 2015; Wilson & DeLucas, 2014).

In addition to self-interaction, there have been measurements made of crossinteraction, that is, the interaction between two different protein molecules. Typically, this is done using crossinteraction chromatography (CIC) (Teske, Blanch, & Prausnitz, 2004; Tessier, Sandler, & Lenhoff, 2004), which has been used to identify soluble mAb species (Jacobs, Wu, Feng, Bethea, & O'Neil, 2010; Spencer, Bethea, Raju, Giles-Komar, & Feng, 2012; Sun et al., 2013). Analogous studies to identify soluble antibody candidates have been performed using affinity capture-SINS (Sun et al., 2013; Wu et al., 2015) and mixed-mode chromatography (Kohli et al., 2015; Yang et al., 2015). Of note, the results from CIC studies do correlate with SIC data (Sun et al., 2013). In fact, one can extract B_{22} and B_{23} values from a single set of CIC experiments (Quigley & Williams, 2016). In addition, osmotic second virial coefficients have been determined for both the native and denatured state of lysozyme (Dong, Liu, Liu, & Sun, 2009). Thus, it is possible to investigate the colloidal stability of both the folded and unfolded states of a protein.

There has been some discussion in the literature about the correlation of A_{22} and B_{22} values (Wills, Scott, & Winzor, 2015), but for the formulation scientist, both can be considered as reliable indicators of the overall colloidal stability of a protein in aqueous solution. Note that the exact designation of osmotic virial coefficients can vary depending on the discipline, the method used to obtain the value, and whether it has been corrected for hard-sphere interactions.

3.2 k_D Values

Another measure of colloidal stability that has emerged is the determination of k_D values from dynamic light scattering (DLS) experiments (He et al., 2010; He, Woods, Becker, Narhi, & Razinkov, 2011; Lehermayr, Mahler, Mader, & Fischer, 2011; Yadav, Scherer, Shire, & Kalonia, 2011). k_D values can be obtained from a plot of diffusion coefficient vs concentration (Lehermayr et al., 2011; Neergaard et al., 2013). Deviation from linearity has been observed and appears to correlate with increased viscosity and aggregation potential (Bauer, Gobel, Schwab, Schermeyer, & Hubbuch, 2016). The k_D value is a primary component of the osmotic second virial coefficient (Arzensek, Kuzman, & Podgornik, 2012; Nicoud, Lattuada, Yates, & Morbidelli, 2015), and the correlation between k_D values and B_{22} values has been widely investigated (Menzen & Friess, 2014; Roberts et al., 2015; Saito et al., 2012). For the most part, the two measures of colloidal stability agree reasonably well, although some sizable differences have been reported (Kumar et al., 2011). In any case, both metrics can be regarded as providing a quantitative measure of protein colloidal stability.

There are many examples of k_D and/or other measures of colloidal stability correlating to aggregation propensity, especially for mAbs at higher concentrations (Banks et al., 2012; Ghosh, Calero-Rubio, Saluja, & Roberts, 2016; Kim et al., 2013; Kumar et al., 2011; Rubin, Linden, Coco, Bommarius, & Behrens, 2013; Saito et al., 2012, 2013; Zheng et al., 2017). In addition, it is worth noting that both colloidal and conformational stability can have comparable effects on aggregation behavior as well (Chou, Krishnamurthy, Manning, Randolph, & Carpenter, 2012; Goldberg, Bishop, Shah, & Sathish, 2011; Kalonia et al., 2016; Yageta et al., 2015). Thus, both conformational and colloidal aspects of stability play a critical role in controlling aggregation behavior. However, one must exercise caution when using DLS for assessing colloidal stability (Fischer & Schmidt, 2016). Hydrodynamic nonidealities may produce incorrect k_D values by DLS (Blanco, Perevozchikova, Martorana, Manno, & Roberts, 2014). An example of this is the incorrect estimate of hydrodynamic radius, and consequently k_D values (Sorret, DeWinter, Schwartz, & Randolph, 2016), in solutions of low ionic strength and/or low pH (Arakawa, Ejima, & Akuta, 2017; Arakawa, Philo, Ejima, Tsumoto, & Arisaka, 2007).

3.3 Colloidal Stability and Solubility

The correlation between solubility and B_{22} is sigmoidal rather than linear (Franco & Pessôa Filho, 2013; Wilson & DeLucas, 2014). Such a correlation has been documented in a number of studies (Ahamed et al., 2007;

Demoruelle et al., 2002; Guo et al., 1999; Mehta, White, & Litster, 2012; Rakel et al., 2015). In general, it has been shown that as B_{22} values become more negative, solubility starts to decrease, often precipitously, and then appears to plateau at values near -8 or -10×10^{-4} mol mL g^{-2} (for example, see Table 4).

The Haas–Drenth–Wilson theoretical model correlates solubility with B_{22} values (Haas, Drenth, & Wilson, 1999). A similar model was derived by Lenhoff and coworkers (Ruppert, Sandler, & Lenhoff, 2001). With this framework, one can effectively use measures of colloidal stability to guide formulation development toward conditions of greater solubility. This is important in the early stage development of many proteins, where a certain concentration must be achieved to be clinically beneficial. In other words, if one cannot adjust the formulation to obtain a protein concentration that can allow proper dosing, then stability issues become a moot point. It is important to note that each protein has an intrinsic solubility limit. Formulation

Table 4 Relationship of Solubilities and B_{22} Values of Three Model Proteins

Protein	B_{22} ($\times 10^4$ mol mL g^{-2})	Solubility (mg/mL)	References
Equine serum albumin	-1.0 ± 0.3	35.7 ± 1.5	Demoruelle et al. (2002)
	-1.7 ± 0.1	15.0 ± 0.7	
	-2.4 ± 0.3	6.3 ± 0.1	
	-4.2 ± 0.3	1.6 ± 0.3	
Ovalbumin	-0.5 ± 0.6	46.4 ± 2.3	Demoruelle et al. (2002)
	-1.9 ± 0.4	17.1 ± 0.9	
	-3.0 ± 0.4	18.8 ± 0.8	
	-2.3 ± 0.4	15.3 ± 0.9	
	-3.3 ± 0.5	15.0 ± 0.8	
	-3.5 ± 0.4	10.3 ± 0.5	
	-4.2 ± 0.2	6.9 ± 0.2	
Lysozyme	$+0.1 \pm 0.5$	55.8 ± 2.6	Guo et al. (1999)
	-3.2 ± 0.5	14.5 ± 0.8	
	-4.4 ± 0.3	7.2 ± 0.3	
	-5.3 ± 0.4	3.8 ± 0.1	
	-7.4 ± 0.3	2.1 ± 0.0	

development can only move the protein concentration toward that upper limit, and will not be able to overcome intrinsic limitations of aqueous solubility.

3.4 Colloidal Stability and Viscosity

As more groups seek to develop high concentration (HC) formulations, particularly of mAbs, the limitations of increased viscosity have become apparent. Over the past decade or more, a number of articles have appeared which demonstrate that viscosity behavior is directly related to colloidal stability, although it is often referred to in other terms such as self-interaction, cluster formation, etc., as described below.

The behavior of HC protein formulations, particularly with respect to viscosity behavior, has been reviewed recently (Jezek et al., 2011; Tomar, Kumar, Singh, Goswami, & Li, 2016), although some reports discussing viscosity of HC protein solutions go back more than 30 years (Lefebvre, 1982).

There have been numerous recent studies that have found a correlation between colloidal stability, usually in terms of k_D values, and the viscosity of concentrated protein solutions, especially mAbs (Chow, Allan, Chai, Atwell, & Lu, 2016; Connolly et al., 2012; Lorenz et al., 2014; Saito et al., 2012; Yadav, Shire, & Kalonia, 2012). However, there is still some disagreement as to whether colloidal stability measurements made at low concentrations (as is usually done to determine k_D values by DLS) will accurately reflect the behavior at high concentration (Dear, Hung, Truskett, & Johnston, 2017). In the case of nonspherical proteins, such as antibodies and fusion proteins, the PPIs that seem to dominate colloidal stability appear to be dipolar in nature (Fast, Cordes, Carpenter, & Randolph, 2009; Singh, Yadav, Shire, & Kalonia, 2014; Yadav, Liu, Shire, & Kalonia, 2010; Yadav, Shire, & Kalonia, 2010), although some correlations to zeta potential have also been noted (Li, Kumar, et al., 2014; Neergaard et al., 2013).

The concentration-dependent viscosity behavior of proteins has often been associated with reversible self-association (RSA), which is another physical manifestation of colloidal instability (Arora et al., 2016; Fukuda et al., 2017; Lilyestrom, Yadav, Shire, & Scherer, 2013; Sarangapani et al., 2016). Another recent development has been the correlation of increased viscosity with cluster formation (Castellanos, Pathak, & Colby, 2014; Fukuda, Moriyama, Yamazaki, Imaeda, & Koga, 2015; Lilyestrom et al., 2013; Yearley et al., 2014, 2013; Zarraga et al., 2013), whether this be a transient association (e.g., RSA) or the actual presence of cluster aggregates (irreversible clusters). Typically, rigorous removal of aggregates will lower the viscosity of a given HC solution (Castellanos, Pathak, Leach, Bishop, & Colby, 2014; Nicoud, Lattuada, et al., 2015).

One final point worth noting is that some of the approaches listed above for determining colloidal stability allow one to evaluate colloidal stability directly at higher protein concentrations. Even though osmotic virial coefficients are defined for dilute solutions, it is clear that an effective estimate of PPIs can be obtained and that these values can guide formulation development, whether one seeks to lower viscosity or hinder aggregation. This has been shown to be an effective approach for assessing the concentration-dependent solution behavior of mAbs (Hofmann et al., 2016; Inouye et al., 2016; Mosbaek et al., 2012; Sule et al., 2011).

3.5 Colloidal Instability and Phase Separation

In addition to facilitating RSA, colloidal instability is also the basis for other types of physical instability, such as opalescence and liquid–liquid phase separation (LLPS). While these phenomena may or may not be directly related to aggregation, they are examples of physical instability in and of themselves (Vekilov, 2010). In the case of LLPS, the relationship between B_{22} values (or k_D values) and phase separation has been investigated in some detail (Liu, Wang, & Ching, 2010; Petsev, Wu, Galkin, & Velikov, 2003; Rozhkov & Goryunov, 2010; Wolf et al., 2014), including for mAbs (Lewus, Darcy, Lenhoff, & Sandler, 2011; Lewus, Levy, Lenhoff, & Sandler, 2015; Raut & Kalonia, 2015). In some cases, opalescence and LLPS seem to be interrelated (Mason, Zhang, Remmele, & Zhang, 2011; Mason, Zhang-van Enk, Zhang, Remmele, & Zhang, 2010; Reiche, Hartl, Blume, & Garidel, 2017), but both are governed by changes in colloidal stability. As with viscosity behavior, there has been a report of cluster formation facilitating LLPS (Zhang et al., 2012). In addition, salts primarily appear to promote LLPS via a salting-out mechanism (Rozhkov & Goryunov, 2012).

3.6 Formulation Strategies for Increasing Colloidal Stability

Of all the classes of excipients, the effects of salts have been most widely investigated for their effects on the colloidal stability of proteins. This is due to the fact that favorable PPIs are primarily due to electrostatic repulsion. Addition of salts can lead to charge shielding and reduced electrostatic repulsion, thus leading to poorer colloidal stability. In the extreme, this will lead to salting-out or precipitation of the native protein.

Some systems respond favorably to the addition of NaCl or other salts, as was shown in the case of reduced viscosity for several mAbs (Kanai, Liu, Patapoff, & Shire, 2008; Liu, Nguyen, Andya, & Shire, 2005; Zarraga et al., 2013). However, it should be stated that many mAbs do not exhibit decreased viscosity (or very little change) in the presence of increased NaCl concentrations (Binabaji, Ma, & Zydney, 2015; Liu, Nguyen, et al., 2005). The basis for viscosity reduction for some mAbs in the presence of salt is attributed to decreased RSA, which is driven by electrostatic interactions. Lower levels of RSA can occur in the presence of salts, with chaotropic anions being more effective than kosmotropic ones (Kanai et al., 2008). For salts, reduced solubility is usually observed as well. A series of studies on hydrophobic salts and their ability to reduce viscosity have appeared (Du & Klibanov, 2011; Guo et al., 2012; Larson et al., 2017). These studies seem to indicate that aliphatic, bulky salts are the most effective in this regard (Guo et al., 2012).

Buffers can also impact colloidal stability (Zbacnik et al., 2017). For example, acetate and citrate were found to yield higher solubilities for an antibody compared to phosphate (Parkins & Lashmar, 2000). McIlvaine's buffer, a mixture of citrate and phosphate, has been reported to have a salting-in effect, resulting in increased solubility (Li, Kasha, Late, & Banga, 2009). Whether these cases represent exceptions to the usual salting-out behavior or are examples of solubilization by modulation of colloidal stability by some other mechanism is not clearly understood.

There are some reports which state that sugars and polyols increase solubility (Antipova & Semenova, 1995, 1997; Arakawa & Timasheff, 1985; Esteve-Romero, Bossi, & Righetti, 1996), and at least one issued patent on the use of starches to increase protein solubility (Brewer, Kim, Montgomery, & Sheety, 1988). Conversely, one study found decreased solubility of an IgG1 in the presence of sugars (Rubin, Sharma, Linden, Bommarius, & Behrens, 2014). Sugars have also been shown to directly increase B_{22} values (Baek, Singh, Arunkumar, & Zydney, 2017; James & McManus, 2012; Valente, Fryksdale, Dale, Gaertner, & Henry, 2006), especially when used at pH values near the isoelectric point of the protein (Valente et al., 2006). Polyols, such as glycerol, have been shown to be effective at increasing solubility (Charron, Kadri, Robert, Giege, & Lorber, 2002; Farnum & Zukoski, 1999; Kulkarni & Zukoski, 2002; Sleutel et al., 2012).

A number of reports can be found stating that amino acids, especially basic ones (Arg, Lys, His), reduce viscosity of HC protein solutions, presumably by improving colloidal stability (Dear et al., 2017; Inoue, Takai, Arakawa, & Shiraki, 2014a; Wang et al., 2015). Interestingly, it was discovered that while basic amino acids were effective at reducing viscosity for HC antibody solutions, they were not helpful in reducing the viscosity of other highly concentrated globular proteins (Inoue, Takai, Arakawa, & Shiraki, 2014b). The use of His to reduce the viscosity of HC antibody solutions is the subject of patents by Liu and Shire (2005, 2012, 2014), while the use of amino acid combinations to reduce viscosity has been recently patented as well (Chang, 2016). Besides basic amino acids, proline (Pro) has also been reported to be effective at lowering viscosity for concentrated IVIG products (Maeder et al., 2011).

Another amino acid-based system that appears to be effective at lowering viscosity (Borwankar et al., 2016; Kheddo, Cliff, Uddin, van der Walle, & Golovanov, 2016), as well as improving solubility (Golovanov, Hautbergue, Wilson, & Lian, 2004) and stability (Golovanov et al., 2004; Kheddo et al., 2014), is the combination of Arg and Glu (Arg-Glu). This equal stoichiometric mixture, usually employed at 50 mM, appears to reduce unfavorable PPIs thereby increasing solubility and decreasing viscosity.

3.6.1 Summary

Colloidal stability of proteins, that is, the nature of PPIs, can be quantified using A_{22}/B_{22} or k_D values. It is clear that colloidal stability can directly impact aggregation behavior of proteins. Electrostatic interactions dominate colloidal stability, especially at lower concentrations. In general, pH is the factor that has the largest impact, as it controls the overall charge state of the protein molecule. Certain excipients also have marked effects on PPIs. For the most part, addition of ionic compounds, like NaCl, lead to charge shielding, resulting in diminished electrostatic repulsion and poorer colloidal stability. There are number of reports of compounds that improve colloidal stability, including sugars, amino acids, and cyclodextrins.

In addition to the impact of PPIs on storage stability, colloidal stability is integral for crystallization, solubility, and phase separation of proteins as well. Consequently, it is an essential element for obtaining a formulation that is manufacturable as well as stable over an extended period of time. Of the excipients that have been shown to modulate colloidal stability, many, if not all, will also impact conformational stability. Therefore, it is important to identify how to achieve maximal conformational as well as colloidal stability for any given liquid protein composition.

4. INTERFACIAL INSTABILITY

Protein molecules are only marginally stable in solution and are susceptible to stress conditions during production, distribution, storage, and delivery which can adversely affect their stability (Pinholt, Hartvig, Medlicott, & Jorgensen, 2011). For example, it is well known that common manufacturing processes such as filtration, mixing, storage, freeze–thaw, lyophilization, nebulization, and transportation can impart interfacial stress to a protein that can change its structure and cause aggregation (Randolph & Jones, 2002). Thus, to maintain product quality, the interfacial stability of a given protein should be carefully evaluated during the product development phase. Interfacial stress commonly occurs at air–liquid, liquid–liquid, and liquid–solid interfaces. The direct exposure of proteins to glass, plastic polymers, metal, air, ice crystals, and silicone oil can lead to adsorption losses and degradation (Pinholt et al., 2011; Rabe, Verdes, & Seeger, 2011). This degradation can be further exacerbated by mechanical stress, such as shearing and agitation (Hawe et al., 2012; Tamizi & Jouyban, 2016). Degradation resulting from interfacial stress has been studied extensively for protein drugs under different stress conditions (Martos et al., 2017). Depending on the mechanism of degradation, robust process and formulation strategies have been developed, and these will be reviewed and discussed in the following section. Armed with this information, the formulation scientist will be well equipped to design formulations rationally that will mitigate or prevent protein loss or damage resulting from interfacial processes.

4.1 Mechanisms of Interfacial Degradation

Although the molecular details of interfacial stress on protein stability are not fully understood, there are a number of mechanisms by which interfacial stress can cause damage to proteins (Lee, McAuley, Schilke, & McGuire, 2011; Randolph & Jones, 2002). These include (1) preferential accumulation of protein at the liquid–air interface resulting in the alteration of protein structure, (2) direct binding of protein to the contact surface resulting in loss of the protein molecule and a conformation change, and (3) leachables from a contact surface causing protein degradation. The details of these mechanisms are discussed below.

Degradation resulting from preferential accumulation of protein at an interface is, for the most part, caused by the amphiphilic and surface active nature of proteins. Without the presence of a more surface active molecule (i.e., a surfactant), proteins can accumulate at an air–liquid interface and over time can undergo conformational changes or partial unfolding (Haynes & Norde, 1995; Lee et al., 2011; Norde, 2008; Vidanovic, Askrabic, Stankovic, & Poprzen, 2003). The extent of structural rearrangement appears to be correlated with the intrinsic conformational stability of the protein in solution (Haynes & Norde, 1995). Increased colloidal stability can also slow aggregation due to interfacial stress (Le Brun, Friess, Bassarab, & Garidel, 2010; Thirumangalathu et al., 2009).

These structurally altered proteins have a higher tendency to interact irreversibly with each other (compared to native state), mainly through hydrophobic interactions (Alves, Reis, & Hunt, 2010; Norde, 2008). In turn, these irreversible interactions generally lead to aggregation and/or particulate formation. Surface-mediated protein degradation can be further enhanced by mechanical stress, such as shearing, shaking, mixing, and dropping. Nonionic surfactants, such as polysorbate 20 (PS 20) and polysorbate 80 (PS 80), are widely used to protect proteins against surface-mediated aggregation and particulate formation. The effect of nonionic surfactants in protecting protein from interfacial stress has been widely studied. Indeed, it has been reported that addition of a surfactant to most antibody formulations can reduce aggregate or particulate formation, in particular under agitation conditions (Wang et al., 2007). Regarding the mechanism of stabilization, it is thought that surfactants protect against aggregation by directly binding to the protein, thus minimizing PPIs, or by accumulating at the air–liquid interface thereby preventing protein accumulation at the interface. Furthermore, it is known that addition of a surfactant to a solution reduces its surface tension, thereby decreasing the overall free energy of the system. As the free energy of the system is reduced, protein–protein and protein–surface interactions are reduced as well (Chang, Kendrick, & Carpenter, 1996).

Interfacial-mediated protein degradation can also occur at a liquid–liquid interface, in particular at an interface containing a hydrophobic organic compound. The strong hydrophobic nature of certain organic compounds can facilitate interaction with proteins and induce conformational changes. This in turn can lead to the formation of protein aggregates and particulates. For example, adsorption of a mAb to silicone oil droplets has been related to protein aggregation (Basu, Krishnan, Thirumangalathu, Randolph, & Carpenter, 2013; Britt et al., 2012), especially under agitation conditions (Thirumangalathu et al., 2009). In fact, exposure to silicone oil without agitation may not cause any adverse effects, in terms of aggregation. An approach toward mitigating interfacial aggregation of a protein is to incorporate a nonionic surfactant into the final formulation (Basu, Blake-Haskins,

O'Berry, Randolph, & Carpenter, 2014; Britt et al., 2012; Ludwig, Carpenter, Hamel, & Randolph, 2010).

Besides the potential of forming altered structures/conformations at liquid–air and liquid–liquid interfaces, proteins can also interact with a solid surface which can lead to the adsorption and reduction of protein concentration in solution. The interaction of a protein with a solid surface is thought to be primarily hydrophobic in nature. These types of surface interactions are particularly problematic for low concentration protein formulations, as significant amounts of protein can be lost to the container/closure surface, as was demonstrated for a low concentration antibody formulation (Morar-Mitrica et al., 2015). Addition of a surfactant to low concentration protein formulations has been shown in several examples to be an effective strategy to prevent loss from irreversible adsorption of the protein to the container/closure surface (Rabe et al., 2011; Wang et al., 2017).

For long-term storage, the potential leachables from the contact surface should also been carefully evaluated, especially for plastic or polymer materials (Rabe et al., 2011). Although a surfactant can protect proteins from interfacial-mediated degradation, it may also enhance the extraction of leachables into solution. These leachables can not only lead to safety concerns but can also very often result in physical and chemical degradation of the protein product as well. Control of surfactant concentration in a formulation is an effective approach to reduce the leachable content in solution. In addition, the use of inert coating materials, such as Fluortec on butyl stoppers, can effectively reduce extraction of leachables from these types of materials. On the other hand, poly(tetrafluoroethylene) (PTFE, Teflon) is extremely hydrophobic and can be the site of interfacial damage to proteins as well (Jorgensen et al., 2011). The negative effect of a leachable on product quality/safety was demonstrated previously for Eprex (epoetin alpha) syringes with uncoated rubber stoppers. The leachable from the uncoated stopper was thought to have acted as an adjuvant and caused an increase in the immune response against the API. This immune response was largely minimized by using a coated rubber stopper (Boven et al., 2005; McCormick, 2004).

4.2 Formulation Strategies for Addressing Interfacial Instability

Nonionic surfactants are often used in protein formulations to reduce interfacial stress and prevent protein denaturation, aggregation, and adsorption loss (Lee et al., 2011). Thus, if interfacial stress is a stability concern for a particular protein molecule, then use of a nonionic surfactant is a good first step in mitigating this problem. For example, PS 20 (also known as Tween 20) and PS 80 (also known as Tween 80) are the two most commonly used nonionic surfactants in biopharmaceutical products (Khan, Mahler, & Kishore, 2015). Both compounds have shown low toxicity and high surface activity, even at low concentrations, although some issues with polysorbate toxicity have been raised (Maggio, 2012a, 2012b). The typical polysorbate (PS) concentration contained within therapeutic protein products ranges from 0.0001% to 0.1% (w/v). For example, the polysorbate concentration in Neulasta is 0.004%, while that of HUMIRA is 0.1%. Besides polysorbates, other alternative surfactants have also been used successfully in commercial products. Two such examples include poloxamer (Gazyva, Orencia, and Rebif) and sodium dodecyl sulfate (PROLEUKIN) (Martos et al., 2017). A list of the surfactants approved for use in parenteral protein products is provided in Table 5.

Generally, the amount of polysorbate included in a product should only be enough to achieve the desired protective effect, which is usually at or above the critical micelle concentration (cmc), although examples are known where the protein is protected by surfactant levels below the cmc (Mahler, Senner, Maeder, & Mueller, 2009; Parkins & Lashmar, 2000). More recently, a study on two mAbs indicated that the concentration of the surfactant and its corresponding cmc is not relevant to stabilization, provided the amounts exceed a critical threshold level near 0.02% (Wang et al., 2017).

Table 5 Surfactants Found in Approved Parenteral Protein Products (in the United States)

Generic Name	Trade Names
Polysorbate 20	Tween 20
Polysorbate 40	Tween 40
Polysorbate 80	Tween 80
Poloxamer 188	Pluronic F-68
Sodium dodecyl sulfate (SDS)	
Sorbitan monopalmitate	Span 40[a]
Lecithin	
Sodium glycocholate	
Polyoxyethylated castor oil	Cremophor EL

[a]For intramuscular administration only.

The selection and concentration of a nonionic surfactant is often determined via agitation (Bai, Bee, Biddlecombe, Chen, & Leach, 2012; Eppler, Weigandt, Hanefeld, & Bunjes, 2010) and repeated freeze–thaw studies (Radmanovic, Serno, Jörg, & Germerhaus, 2013). These studies are usually conducted as part of formulation screening in early development. It is generally recommended that such studies are conducted using the same container closure system that will be employed for the drug product and/or drug substance. Additionally, the conditions tested should mimic the worst stress conditions the protein will be exposed to during production, transportation, and delivery. In the case of prefilled syringes (PFS), addition of a surfactant and/or removal of the air bubble, if possible, can significantly reduce the formation of aggregates and particles upon agitation (Gerhardt et al., 2014, 2015).

The assessment of product quality after stress testing should include assays that not only detect soluble aggregates, but subvisible and visible particles as well. Tables 6 and 7 give the results of an agitation screening study conducted for a mAb in PFS. For this study, samples were formulated in the same base formulation (F1) with and without PS 80. Mechanical stress

Table 6 Summary of SEC Results After Stress Testing

	***t*0**		**24 h agitation, 500 rpm**		**12 × Dropping, Ambient**	
Formulation	**Main (%)**	**HMWS (%)**	**Main (%)**	**HWMS (%)**	**Main (%)**	**HMWS (%)**
F1 without PS 80	99.8	0.1	99.8	0.1	99.8	0.1
F1 with 0.1% PS 80	99.5	0.3	99.5	0.3	99.5	0.3

HWMS, high-molecular weight species; main, main peak by size exclusion chromatography.

Table 7 Summary of Subvisible (≥10 μm) and Visible Particle Results After Stress Testing

	***t*0**		**24 h Agitation, 500 rpm**		**12 × Dropping, Ambient**	
Formulation	**SbVP, Counts/mL**	**VP**	**SbVP, Counts/mL**	**VP**	**SbVP, Counts/mL**	**VP**
F1 without PS 80	617	No	1420	Yes	14,865	Yes
F1 with 0.1% PS 80	395	No	549	No	1754	No

VP, visible particle; SbVP, subvisible particles.

testing was performed by both agitation on a rotary shaker and by drop testing (Randolph et al., 2015). The drop test included 12 drops from a bench top (standard height) with the syringe packaged in a rectangular box (i.e., primary packaging). Although both agitation stresses did not appear to impact soluble aggregate formation under the conditions investigated, both were shown to have a significant impact on the levels of subvisible particles and visible particles measured. In addition, it appears that the dropping stress had a more adverse impact on product quality for this particular protein than shaking. For both stress conditions, the addition of surfactant clearly helped to minimize the formation of both visible and subvisible particles.

For formulation development, it is important that mechanical stress studies be performed in the container/closure system that will be used for storage. Moreover, there should be a well-defined protocol that allows adequate exposure to the air–water interface (Eppler et al., 2010). Usually, this is done on an orbital shaker, as this produces the most consistent results (Bai et al., 2012). In addition, product quality should also be assessed for the container system that will be used for the product administration. For example, a product intended for IV delivery should be evaluated in IV bags using the appropriate diluent at the projected dose levels. The concentration of surfactant in the formulation should be sufficient to protect the drug product from degradation under these potential stress conditions (Wang et al., 2017). Several recent examples show the importance of performing agitation and compatibility studies in IV bags to ensure product strength and quality upon administration (Glover et al., 2013; Kumru et al., 2012; Morar-Mitrica et al., 2015; Sreedhara et al., 2012). Adsorption to the in-line filters can also be problematic and should be considered as well (Besheer, 2017).

Despite the successful use of polysorbate in therapeutic protein formulations to prevent protein aggregation, surfactants, such as PS 20 and PS 80, can undergo chemical degradation reactions (Kerwin, 2008). This can not only reduce their effectiveness at protecting protein from interfacial degradation but can also result in new unwanted degradants in the protein product. It has been shown previously that residual levels of hydrogen peroxide from surfactant raw materials can lead to the oxidative degradation of both surfactant and protein during manufacturing and storage (Martos et al., 2017). This risk factor usually can be minimized by reducing surfactant concentration and/or by selecting high-quality raw materials. Additionally, implementing strict storage procedures, such as a proper cold chain, as well as including a nitrogen overlay in the product headspace can help to minimize this risk factor (Wang et al., 2007).

The degradation of surfactant can also be mediated by process impurities contained within the drug substance. Recent examples have shown that host cell protein and lipase activity from the upstream process can lead to the hydrolysis of the ester bonds contained in polysorbate surfactants (Labrenz, 2014; Li, Hewitt, et al., 2014; McShan, Kei, Ji, Kim, & Wang, 2016). In one example, subsequent improvements in the purification process appeared to prevent the degradation of surfactant, which was seen earlier when these process impurities were present in the drug substance (Dixit, Salamat-Miller, Salinas, Taylor, & Basu, 2016; Hall, Sandefur, Frye, Tuley, & Huang, 2016). Not only are the degraded surfactants less protective (Labrenz, 2014), but the free fatty acids that are released can be nucleation sites for particulate formation, due to their poor aqueous solubility (Saggu, Liu, & Patel, 2015).

Other excipients have been investigated for their ability to reduce interfacial damage. In the case of silicone oil adsorption, NaCl appears to accelerate adsorption to the oil droplet and lead to increased aggregation (Britt et al., 2012; Ludwig et al., 2010). Likely, this is related to diminished colloidal stability, as discussed above, but the overall process appears to be fairly complicated.

The concept of increasing conformational stability in order to inhibit structural rearrangement on the surface has been explored in some detail. The approach has rarely been successful for a number of reasons. First, sugars appear to accelerate the rate of surface adsorption (Guzey, McClements, & Weiss, 2003). Second, they can also impact colloidal stability and interfacial tension, making interfacial interactions more difficult to predict (Ruíz-Henestrosa, Sánchez, & Patino, 2008). In any case, there are number of examples of polyols and sugars increasing the extent of interfacially induced aggregation (Abbas, Sharma, Patapoff, & Kalonia, 2012; Lim et al., 2016; Liu, Qi, Schwartz, Randolph, & Carpenter, 2013; Wen, Zheng, Wang, Lan, & Yin, 2017). However, even in the presence of these excipients, surfactants have been shown to be effective at protecting the protein (Liu et al., 2013).

4.2.1 Summary

Protein will contact a variety of interfaces during production, storage, and use. Each interface has the potential to cause damage to the protein by adsorption (in and of itself a physical instability), structural rearrangement, and subsequent desorption in a conformationally altered state. The primary approach is to add a surfactant that will prevent or limit exposure of the protein to the interface, thereby reducing the extent of interfacial damage (Lee et al., 2011; Randolph & Jones, 2002). In addition, increasing the colloidal or conformational stability will also reduce interfacial damage, although some excipients, like sugars and polyols, can produce complex behavior with respect to interactions with surfaces (Abbas et al., 2012; Lim et al., 2016; Liu et al., 2013; Wen et al., 2017). A number of surfactants are found in approved injectable protein products. These excipients may not only reduce formation of soluble aggregates but also block the progression from soluble aggregates to particulates (Arakawa & Kita, 2000; Respaud et al., 2014). Modification of the container surface and other packaging strategies may also be beneficial.

5. CONCLUSIONS

For the first time, a rational approach to developing stable liquid formulations of therapeutic proteins has been delineated. While these strategies do not guarantee success, they offer the best chance of obtaining a stable liquid product, given the current state of knowledge. One must seek to maintain or optimize conformational, colloidal, and interfacial stability, while not compromising the chemical stability of the polypeptide. Together, these four aspects (conformational, colloidal, interfacial, and chemical) are like four legs of a table. Loss of any one will greatly reduce the stability of the drug product.

Approaches for achieving maximal stability in each one of these areas have been described in detail. In most cases, the pH of the solution plays a central or dominant role. However, for each of these areas certain excipients can be highly influential in improving the stability of the protein. Not only is excipient selection important, but the nature of the container closure system will affect the stability, especially, with respect to interfacial stability. Moreover, handling considerations must also be taken into account. The formulation strategies described herein should allow one to identify formulations that are stable for an extended period of time (e.g., 24 months) when stored at 2–8°C, which is the usual shelf-life and storage condition for a marketed biopharmaceutical product. On the other hand, most proteins are not stable in aqueous solution when stored at room temperature or higher. However, optimized formulations may be stable for a period of days or weeks when held at 25°C.

REFERENCES

Abaturov, L. V., Burstein, E. A., & Nosova, N. G. (2008). Mechanisms involved in proteolytic degradation of globular proteins: The role of local and global fluctuations in the native structure. *Molecular Biology*, *42*(2), 289–301.

Abbas, S. A., Sharma, V. K., Patapoff, T. W., & Kalonia, D. S. (2011). Solubilities and transfer free energies of hydrophobic amino acids in polyol solutions: Importance of the hydrophobicity of polyols. *Journal of Pharmaceutical Sciences*, *100*(8), 3096–3104.

Abbas, S. A., Sharma, V. K., Patapoff, T. W., & Kalonia, D. S. (2012). Opposite effects of polyols on antibody aggregation: Thermal versus mechanical stresses. *Pharmaceutical Research*, *29*, 683–694.

Ahamed, T., Esteban, B. N. A., Ottens, M., van Dedem, G. W. K., van der Wielen, L. A. M., Bisschops, M. A. T., et al. (2007). Phase behavior of an intact monoclonal antibody. *Biophysical Journal*, *93*(2), 61–619.

Ahamed, T., Ottens, M., van Dedem, G. W. K., & van der Wielen, L. A. M. (2005). Design of self-interaction chromatography as an analytical tool for predicting protein phase behavior. *Journal of Chromatography A*, *1089*(1–2), 111–124.

Aledo, J. C., Canton, F. R., & Veredas, F. J. (2017). A machine learning approach for predicting methionine oxidation sites. *BMC Bioinformatics*, *18*, 430.

Alves, C. M., Reis, R. L., & Hunt, J. A. (2010). The dynamics, kinetics, and reversibility of protein adsorption onto the surface of biodegradable materials. *Soft Matter*, *6*, 4135–4143.

Andersen, C. B., Manno, M., Rischel, C., Thorolfsson, M., & Martorana, V. (2010). Aggregation of a multidomain protein: A coagulation mechanism governs aggregation of a model IgG1 antibody under weak thermal stress. *Protein Science*, *19*(2), 279–290.

Antipova, A. S., & Semenova, M. G. (1995). Effect of sucrose on the thermodynamic incompatibility of different biopolymers. *Carbohydrate Polymers*, *28*(4), 359–365.

Antipova, A. S., & Semenova, M. G. (1997). Influence of sucrose on the thermodynamic properties of the 11S globulin of Vicia faba-dextran-aqueous solvent system. *Food Hydrocolloids*, *11*(4), 415–421.

Arakawa, T., Ejima, D., & Akuta, T. (2017). Protein aggregation under high concentration/density state during chromatographic and ultrafiltration processes. *International Journal of Biological Macromolecules*, *95*, 1153–1158.

Arakawa, T., Ejima, D., Tsumoto, K., Obeyama, N., Tanaka, Y., Kita, Y., et al. (2007). Suppression of protein interactions by arginine: A proposed mechanism of the arginine effects. *Biophysical Chemistry*, *127*, 1–8.

Arakawa, T., & Kita, Y. (2000). Protection of bovine serum albumin form aggregation by Tween 80. *Journal of Pharmaceutical Sciences*, *89*(5), 646–651.

Arakawa, T., Philo, J. S., Ejima, D., Tsumoto, K., & Arisaka, F. (2007). Aggregation analysis of therapeutic proteins, part 2. *Bioprocess International*, *5*(4), 36+.

Arakawa, T., & Timasheff, S. N. (1985). Theory of protein solubility. *Methods in Enzymology*, *114*, 49–77.

Arakawa, T., Tsumoto, K., Kita, Y., Chang, B., & Ejima, D. (2007). Biotechnology applications of amino acids in protein purification and formulations. *Amino Acids*, *33*(4), 587–605.

Arora, J., Hu, Y., Esfandiary, R., Sathish, H. A., Bishop, S. M., Joshi, S. B., et al. (2016). Charge-mediated Fab-Fc interactions in an IgG1 antibody induce reversible self-association, cluster formation, and elevated viscosity. *MAbs*, *8*(8), 1561–1574.

Arzensek, D., Kuzman, D., & Podgornik, R. (2012). Colloidal interactions between monoclonal antibodies in aqueous solutions. *Journal of Colloid and Interface Science*, *384*, 207–216.

Aswad, D. W., Paranandi, M. V., & Schurter, B. T. (2000). Isoaspartate in peptide and proteins: Formation, significance and analysis. *Journal of Pharmaceutical and Biomedical Analysis*, *21*(6), 1129–1136.

Austerberry, J. I., Dajani, R., Panova, S., Roberts, D., Golovanov, A. P., Pluen, A., et al. (2017). The effect of charge mutations on the stability and aggregation of a human single chain Fv fragment. *European Journal of Pharmaceutics and Biopharmaceutics*, *115*, 18–30.

Auton, M., Rosgen, J., Sinev, M., Holthauzen, L. M. F., & Bolen, D. W. (2011). Osmolyte effects on protein stability and solubility: A balancing act between backbone and side-chains. *Biophysical Chemistry*, *159*(1), 90–99.

Baek, Y., Singh, N., Arunkumar, A., & Zydney, A. (2017). Ultrafiltration behavior of an Fc-fusion protein: Filtrate flux data and modeling. *Journal of Membrane Science*, *528*, 171–177.

Bai, G., Bee, J. S., Biddlecombe, J. G., Chen, Q. M., & Leach, W. T. (2012). Computational fluid dynamics (CFD) insights into agitation stress methods in biopharmaceutical development. *International Journal of Pharmaceutics*, *423*(2), 264–280.

Bajaj, H., Sharma, V. K., & Kalonia, D. S. (2004). Determination of second virial coefficient of protein using a dual-detector cell for simultaneous measurement of scattered light intensity and concentration in SEC-HPLC. *Biophysical Journal*, *87*(6), 4048–4055.

Bam, N. B., Cleland, J. L., Yang, J., Manning, M. C., Carpenter, J. F., Kelley, R. F., et al. (1998). Tween protects recombinant human growth hormone against agitation-induced damage via hydrophobic interactions. *Journal of Pharmaceutical Sciences*, *87*(12), 1554–1559.

Banks, D. D., Latypov, R. F., Ketchem, R. R., Woodard, J., Scavezze, J. L., Siska, C. C., et al. (2012). Native-state solubility and transfer free energy as predictive tools for selecting excipients to include in protein formulation development studies. *Journal of Pharmaceutical Sciences*, *101*(8), 2720–2732.

Barnett, G. V., Razinkov, V. I., Kerwin, B. A., Blake, S., Qi, W., Curtis, R. A., et al. (2016). Osmolyte effects on monoclonal antibody stability and concentration-dependent protein interactions with water and common osmolytes. *Journal of Physical Chemistry B*, *120*(13), 3318–3330.

Barth, A. (2007). Infrared spectroscopy of proteins. *Biochimica et Biophysica Acta*, *1767*, 1073–1101.

Basu, P., Blake-Haskins, A. W., O'Berry, K. B., Randolph, T. W., & Carpenter, J. F. (2014). Albinterferon a_{2b} adsorption to silicone oil-water interfaces: Effects on protein conformation, aggregation, and subvisible particle formation. *Journal of Pharmaceutical Sciences*, *103*(2), 427–436.

Basu, P., Krishnan, S., Thirumangalathu, R., Randolph, T. W., & Carpenter, J. F. (2013). IgG_1 aggregation and particle formation induced by silicone-water interfaces on siliconized borosilicate glass beads: A model for siliconized primary containers. *Journal of Pharmaceutical Sciences*, *102*(3), 852–865.

Bauer, K. C., Gobel, M., Schwab, M.-L., Schermeyer, M.-T., & Hubbuch, J. (2016). Concentration-dependent changes in apparent diffusion coefficients as indicator for colloidal stability of protein solutions. *International Journal of Pharmaceutics*, *511*, 276–287.

Baynes, B. M., & Trout, B. L. (2003). Proteins in mixed solvents: A molecular-level perspective. *Journal of Physical Chemistry B*, *107*(50), 14058–14067.

Baynes, B. M., Wang, D. I. C., & Trout, B. L. (2005). Role of arginine in the stabilization of protein against aggregation. *Biochemistry*, *44*, 4919–4925.

Besheer, A. (2017). Protein adsorption to in-line filters of intravenous administration sets. *Journal of Pharmaceutical Sciences*, *106*(10), 2959–2965.

Bhat, R., & Timasheff, S. N. (1992). Steric exclusion is the principal source of the preferential hydration of proteins in the presence of polyethylene glycol. *Protein Science*, *1*(9), 1133–1143.

Binabaji, E., Ma, J. F., & Zydney, A. L. (2015). Intermolecular interactions and the viscosity of highly concentrated monoclonal antibody solutions. *Pharmaceutical Research*, *32*(9), 3102–3109.

Blanco, M. A., Perevozchikova, T., Martorana, V., Manno, M., & Roberts, C. J. (2014). Protein-protein interactions in dilute to concentrated solutions: Alpha-chymotrypsinogen in acidic conditions. *Journal of Physical Chemistry B*, *118*(22), 5817–5831.

Blanco, M. A., Sahin, E., & Roberts, C. J. (2011). Reexamining protein-protein and protein-solvent interactions from Kirkwood-Buff analysis of light scattering in multi-component solutions. *Journal of Chemical Physics*, *134*, 225103.

Blobel, J., Brath, U., Bernado, P., Diehl, C., Ballester, L., Sornosa, A., et al. (2011). Protein loop compaction and the origin of the effect of arginine and glutamic acid mixtures on solubility, stability and transient oligomerization of proteins. *European Biophysics Journal With Biophysics Letters*, *40*(12), 1327–1338.

Bloustine, J., Berejnov, V., & Fraden, S. (2003). Measurements of protein-protein interactions by size exclusion chromatography. *Biophysical Journal*, *85*(4), 2619–2623.

Borwankar, A. U., Dear, B. J., Twu, A., Hung, J. J., Dinin, A. K., Wilson, B. K., et al. (2016). Viscosity reduction of a concentrated monoclonal antibody with arginine-HCl and arginine-glutamate. *Industrial and Engineering Chemistry Research*, *55*, 11225–11234.

Boven, K., Stryker, S., Knight, J., Thomas, A., van Regenmortel, M., Kemeny, D. M., et al. (2005). The increased incidence of pure red cell aplasia with an Eprex formulation in uncoated rubber stopper syringes. *Kidney International*, *67*(6), 2346–2353.

Box, V. G. S., & Smith, W. (2000). The chemical significance of the tertiary structures of polypeptides and proteins. *Journal of Molecular Structure*, *520*, 229–235.

Brader, M. L., Estey, T., Bai, S. J., Alston, R. W., Lucas, K. K., Lantz, S., et al. (2015). Examination of thermal unfolding and aggregation profiles of a series of developable therapeutic monoclonal antibodies. *Molecular Pharmaceutics*, *12*(4), 1005–1017.

Brewer, J. W., C. Y. Kim, C. J. Montgomery and J. K. Sheety (1988). Process for the production of solution stable alpha-amylase and liquid alpha-amylase produced thereby. US Patent 4,724,208, issued 9 February 1988.

Bridgewater, J. D., & Vachet, R. W. (2005). Metal-catalyzed oxidation reactions and mass spectrometry: The roles of ascorbate and different oxidizing agents in determining Cu-protein-binding sites. *Analytical Biochemistry*, *341*, 122–130.

Britt, K. A., Schwartz, D. K., Wurth, C., Mahler, H.-C., Carpenter, J. F., & Randolph, T. W. (2012). Excipient effects on humanized monoclonal antibody interactions with silicone oil emulsions. *Journal of Pharmaceutical Sciences*, *101*(12), 4419–4432.

Broecker, J., Klingel, V., Ou, W.-L., Balo, A. R., Kissick, D. J., Ogata, C. M., et al. (2016). A versatile system for high-throughput in situ x-ray screening and data collection of soluble and membrane-protein crystals. *Crystal Growth & Design*, *16*, 6318–6326.

Bruzdziak, P., Panuszko, A., Jourdan, M., & Stangret, J. (2016). Protein thermal stabilization in aqueous solutions of osmolytes. *Acta Biochimica Polonica*, *63*(1), 65–70.

Bruździak, P., Panuszko, A., & Stangret, J. (2013). Influence of osmolytes on protein and water structure: A step to understanding the mechanism of protein stabilization. *Journal of Physical Chemistry B*, *117*, 11502–11508.

Carpenter, J. F., Kendrick, B. S., Chang, B. S., Manning, M. C., & Randolph, T. W. (1999). Inhibition of stress-induced aggregation of protein therapeutics. *Methods in Enzymology*, *309*, 236–255.

Carpenter, J. F., Pikal, M. J., Chang, B. S., & Randolph, T. W. (1997). Rational design of stable lyophilized protein formulations: Some practical advice. *Pharmaceutical Research*, *14*(8), 969–975.

Carpenter, J. F., Randolph, T. W., Jiskoot, W., Crommelin, D. J. A., Middaugh, C. R., Winter, G., et al. (2009). Overlooking subvisible particles in therapeutic protein

products: Gaps that may compromise product quality. *Journal of Pharmaceutical Sciences, 98*(4), 1201–1205.
Castellanos, M. M., Pathak, J. A., & Colby, R. H. (2014). Both protein adsorption and aggregation contribute to shear yielding and viscosity increase in protein solutions. *Soft Matter, 10*(1), 122–131.
Castellanos, M. M., Pathak, J. A., Leach, W., Bishop, S. M., & Colby, R. H. (2014). Explaining the non-Newtonian character of aggregating monoclonal antibody solutions using small-angle neutron scattering. *Biophysical Journal, 107*(2), 469–476.
Cerdá-Costa, N., De la Arada, I., Avilés, F. X., Arrondo, J. L. R., & Villegas, S. (2009). Influence of aggregation propensity and stability on amyloid fibril formation as studied by Fourier transform infrared spectroscopy and two-dimensional COS analysis. *Biochemistry, 48*, 10582–10590.
Chakravarti, B., & Chakravarti, D. N. (2007). Oxidative modification of proteins: Age-related changes. *Gerontology, 53*, 128–139.
Chakroun, N., Hilton, D., Ahmad, S. S., Platt, G. W., & Dalby, P. A. (2016). Mapping the aggregation kinetics of a therapeutic antibody fragment. *Molecular Pharmaceutics, 13*(2), 307–319.
Chalikian, T. V. (2016). Excluded volume contributions to cosolvent-mediated modulation of macromolecular folding and binding reactions. *Biophysical Chemistry, 209*, 1–8.
Chang, B. S. (2016). Protein formulations containing amino acids. US Patent 9,364,542, issued 14 June 2016.
Chang, B. S., Kendrick, B. S., & Carpenter, J. F. (1996). Surface-induced denaturation of proteins during freezing and its inhibition by surfactants. *Journal of Pharmaceutical Sciences, 85*(12), 1325–1330.
Charron, C., Kadri, A., Robert, M. C., Giege, R., & Lorber, B. (2002). Crystallization in the presence of glycerol displaces water molecules in the structure of thaumatin. *Acta Crystallographica Section D, 58*, 2060–2065.
Chebotareva, N. A., Kurganov, B. I., & Livanova, N. B. (2004). Biochemical effects of molecular crowding. *Biochemistry (Moscow), 69*(11), 1239–1251.
Chen, X. B., Zeng, F., Huang, T., Cheng, L., Liu, H., & Gong, R. (2016). Optimization on Fc for improvement of stability and aggregation resistance. *Current Pharmaceutical Biotechnology, 17*(15), 1353–1359.
Cheng, Y. C., Bianco, C. L., Sandler, S. I., & Lenhoff, A. M. (2008). Salting-out lysozyme and ovalbumin from mixtures: Predicting precipitation performance from protein-protein interactions. *Industrial and Engineering Chemistry Research, 47*(15), 5203–5213.
Chennamsetty, N., Quan, Y., Nashine, V., Sadineni, V., Lyngberg, O., & Krystek, S. (2015). Modeling the oxidation of methionine residues by peroxides in proteins. *Journal of Pharmaceutical Sciences, 104*(4), 1246–1255.
Chi, E. Y., Krishnan, S., Kendrick, B. S., Chang, B. S., Carpenter, J. F., & Randolph, T. W. (2003). Roles of conformational stability and colloidal stability in the aggregation of recombinant human granulocyte colony-stimulating factor. *Protein Science, 12*, 903–913.
Chiti, F., & Dobson, C. M. (2009). Amyloid formation of globular proteins under native conditions. *Nature Chemical Biology, 5*(1), 15–22.
Chiti, F., & Dobson, C. M. (2017). Protein misfolding, amyloid formation, and human disease: A summary of progress over the last decade. *Annual Review of Biochemistry, 86*, 27–68.
Chou, D. K., Krishnamurthy, R., Manning, M. C., Randolph, T. W., & Carpenter, J. F. (2012). Physical stability of albinterferon-alpha$_{2b}$ in aqueous solution: Effects of conformational stability and colloidal stability on aggregation. *Journal of Pharmaceutical Sciences, 101*(8), 2702–2719.
Chow, C. K., Allan, B. W., Chai, Q., Atwell, S., & Lu, J. R. (2016). Therapeutic antibody engineering to improve viscosity and phase separation guided by crystal structure. *Molecular Pharmaceutics, 13*(3), 915–923.

Chu, J.-W., & Trout, B. L. (2004). On the mechanisms of oxidation of organic sulfides by H_2O_2 in aqeuous solutions. *Journal of the American Chemical Society, 126*, 900–908.
Chu, J.-W., Yin, J., Brooks, B. R., Wang, D. I. C., Ricci, M. S., Brems, D. N., et al. (2004). A comprehensive picture of non-site specific oxidation of methionine residues by peroxides in protein pharmaceuticals. *Journal of Pharmaceutical Sciences, 93*(12), 3096–3102.
Chu, J.-W., Yin, J., Wang, D. I. C., & Trout, B. L. (2004b). Molecular dynamics simulations and oxidation rates of methionine residues of granulocyte colony-stimulating factor at different pH values. *Biochemistry, 43*, 1019–1029.
Chu, J.-W., Yin, J., Wang, D. I. C., & Trout, B. L. (2004c). A structural and mechanistic study of the oxidation of methionine residues in hPTH(1-34) via experiments and simulations. *Biochemistry, 43*, 14139–14148.
Cimmperman, P., Baranauskienė, L., Jachimovičiūtė, S., Jachno, J., Torresan, J., Michailovienė, V., et al. (2008). A quantitative model of thermal stabilization and destabilization of proteins by ligands. *Biophysical Journal, 95*(7), 3222–3231.
Connolly, B. D., Petry, C., Yadav, S., Demeule, B., Ciaccio, N., Moore, J. M. R., et al. (2012). Weak interactions govern the viscosity of concentrated antibody solutions: High-throughput analysis using the diffusion interaction parameter. *Biophysical Journal, 103*(1), 69–78.
Cordoba, A. J., Shyong, B.-J., Breen, D., & Harris, R. J. (2005). Non-enzymatic hinge region fragmentation of antibodies in solution. *Journal of Chromatography B, 818*, 115–121.
Costanzo, J. A., O'Brien, C. J., Tiller, K., Tamargo, E., Robinson, A. S., Roberts, C. J., et al. (2014). Conformational stability as a design target to control protein aggregation. *Protein Engineering, Design & Selection*, 27(5), 157–167.
Curnow, P., & Booth, P. J. (2010). The contribution of a covalently bound cofactor to the folding and thermodynamic stability of an integral membrane protein. *Journal of Molecular Biology, 403*(4), 630–642.
Dear, B. J., Hung, J. J., Truskett, T. M., & Johnston, K. P. (2017). Contrasting the influence of cationic amino acids on the viscosity and stability of a highly concentrated monoclonal antibody. *Pharmaceutical Research, 34*, 193–207.
Demoruelle, K., Guo, B., Kao, S., McDonald, H., Nikic, D. B., Holman, S. C., et al. (2002). Correlation between the osmotic second virial coefficient and solubility for equine serum albumin and ovalbumin. *Acta Crystallographica, Section D: Biological Crystallography, 58*, 1544–1548.
DePaz, R. A., Barnett, C. C., Dale, D. A., Carpenter, J. F., Gaertner, A. L., & Randolph, T. W. (2000). The excluding effects of sucrose on a protein chemical degradation pathway: Methionone oxidation in subtilisin. *Archives of Biochemistry and Biophysics, 384*(1), 123–132.
Derrick, T., Grillo, A. O., Vitharana, S. N., Jones, L., Rexroad, J., Shah, A. U., et al. (2007). Effect of polyanions on the structure and stability of Repifermin™ (keratinocyte growth factor-2). *Journal of Pharmaceutical Sciences, 96*(4), 761–776.
Deshpande, K. S., Ahamed, T., Horst, J. H., Jansens, P. J., van der Wielen, L. A. M., & Ottens, M. (2009). The use of self-interaction chromatography in stable formulation and crystallization of proteins. *Biotechnology Journal, 4*, 1266–1277.
Dieterle, M., Blaschke, T., & Hasse, H. (2013). Second osmotic virial coefficients and aggregation of monoclonal antibodies by static laser light scattering. *Zeitschrift für Physikalische Chemie*, 227(203), 333–343.
Dixit, N., Salamat-Miller, N., Salinas, P. A., Taylor, K. D., & Basu, S. K. (2016). Residual host cell protein promotes polysorbate 20 degradation in a sulfatase drug product leading to free fatty acid particles. *Journal of Pharmaceutical Sciences, 105*(5), 1657–1666.
Dong, X.-Y., Liu, J.-H., Liu, F.-F., & Sun, Y. (2009). Self-interaction of native and denatured lysozyme in the presence of osmolytes, L-arginine and guanidone hydrochloride. *Biochemical Engineering Journal, 43*, 321–326.

Drenski, M. F., Brader, M. L., Alston, R. W., & Reed, W. F. (2013). Monitoring protein aggregation kinetics with simultaneous multiple sample light scattering. *Analytical Biochemistry, 437*, 185–197.

Du, W., & Klibanov, A. M. (2011). Hydrophobic salts markedly diminish viscosity of concentrated protein solutions. *Biotechnology and Bioengineering, 108*(3), 632–636.

Duenas, E. T., Keck, R., DeVos, A., Jones, A. J. S., & Cleland, J. L. (2001). Comparison between light induced and chemically induced oxidation of rhVEGF. *Pharmaceutical Research, 18*(10), 1455–1460.

Dunkelberger, E. B., Buchanan, L. E., Marek, P., Cao, P., Raleigh, D. P., & Zanni, M. T. (2012). Deamidation accelerates amyloid formation and alters amylin fiber structure. *Journal of the American Chemical Society, 134*(30), 12658–12667.

Ellis, R. J., & Minton, A. P. (2006). Protein aggregation in crowded environments. *Biological Chemistry, 387*, 485–497.

Eng, M., Ling, V., Briggs, J. A., Souza, K., Canova-Davis, E., Powell, M. F., et al. (1997). Formulation development and primary degradation pathways for recombinant human nerve growth factor. *Analytical Chemistry, 69*(20), 4184–4190.

Eppler, A., Weigandt, M., Hanefeld, A., & Bunjes, H. (2010). Relevant shaking stress conditions for antibody preformulation development. *European Journal of Pharmaceutics and Biopharmaceutics, 74*(2), 139–147.

Esteve-Romero, J. S., Bossi, A., & Righetti, P. G. (1996). Purification of thermamylase in multicompartment electrolyzers with isoelectric membranes: The problem of protein solubility. *Electrophoresis, 17*(7), 1241–1247.

Farnum, M., & Zukoski, C. (1999). Effect of glycerol on the interactions and solubility of bovine pancreatic trypsin inhibitor. *Biophysical Journal, 76*(5), 2716–2725.

Fast, J. L., Cordes, A. A., Carpenter, J. F., & Randolph, T. W. (2009). Physical instability of a therapeutic Fc fusion protein: Domain contributions to conformational and colloidal stability. *Biochemistry, 48*(49), 11724–11736.

Fernández, C., & Minton, A. P. (2008). Automated measurement of the static light scattering of macromolecular solutions over a broad range of concentrations. *Analytical Biochemistry, 381*, 254–257.

Fischer, K., & Schmidt, M. (2016). Pitfalls and novel applications of particle sizing by dynamic light scattering. *Biomaterials, 98*, 79–91.

Franco, L. F. M., & Pessôa Filho, P. D. A. (2013). On the relationship between the solubility of proteins and the osmotic second virial coefficient. *Brazilian Journal of Chemical Engineering, 30*(1), 95–104.

Fransson, J., Florin-Robertsson, E., Axelsson, K., & Nyhlén, C. (1996). Oxidation of human insulin-like growth factor I in formulation studies: Kinetics of oxidation in aqueous solution and in solid state. *Pharmaceutical Research, 13*(8), 1252–1257.

Freire, E., Schön, A., Hutchins, B. M., & Brown, R. K. (2013). Chemical denaturation as a tool in the formulation optimization of biologics. *Drug Discovery Today, 18*(19–20), 1007–1013.

Fukuda, M., Moriyama, C., Yamazaki, T., Imaeda, Y., & Koga, A. (2015). Quantitative correlation between viscosity of concentrated MAb solutions and particle size parameters obtained from small-angle x-ray scattering. *Pharmaceutical Research, 32*, 3803–3812.

Fukuda, M., Watanabe, A., Hayasaka, A., Muraoka, M., Hori, Y., Yamazaki, T., et al. (2017). Small-scale screening method for low-viscosity antibody solutions using small-angle X-ray scattering. *European Journal of Pharmaceutics and Biopharmaceutics, 112*, 132–137.

Gadgil, H. S., Bondarenko, P. V., Pipes, G., Rehder, D., McAuley, A., Perico, N., et al. (2007). The LC/MS analysis of glycation of IgG molecules in sucrose containing formulations. *Journal of Pharmaceutical Sciences, 96*(10), 2607–2621.

Gaudiano, M. C., Colone, M., Bombelli, C., Chistolini, P., Valvo, L., & Diociaiuti, M. (2005). Early stages of salmon calcitonin aggregation: Effect induced by ageing and oxidation processes in water and in the presence of model membranes. *Biochimica et Biophysica Acta*, *1750*, 134–145.

Gebicki, J. M., Nauser, T., Domazou, A., Steinmann, D., Bounds, P. L., & Koppenol, W. H. (2010). Reduction of protein radicals by GSH and ascorbate: Potential biological significance. *Amino Acids*, *39*, 1131–1137.

Geng, S. B., Cheung, J. K., Narasimhan, C., Shameem, M., & Tessier, P. M. (2014). Improving monoclonal antibody selection and engineering using measurements of colloidal protein interactions. *Journal of Pharmaceutical Sciences*, *103*(11), 3356–3363.

Geng, S. B., Wittekind, M., Vigil, A., & Tessier, P. M. (2016). Measurements of monoclonal antibody self-association are correlated with complex biophysical properties. *Molecular Pharmaceutics*, *13*(5), 1636–1645.

Gerhardt, A., McGraw, N. R., Schwartz, D. K., Bee, J. S., Carpenter, J. F., & Randolph, T. W. (2014). Protein aggregation and particle formation in prefilled glass syringes. *Journal of Pharmaceutical Sciences*, *103*(6), 1601–1612.

Gerhardt, A., Mcumber, A. C., Nguyen, B. H., Lewus, R., Schwartz, D. K., Carpenter, J. F., et al. (2015). Surfactant effects on particle generation in antibody formulations in prefilled syringes. *Journal of Pharmaceutical Sciences*, *104*(12), 4056–4064.

Ghosh, R., Calero-Rubio, C., Saluja, A., & Roberts, C. J. (2016). Relating protein-protein interactions and aggregation rates from low to high cocentrations. *Journal of Pharmaceutical Sciences*, *105*(3), 1086–1096.

Glover, Z. W., Gennaro, L., Yadav, S., Demeule, B., Wong, P. Y., & Sreedhara, A. (2013). Compatibility and stability of pertuzumab and trastuzumab admixtures in i.v. infusion bags for coadministration. *Journal of Pharmaceutical Sciences*, *102*(3), 794–812.

Goldberg, D. S., Bishop, S. M., Shah, A. U., & Sathish, H. A. (2011). Formulation development of therapeutic monoclonal antibodies using high-throughput fluorescence and static light scattering techniques: Role of conformational and colloidal stability. *Journal of Pharmaceutical Sciences*, *100*(4), 1306–1315.

Goldberg, D. S., Lewus, R. A., Esfandiary, R., Farkas, D. C., Mody, N., Day, K. J., et al. (2017). Utility of high throughput screening techniques to predict stability of monoclonal antibody formulations during early stage development. *Journal of Pharmaceutical Sciences*, *106*(6), 1971–1977.

Golovanov, A. P., Hautbergue, G. M., Wilson, S. A., & Lian, L. Y. (2004). A simple method for improving protein solubility and long-term stability. *Journal of the American Chemical Society*, *126*(29), 8933–8939.

Griffiths, S. W., & Cooney, C. L. (2002). Relationship between protein structure and methionine oxidation in recombinant human alpha-1-antitrypsin. *Biochemistry*, *41*(20), 6245–6252.

Groenning, M. (2010). Binding mode of thioflavin T and other molecular probes in the context of amyloid fibrils-current status. *Journal of Chemical Biology*, *3*, 1–18.

Gruia, F., Parupudi, A., Baca, M., Ward, C., Nyborg, A., Remmele, R. L., Jr., et al. (2017). Impact of mutations on the higher order structure and activity of a recombinant uricase. *Journal of Pharmaceutical Sciences*, *106*, 1018–1024.

Guo, Z., Chen, A., Nassar, R., Helk, B., Mueller, C., Tang, Y., et al. (2012). Structure-activity relationship for hydrophobic salts as viscosity-lowering excipients for concentrated solutions of monoclonal antibodies. *Pharmaceutical Research*, *29*(11), 3102–3109.

Guo, B., Kao, S., McDonald, H., Asanov, A., Combs, L. L., & Wilson, W. W. (1999). Correlation of second virial coefficients and solubilities useful in protein crystal growth. *Journal of Crystal Growth*, *196*, 424–433.

Guzey, D., McClements, D. J., & Weiss, J. (2003). Adsorption kinetics of BSA at air-water solution interfaces as affected by sugar type and concentration. *Food Research International*, *36*, 649–660.

Haas, C., Drenth, J., & Wilson, W. W. (1999). Relation between the solubility of proteins in aqueous solutions and the second virial coefficient of the solution. *Journal of Physical Chemistry B*, *103*(14), 2808–2811.

Haidekker, M. A., & Theodorakis, E. A. (2005). Molecular rotors—Fluorescent biosensors for viscosity and flow. *Organic and Biomolecular Chemistry*, *5*, 1669–1678.

Hall, T., Sandefur, S. L., Frye, C. C., Tuley, T. L., & Huang, L. (2016). Polysorbates 20 and 80 degradation by group XV lysosomal phospholipase A2 isomer X1 in monoclonal antibody formulations. *Journal of Pharmaceutical Sciences*, *105*(5), 1633–1642.

Hamada, H., Arakawa, T., & Shiraki, K. (2009). Effect of additives on protein aggregation. *Current Pharmaceutical Biotechnology*, *10*, 400–407.

Harms, M. J., Wilmarth, P. A., Kapfer, D. M., Steel, E. A., David, L. L., Bächinger, H. P., et al. (2004). Laser light-scattering evidence for an altered association of bB1-crystallin deamidated in the connecting peptide. *Protein Science*, *13*, 678–686.

Hartmann, W. K., Saptharishi, N., Yang, X. Y., Mitra, G., & Soman, G. (2004). Characterization and analysis of thermal denaturation of antibodies by size exclusion high-performance liquid chromatography with quadruple detection. *Analytical Biochemistry*, *325*, 227–239.

Hawe, A., Wiggenhorn, M., Van De Weert, M., Garbe, J. H. O., Mahler, H.-C., & Jiskoot, W. (2012). Forced degradation of therapeutic proteins. *Journal of Pharmaceutical Sciences*, *101*(3), 895–913.

Hawkins, C. L., & Davies, M. J. (2001). Generation and propagation of radical reactions on proteins. *Biochimica et Biophysica Acta*, *1504*, 196–219.

Haynes, C. A., & Norde, W. (1995). Structures and stabilities of adsorbed proteins. *Journal of Colloid and Interface Science*, *169*, 313–328.

He, F., Becker, G. W., Litowski, J. R., Narhi, L. O., Brems, D. N., & Razinkov, V. I. (2010). High-throughput dynamic light scattering method for measuring viscosity of concentrated protein solutions. *Analytical Biochemistry*, *399*(1), 141–143.

He, F., Woods, C. E., Becker, G. W., Narhi, L. O., & Razinkov, V. I. (2011). High-throughput assessment of thermal and colloidal stability parameters for monoclonal antibody formulations. *Journal of Pharmaceutical Sciences*, *100*(12), 5126–5141.

Hedberg, S. H. M., Heng, J. Y. Y., Williams, D. R., & Liddell, J. M. (2015). Self-interaction chromatography of mAbs: Accurate measurement of dead volumes. *Pharmaceutical Research*, *32*(12), 3975–3985.

Hedberg, S. H. M., Heng, J. Y. Y., Williams, D. R., & Liddell, J. M. (2016). Micro scale self-interaction chromatography of proteins: A mAb case-study. *Journal of Chromatography A*, *1434*, 57–63.

Ho, J. G. S., Middelberg, A. P. J., Ramage, P., & Kocher, H. P. (2003). The likelihood of agrgegation during protein renaturation can be assessed using the second virial coefficient. *Protein Science*, *12*(4), 708–716.

Hofmann, M., Winzer, M., Weber, C., & Gieseler, H. (2014). A low-volume screening tool to optimize colloidal stability for highly concentrated protein formulations. *American Pharmaceutical Review*, *17*(6), 36–42.

Hofmann, M., Winzer, M., Weber, C., & Gieseler, H. (2016). Prediction of protein aggregation in high concentration protein solutions utilizing protein-protein interactions determined by low volume static light scattering. *Journal of Pharmaceutical Sciences*, *105*, 1819–1828.

Holthauzen, L. M. F., Auton, M., Sinev, M., & Rosgen, J. (2011). Protein stability in the presence of cosolutes. *Methods in Enzymology*, *492*, 61–125.

Holthauzen, L. M. F., & Bolen, D. W. (2007). Mixed osmolytes: The degree to which one osmolyte affects the protein stabilizing ability of another. *Protein Science, 16*(2), 293–298.

Inglis, A. S. (1983). Single hydrolysis method for all amino-acids, including cysteine and tryptophan. *Methods in Enzymology, 91*, 26–36.

Inoue, N., Takai, E., Arakawa, T., & Shiraki, K. (2014a). Arginine and lysine reduce the high viscosity of serum albumin solutions, for pharmaceutical injection. *Journal of Bioscience and Bioengineering, 117*(5), 539–543.

Inoue, N., Takai, E., Arakawa, T., & Shiraki, K. (2014b). Specific decrease in solution viscosity of antibodies by arginine for therapeutic formulations. *Molecular Pharmaceutics, 11*(6), 1889–1896.

Inouye, H., Houde, D., Temel, D. B., & Makowski, L. (2016). Utility of solution x-ray scattering for the development of antibody biopharmaceuticals. *Journal of Pharmaceutical Sciences, 105*, 3278–3289.

Jacobs, S. A., Wu, S. J., Feng, Y. Q., Bethea, D., & O'Neil, K. T. (2010). Cross-interaction chromatography: A rapid method to identify highly soluble monoclonal antibody candidates. *Pharmaceutical Research, 27*(1), 65–71.

James, S., & McManus, J. J. (2012). Thermal and solution stability of lysozyme in the presence of sucrose, glucose, and trehalose. *Journal of Physical Chemistry B, 116*(34), 10182–10188.

Jayaraman, J., Wu, J. M., Brunelle, M. C., Cruz, A. M. M., Goldberg, D. S., Lobo, B., et al. (2014). Plasmonic measurements of monoclonal antibody self-association using self-interaction nanoparticle spectroscopy. *Biotechnology and Bioengineering, 111*(8), 1513–1520.

Jensen, J. L., Kolvenbach, C., Roy, S., & Schöneich, C. (2000). Metal-catalyzed oxidation of brain-derived neurotrophic factor (BDNF): Analytical challenges for the identification of modified sites. *Pharmaceutical Research, 17*(2), 190–196.

Jezek, J., Rides, M., Derham, B., Moore, J., Cerasoli, E., Simler, R., et al. (2011). Viscosity of concentrated therapeutic protein compositions. *Advanced Drug Delivery Reviews, 63*(13), 1107–1117.

Ji, J. A., Zhang, B. Y., Cheng, W., & Wang, Y. J. (2009). Methionine, tryptophan, and histidine oxidation in a model protein, PTH: Mechanisms and stabilization. *Journal of Pharmaceutical Sciences, 98*(12), 4485–4500.

Jia, L., & Sun, Y. (2017). Protein asparagine deamidation prediction based on structures with machine learning methods. *PLoS One, 12*(7), e0181347.

Johnson, D. H., Parupudi, A., Wilson, W. W., & DeLucas, L. J. (2009). High-throughput self-interaction chromatography: Applications in protein formulation prediction. *Pharmaceutical Research, 26*(2), 296–305.

Jorgensen, L., Bennedsen, P., Hoffmann, S. V., Krogh, R. L., Pinholt, C., Groenning, M., et al. (2011). Adsorption of insulin with varying self-association profiles to a solid teflon surface—Influence on protein structure, fibrillation tendency and thermal stability. *European Journal of Pharmaceutical Sciences, 42*, 509–516.

Joshi, V., Shivach, T., Yadav, N., & Rathore, A. S. (2014). Circular dichroism spectroscopy as a tool for monitoring aggregation in monoclonal antibody therapeutics. *Analytical Chemistry, 86*, 11606–11613.

Kaieda, S., Lund, M., Plivelic, T. S., & Halle, B. (2014). Weak self-interactions of globular proteins studied by small-angle x-ray scattering and structure-based modeling. *Journal of Physical Chemistry B, 118*(34), 10111–10119.

Kalonia, C., Toprani, V., Toth, R., Wahome, N., Gabel, I., Middaugh, C. R., et al. (2016). Effects of protein conformation, apparent solubility, and protein-protein interactions on the rates and mechanisms of aggregation for an IgG1Monoclonal antibody. *Journal of Physical Chemistry B, 120*(29), 7062–7075.

Kamal, M. Z., Ahmad, S., & Rao, N. M. (2011). Stabilizing effect of polyols is sensitive to inherent stability of protein. *Biophysical Chemistry*, *156*, 68–71.

Kameoka, D., Ueda, T., & Imoto, T. (2011). Effect of the conformational stability of the CH2 domain on the aggregation and peptide cleavage of a humanized IgG. *Applied Biochemistry and Biotechnology*, *164*, 642–654.

Kamerzell, T. J., Esfandiary, R., Joshi, S. B., Middaugh, C. R., & Volkin, D. B. (2011). Protein-excipient interactions: Mechanisms and biophysical characterization applied to protein formulation development. *Advanced Drug Delivery Reviews*, *63*(13), 1118–1159.

Kanai, S., Liu, J., Patapoff, T. W., & Shire, S. J. (2008). Reversible self-association of a concentrated monoclonal antibody solution mediated by Fab-Fab interaction that impacts solution viscosity. *Journal of Pharmaceutical Sciences*, *97*(10), 4219–4227.

Karamitros, C. S., & Labrou, N. E. (2017). Preserving enzymatic activity and enhancing biochemical stability of glutathione transferase by soluble additives under free and tethered conditions. *Biotechnology and Applied Biochemistry*, *64*(5), 754–764.

Kawai, A., Chuang, V. T. G., Kuono, Y., Yamasaki, K., Miyamoto, S., Anraku, M., et al. (2017). Crystallographic analysis of the ternary complex of octanoate and N-acetyl-L-methionine with human serum albumin reveals the mode of their stabilizing interactions. *Biochimica et Biophysica Acta*, *1865*, 979–984

Kayser, V., Chennamsetty, N., Voynov, V., Helk, B., Forrer, K., & Trout, B. L. (2011). Evaluation of a non-Arrhenius model for therapeutic monoclonal antibody aggregation. *Journal of Pharmaceutical Sciences*, *100*(7), 2526–2542.

Kayser, V., Chennamsetty, N., Voynov, V., Helk, B., & Trout, B. L. (2011). Conformational stability and aggregation of therapeutic monoclonal antibodies studied with ANS and thioflavin T binding. *MAbs*, *3*(4), 408–411.

Kelly, R. L., Sun, T. W., Jain, T., Caffry, I., Yu, Y., Cao, Y., et al. (2015). High throughput cross-interaction measures for human IgG1 antibodies correlate with clearance rates in mice. *MAbs*, 7(4), 770–777.

Kerwin, B. A. (2008). Polysorbates 20 and 80 used in the formulation of protein biotherapeutics: Structure and degradation pathways. *Journal of Pharmaceutical Sciences*, *97*(8), 2924–2935.

Kerwin, B. A., & Remmele, R. L., Jr. (2007). Protect from light: Photodegradation and protein biologics. *Journal of Pharmaceutical Sciences*, *96*(6), 1468–1479.

Khan, T. A., Mahler, H. C., & Kishore, R. S. (2015). Key interactions of surfactants in therapeutic protein formulations: A review. *European Journal of Pharmaceutics and Biopharmaceutics*, *97*(Pt. A), 60–67.

Kheddo, P., Cliff, M. J., Uddin, S., van der Walle, C. F., & Golovanov, A. P. (2016). Characterizing monoclonal antibody formulations in arginine glutamate solutions using H-1 NMR spectroscopy. *MAbs*, *8*(7), 1245–1258.

Kheddo, P., Tracka, M., Armer, J., Dearman, R. J., Uddin, S., van der Walle, C. F., et al. (2014). The effect of arginine glutamate on the stability of monoclonal antibodies in solution. *International Journal of Pharmaceutics*, *473*, 126–133.

Kim, N., Remmele, R. L., Jr., Liu, D., Razinkhov, V. L., Fernandez, E. J., & Roberts, C. J. (2013). Aggregation of anti-streptavidin immunoglobulin gamma-1 involves Fab unfolding and competing growth pathways mediated by pH and salt concentration. *Biophysical Chemistry*, *172*, 26–36.

Kohli, N., Jain, N., Geddie, M. L., Razlog, M., Xu, L. H., & Lugovskoy, A. A. (2015). A novel screening method to assess developability of antibody-like molecules. *MAbs*, 7(4), 752–758.

Kosky, A. A., Dharmavaram, V., Ratnaswamy, G., & Manning, M. C. (2009). Multivariate analysis of the sequence dependence of asparagine deamidation rates in peptides. *Pharmaceutical Research*, *26*(11), 2417–2428.

Kossiakoff, A. A. (1990). Tertiary structure is a principal determinant to protein deamidation. *Science*, *240*(4849), 191–194.

Koudelka, T., Dehle, F. C., Musgrave, I. F., Hoffmann, P., & Carver, J. A. (2012). Methionine oxidation enhances k—Casein amyloid fibril formation. *Journal of Agricultural and Food Chemistry*, *60*, 4144–4155.

Kress, C., & Brandenbusch, C. (2015). Osmotic virial coefficients as access to the partitioning in aqueous two-phase systems. *Journal of Pharmaceutical Sciences*, *104*, 3703–3709.

Krishnan, S., Chi, E. Y., Webb, J. N., Chang, B. S., Shan, D., Goldenberg, M., et al. (2002). Aggregation of granulocyte colony stimulating factor under physiological conditions: Characterization and thermodynamic inhibition. *Biochemistry*, *41*, 6422–6431.

Kroon, D. J., Baldwin-Ferro, A., & Lalan, P. (1992). Identification of sites of degradation in a therapeutic monoclonal antibody by peptide mapping. *Pharmaceutical Research*, *9*(11), 1386–1393.

Kulkarni, A. M., & Zukoski, C. F. (2002). Nanoparticle crystal nucleation: Influence of solution conditions. *Langmuir*, *18*(8), 3090–3099.

Kumar, V., Dixit, N., Zhou, L., & Fraunhofer, W. (2011). Impact of short range hydrophobic interactions and long range electrostatic forces on the aggregation kinetics of a monoclonal antibody and a dual-variable domain immunoglobulin at low and high concentrations. *International Journal of Pharmaceutics*, *421*, 82–93.

Kumar, V., Singh, S. N., & Kalonia, D. S. (2012). Mechanism of stabilization of proteins by poly-hydroxy co-solvents: Concepts and implications in formulation development. *American Pharmaceutical Review*, *15*(3), 99+.

Kumari, A., Rosenkranz, T., Fitter, J., & Kayastha, A. M. (2011). Structural stability of soybean (Glycine max) a-amylase: Properties of the unfolding transition studied with fluorescence and CD spectroscopy. *Protein & Peptide Letters*, *18*, 253–260.

Kumru, O. S., Liu, J., Ji, J. A., Cheng, W., Wang, Y. J., Wang, T., et al. (2012). Compatibility, physical stability, and characterization of an IgG4 monoclonal antibody after dilution into different intravenous administration bags. *Journal of Pharmaceutical Sciences*, *101*(10), 3636–3650.

Laber, J. R., Dear, B. J., Martins, M. L., Jackson, D. E., DiVenere, A., Gollihar, J. D., et al. (2017). Charge shielding prevents aggregation of supercharged GFP variants at high protein concentration. *Molecular Pharmaceutics*, *14*, 3269–3280.

Labrenz, S. R. (2014). Ester hydrolysis of polysorbate 80 in mAb drug product: Evidence in support of the hypothesized risk after the observation of visible particulate in mAb formulations. *Journal of Pharmaceutical Sciences*, *103*, 2268–2277.

Larson, A. M., Weight, A. K., Love, K., Bonificio, A., Westcott, C. R., & Klibanov, A. M. (2017). Bulky polar additives that greatly reduce the viscosity of concentrated solutions of therapeutic monoclonal antibodies. *Journal of Pharmaceutical Sciences*, *106*(5), 1211–1217.

Lauer, T. M., Wood, G. P. F., Farkas, D., Sathish, H. A., Samra, H. S., & Trout, B. L. (2016). Moelcular investigation of the mechanism of non-enzymatic hydrolysis of proteins and the predictive algorithm for susceptibility. *Biochemistry*, *55*(23), 3315–3328.

Le Brun, V., Friess, W., Bassarab, S., & Garidel, P. (2010). Correlation of protein-protein interactions as assessed by affinity chromatography with colloidal protein stability: A case study with lysozyme. *Pharmaceutical Development and Technology*, *15*(4), 421–430.

Le Brun, V., Friess, W., Bassarab, S., Muhlau, S., & Garidel, P. (2010). A critical evaluation of self-interaction chromatography as a predictive tool for the assessment of protein-protein interactions in protein formulation development: A case study of a therapeutic monoclonal antibody. *European Journal of Pharmaceutics and Biopharmaceutics*, *75*, 16–25.

Lecour, S., Ben Baouali, A., Maupoil, V., Chahine, R., Abadie, C., Jahouhey-Donzel, A., et al. (1998). Demonstration of the production of oxygen-centered free radicals during electrolysis using E.S.R. spin-trapping techniques: Effects on cardiac function in the isolated rat heart. *Free Radical Biology & Medicine*, *24*(4), 573–579.

Ledl, F., & Schleicher, E. (1990). New aspects of the Maillard reaction in foods and in the human body. *Angewandte Chemie International Edition in English*, *29*, 565–594.
Lee, H. J., McAuley, A., Schilke, K. F., & McGuire, J. (2011). Molecular origins of surfactant-mediated stabilization of protein drugs. *Advanced Drug Delivery Reviews*, *63*(13), 1160–1171.
Lefebvre, J. (1982). Viscosity of concentrated protein solutions. *Rheologica Acta*, *21*(4–5), 620–625.
Lehermayr, C., Mahler, H.-C., Mader, K., & Fischer, S. (2011). Assessment of net charge and protein-protein interactions of different monoclonal antibodies. *Journal of Pharmaceutical Sciences*, *100*(7), 2551–2562.
Lewus, R. A., Darcy, P. A., Lenhoff, A. M., & Sandler, S. I. (2011). Interactions and phase behavior of a monoclonal antibody. *Biotechnology Progress*, *27*(1), 280–289.
Lewus, R. A., Levy, N. E., Lenhoff, A. M., & Sandler, S. I. (2015). A comparative study of monoclonal antibodies. 1. Phase behavior and protein-protein interactions. *Biotechnology Progress*, *31*(1), 268–276.
Li, B., Borchardt, R. T., Topp, E. M., Vander Velde, D. G., & Schowen, R. L. (2003). Racemization of an asparagine residue during peptide deamidation. *Journal of the American Chemical Society*, *125*, 11486–11487.
Li, N., Fort, F., Kessler, K., & Wang, W. (2009). Factors affecting cleavage at aspartic residues in model decapeptides. *Journal of Pharmaceutical and Biomedical Analysis*, *50*(1), 73–78.
Li, R., Hageman, M. J., & Topp, E. M. (2002). Effect of viscosity on the deamidation rate of a model Asn-hexapeptide. *Journal of Peptide Research*, *59*, 211–220.
Li, Y., Hewitt, D., Lentz, Y. K., Ji, J. Y. A., Zhang, T. Y., & Zhang, K. (2014). Characterization and stability study of polysorbate 20 in therapeutic monoclonal antibody formulation by multidimensional ultrahigh-performance liquid chromatography-charged aerosol detection-mass spectrometry. *Analytical Chemistry*, *86*(10), 5150–5157.
Li, G., Kasha, P. C., Late, S., & Banga, A. K. (2009). Application of hanging drop technique to optimize human IgG formulations. *Journal of Pharmacy and Pharmacology*, *62*, 125–131.
Li, L., Kumar, S., Buck, P. M., Burns, C., Lavoie, J., Singh, S. K., et al. (2014). Concentration dependent viscosity of monoclonal antibody solutions: explaining experimental behavior in terms of molecular properties. *Pharmaceutical Research*, *31*, 3161–3178.
Li, Y., Ogunnaike, B. A., & Roberts, C. J. (2010). Multi-variate approach to global protein aggregation behavior and kinetics: Effects of pH, NaCl, and temperature for alpha-chymotrypsinogen A. *Journal of Pharmaceutical Sciences*, *99*(2), 645–662.
Li, S. H., Patapoff, T. W., Nguyen, T. H., & Borchardt, R. T. (1996). Inhibitory effects of sugars and polyols on the metal-catalyzed oxidation of human relaxin. *Journal of Pharmaceutical Sciences*, *85*(8), 868–872.
Liang, Y. R., Zhu, Y., & Fang, Q. (2016). Advances in high-throughput, automated techniques and instrumentation for protein crystallization screening. *Chinese Journal of Chromatography*, *34*(12), 1137–1144.
Lilyestrom, W. G., Yadav, S., Shire, S. J., & Scherer, T. M. (2013). Monoclonal antibody self-association, cluster formation, and rheology at high concentrations. *Journal of Physical Chemistry B*, *117*(21), 6373–6384.
Lim, J. Y., Kim, N. A., Lim, D. G., Eun, C., Choi, D., & Jeong, S. H. (2016). Biophysical stability of hyFc fusion protein with regards to buffers and various excipients. *International Journal of Biological Macromolecules*, *86*, 622–629.
Liu, Y., Caffry, I., Wu, J., Geng, S. B., Jain, T., Sun, T., et al. (2014). High-throughput screening for developability during early-stage antibody discovery using self-interaction nanoparticle spectroscopy. *MAbs*, *6*(2), 483–492.
Liu, Y., Fratini, E., Baglioni, P., Chen, W.-R., & Chen, S.-H. (2005). Effective long-range attraction between protein molecules in solutions studied by small angle neutron scattering. *Physical Review Letters*, *95*, 118102.

Liu, Z., Gong, Z., Dong, X. Y., & Tang, C. (2016). Transient protein-protein interactions visualized by solution NMR. *Biochimica et Biophysica Acta*, *1864*, 115–122.
Liu, J., Nguyen, M. D. H., Andya, J. D., & Shire, S. J. (2005). Reversible self-association increases the viscosity of a concentrated monoclonal antibody in aqueous solution. *Journal of Pharmaceutical Sciences*, *94*(9), 1928–1940.
Liu, L., Qi, W., Schwartz, D. K., Randolph, T. W., & Carpenter, J. F. (2013). The effects of excipients on protein aggregation during agitation: An interfacial shear rheology study. *Journal of Pharmaceutical Sciences*, *102*(8), 2460–2470.
Liu, J. and S. J. Shire (2005). Reduced-viscosity concentrated protein formulations. US Patent 6,875,432, issued 5 April 2005.
Liu, J. and S. J. Shire (2012). Reduced-viscosity concentrated protein formulations. US Patent 8,142,776, issued 27 March 2012.
Liu, J. and S. J. Shire (2014). Reduced-viscosity concentrated protein formulations. US Patent 8,703,126, issued 22 April 2014.
Liu, Y. X., Wang, X. J., & Ching, C. B. (2010). Toward further understanding of lysozyme crystallization: Phase diagram, protein-protein interaction, nucleation kinetics, and growth kinetics. *Crystal Growth & Design*, *10*(2), 548–558.
Lorenz, T., Fiaux, J., Heitmann, D., Gupta, K., Kocher, H. P., Knopf, H.-P., et al. (2014). Developability assessment of biologics by integrated biologics profiling. *American Pharmaceutical Review*, *17*(5), 71–75.
Lorenzo, J. R., Alonso, L. G., & Sanchez, I. E. (2015). Prediction of spontaneous protein deamidation from sequence-derived secondary structure and intrinsic disorder. *PLoS One*, *10*(12), e145186.
Ludwig, D. B., Carpenter, J. F., Hamel, J.-B., & Randolph, T. W. (2010). Protein adsorption and excipient effects on kinetic stability of silicone oil emulsions. *Journal of Pharmaceutical Sciences*, *99*(4), 1721–1733.
Luo, Q., Joubert, M. K., Stevenson, R., Ketchem, R. R., Narhi, L. O., & Wypych, J. (2011). Chemical modifications in therapeutic protein aggregates generated under different stress conditions. *Journal of Biological Chemistry*, *285*(28), 25134–25144.
Maeder, W., Lieby, P., Sebald, A., Spycher, M., Pedrussio, R., & Bolli, R. (2011). Local tolerance and stability up to 24 months of a new 20% proline-stabilized polyclonal immunoglobulin for subcutaneous administration. *Biologicals*, *39*, 43–49.
Maggio, E. T. (2012a). Polysorbates, peroxides, protein aggregation, and immunogenicity—A growing concern. *Journal of Excipients and Food Chemistry*, *3*(2), 45–53.
Maggio, E. T. (2012b). Polysorbates, immunogenicity, and the totality of the evidence. *Bioprocess International*, *10*(10), 44–49.
Mahler, H. C., Senner, F., Maeder, K., & Mueller, R. (2009). Surface activity of a monoclonal antibody. *Journal of Pharmaceutical Sciences*, *98*(12), 4525–4533.
Malek-Sabet, N., Masaumian, M. R., Zeinali, M., Khalilzadeh, R., & Mousaabadi, J. M. (2013). Production, purification, and chemical stability of recombinant human interferon-γ in low oxygen tension condition: A formulation approach. *Preparative Biochemistry & Biotechnology*, *43*, 586–600.
Manning, M. C. (2005). Use of infrared spectroscopy to monitor protein structure and stability. *Expert Reviews in Proteomics*, *2*(5), 731–743.
Manning, M. C., Chou, D. K., Murphy, B. M., Payne, R. W., & Katayama, D. S. (2010). Stability of protein pharmaceuticals: An update. *Pharmaceutical Research*, *27*(4), 544–575.
Manning, M. C., Patel, K., & Borchardt, R. T. (1989). Stability of protein pharmaceuticals. *Pharmaceutical Research*, *6*(11), 903–918.
Martin, C., & Lenhoff, A. M. (2011). Self-interactions chromatography of proteins on a microfluidic monolith. *Biochemical Engineering Journal*, *53*, 216–222.
Martos, A., Koch, W., Jiskoot, W., Wuchner, K., Winter, G., Friess, W., et al. (2017). Trends on analytical characterization of polysorbates and their degradation products in biopharmaceutical formulations. *Journal of Pharmaceutical Sciences*, *106*(7), 1722–1735.

Mason, B. D., Zhang, L., Remmele, R. L., & Zhang, J. (2011). Opalescence of an IgG2 monoclonal antibody solution as it relates to liquid-liquid phase separation. *Journal of Pharmaceutical Sciences*, *100*(11), 4587–4596.

Mason, B. D., Zhang-van Enk, J., Zhang, L., Remmele, R. L., Jr., & Zhang, J. (2010). Liquid-liquid phase separation of a monoclonal antibody and nonmonotonic influence of Hofmeister anions. *Biophysical Journal*, *99*, 3792–3800.

McCormick, D. (2004). Small changes, big effects in biological manufacturing. *Pharmaceutical Technology*, *28*, 16+.

McShan, A. C., Kei, P., Ji, J. A., Kim, D. C., & Wang, Y. J. (2016). Hydrolysis of polysorbate 20 and 80 by a range of carboxylester hydrolases. *PDA Journal of Pharmaceutical Science and Technology*, *70*(4), 332–345.

Mehta, C. M., White, E. T., & Litster, J. D. (2012). Correlation of second virial coefficient with solubility for proteins in salt solutions. *Biotechnology Progress*, *28*, 163–170.

Menzen, T., & Friess, W. (2014). Temperature-ramped studies on the aggregation, unfolding, and interaction of a therapeutic monoclonal antibody. *Journal of Pharmaceutical Sciences*, *103*, 445–455.

Minton, A. P. (2016). Recent applications of light scattering measurement in the biological and biopharmaceutical sciences. *Analytical Biochemistry*, *501*, 4–22.

Miroliaei, M., & Nemat-Gorgani, M. (2001). Sugars protect native and apo yeast alcohol dehydrogenase against irreversible thermoinactivation. *Enzyme and Microbial Technology*, *29*, 554–559.

Morar-Mitrica, S., Puri, M., Beumer Sassi, A., Fuller, J., Hu, P., Crotts, G., et al. (2015). Development of a stable low-dose aglycosylated antibody formulation to minimize protein loss during intravenous administration. *MAbs*, 7(4), 792–803.

Morris, A. M., Watzky, M. A., & Finke, R. G. (2009). Protein aggregation kinetics, mechanism, and curve-fitting: A review of the literature. *Biochimica et Biophysica Acta*, *1794*, 375–397.

Mosbaek, C. R., Konarev, P. V., Svergun, D. I., Rischel, C., & Vestergaard, B. (2012). High concentration formulation studies of an IgG2 antibody using small angle X-ray scattering. *Pharmaceutical Research*, *29*(8), 2225–2235.

Mueller, M., Loh, M. Q. T., Tscheliessnig, R., Tee, D. H. Y., Tan, E., Bardor, M., et al. (2013). Liquid formulations for stabilizing IgMs during physical stress and long-term storage. *Pharmaceutical Research*, *30*, 735–750.

Mulinacci, F., Poirier, E., Capelle, M. A. H., Gurny, R., & Arvinte, T. (2011). Enhanced physical stability of human calcitonin after methionine oxidation. *European Journal of Pharmaceutics and Biopharmaceutics*, *78*, 229–238.

Muzammil, S., Kumar, Y., & Tayyab, S. (2000). Anion-induced stabilization of human serum albumin prevents the formation of intermediate during urea denaturation. *Proteins: Structure, Function, and Genetics*, *40*, 29–38.

Neal, B. L., Asthagiri, D., & Lenhoff, A. M. (1998). Molecular origins of osmotic second virial coefficient of proteins. *Biophysical Journal*, *75*, 2469–2477.

Neergaard, M. S., Kalonia, D. S., Parshad, H., Nielsen, A. D., Moller, E. H., & van de Weert, M. (2013). Viscosity of high concentration protein formulations of monoclonal antibodies of the IgG1 and IgG4 subclass—Prediction of viscosity through protein-protein interaction measurements. *European Journal of Pharmaceutical Sciences*, *49*, 400–410.

Nguyen, T. H. (1994). Oxidation degradation of protein pharmaceuticals. *ACS Symposium Series*, *567*, 59–71.

Nguyen, L. T., Wiencek, J. M., & Kirsch, L. E. (2003). Characterization methods for the physical stability or biopharmaceuticals. *PDA Journal of Pharmaceutical Science and Technology*, *57*(6), 429–445.

Nicolai, T., Britten, M., & Schmitt, C. (2011). Beta-lactoglobulin and WPI aggregates: Formation, structure and applications. *Food Hydrocolloids*, *25*, 1945–1962.
Nicoud, L., Cohrs, N., Arosio, P., Norrant, E., & Morbidelli, M. (2015). Effect of polyol sugars on the stabilization of monoclonal antibodies. *Biophysical Chemistry*, *197*, 40–46.
Nicoud, L., Lattuada, M., Yates, A., & Morbidelli, M. (2015). Impact of aggregate formation on the viscosity of protein solutions. *Soft Matter*, *11*(27), 5513–5522.
Nilsson, M. R., Driscoll, M., & Raleigh, D. P. (2002). Low levels of asparagine deamidation can have a dramatic effect on aggregation of amyloidogenic peptides: Implication for the study of amyloid formation. *Protein Science*, *11*, 342–349.
Niu, W. L., Shu, Q., Chen, Z. W., Mathews, S., Di Cera, E., & Frieden, C. (2010). The role of Zn2 + on the structure and stability of murine adenosine deaminase. *Journal of Physical Chemistry B*, *114*(49), 16156–16165.
Norde, W. (2008). My voyage of discovery to proteins in flatland... and beyond. *Colloids and Surfaces B: Biointerfaces*, *61*, 1–9.
Obrezanova, O., Arnell, A., de la Cuesta, R. G., Berthelot, M. E., Gallagher, T. R. A., Zurdo, J., et al. (2015). Aggregation risk prediction for antibodies ad its application to biotherapeutic development. *MAbs*, 7(2), 352–363.
Olsen, S. N., Andersen, K. B., Randolph, T. W., Carpenter, J. F., & Westh, P. (2009). Role of electrostatic repulsion on colloidal stability of *Bacillus halmapalus* alpha-amylase. *Biochimica et Biophysica Acta*, *1794*, 1058–1065.
Østdal, H., Weisbjerg, M. R., Skibsted, L. H., & Nielsen, J. H. (2008). Protection against photooxidation of milk by high urate content. *Milchwissenschaft*, *63*(2), 119–122.
Ouellette, D., Alessandri, L., Piparia, R., Aikhoje, A., Chin, A., Radziejewski, C., et al. (2009). Elevated cleavage of human immunoglobulin gamma molecules containing a lambda light chain mediated by iron and histidine. *Analytical Biochemistry*, *389*, 107–117.
Palmblad, M., Westlind-Danielsson, A., & Bergquist, J. (2002). Oxidation of methionine 35 attenuates formation of amyloid β-peptide 1-40 oligomers. *Journal of Biological Chemistry*, *277*(22), 19506–19510.
Pan, B., Abel, J., Ricci, M. S., Brems, D. N., Wang, D. I. C., & Trout, B. L. (2006). Comparative oxidation studies of methionone residues reflect a structural effect on chemical kinetics in rhG-CSF. *Biochemistry*, *45*(51), 15430–15443.
Pan, H., Chen, K., Chu, L., Kinderman, F., Apostol, I., & Huang, G. (2009). Methionine oxidation in human IgG2 Fc decreases binding affinities to Protein A and FcRn. *Protein Science*, *19*, 424–433.
Pan, B., Ricci, M. S., & Trout, B. L. (2011). A molecular mechanism of hydrolysis of peptide bonds at neutral pH using a model compound. *Journal of Physical Chemistry B*, *115*, 5958–5970.
Panuszko, A., Bruzdziak, P., Kaczkowska, E., & Stangret, J. (2016). General mechanism of osmolytes' influence on protein stability irrespective of the type of osmolyte cosolvent. *Journal of Physical Chemistry B*, *120*(43), 11159–11169.
Parkins, D. A., & Lashmar, U. T. (2000). The formulation of biopharmaceutical products. *Pharmaceutical Science & Technology Today*, *3*(4), 129–137.
Patel, K., & Borchardt, R. T. (1990). Chemical pathways of peptide degradation. II. Kinetics of deamidation of an asparaginyl residue in a model hexapeptide. *Pharmaceutical Research*, 7(7), 703–711.
Patro, S. Y., & Przbycien, T. M. (1996). Self-interaction chromatography: A tool for the study of protein-protein interactions in bioprocessing environments. *Biotechnology and Bioengineering*, *52*, 193–203.
Pellicane, G., Costa, D., & Caccamo, C. (2004). Microscopic determination of the phase diagrams of lysozyme and gamma-crystallin solutions. *Journal of Physical Chemistry B*, *108*(23), 7538–7541.

Petsev, D. N., Thomas, B. R., Yau, S.-T., & Vekilov, P. G. (2000). Interactions and aggregation of apoferritin molecules in solution: Effects of added electrolytes. *Biophysical Journal*, *78*, 2060–2069.

Petsev, D. N., Wu, X. X., Galkin, O., & Velikov, P. G. (2003). Thermodynamic functions of concentrated protein solutions from phase equilibria. *Journal of Physical Chemistry B*, *107*(16), 3921–3926.

Pham, N., Radajewski, D., Round, A., Brennich, M., Pernot, P., Biscans, B., et al. (2017). Coupling high throughput microfluidics and small-angle x-ray scattering to study protein crystallization. *Analytical Chemistry*, *89*(4), 2242–2287.

Philo, J. S., & Arakawa, T. (2009). Mechanisms of protein aggregation. *Current Pharmaceutical Biotechnology*, *10*, 348–351.

Pinholt, C., Hartvig, R. A., Medlicott, N. J., & Jorgensen, L. (2011). The importance of interfaces in protein drug delivery—Why is protein adsorption of interest in pharmaceutical formulations? *Expert Opinion in Drug Delivery*, *8*(7), 949–964.

Poddar, N. K., Ansari, Z. A., Singh, R. K. B., Movahedi, A. A. M., & Ahmad, F. (2010). Effect of oligosaccharides and their monosaccharide mixtures on the stability of proteins: A scaled particle study. *Journal of Biomolecular Structure and Dynamics*, *28*(3), 331–341.

Poshusta, T. L., Katoh, N., Gertz, M. A., Dispenzieri, A., & Ramirez-Alvarado, M. (2013). Thermal stability threshold for amyloid formation in light chain amyloidosis. *International Journal of Molecular Sciences*, *14*, 22604–22617.

Printz, M., Kalonia, D. S., & Friess, W. (2012). Individual second virial coefficient determination of monomer and oligomers in heat-stressed protein samples using size-exclusion chromatography-light scattering. *Journal of Pharmaceutical Sciences*, *101*(1), 363–372.

Qi, P., Volkin, D. B., Zhao, H., Nedved, M. L., Hughes, R., Bass, R., et al. (2009). Characterization of the photodegradation of a human IgG1 monoclonal antibody formulated as a high-concentration liquid dosage form. *Journal of Pharmaceutical Sciences*, *98*(9), 3117–3130.

Quigley, A., Heng, J. Y. Y., Liddell, J. M., & Williams, D. R. (2013). The accurate measurement of second virial coefficients using self-interaction chormatography: Experimental considerations. *European Journal of Pharmaceutics and Biopharmaceutics*, *85*, 1103–1111.

Quigley, A., & Williams, D. R. (2015). The second virial coefficient as a predictor of protein aggregation propensity: A self-interaction chromatography. *European Journal of Pharmaceutics and Biopharmaceutics*, *96*, 282–290.

Quigley, A., & Williams, D. R. (2016). Similar interaction chromatography of proteins: A cross interaction chromatographic approach to estimate the osmotic second virial coefficient. *Journal of Chromatography A*, *1459*, 47–56.

Rabe, M., Verdes, D., & Seeger, S. (2011). Understanding protein adsorption phenomena at solid interfaces. *Advances in Colloid and Interface Science*, *162*, 87–106.

Radmanovic, N., Serno, T., Jörg, S., & Germerhaus, O. (2013). Understanding the freezing of biopharmaceuticals: First-principle modeling of the process and evaluation of its effects on product quality. *Journal of Pharmaceutical Sciences*, *102*(8), 2495–2507.

Raibekas, A. A., Bures, E. J., Siska, C. C., Kohno, T., Latypov, R. F., & Kerwin, B. A. (2005). Anion binding and controlled aggregation of human interleukin-1 receptor antagonist. *Biochemistry*, *44*, 9871–9879.

Rakel, N., Galm, L., & Bauer, K. C. (2015). Influence of macromolecular precipitants on phase behavior of monoclonal antibodies. *Biotechnology Progress*, *31*(1), 145–153.

Ramos, C. H. I., & Baldwin, R. L. (2002). Sulfate anion stabilization of native ribonuclease A both by anion binding and by the Hofmeister effect. *Protein Science*, *11*, 1771–1778.

Randolph, T. W., & Jones, L. S. (2002). Surfactant-protein interactions. *Pharmaceutical Biotechnology*, *13*, 159–175.

Randolph, T. W., Schiltz, E., Sederstrom, D., Steinmann, D., Mozziconacci, O., Schöneich, C., et al. (2015). Do not drop: Mechanical shock in vials causes cavitation, protein aggregation, and particle formation. *Journal of Pharmaceutical Sciences*, *104*, 602–611.

Raso, S. W., Abel, J., Barnes, J. M., Maloney, K. M., Pipes, G., Treuheit, M. J., et al. (2005). Aggregation of granulocyte-colony stimulating factor in vitro involves a conformationally altered aggregation state. *Protein Science*, *14*, 2246–2257.

Raut, A. S., & Kalonia, D. S. (2015). Liquid-liquid phase separation in a dual variable domain immunoglobulin protein solution: Effect of formulation factors and protein-protein interactions. *Molecular Pharmaceutics*, *12*, 3261–3271.

Receveur, V., Durand, D., Desmadril, M., & Calmettes, P. (1998). Repulsive interparticle interactions in a denatured protein solution revealed by small angle neutron scattering. *FEBS Letters*, *426*, 57–61.

Reddy, R. C. K., Lilie, H., Rudolph, R., & Lange, C. (2005). L-arginine increases the solubility of unfolded species of hen egg white lysozyme. *Protein Science*, *14*, 929–935.

Reiche, K., Hartl, J., Blume, A., & Garidel, P. (2017). Liquid-liquid phase separation of a monoclonal antibody at low ionic strength: Influence of anion charge and concentration. *Biophysical Chemistry*, *220*, 7–19.

Remmele, R. L., Jr., Nightlinger, N. S., Srinivasan, S., & Gombotz, W. R. (1998). Interleukin-1 receptor (IL-1R) liquid formulation development using differential scanning calorimetry. *Pharmaceutical Research*, *15*(2), 200–208.

Respaud, R., Marchand, D., Parent, C., Pelat, T., Thullier, P., Tournamile, J.-F., et al. (2014). Effect of formulation on the stability and aersol performance of a nebulized antibody. *MAbs*, *6*(5), 1347–1355.

Roberts, D., Keeling, R., Tracka, M., van der Walle, C. F., Uddin, S., Warwicker, J., et al. (2015). Specific ion and buffer effects on protein-protein interactions of a monoclonal antibody. *Molecular Pharmaceutics*, *12*(1), 179–193.

Robinson, A. B., & Rudd, C. J. (1974). Deamidation of glutaminyl and asparaginyl residues in peptides and proteins. *Current Topics in Cellular Regulation*, *8*, 247–295.

Rosenberg, A. S. (2006). Effects of protein aggregates: An immunologic perspective. *The AAPS Journal*, *8*(3), E501–E507.

Rosenberg, A. S., Verthelyi, D., & Cherney, B. W. (2012). Managing uncertainty: A perspective on risk pertaining to product quality attributes as they bear on immunogenicity of therapeutic proteins. *Journal of Pharmaceutical Sciences*, *101*(10), 3560–3567.

Rozhkov, S. P., & Goryunov, A. S. (2010). Thermodynamic study of protein phases formation and clustering in model water-protein-salt solutions. *Biophysical Chemistry*, *151*(1–2), 22–28.

Rozhkov, S. P., & Goryunov, A. S. (2012). Salt induced thermodynamic instability, concentration heterogeneity and phase transitions in lysozyme solutions. *Biophysical Chemistry*, *170*, 34–41.

Rubin, J., Linden, L., Coco, W. M., Bommarius, A. S., & Behrens, S. H. (2013). Salt-induced aggregation of a monoclonal human immunoglobulin G1. *Journal of Pharmaceutical Sciences*, *102*(2), 377–386.

Rubin, J., Sharma, A., Linden, L., Bommarius, A. S., & Behrens, S. H. (2014). Gauging colloidal and thermal stability in human IgG1-sugar solutions through diffusivity measurements. *Journal of Physical Chemistry B*, *118*(11), 2803–2809.

Ruíz-Henestrosa, V. P., Sánchez, C. C., & Patino, J. M. R. (2008). Effect of sucrose on functional properties of soy globulins: Adsorption and foam characteristics. *Journal of Agricultural and Food Chemistry*, *56*, 2512–2521.

Ruppert, S., Sandler, S. I., & Lenhoff, A. M. (2001). Correlation between the osmotic second virial coefficent and the solubility of proteins. *Biotechnology Progress*, *17*, 182–187.

Rustandi, R. R., & Wang, Y. (2011). Use of CE-SDS gel for characterization of monoclonal antibody hinge region clipping due to copper and high pH stress. *Electrophoresis*, *32*, 3078–3084.

Sadineni, V., Chandrasekharan, S., & Nassar, M. N. (2014). Implications of trace levels of redox-active metals in drug-product formulation. *BioPharm International*, *27*(4), 30+.

Saggu, M., Liu, J., & Patel, A. (2015). Identification of subvisible particles in biopharmaceutical formulations using Raman spectroscopy provides insight into polysorbate 20 degradation pathway. *Pharmaceutical Research*, *32*, 2877–2888.

Sahin, E., Jordan, J. L., Spatara, M. L., Naranjo, A., Costanzo, J. A., Weiss, W. F., et al. (2011). Computational design and biophysical characterization of aggregation-resistant point mutations for gamma D crystallin illustrate a balance of conformational stability and intrinsic aggregation propensity. *Biochemistry*, *50*(5), 628–639.

Saito, S., Hasegawa, J., Kobayashi, N., Kishi, N., Uchiyama, S., & Fukui, K. (2012). Behavior of monoclonal antibodies: Relation between the second virial coefficient (B(2)) at low concentrations and aggregation propensity and viscosity at high concentrations. *Pharmaceutical Research*, *29*(2), 397–410.

Saito, S., Hasegawa, J., Kobayashi, N., Tomitsuka, T., Uchiyama, S., & Fukui, K. (2013). Effects of ionic strength and sugars on the aggregation propensity of monoclonal antibodies: Influence of colloidal and conformational stabilities. *Pharmaceutical Research*, *30*, 1263–1280.

Salinas, B. A., Sathish, H. A., Shah, A. U., Carpenter, J. F., & Randolph, T. W. (2010). Buffer-dependent fragmentation of a humanized full-length monoclonal antibody. *Journal of Pharmaceutical Sciences*, *99*(7), 2962–2974.

Sancataldo, G., Vetri, V., Foderá, V., Di Cara, G., Militello, V., & Leone, M. (2014). Oxidation enhances human serum albumin thermal stability and changes the routes of amyloid fibril formation. *PLoS One*, *9*(1), e84552.

Sarangapani, P. S., Weaver, J., Parupudi, A., Besong, T. M. D., Adams, G. G., Harding, S. E., et al. (2016). Both reversible self-association and structural changes underpin molecular viscoelasticity of mAb solutions. *Journal of Pharmaceutical Sciences*, *105*, 3496–3506.

Schaink, H. M., & Smit, J. A. M. (2007). Protein-polysaccharide interactions: The determination of the osmotic second virial coefficients in aqueous solution of β-lactoglobulin and dextran. *Food Hydrocolloids*, *21*, 1389–1396.

Schöneich, C. (2000). Mechanisms of metal-catalyzed oxidation of histidine to 2-oxo-histidine in peptides and proteins. *Journal of Pharmaceutical and Biomedical Analysis*, *21*, 1093–1097.

Shiraki, K., Tomita, S., & Inoue, N. (2016). Small amine molecules: Solvent design towards facile improvement of protein stability against agrgegation and inactivation. *Current Pharmaceutical Biotechnology*, *17*, 116–125.

Shojaee, H., Sabbaghian, M., & Ebrahim-Habibi, A. (2016). Acarbose and the thermal aggregation of Bacillus amyloliquefaciens alpha-amylase (BAA): Protective effect of an inhibitor. *Journal of Chemical Technology and Biotechnology*, *91*, 1397–1402.

Shukla, D., & Trout, B. L. (2011). Perferential interaction coefficients of proteins in aqueous arginine solutions and their moelcular origins. *Journal of Physical Chemistry B*, *115*, 1243–1253.

Siderius, D. W., Krekelberg, W. P., Roberts, C. J., & Shen, V. K. (2012). Osmotic virial coefficients for model protein and colloidal solutions: Importance of ensemble constraints in the analysis of light scattering data. *Journal of Chemical Physics*, *136*, 175102.

Singh, S., & Singh, J. (2003). Effect of polyols on the conformational stability and biological activity of a model protein lyoszyme. *AAPS PharmSciTech*, *4*(3).

Singh, S. N., Yadav, S., Shire, S. J., & Kalonia, D. S. (2014). Dipole-dipole interaction in antibody solutions: Correlation with viscosity behavior at high concentration. *Pharmaceutical Research*, *31*, 2549–2558.

Sleutel, M., Van Driessche, A. E. S., Pan, W., Reichel, E. K., Maes, D., & Vekilov, P. G. (2012). Does solution viscosity scale the rate of aggregation of folded proteins? *Journal of Physical Chemistry Letters*, *3*, 1258–1263.

Smith, M. A., Easton, M., Everett, P., Lewis, G., Payne, M., Riveros-Moreno, V., et al. (1996). Specific cleavage of immunoglobulin G by copper ions. *International Journal of Peptide and Protein Research*, *48*, 48–55.

Smith, R. M., & Hansen, D. E. (1998). The pH-rate profile for the hydrolysis of a peptide bond. *Journal of the American Chemical Society*, *120*, 8910–8913.

Sorret, L. L., DeWinter, M. A., Schwartz, D. K., & Randolph, T. W. (2016). Challenges in predicting protein-protein interactions from measurements of molecular diffusivity. *Biophysical Journal*, *111*(9), 1831–1842.

Spencer, S., Bethea, D., Raju, T. S., Giles-Komar, J., & Feng, Y. Q. (2012). Solubility evaluation of murine hybridoma antibodies. *MAbs*, *4*(3), 319–325.

Sreedhara, A., Glover, Z. K., Piros, N., Xiao, N., Patel, A., & Kabakoff, B. (2012). Stability of IgG1 monoclonal antibodies in intravenous infusion bags under clinical in-use conditions. *Journal of Pharmaceutical Sciences*, *101*(1), 21–30.

Sreedhara, A., Yin, J., Joyce, M., Lau, K., Wecksler, A. T., Deperalta, G., et al. (2016). Effect of ambient light on IgG1 monoclonal antibodies during drug product processing and development. *European Journal of Pharmaceutics and Biopharmaceutics*, *100*, 38–46.

Stratton, L. P., Kelly, R. M., Rowe, J., Shively, J. E., Smith, D. D., Carpenter, J. F., et al. (2001). Controlling deamidation rates in a model peptide: Effects of temperature, peptide concentration, and additives. *Journal of Pharmaceutical Sciences*, *90*(12), 2141–2148.

Street, T. O., Bolen, D. W., & Rose, G. D. (2006). A molecular mechanism for osmolyte-induced protein stability. *Proceedings of the National Academy of Sciences of the United States of America*, *103*(38), 13997–14002.

Sule, S. V., Sukumar, M., Weiss, W. F., Marcelino-Cruz, A. M., Sample, T., & Tessier, P. M. (2011). High-throughput analysis of concentration-dependent antibody self-association. *Biophysical Journal*, *101*(7), 1749–1757.

Sun, T., Reid, F., Liu, Y., Cao, Y., Estep, P., Nauman, C., et al. (2013). High throughput detection of antibody self-interaction by bio-layer interferometry. *MAbs*, *5*(6), 838–841.

Sydow, J. F., Lipsmeier, F., Larraillet, V., Hilger, M., Mautz, B., Molhoj, M., et al. (2014). Structure-based prediction of asparagine and aspartate degradation sites in antibody variable regions. *PLoS One*, *9*(6), e100736.

Taddei, N., Capanni, C., Chiti, F., Stefani, M., Dobson, C. M., & Ramponi, G. (2001). Folding and aggregation are selectively influenced by the conformational preferences of the alpha-helices of muscle acylphosphatase. *Journal of Biological Chemistry*, *276*(40), 37149–37154.

Tamizi, E., & Jouyban, A. (2016). Forced degradation studies of biopharmaceuticals: Selection of stress conditions. *European Journal of Pharmaceutics and Biopharmaceutics*, *98*, 26–46.

Teh, L. C., Murphy, L. J., Huq, N. L., Surus, A. S., Friesen, H. G., Lazarus, L., et al. (1987). Methionine oxidation in human growth hormone and human chorionic somatotropin. Effects on receptor binding and biological activity. *Journal of Biological Chemistry*, *262*(14), 6472–6477.

Teske, C. A., Blanch, H. W., & Prausnitz, J. M. (2004). Chromatographic measurement of interactions between unlike proteins. *Fluid Phase Equilibria*, *219*, 139–148.

Tessier, P. M., Lenhoff, A. M., & Sandler, S. I. (2002). Rapid measurement of protein osmotic second virial coefficients by self-interaction chromatography. *Biophysical Journal*, *82*(3), 1620–1631.

Tessier, P. M., Sandler, S. I., & Lenhoff, A. M. (2004). Direct measurement of protein osmotic second virial cross coefficients by cross-interaction chromatography. *Protein Science*, *13*(5), 1379–1390.

Tessier, P. M., Vandrey, S. D., Berger, B. W., Pazhianur, R., Sandler, S. I., & Lenhoff, A. M. (2002). Self-interaction chromatography: A novel screening method for rational protein crystallization. *Acta Crystallographica, Section D: Biological Crystallography*, *58*, 1531–1535.

Thirumangalathu, R., Krishnan, S., Ricci, M. S., Brems, D. N., Randolph, T. W., & Carpenter, J. F. (2009). Silicone oil- and agitation-induced aggregation of a monoclonal antibody in aqueous solution. *Journal of Pharmaceutical Sciences*, *98*(9), 3167–3181.

Tidy, R. J., Lam, V., Fimognari, N., Mamo, J. C., & Hackett, M. J. (2017). FTIR studies of the similarities between pathology induced protien aggregation *in vivo* and chemically induced protein aggregation *in vitro*. *Vibrational Spectroscopy*, *91*, 68–76.

Timasheff, S. N. (1998). Control of protein stability and reactions by weakly interacting cosolvents: The simplicity of the complicated. *Advances in Protein Chemistry*, *51*, 352–432.

Tiwari, A., & Bhat, R. (2006). Stabilization of yeast hexokinase A by polyol osmolytes: Correlation with the physicochemical properties of aqueous solutions. *Biophysical Chemistry*, *124*, 90–99.

To, B. C. S., & Lenhoff, A. M. (2007). Hydrophobic interaction chromatography of proteins II. Solution thermodynamic properties as a determinant of retention. *Journal of Chromatography A*, *1141*(2), 235–243.

Toews, J., Rogalski, J. C., & Kast, J. (2010). Accessibility governs the relative reactivity of basic residues in formaldehyde-induced protein modifications. *Analytica Chimica Acta*, *676*, 60–67.

Tomar, D. S., Kumar, S., Singh, S. K., Goswami, S., & Li, L. (2016). Molecular basis of high viscosity in concentrated antibody solutions: Strategies for high concentration drug product development. *MAbs*, *8*(2), 216–228.

Tomita, S., & Shiraki, K. (2011). Why do solution additives suppress the heat-induced inactivation of proteins? Inhibition of chemical modifications. *Biotechnology Progress*, *27*(3), 855–862.

Topp, E. M., Zhang, L., Zhao, H., Payne, R. W., Evans, G. J., & Manning, M. C. (2010). Chemical instability in peptide and protein pharmaceuticals. In F. Jameel & S. Hershenson (Eds.), *Formulation and process development strategies for manufacturing biopharmaceuticals* (pp. 41–67). New York: John Wiley and Sons.

Torosantucci, R., Schöneich, C., & Jiskoot, W. (2014). Oxidation of therapeutic proteins and peptides: Structural and biological consequences. *Pharmaceutical Research*, *31*, 541–553.

Trnková, L., Dršata, J., & Boušová, I. (2015). Oxidation as an important factor of protein damage: Implications for Maillard reaction. *Journal of Biosciences*, *40*(2), 419–439.

Uchiyama, S. (2014). Liquid formulation for antibody drugs. *Biochimica et Biophysica Acta*, *1844*, 2041–2052.

Valente, J. J., Fryksdale, B. G., Dale, D. A., Gaertner, A. L., & Henry, C. S. (2006). Screening for physical stability of a Pseudomonas amylase using self-interaction chromatography. *Analytical Biochemistry*, *357*(1), 35–42.

Valente, J. J., Payne, R. W., Manning, M. C., Wilson, W. W., & Henry, C. S. (2005). Colloidal behavior of proteins: Effects of the second virial coefficient on solubility, crystallization and aggregation of proteins in aqueous solution. *Current Pharmaceutical Biotechnology*, *6*, 427–436.

Vekilov, P. G. (2010). Phase transitions of folded proteins. *Soft Matter*, *6*(21), 5254–5272.

Vidanovic, D., Askrabic, J. M., Stankovic, M., & Poprzen, V. (2003). Effects of nonionic surfactants on the physical stability of immunoglobulin G in aqueous solution during mechanical agitation. *Pharmazie*, *58*, 399–404.

Vlasak, J., & Ionescu, R. (2011). Fragmentation of monoclonal antibodies. *MAbs*, *3*(3), 253–263.

Wakankar, A. A., & Borchardt, R. T. (2006). Formulation considerations for proteins susceptible to asparagine deamidation and aspartate isomerization. *Journal of Pharmaceutical Sciences, 95*(11), 2321–2336.

Wang, W., Ignatius, A. A., & Thakkar, S. V. (2014). Impact of residual impurities and contaminants on protein stability. *Journal of Pharmaceutical Sciences, 103*, 1315–1330.

Wang, W., Singh, S., Zeng, D. L., King, K., & Nema, S. (2007). Antibody structure, instability, and formulation. *Journal of Pharmaceutical Sciences, 96*(1), 1–26.

Wang, S., Wu, G., Zhang, X., Tian, Z., Zhang, N., Hu, T., et al. (2017). Stabilizing two IgG1 monoclonal antibodies by surfactants: Balance between aggregation prevention and structure perturbation. *European Journal of Pharmaceutics and Biopharmaceutics, 114*, 263–277.

Wang, S. J., Zhang, N., Hu, T., Dai, W. G., Feng, X. Y., Zhang, X. Y., et al. (2015). Viscosity-lowering effect of amino acids and salts on highly concentrated solutions of two IgG1 monoclonal antibodies. *Molecular Pharmaceutics, 12*(12), 4478–4487.

Waterman, K. C., Adami, R. C., Alsante, K. M., Hong, J., Landis, M. S., Lombardo, F., et al. (2002). Stabilization of pharmaceuticals to oxidative degradation. *Pharmaceutical Development and Technology*, 7(1), 1–32.

Wen, L., Zheng, X., Wang, X., Lan, H., & Yin, Z. (2017). Bilateral effects of excipients on protein stability: preferential interaction type of excipient and surface aromatic hydrophobicity of protein. *Pharmaceutical Research, 34*, 1378–1390.

Wills, P. R., Scott, D. J., & Winzor, D. J. (2015). The osmotic second virial coefficient for protein self-interaction: Use and misuse to describe thermodynamic nonideality. *Analytical Biochemistry, 490*, 55–65.

Wills, P. R., & Winzor, D. J. (2017). Rigorous analysis of static light scattering measurements on buffered protein solutions. *Biophysical Chemistry, 228*, 108–113.

Wilson, W. W., & DeLucas, L. J. (2014). Applications of the second virial coefficient: Protein crystallization and solubility. *Acta Crystallographica Section F: Structural Biology Communications, F70*, 543–554.

Winzor, D. J. (2016). Measurement of osmotic second virial coefficients by zonal size exclusion chromatography. *Analytical Biochemistry, 504*, 59–63.

Winzor, D. J., Scott, D. J., & Wills, P. R. (2007). A simpler analysis for the measurement of second virial coefficients by self-interaction chromatography. *Analytical Biochemistry, 371*(1), 21–25.

Wolf, M., Roosen-Runge, F., Zhang, F. J., Roth, R., Skoda, M. W. A., Jacobs, R. M. J., et al. (2014). Effective interactions in protein-salt solutions approaching liquid-liquid phase separation. *Journal of Molecular Liquids, 200*, 20–27.

Wu, Y., Levons, J., Narang, A. S., Raghavan, K., & Rao, V. M. (2011). Reactive impurities in excipients: Profiling, identification and mitigation of drug-excipient incompatibility. *AAPS PharmSciTech, 12*(4), 1248–1263.

Wu, J., Schultz, J. S., Weldon, C. L., Sule, S. V., Chai, Q., Geng, S. B., et al. (2015). Discovery of highly soluble antibodies prior to purification using affinity-capture self-interaction nanoparticle spectroscopy. *Protein Engineering, Design & Selection, 28*(10), 403–414.

Yadav, S., Liu, J., Shire, S. J., & Kalonia, D. S. (2010). Specific interactions in high concentration antibody solutions resulting in high viscosity. *Journal of Pharmaceutical Sciences, 99*(3), 1152–1168.

Yadav, S., Scherer, T. M., Shire, S. J., & Kalonia, D. S. (2011). Use of dynamic light scattering to determine second virial coefficient in a semidilute concentration regime. *Analytical Biochemistry, 411*(2), 292–296.

Yadav, S., Shire, S. J., & Kalonia, D. S. (2010). Factors affecting the viscosity in high concentration solutions of different monoclonal antibodies. *Journal of Pharmaceutical Sciences, 99*(12), 4812–4829.

Yadav, S., Shire, S. J., & Kalonia, D. S. (2012). Viscosity behavior of high-concentration monoclonal antibody solutions: Correlation with interaction parameter and electroviscous effects. *Journal of Pharmaceutical Sciences*, *101*(3), 998–1011.

Yageta, S., Lauer, T. M., Trout, B. L., & Honda, S. (2015). Conformational and colloidal stabilities of isolated constant domains of human immunoglobulin G and their impact on antibody aggregation under acidic conditions. *Molecular Pharmaceutics*, *12*(5), 1443–1455.

Yan, B., & Boyd, D. (2011). Breaking the light and heavy chain linkage of human immunoglobulin G1 (IgG1) by radical reactions. *Journal of Biological Chemistry*, *286*(28), 24674–24684.

Yang, X., Zhang, Y., Wang, F., Wang, J., Richardson, D., Shameem, M., et al. (2015). Analysis and purification of IgG4 bispecific antibodies by a mixed-mode chromatography. *Analytical Biochemistry*, *484*, 173–179.

Yates, Z., Gunasekaran, K., Zhou, H., Hu, Z., Liu, Z., Ketchem, R. R., et al. (2010). Histidine residue mediates radical-induced hinge cleavage of human IgG1. *Journal of Biological Chemistry*, *285*(24), 18662–18671.

Yearley, E. J., Godfrin, P. D., Perevozchikova, T., Zhang, H., Falus, P., Porcar, L., et al. (2014). Observation of small cluster formation in concentrated monoclonal antibody solutions and its implications to solution viscosity. *Biophysical Journal*, *106*, 1763–1770.

Yearley, E. J., Zarraga, I. E., Shire, S. J., Scherer, T. M., Gokarn, Y., Wagner, N. J., et al. (2013). Small-angle neutron scattering characterization of monoclonal antibody conformations and interactions at high concentrations. *Biophysical Journal*, *105*(3), 720–731.

Yin, J., Chu, J.-W., Ricci, M. S., Brems, D. N., Wang, D. I. C., & Trout, B. L. (2004). Effects of antioxidants on the hydrogen peroxide-mediated oxidation of methionine residues in granulocyte colony-stimulating factor and human parathyroid hormone fragment 13-34. *Pharmaceutical Research*, *21*(12), 2377–2382.

Yin, S., Zhang, C., Li, Z., Wang, Q., Shi, H., Yu, R., et al. (2017). Identification, characterization, and stabilization of the deamidation degradation of recombinant human tumor necrosis factor-α. *Process Biochemistry*, *53*, 216–223.

Yokota, H., Saito, H., Masuoka, K., Kaniwa, H., & Shibanuma, T. (2000). Reversed phase HPLC of Met(58) oxidized rhIL-11: Oxidation enhanced by plastic tubes. *Journal of Pharmaceutical and Biomedical Analysis*, *24*(2), 317–324.

Youssef, A. M. K., & Winter, G. (2013). A critical evaluation of microcalorimetry as a predictive tool for long term stability of liquid protein formulations: Granulocyte colony stimulating factor (GCSF). *European Journal of Pharmaceutics and Biopharmaceutics*, *84*(1), 145–155.

Zakrzewska, M., Krowarsch, D., Wiedlocha, A., Olsnes, S., & Otlewski, J. (2005). Highly stable mutants of human fibroblast growth factor-1 exhibit prolonged biological action. *Journal of Molecular Biology*, *352*, 860–875.

Zarraga, I. E., Taing, R., Zarzar, J., Luoma, J., Hsiung, J., Patel, A., et al. (2013). High shear rheology and anisotropy in concentrated solutions of monoclonal antibodies. *Journal of Pharmaceutical Sciences*, *102*(8), 2538–2549.

Zbacnik, T. J., Holcomb, R. E., Katayama, D. S., Murphy, B. M., Payne, R. W., Coccaro, R. C., et al. (2017). Role of buffers in protein formulations. *Journal of Pharmaceutical Sciences*, *106*(3), 713–733.

Zhang, J., & Liu, X. Y. (2003). Effect of protein-protein interactions on protein aggregation kinetics. *Journal of Chemical Physics*, *119*(20), 10972–10976.

Zhang, F. J., Roosen-Runge, F., Sauter, A., Roth, R., Skoda, M. W. A., Jacobs, R. M. J., et al. (2012). The role of cluster formation and metastable liquid-liquid phase separation in protein crystallization. *Faraday Discussions*, *159*, 313–325.

Zheng, S., Qiu, D., Adams, M., Li, J., Mantri, R. V., & Gandhi, R. (2017). Investigating the degradation behaviors of a therapeutic monoclonal antibody associated with pH and buffer species. *AAPS PharmSciTech*, *18*(1), 42–48.

Zhou, S., Zhang, B., Sturm, E., Teagarden, D. L., Schöneich, C., Kolhe, P., et al. (2010). Comparative evaluation of disodium edetate and diethylenetriaminepentaacetic acid as iron chelators to prevent metal-catalyzed destabilization of a therapeutic monoclonal antibody. *Journal of Pharmaceutical Sciences*, *99*(10), 4239–4250.

Zidar, M., Susteric, A., Ravnik, M., & Kuzman, D. (2017). High throughput prediction approach for monoclonal antibody aggregation at high concentration. *Pharmaceutical Research*, *34*(9), 1831–1839.

CHAPTER TWO

Therapeutic Monoclonal Antibodies Delivery for the Glioblastoma Treatment

Flávia Sousa*[,†,‡,§], Rui P. Moura[§], Elias Moreira[§], Cláudia Martins*[,†], Bruno Sarmento*[,†,§,1]

*i3S—Instituto de Investigação e Inovação em Saúde, Universidade do Porto, Porto, Portugal
[†]INEB—Instituto Nacional de Engenharia Biomédica, Universidade do Porto, Porto, Portugal
[‡]ICBAS—Instituto Ciências Biomédicas Abel Salazar, Universidade do Porto, Porto, Portugal
[§]CESPU—Instituto de Investigação e Formação Avançada em Ciências e Tecnologias da Saúde, Gandra, Portugal
[1]Corresponding author: e-mail address: bruno.sarmento@ineb.up.pt

Contents

Abstract

Glioblastoma multiforme (GBM) is the most common and challenging primary malignant brain tumor, being the median overall survival between 10 and 14 months due to its invasive characteristics. GBM treatment is mainly based on the maximal surgical resection and radiotherapy associated to chemotherapy. Monoclonal antibodies (mAbs) have been used in chemotherapy protocols for GBM treatment in order to improve immunotherapy and antiangiogenic processes. High specificity and affinity of mAbs for biological targets make them highly used for brain tumor therapy. Specifically, antiangiogenic mAbs have been wisely indicated in chemotherapy protocols because GBM is the most vascularized tumors in humans with high expression of cytokines. However, mAb-based therapy is not that effective due to the aggressive spread of the tumor

Advances in Protein Chemistry and Structural Biology, Volume 112
ISSN 1876-1623
https://doi.org/10.1016/bs.apcsb.2018.03.001

associated to the difficulty in the access of mAb into the brain (due to the blood–brain barrier). For that reason, nanobiotechnology has played an important role in the treatment of several tumors, mainly in the tumors of difficult access, such as GBM. In this chapter will be discussed strategies related with nanobiotechnology applied to the mAb delivery and how these therapeutics can improve the GBM treatment and life quality of the patient.

1. INTRODUCTION

Malignant primary and metastatic brain tumors are still the most challenging tumors regarding their treatment since brain tumors remain consistently fatal. It is well known that 80% of malignant brain tumors are mainly gliomas, among them more than half of them are glioblastoma multiforme (GBM) (Jain et al., 2007). Actually, GBM is a high morbidity and mortality brain tumor with less than 15% of patient survival, being the median overall survival between 10 and 14 months, after diagnosis (Wong et al., 1999). The worst characteristic associated to GBM is attributed to its high vascularization, characterized by high expression and secretion of angiogenic cytokines (Batchelor, Reardon, de Groot, Wick, & Weller, 2014). Tumor growth is highly dependent of blood vessels supply in order to acquire all nutrients and oxygen. For that reason, antiangiogenic strategies are totally mandatory due to its potential value for the treatment. Antiangiogenic drugs might be used as a single agent or in combination with other chemotherapeutic drugs in the current chemotherapy protocol after surgical resection of tumor, if possible (Carlsson, Brothers, & Wahlestedt, 2014).

Several antiangiogenic drugs have been used in the GBM treatment, which are responsible to target vascular endothelial growth factor (VEGF), avoiding high vascularization of gliomas and tumor survival (McNamara & Mason, 2012). It has been described that blood vessel normalization attributed to antiangiogenic drugs happens due to an increase of blood flow and oxygenation (Jain et al., 2007; McNamara & Mason, 2012). Different antiangiogenic drugs have been studied in GBM from large macromolecules, such as monoclonal antibodies (mAbs) (e.g., bevacizumab, rilotumumab) or recombinant fusion protein (e.g., aflibercept), to small-molecule kinase or integrin inhibitors (e.g., sunitinib, dasatinib, vatalanib, sorafenib) (Batchelor et al., 2014; McNamara & Mason, 2012). Fig. 1 represents the main mechanism of action of different types of antiangiogenics drugs, resulting in a decrease of vascularization, vascular permeability, endothelial cell survival, and cell migration.

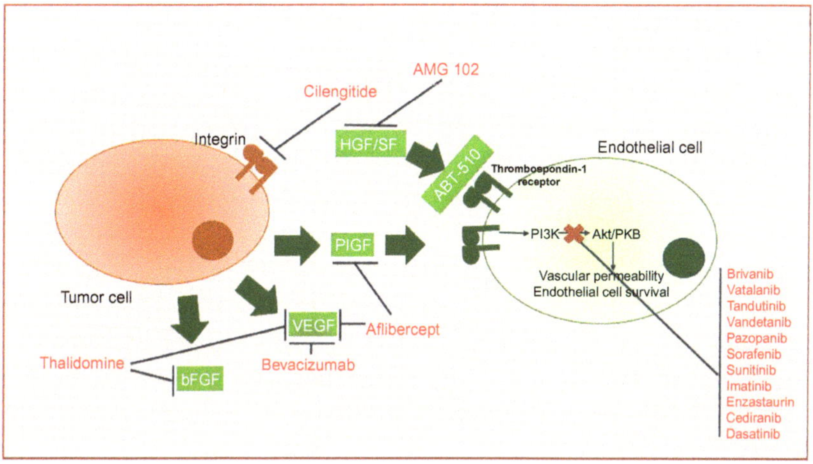

Fig. 1 Mechanism of action of antiangiogenic drugs for the glioblastoma multiforme treatment. Antiangiogenic drugs are represented at *red* color. *bFGF*, basic fibroblast growth factor; *HGF/SF*, hepatocyte growth factor; *PIGF*, placental growth factor; *VEGF*, vascular endothelial growth factor.

However, bevacizumab is the first antiangiogenic drug that has shown promising results in the progression-free survival for the GBM treatment, either as a single agent or in combination with chemotherapy (Blesa et al., 2012; Gil et al., 2012; Gil-Gil, Mesia, Rey, & Bruna, 2013). Recently, a systematic review of randomized clinical trials showed its potential role in the GBM treatment compared with other 13 antiangiogenic drugs, being the only one that improved progression survival ratio (Lombardi et al., 2017). Bevacizumab is an IgG1 mAb, who acts against VEGF, avoiding its interaction with its receptors (VEGFR). The inhibition of the VEGF binding to its VEGFR results in the cell proliferation inhibition, decrease of vascular permeability, and cell migration (Chamberlain, 2011; Ferrara, Hillan, Gerber, & Novotny, 2004; Wang, Fei, Vanderlaan, & Song, 2004). Despite the high specificity and high affinity of mAbs to their target, the access of these biomacromolecules to the brain tumor is limited due to the blood–brain barrier (BBB) (Jain, 2007). Low permeability of BBB becomes an obstacle for the GBM treatment, requiring higher doses of mAbs and consequently, more damage due to off-targets toxicity (Glaser, Han, Wu, & Zeng, 2017). On the other hand, poor tissue distribution of mAbs in combination with the aggressive local spread of tumor becomes the efficacy of this antibody-based therapy which is very limited. Furthermore, malignant brain tumors have the ability to induce severe modification in the BBB, developing a new barrier called blood–brain tumor barrier (BBTB), characterized by disorganized and hyperpermeable tumor blood vessels (Kim & Lee, 2009).

Therefore, in order to overcome these limitations and achieve a high efficacy for the GBM treatment, researchers have started using nanobiotechnology as a tool to improve drug delivery in cancer. For the delivery of mAbs through nanobiotechnology, mAbs have been encapsulated into organic nanoparticles (e.g., poly(propyleneglycol) nanoparticles, poly(ethylene glycol) nanoparticles, poly(lactic-co-glycolic) acid nanoparticles, polymeric micelles, liposomes, dendrimers) or inorganic nanoparticles (iron, gold nanoparticles). On the other hand, biochemical modifications might still be done in the surface of nanoparticles in order to enable the delivery in a specific target, avoiding side effects.

The aim of this chapter is to explain the advantages of using therapeutic mAbs delivery for the GBM treatment and how can nanotechnology allow the improvement of mAbs pharmacokinetics and pharmacodynamics. The importance of glioblastoma pathophysiology, glioblastoma diagnosis, and current antibody-based and nonbased therapy will also be addressed.

2. GLIOBLASTOMA PATHOPHYSIOLOGY

Glioblastoma is the most prevalent and lethal type of primary brain tumors (D'Alessandris & Ricci-Vitiani, 2016; Tian et al., 2017). According to the American Brain Tumor Association, 14.9% of all primary brain tumors and 55.4% of all gliomas consist on GBMs. The high incidence of this pathology leads to an estimation of 12,390 new clinical cases in the year 2017.

GBM is a glioma, which is a tumor raised from glial cells or their precursors, classified as grade IV in the current World Health Organization (WHO) system (Tian et al., 2017). This system denotes lesions with low proliferative capacity, surgically removable, as grade I (low-grade glioma); infiltrating and possibly progressive lesions as grade II (low-grade glioma); lesions with high mitotic activity and nuclear atypia or anaplasia, possibly progressive, as grade III (high-grade or malignant glioma); and extremely hostile tumors, capable of infiltrating into neighboring tissues, with high mitotic activity, necrosis events, widespread microvascular proliferations, and a fatal outcome as grade IV (high-grade or malignant glioma) (Louis et al., 2007; Nazarenko et al., 2012). Moreover, gliomas can be divided into astrocytomas (originated from astroglial cells), oligodendrogliomas (originated from oligodendroglial cells), oligoastrocytomas (originated from both astrocytes and oligodendrocytes), and ependymomas (originated from ependymal cells) (Ho et al., 2014). According to this classification, GBM is considered an astrocytoma.

Most part of the GBMs are located in the cerebral hemispheres, especially on the temporal, parietal, and frontal lobes, where they proliferate from irregular-shaped lesions usually originated in the white matter (Nelson & Cha, 2003). GBMs are characterized by a heterogeneous environment, simultaneously demonstrating different cystic and gelatinous areas, as well as hemorrhage and necrosis events (Agnihotri et al., 2013). In what concerns cellular development, these tumors are also characterized by the presence of cells in different differentiation stages and with variable tumoral activity. As usual in gliomas biology, the GBM-associated malignancy arises from the accumulation of genetic alterations and abnormal regulation of the signaling pathways related to growth factors, consequently leading to a tumoral environment enriched in proinflammatory cytokines and chemokines (Alifieris & Trafalis, 2015). Invasion is the key cellular feature of GBM during the initial stage of the pathology. The malignant cells are capable of extensively migrating through the brain tissue, infiltrating and destroying physical barriers, which include basement membranes, cell junctions, and extracellular matrix. This may be attributed to the overexpression of several elements of the metzincin family (zinc-based proteinases), as well as the action of different integrins (e.g., $apha_v$–$beta_3$ integrin) and transforming growth factor-β (TGF-β) (Alves et al., 2011). Although GBMs can easily proliferate, they rarely metastasize to other organs (Lathia, Mack, Mulkearns-Hubert, Valentim, & Rich, 2015).

GBMs are classified as primary and secondary. Primary GBMs are the most common ones and arise de novo without prior clinical antecedents (within 3–6 months), whereas secondary ones slowly accrue from already existing lower-grade astrocytomas (over 10–15 years) (Agnihotri et al., 2013; Alifieris & Trafalis, 2015). Primary GBMs generally manifest in older patients, while secondary GBMs are most common on the younger ones (Aldape, Zadeh, Mansouri, Reifenberger, & von Deimling, 2015). Despite the similarities between the pathophysiology involved in primary and secondary GBMs, some molecular differences may be pointed out. In primary GBMs, more severe mutations occur in the genes related to the epidermal growth factor receptor (EGFR), consequently encoding a mutated type of EGFR, known as epidermal growth factor receptor variant III (EGFRvIII) (Alifieris & Trafalis, 2015). This mutated type, EGFRvIII, not only promotes the proliferation of EGFRvIII-positive cells but also drives the proliferation of their neighboring cells which express the nonmutated type, EGFR (Aldape et al., 2015). In secondary GBMs, the key player is the amplification of the alpha-type of the gene associated with the platelet-derived growth factor receptor (PDGFR), hence resulting in the platelet-derived growth factor receptor alpha (PDGFRA) (Alifieris & Trafalis, 2015). PDGFRA is involved in glial cell proliferation and oligodendrocyte differentiation (Alentorn et al., 2012). Both primary and secondary GBMs-associated mutations culminate in a higher activity of the tyrosine kinase receptor (TKR), hence activating the rat sarcoma (RAS) and phosphatidylinositol 3-kinase (PI3K) signaling pathways (Alifieris & Trafalis, 2015). Biologically, this ultimately results in an aggressive growth of the glioma mass and stimulation of the microenvironment tumorigenicity.

A different GBMs classification was implemented by the Cancer Genome Atlas, which divided them into proneural, mesenchymal, classical, and neural (Verhaak et al., 2010). This gene expression-based molecular classification involved more biological interveners, rather than just EGFR/EGFRvIII and PDGFR/PDGFRA. According to this classification, proneural GBMs are characterized to be highly mutated at the genes encoding the tumor protein p53 (TP53 gene), isocitrate dehydrogenase 1 (IDH1 gene), and PDGFRA; the mesenchymal GBMs are more prone to mutations in the genes encoding the neurofibromatosis type 1 (NF1 gene), phosphatase and tensin homolog (PTEN gene), and also tumor protein p53; in classical GBMs, the main mutations involve the EGFR, whereas no alterations occur in tumor suppressor genes; and neural GBMs are characterized by the expression of noncancerous proteins associated with normal neuronal differentiation (Olar & Aldape, 2014; Verhaak et al., 2010). Table 1 summarizes some of the most common gene mutations in GBM development.

The GBM stemness features are also an important issue in what concerns the development of the malignancy. The cancer stem cells (CSCs), also referred as tumor-initiating cells, were found to be able to stimulate and sustain GBM growth, demonstrating close similarities to the adult normal neural stem cells (NSCs), which include self-renewal and multilineage differentiation abilities (Davis, 2016; Wakimoto et al., 2009). The emergence of CSCs from mutated NSCs was hypothesized as a process involving the deletion of the NF1 and TP53 genes, among others, despite the lack of enough scientific evidences to corroborate this theory (Pérez-Castillo, Aguilar-Morante, Morales-García, & Dorado, 2008; Zhu et al., 2005). CSCs intimately contact with aberrant blood vessel formation during GBM pathogenesis, which

Table 1 Some of the Gene Mutations Frequently Found in GBM Pathology

Gene Symbol	Gene Identification	Encoded Protein Function	References
EGFR	Epidermal growth factor receptor	Regulation of signaling networks, metabolic reprogramming, and cell proliferation	Furnari, Cloughesy, Cavenee, and Mischel (2015) and Liu et al. (2015)
PDGFR	Platelet-derived growth factor receptor	Glial cell proliferation and oligodendrocyte differentiation	Alentorn et al. (2012) and Verhaak et al. (2010)
IDH1	Isocitrate dehydrogenase 1	NADPH production	Olar and Aldape (2014) and SongTao et al. (2012)
NF1	Neurofibromin 1	Regulation of signaling networks (RAS pathway)	Alifieris and Trafalis (2015) and Karsy et al. (2012)
PTEN	Phosphatase and tensin homolog	Regulation of signaling networks (PI3K pathway)	Alifieris and Trafalis (2015) and Molina et al. (2011)
TP53	Tumor protein p53	Induction of cellular apoptosis	Ohgaki et al. (2004) and Wang et al. (2014)
RB1	Retinoblastoma 1	Regulation of cell cycle	Parsons et al. (2008) and TCGAN (2008)
$p16^{INK4a}$	Prototype member of the INK4 family of cyclin-dependent kinase inhibitors	Regulation of cell cycle	Karsy et al. (2012) and Solomon, Kim, Jean, and Waldman (2008)
MGMT	*O*-6-methylguanine-DNA methyltransferase	DNA repair	Aldape et al. (2015) and Karsy et al. (2012)
ERBB2	V-erb-b2 erythroblastic leukemia viral oncogene homolog 2	Regulation of signaling networks and cell proliferation	Parsons et al. (2008) and TCGAN (2008)
PIK3CA	Phosphoinositide-3-kinase	Regulation of signaling networks, cell growth, migration, and survival	Gallia et al. (2006) and Hartmann, Bartels, Gehlhaar, Holtkamp, and von Deimling (2005)
PIK3R1	Phosphoinositide-3-kinase regulatory subunit 1	Regulation of signaling networks, cell growth, migration, and survival	Parsons et al. (2008) and Verhaak et al. (2010)
PTPRD	Protein tyrosine phosphatase receptor type D	Regulation of signaling networks and cell development	Solomon et al. (2008) and Veeriah et al. (2009)

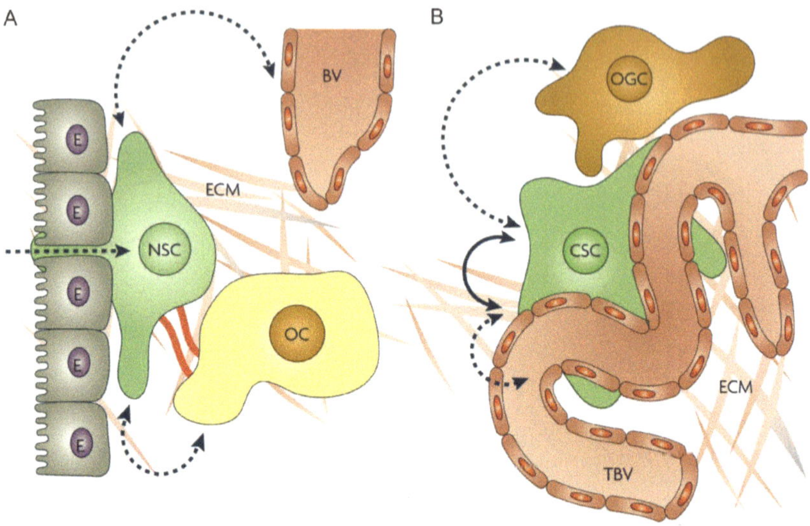

Fig. 2 The normal and malignant neural stem cell niche. (A) Stylized view of the normal neural stem cell (NSC) niche. NSCs intimately interact with ependymal cells (E), blood vessels (BV), and various other cell types including progenitor and support cells (OC). NSC function may be regulated by various diffusible factors derived from the surrounding cells, as well as the cerebrospinal fluid (*broken lines*; e.g., growth factors, pigment epithelium-derived factor (PEDF), and bone morphogenetic proteins (BMPs)). Additional regulation might be provided by direct cell contacts (*red solid lines*; e.g., NOTCH receptor-ligand signals) and the extracellular matrix (ECM). (B) GBM cancer stem cells (CSCs) are found in intimate contact with the aberrant tumor vasculature (TBV). CSCs can secrete diffusible factors such as VEGF, which recruit TBV to the niche. In turn, TBV and other glioma cells (OGC) secrete factors that maintain aberrant CSC self-renewal. Mutations in CSC might also enable aberrant, intrinsic self-renewal (*solid line*). *Adapted with permission from Gilbertson, R. J., & Rich, J. N. (2007). Making a tumour's bed: Glioblastoma stem cells and the vascular niche.* Nature Reviews Cancer, *7*, *733*.

consists of one of the hallmarks of the disease, therefore contributing to fast tumor growth. In fact, it is known that the necrotic and hypoxic microenvironment of GBM strongly promotes angiogenesis through the action of the VEGF, vascular endothelial growth factor receptor 2 (VEGFR2), and angiopoietin 2 (Gilbertson & Rich, 2007). Fig. 2 elucidates the main differences related to the neuromicroenvironment of NSCs and CSCs niche.

3. DIAGNOSIS OF GLIOBLASTOMA

The primary diagnosis of GBM is usually performed through tomographic imaging techniques, in which magnetic resonance imaging is considered the gold standard. Other techniques may be used to diagnose GBM, including perfusion-weighted imaging, diffusion-weighted imaging, diffusion tensor imaging, dynamic contrast-enhanced T1 permeability imaging, and magnetic resonance spectroscopy (Kesari, 2011; Young, 2007). Moreover, an angiography may be useful to detect the highly atypical vasculature of the GBM, usually alongside early venous drainage and arteriovenous shunting (Weller et al., 2014).

The final diagnosis is usually based on a histopathological analysis of the GBM or its sections (Urbańska, Sokołowska, Szmidt, & Sysa, 2014). Central necrosis and perilesional edema are key aspects in what concerns the diagnosis of a GBM scenario (Weller et al., 2014). Also, the immunohistochemical detection of the glial fibrillary acidic protein (GFAP), which is a protein involved in the maturation of astrocytes, is commonly used in this type of diagnostic, since less protein expression is correlated with high malignancy of GBMs (Rutka et al., 1997; Urbańska et al., 2014). The histopathological analysis involves the resection of the tumor mass, or a fine-needle aspiration biopsy, when the previous is not possible (Urbańska et al., 2014).

New noninvasive diagnostic methods have been also explored to diagnose GBM, since the histopathological examination may be challenging due to the tumor location and sensitivity of brain tissue (Carlsson et al., 2014). These strategies involve the study of blood samples of GBM patients, in order to detect disease biomarkers. For example, the blood-circulating GBM-derived microvesicles were pointed out as suitable to provide information about pathological alterations associated with the IDH1, EGFR, and EGFRvIII (Shao et al., 2012).

Other aspects required to complement the evaluation of the GBM stage include the assessment of the morphology, tumor grade (I–IV), degree of proliferation, genetic analysis, behavior of the pathology after treatment, and survival time (Urbańska et al., 2014).

4. TREATMENT OF GLIOBLASTOMA

The deadly profile of GBM leads to a median survival of the patients of only 6–12 months after diagnosis (Urbańska et al., 2014). Most of the currently available treatments are not sufficiently effective against the disease and, undesirably, damage surrounding healthy tissues. Since the malignancy can resist treatments, recurrent tumors often emerge (Birk, Han, & Butowski, 2016). Therefore, novel and targeted treatment approaches are desperately needed to fight GBM pathology.

4.1 Nonantibody-Based Therapies

The current standard treatment for GBM includes surgical resection of the tumor and, further, both radiotherapy and chemotherapy (Carlsson et al., 2014). Complete tumor resection is a critical surgical issue since GBM highly infiltrates through neighboring tissues (Urbańska et al., 2014). Thus, it is usually not possible to proceed with a total removal of the GBM mass. Besides this, the resection involves a careful study of the tumor shape and size, location of blood vessels, and neurological coordinates (Carlsson et al., 2014). Radiotherapy against GBM after resection usually involves a time period of 6 weeks, 5 sessions per week (Carlsson et al., 2014). This therapy is responsible for inducing DNA damages, ultimately leading to cellular apoptosis. However, resistance to radiotherapy is commonly verified in GBM scenarios and this may be attributed to the role of EGFR vIII, which upregulates the mechanisms involved in DNA repair (Mukherjee et al., 2009).

Chemotherapy for GBM treatment uses temozolomide as the standard (Carlsson et al., 2014). The therapy involves the oral administration of this drug, 150–200 mg/m^2, days 1–5, for 4 weeks (Weller et al., 2014). Temozolomide is an alkylating agent which is capable of methylating purines in DNA structure, thus causing cellular apoptosis. However, the presence of O-6-methylguanine-DNA methyltransferase (MGMT) in the tumor microenvironment hinders this type of therapy and promotes malignant cells survival, since it is an enzyme capable of repairing DNA after genomic lesion through methylating agents, specifically restoring guanine from O-6-methylguanine (Aldape et al., 2015).

Despite the aggressiveness of the treatments, the long-term survival of patients remains challenging, mainly due to the intrinsic resistance of the CSCs or the acquired resistance of other GBM cell populations, which are able to repopulate the tumor later on (Ramirez, Weatherbee, Wheelhouse, & Ross, 2013). The treatment of recurrent GBMs is based on the same previously mentioned protocol, thus including re-resection, reirradiation, and rechallenge chemotherapy. Besides this, in recurrent situations, therapies against tumor angiogenesis are also introduced (Weller et al., 2014).

Regarding the critical role of CSCs in GBM development, some authors claim that future directions in treatments should envisage this specific type of cells (Cho et al., 2013; Han, Shi, Gong, Zhang, & Sun, 2013; Kim, Harford, Pirollo, & Chang, 2015; Safa, Saadatzadeh, Cohen-Gadol, Pollok, & Bijangi-Vishehsaraei, 2016). However, it is important to highlight that novel therapeutic strategies must take into account the BBB crossing of molecules, since this biological barrier is one of the major obstacles in what concerns drug delivery to the brain (Kim et al., 2015; Sminia & Westerman, 2016).

4.2 Monoclonal Antibodies-Based Therapies

When the hybridoma method for the production of antibodies was first described, a whole new plethora of treatment options arrived (Kohler & Milstein, 1975): the ability to produce antibodies, extremely powerful therapeutic agents with high receptor-affinity/selectivity and therapeutic prowess, as well as relatively low toxicity and relatively high and stable plasma half-lives (Freskgard & Urich, 2017). However, antibodies that are developed for the treatment of glioblastoma face a worthy opponent in their path to the tumor site, this opponent being the BBB. Due to their considerable molecular weight and poor tissue distribution, their ability to cross the BBB is limited, resulting in lower therapeutic effectiveness (Lampson, 2011). Furthermore, central nervous system tumors have the ability to induct direct changes in the BBB, inducing vascular alterations at the area, essentially developing a new barrier, occasionally labeled BBTB. This will further hinder mAbs bioavailability to the central nervous system, affecting their therapeutic outcome against glioblastoma (van Tellingen et al., 2015). However, considering all of these notions, there are therapeutic options with mAbs aimed at glioblastoma, and surely more will come with clinical research.

As mentioned previously, glioblastoma induces the formation of new vasculature at its site, characterized by high expression of VEGF. Thus, one of the logic approaches when using mAbs was to employ mAbs in that target and block the VEGF receptors. Bevacizumab has been approved for a multitude of cancers, and is currently recognized as a second-line agent for the treatment of GBM (Narita, 2013). It was given fast track approval by the US Food and Drug administration (FDA) as a single-agent treatment option for GBM. However, its European counterpart, the European Medicines Agency, did not grant the indication as it considered there was a lack of evidence regarding bevacizumab's effectiveness. Readers are pointed to an eloquent review by Narita (2013), where studies regarding bevacizumab as a single agent are presented.

Bevacizumab is a humanized monoclonal immunoglobulin (IgG1), and works by binding to VEGF, incapacitating it from acting on its target receptors, thus decreasing new vessel synthesis (Gil-Gil et al., 2013). Most treatment regimens regarding bevacizumab employ it at dosages of 10 mg/kg every 2 weeks, as a single agent, until tumor progression, or in combination with irinotecan 125 mg/m^2 (340 mg/m^2 if patients are on enzyme-inducing antiepileptic drugs, similarly repeating the treatment every 2 weeks) (Gil et al., 2012; Vredenburgh et al., 2007).

Bevacizumab has also been reported to decrease edema in patients with glioblastoma, reducing the necessity of steroid-based treatments (Kreisl et al., 2011). Most studies are concordant in the fact that bevacizumab improves the progression-free survival and overall survival in most patients; however, there is still the necessity to optimize bevacizumab's dosage and the treatment duration. Randomized studies are currently undergoing that aim to answer these questions. Lastly, bevacizumab seems to lack any biomarker or genetic indication that a patient may benefit from its use, making it difficult to predict which patients will have a better than usual response to bevacizumab. Further research should be conducted to attempt to identify potential targets (Gil-Gil et al., 2013).

Although bevacizumab remains the only mAb approved for therapy regarding glioblastoma, other antibodies with different targets than VEGF are being researched, mainly in preclinical and clinical trials. A likely target could be antibodies that block the EGFR and its subsequent aggressive mutation, EGFRvIII, as mentioned previously, such as cetuximab, nimotuzumab, or panitumumab. The use of cetuximab as a single agent was exploited, but was not found to be an improvement over existing therapies (Neyns et al., 2009). A case was reported where the dual therapy of a remission of a glioblastoma was achieved with the combination of cetuximab and bevacizumab, perhaps requiring that further research is conducted into why this particular case worked or how both antibodies potentiated each other's action (Blesa et al., 2012). However, a phase II trial regarding a triple therapy regimen with cetuximab, bevacizumab, and irinotecan, after first-line temozolomide therapy failed, showed that the results with and without cetuximab were not significantly different (Hasselbach et al., 2010). Studies with nimotuzumab showed that drug might be an effective agent regarding overall survival in glioblastoma patients, especially when coupled with radiotherapy. Furthermore, when nimotuzumab was applied in patient EGFR positive, there was a significant increase in overall survival (Ramos et al., 2006; Solomon et al., 2014; Westphal et al., 2015). Thus, further research with EGFR antagonists should be considered as a potential salvage treatment when all else has failed. Thus far, bevacizumab is the only approved antibody for the treatment of glioblastoma, and nimotuzumab, which has an orphan status in both the United States and EU for the treatment of glioblastoma (Yang, Guo, & Chen, 2015).

5. NANOTECHNOLOGY-BASED DRUG DELIVERY SYSTEMS CARRYING ANTIBODIES TO OVERCOME CURRENT THERAPY OF GLIOBLASTOMA

As presented before, antibody-based therapy has a role for the treatment of glioblastoma, and to the point of being approved by the FDA there are obvious limitations that should be addressed. Actually, the low rates of BBB crossing due to the antibodies large molecular size and also the binding of their Fc region to specialized FcγRIIb receptors on the surface of macrophage or lymphocytes can render the antibody's therapeutic action useless (Chekhonin & Gurina, 2015).

Thus, it is vital for researchers to research alternative ways to increase antibody's effectiveness and therapeutic outcomes, as well as increasing their CNS penetration. Nanotechnology here emerges as a very promising strategy. The use of nanocarriers carrying the antibodies within them gives researchers the ability to functionalize the nanocarrier externally, through surface modifications to promote BBB targeting and subsequent crossing, through receptor-mediated transcytosis or adsorptive-mediated transcytosis. These strategies usually give a higher rate of efficiency and effectiveness to therapy as BBB penetration values are usually higher (Glaser et al., 2017). Nanocarriers have also demonstrated clinical safety, allowing better drug pharmacokinetics and pharmacodynamics. Lastly, nanocarriers benefit from a passive targeting effect, labeled the enhanced permeation and retention effect (EPR effect). This happens due to the tumor in questions particularly leaky vasculature (critical in glioblastoma due to the angiogenic prowess of said tumor), which causes a passive accumulation of the nanocarriers in its vasculature, while preventing their return to circulation (Tzeng & Green, 2013). With all these advantages in mind, nanocarriers coupled with antibodies should undoubtedly be researched and perfected with the aim of employing these in human therapy.

5.1 Organic Nanocarriers

Organic nanocarriers include liposomes, dendrimers, micelles, nanoparticles synthesized from synthetic or natural polymers, and many more examples. These compounds are generally well tolerated, biodegradable, and biocompatible, usually showing relatively low toxicity and moderate delivery efficiency. Organic nanocarriers also benefit from the EPR effect and, as mentioned previously, allow surface modification to promote targeted therapy toward the CNS (López-Dávila, Seiifalian, & Loizidou, 2012).

The work carried out by Kito et al. demonstrated that mAbs encapsulated within liposomes, targeting glioma-associated antigen, had selective toxicity against glioblastoma cells, and could in fact be an option regarding chemotherapy of glioblastomas (Kito, Yoshida, Kageyama, Kojima, & Yagi, 1989). Another study by Mortensen et al. showed how liposomes conjugated with cetuximab enjoyed an increase in glioblastoma targeting and accumulation. The key point here is that cetuximab was not encapsulated within the liposome, but used to promote active targeting of glioblastoma cells, due to the high expression of EGFR. Thus, cetuximab could potentially have a dual role of active targeting glioblastoma cells but also exert a therapeutic action on the liposome, as an EGFR antagonist (Mortensen et al., 2013). Battaglia et al. demonstrated promising results when studying in vitro activity of solid lipid nanoparticles carrying bevacizumab. Although the results were promising, further research needed to be conducted regarding solid lipid nanoparticles carrying bevacizumab for glioblastoma therapy, such as BBB penetration values (Battaglia et al., 2015).

5.2 Inorganic Nanocarriers

Inorganic nanocarriers are nanocarriers that are usually derived from metallic compounds, such as iron-oxide nanoparticles, gold nanoparticles, and titanium oxide nanoparticles. These nanocarriers can employ certain benefits that organic nanocarriers hold, such as surface functionalization, but some can also benefit from alternative targeting mechanism, such as magnetism. Some of these nanocarriers can also hold imaging properties, which can also be key when dealing with glioblastoma (Lin, Mi, Chu, Zhang, & Liu, 2016).

Similar to what was employed with liposomes, Hadjipanavis et al. employed iron-oxide nanoparticles conjugated with an EGFRvIII-targeting antibody, coupled with convention-enhanced delivery to target glioblastoma. Little to no toxicity was demonstrated, and a significant decrease in glioblastoma cell survivability was demonstrated (Hadjipanavis et al., 2010).

Also, Kaluzova et al. used iron-oxide nanoparticles conjugated with cetuximab for the therapy of glioblastoma, an EGFR antagonist as well. The results were promising, with significant antitumor effect, with better results than a single-agent delivery of cetuximab (Kaluzova, Bouras, Machaidze, & Hadjipanavis, 2015).

Using nanocarriers as antibody carriers is a promising strategy. However, due to the antibody's size and hydrophobicity, encapsulating an antibody within a nanocarrier is a great challenge, which explains why so little studies have been employed on nanocarriers carrying antibodies within them. Antibodies hold great promise when used as surface modifiers, allowing for active targeting of overexpressed receptors regarding either the tumor microenvironment or the BBB to increase penetration efficiency. Further research should be conducted in alternative ways to promote antibody encapsulation, as the potential therapeutic benefits of an antibody coupled with the improved delivery benefits of a nanocarrier are surely a very powerful and dynamic upcoming combination.

6. CONCLUSION

In last years, representative and significant efforts have been done for the development of new antiangiogenic therapies for glioblastoma. In fact, glioblastoma is one of the most fatal brain tumor in the world, making their treatment an emergency for the scientific and medicinal world.

Nanotechnology applied to drug delivery for GBM has started to be done to overcome some limitations regarding to the lack of drug efficacy and effectiveness. Despite the great results, organic and inorganic nanoparticles have been only applied to the anticancer drugs delivery to GBM (Pourgholi, Hajivalili, Farhad, Kafil, & Yousefi, 2016). On the other hand, functionalization of nanoparticles has done its job, improving drug pharmacokinetics and pharmacodynamics (Xu, Gao, Sun, & Duan, 2015). The same does not happen when it is intended just the delivery of a full mAb for GBM treatment. There is still a failure in mAb delivery through organic or inorganic nanoparticles. No pharmacokinetics and pharmacodynamics study have been done in mouse models to test the advantages of nanotechnology applied to antiangiogenic drugs.

During the next years, there are reasonable aspects to improve the GBM treatment with more effective therapies, mainly concerning antiangiogenic therapies. Nowadays, there are some innovative nanoparticles formulations in the market of anticancer drug, which can easily be translated into mAbs.

ACKNOWLEDGMENTS

F.S. would like to thank to FCT for funding the PhD scholarship (SFRH/BD/112201/2015). This chapter is a result of the project NORTE-01-0145-FEDER-000012, supported by Norte Portugal Regional Operational Programme (NORTE 2020), under the PORTUGAL 2020 Partnership Agreement, through the European Regional Development Fund (ERDF). This work was also financed by FEDER—Fundo Europeu de Desenvolvimento Regional funds through the COMPETE 2020—Operacional Programme for Competitiveness and Internationalisation (POCI), Portugal 2020, and by Portuguese funds through FCT—Fundação para a Ciência e a Tecnologia/Ministério da Ciência, Tecnologia e Ensino Superior in the framework of the project "Institute for Research and Innovation in Health Sciences" (POCI-01-0145-FEDER-007274).

REFERENCES

Agnihotri, S., Burrell, K. E., Wolf, A., Jalali, S., Hawkins, C., Rutka, J. T., et al. (2013). Glioblastoma, a brief review of history, molecular genetics, animal models and novel therapeutic strategies. *Archivum Immunologiae et Therapiae Experimentalis*, *61*(1), 25–41.

Aldape, K., Zadeh, G., Mansouri, S., Reifenberger, G., & von Deimling, A. (2015). Glioblastoma: Pathology, molecular mechanisms and markers. *Acta Neuropathologica*, *129*(6), 829–848.

Alentorn, A., Marie, Y., Carpentier, C., Boisselier, B., Giry, M., Labussière, M., et al. (2012). Prevalence, clinico-pathological value, and co-occurrence of PDGFRA abnormalities in diffuse gliomas. *Neuro-Oncology*, *14*(11), 1393–1403.

Alifieris, C., & Trafalis, D. T. (2015). Glioblastoma multiforme: Pathogenesis and treatment. *Pharmacology and Therapeutics*, *152*, 63–82.

Alves, T. R., Lima, F. R. S., Kahn, S. A., Lobo, D., Dubois, L. G. F., Soletti, R., et al. (2011). Glioblastoma cells: A heterogeneous and fatal tumor interacting with the parenchyma. *Life Sciences*, *89*(15), 532–539.

Batchelor, T. T., Reardon, D. A., de Groot, J. F., Wick, W., & Weller, M. (2014). Antiangiogenic therapy for glioblastoma: Current status and future prospects. *Clinical Cancer Research*, *20*(22), 5612–5619.

Battaglia, L., Gallarate, M., Peira, E., Chirio, D., Solazzi, I., Giordano, S., et al. (2015). Bevacizumab loaded solid lipid nanoparticles prepared by the coacervation technique: Preliminary in vitro studies. *Nanotechnology*, *26*(25), 255102.

Birk, H. S., Han, S. J., & Butowski, N. A. (2016). Treatment options for recurrent high-grade gliomas. *CNS Oncology*, *6*(1), 61–70.

Blesa, J., Mollá, S., Esparcia, M., Ortells, J., Godoy, M., Das, A., et al. (2012). Durable complete remission of a brainstem glioma treated with a combination of bevacizumab and cetuximab. *Case Reports in Oncology*, *5*(3), 676–681.

Carlsson, S. K., Brothers, S. P., & Wahlestedt, C. (2014). Emerging treatment strategies for glioblastoma multiforme. *EMBO Molecular Medicine*, *6*(11), 1359–1370.

Chamberlain, M. C. (2011). Bevacizumab for the treatment of recurrent glioblastoma. *Clinical Medicine Insights. Oncology*, *5*, 117–129.

Chekhonin, I., & Gurina, O. (2015). Trends in malignant glioma monoclonal antibody therapy. *Current Medicinal Chemistry*, *11*(3), 102–118.

Cho, D. Y., Lin, S. Z., Yang, W. K., Lee, H. C., Hsu, D. M., Lin, H. L., et al. (2013). Targeting cancer stem cells for treatment of glioblastoma multiforme. *Cell Transplantation*, *22*(4), 731–739.

D'Alessandris, Q. G., & Ricci-Vitiani, L. (2016). Emerging role for USP1 in glioblastoma stem cell maintenance and radioresistance: A potential target for glioblastoma therapy. *Translational Cancer Research*, *5*(4), S751–S754.

Davis, M. E. (2016). Glioblastoma: Overview of disease and treatment. *Clinical Journal of Oncology Nursing*, *20*(5), S2–S8.

Ferrara, N., Hillan, K. J., Gerber, H.-P., & Novotny, W. (2004). Discovery and development of bevacizumab, an anti-VEGF antibody for treating cancer. *Nature Reviews Drug Discovery*, *3*(5), 391–400.

Freskgard, P., & Urich, E. (2017). Antibody therapies in CNS diseases. *Neuropharmacology, 120*, 38–55.

Furnari, F. B., Cloughesy, T. F., Cavenee, W. K., & Mischel, P. S. (2015). Heterogeneity of epidermal growth factor receptor signalling networks in glioblastoma. *Nature Reviews Cancer, 15*(5), 302–310.

Gallia, G. L., Rand, V., Siu, I. M., Eberhart, C. G., James, C. D., Marie, S. K., et al. (2006). PIK3CA gene mutations in pediatric and adult glioblastoma multiforme. *Molecular Cancer Research, 4*(10), 709–714.

Gil, M., de Las Peñas, R., Reynés, G., Balañá, C., Peréz-Segura, P., García-Velasco, A., et al. (2012). Bevacizumab plus irinotecan in recurrent malignant glioma shows high overall survival in a multicenter retrospective pooled series of the Spanish Neuro-Oncology Research Group (GEINO). *Anti-Cancer Drugs, 23*(6), 659–665.

Gilbertson, R. J., & Rich, J. N. (2007). Making a tumour's bed: Glioblastoma stem cells and the vascular niche. *Nature Reviews Cancer, 7*, 733.

Gil-Gil, M., Mesia, C., Rey, M., & Bruna, J. (2013). Bevacizumab for the treatment of glioblastoma. *Clinical Medicine Insights. Oncology, 7*, 123–145.

Glaser, T., Han, I., Wu, L., & Zeng, X. (2017). Targeted nanotechnology in glioblastoma multiforme. *Frontiers in Pharmacology, 8*, 166.

Hadjipanavis, C., Machaidze, R., Kaluzova, M., Wang, L., Schuette, A., Chen, H., et al. (2010). EGFRvIII antibody-conjugated iron oxide nanoparticles for magnetic resonance imaging-guided convection-enhanced delivery and targeted therapy of glioblastoma. *Cancer Research, 70*(15), 6303–6312.

Han, L., Shi, S., Gong, T., Zhang, Z., & Sun, X. (2013). Cancer stem cells: Therapeutic implications and perspectives in cancer therapy. *Acta Pharmaceutica Sinica B, 3*(2), 65–75.

Hartmann, C., Bartels, G., Gehlhaar, C., Holtkamp, N., & von Deimling, A. (2005). PIK3CA mutations in glioblastoma multiforme. *Acta Neuropathologica, 109*(6), 639–642.

Hasselbach, B., Lassen, U., Hansen, S., Holmberg, M., Sørensen, M., Kosteljanetz, M., et al. (2010). Cetuximab, bevacizumab, and irinotecan for patients with primary glioblastoma and progression after radiation therapy and temozolomide: A phase II trial. *Neuro-Oncology, 12*(5), 508–516.

Ho, V. K., Reijneveld, J. C., Enting, R. H., Bienfait, H. P., Robe, P., Baumert, B. G., et al. (2014). Changing incidence and improved survival of gliomas. *European Journal of Cancer, 50*(13), 2309–2318.

Jain, K. K. (2007). Use of nanoparticles for drug delivery in glioblastoma multiforme. *Expert Review of Neurotherapeutics*, 7(4), 363–372.

Jain, R. K., di Tomaso, E., Duda, D. G., Loeffler, J. S., Sorensen, A. G., & Batchelor, T. T. (2007). Angiogenesis in brain tumours. *Nature Reviews Neuroscience, 8*, 610.

Kaluzova, M., Bouras, A., Machaidze, R., & Hadjipanavis, C. (2015). Targeted therapy of glioblastoma stem-like cells and tumor non-stem cells using cetuximab-conjugated iron-oxide nanoparticles. *Oncotarget, 6*(11), 8788–8806.

Karsy, M., Gelbman, M., Shah, P., Balumbu, O., Moy, F., & Arslan, E. (2012). Established and emerging variants of glioblastoma multiforme: Review of morphological and molecular features. *Folia Neuropathologica, 50*(4), 301–321.

Kesari, S. (2011). Understanding glioblastoma tumor biology: The potential to improve current diagnosis and treatments. *Seminars in Oncology, 38*(Suppl. 4), S2–S10.

Kim, S.-S., Harford, J. B., Pirollo, K. F., & Chang, E. H. (2015). Effective treatment of glioblastoma requires crossing the blood–brain barrier and targeting tumors including cancer stem cells: The promise of nanomedicine. *Biochemical and Biophysical Research Communications, 468*(3), 485–489.

Kim, W.-Y., & Lee, H.-Y. (2009). Brain angiogenesis: Mechanism and therapeutic intervention in brain tumors. *The FEBS Journal, 276*(17), 4653–4664.

Kito, A., Yoshida, J., Kageyama, N., Kojima, N., & Yagi, K. (1989). Liposomes coupled with monoclonal antibodies against glioma-associated antigen for targeting chemotherapy of glioma. *Journal of Neurosurgery*, *71*(3), 382–387.

Kohler, G., & Milstein, C. (1975). Continuous cultures of fused cells secreting antibody of predefined specificity. *Nature*, *256*, 495–497.

Kreisl, T., Zhang, W., Odia, Y., Shih, J., Butman, J., Hammoud, D., et al. (2011). A phase II trial of single-agent bevacizumab in patients with recurrent anaplastic glioma. *Neuro-Oncology*, *13*(10), 1143–1150.

Lampson, L. (2011). Monoclonal antibodies in neuro-oncology. *MAbs*, *3*(2), 153–160.

Lathia, J. D., Mack, S. C., Mulkearns-Hubert, E. E., Valentim, C. L. L., & Rich, J. N. (2015). Cancer stem cells in glioblastoma. *Genes and Development*, *29*(12), 1203–1217.

Lin, G., Mi, P., Chu, C., Zhang, J., & Liu, G. (2016). Inorganic nanocarriers overcoming multidrug resistance for cancer theranostics. *Advanced Science (Weinheim, Baden-Württemberg, Germany)*, *3*(11), 1600134.

Liu, F., Hon, G. C., Villa, G. R., Turner, K. M., Ikegami, S., Yang, H., et al. (2015). EGFR mutation promotes glioblastoma through epigenome and transcription factor network remodeling. *Molecular Cell*, *60*(2), 307–318.

Lombardi, G., Pambuku, A., Bellu, L., Farina, M., Della Puppa, A., Denaro, L., et al. (2017). Effectiveness of antiangiogenic drugs in glioblastoma patients: A systematic review and meta-analysis of randomized clinical trials. *Critical Reviews in Oncology/Hematology*, *111*, 94–102.

López-Dávila, V., Seiifalian, A., & Loizidou, M. (2012). Organic nanocarriers for cancer drug delivery. *Current Opinion in Pharmacology*, *12*(4), 414–419.

Louis, D. N., Ohgaki, H., Wiestler, O. D., Cavenee, W. K., Burger, P. C., Jouvet, A., et al. (2007). The 2007 WHO classification of tumours of the central nervous system. *Acta Neuropathologica*, *114*(2), 97–109.

McNamara, M. G., & Mason, W. P. (2012). Antiangiogenic therapies in glioblastoma multiforme. *Expert Review of Anticancer Therapy*, *12*(5), 643–654.

Molina, J. R., Agarwal, N. K., Morales, F. C., Hayashi, Y., Aldape, K. D., Cote, G., et al. (2011). PTEN, NHERF1 and PHLPP form a tumor suppressor network that is disabled in glioblastoma. *Oncogene*, *31*(10), 1264–1274.

Mortensen, J., Jeppesen, M., Pilgaard, L., Agger, R., Duroux, M., Zachar, V., et al. (2013). Targeted antiepidermal growth factor receptor (cetuximab) immunoliposomes enhance cellular uptake in vitro and exhibit increased accumulation in an intracranial model of glioblastoma multiforme. *Journal of Drug Delivery*, *2013*, 13.

Mukherjee, B., McEllin, B., Camacho, C. V., Tomimatsu, N., Sirasanagandala, S., Nannepaga, S., et al. (2009). EGFRvIII and DNA double-strand break repair: A molecular mechanism for radioresistance in glioblastoma. *Cancer Research*, *69*(10), 4252–4259.

Narita, Y. (2013). Drug review: Safety and efficacy of bevacizumab for glioblastoma and other brain tumors. *Japanese Journal of Clinical Oncology*, *43*(6), 587–595.

Nazarenko, I., Hede, S.-M., He, X., Hedrén, A., Thompson, J., Lindström, M. S., et al. (2012). PDGF and PDGF receptors in glioma. *Upsala Journal of Medical Sciences*, *117*(2), 99–112.

Nelson, S. J., & Cha, S. (2003). Imaging glioblastoma multiforme. *Cancer Journal*, *9*(2), 134–145.

Neyns, B., Sadones, J., Joosens, E., Bouttens, F., Verbeke, L., Baurain, J., et al. (2009). Stratified phase II trial of cetuximab in patients with recurrent high-grade glioma. *Annals of Oncology*, *20*(9), 1596–1603.

Ohgaki, H., Dessen, P., Jourde, B., Horstmann, S., Nishikawa, T., Di Patre, P. L., et al. (2004). Genetic pathways to glioblastoma: A population-based study. *Cancer Research*, *64*(19), 6892–6899.

Olar, A., & Aldape, K. D. (2014). Using the molecular classification of glioblastoma to inform personalized treatment. *The Journal of Pathology*, *232*(2), 165–177.

Parsons, D. W., Jones, S., Zhang, X., Lin, J. C., Leary, R. J., Angenendt, P., et al. (2008). An integrated genomic analysis of human glioblastoma multiforme. *Science*, *321*(5897), 1807–1812.

Pérez-Castillo, A., Aguilar-Morante, D., Morales-García, J. A., & Dorado, J. (2008). Cancer stem cells and brain tumors. *Clinical and Translational Oncology*, *10*(5), 262–267.

Pourgholi, F., Hajivalili, M., Farhad, J.-N., Kafil, H. S., & Yousefi, M. (2016). Nanoparticles: Novel vehicles in treatment of glioblastoma. *Biomedicine and Pharmacotherapy*, *77*, 98–107.

Ramirez, Y. P., Weatherbee, J. L., Wheelhouse, R. T., & Ross, A. H. (2013). Glioblastoma multiforme therapy and mechanisms of resistance. *Pharmaceuticals (Basel, Switzerland)*, *6*(12), 1475–1506.

Ramos, T., Figuerede, J., Catala, M., González, S., Selva, J., Cruz, T., et al. (2006). Treatment of high-grade glioma patients with the humanized anti-epidermal growth factor receptor (EGFR) antibody h-R3: Report from a phase I/II trial. *Cancer Biology & Therapy*, *5*(4), 375–379.

Rutka, J. T., Murakami, M., Dirks, P. B., Hubbard, S. L., Becker, L. E., Fukuyama, K., et al. (1997). Role of glial filaments in cells and tumors of glial origin: A review. *Journal of Neurosurgery*, *87*(3), 420–430.

Safa, A. R., Saadatzadeh, M. R., Cohen-Gadol, A. A., Pollok, K. E., & Bijangi-Vishehsaraei, K. (2016). Emerging targets for glioblastoma stem cell therapy. *Journal of Biomedical Research*, *30*(1), 19–31.

Shao, H., Chung, J., Balaj, L., Charest, A., Bigner, D. D., Carter, B. S., et al. (2012). Protein typing of circulating microvesicles allows real-time monitoring of glioblastoma therapy. *Nature Medicine*, *18*(12), 1835–1840.

Sminia, P., & Westerman, B. A. (2016). Blood–brain barrier crossing and breakthroughs in glioblastoma therapy. *British Journal of Clinical Pharmacology*, *81*(6), 1018–1020.

Solomon, D. A., Kim, J.-S., Cronin, J. C., Sibenaller, Z., Ryken, T., Rosenberg, S. A., et al. (2008). Mutational inactivation of PTPRD in glioblastoma multiforme and malignant melanoma. *Cancer Research*, *68*(24), 10300–10306.

Solomon, D. A., Kim, J.-S., Jean, W., & Waldman, T. (2008). Conspirators in a capital crime: Co-deletion of p18INK4c and p16INK4a/p14ARF/p15INK4b in glioblastoma multiforme. *Cancer Research*, *68*(21), 8657–8660.

Solomon, M., Miranda, N., Jorrín, E., Chon, I., Marinello, J., Alert, J., et al. (2014). Nimotuzumab in combination with radiotherapy in high grade glioma patients: A single institution experience. *Cancer Biology & Therapy*, *15*(5), 504–509.

SongTao, Q., Lei, Y., Si, G., YanQing, D., HuiXia, H., XueLin, Z., et al. (2012). IDH mutations predict longer survival and response to temozolomide in secondary glioblastoma. *Cancer Science*, *103*(2), 269–273.

TCGAN. (2008). Comprehensive genomic characterization defines human glioblastoma genes and core pathways. *Nature*, *455*(7216), 1061–1068.

Tian, Y., Bresenitz, P., Reska, A., El Moussaoui, L., Beier, C. P., & Gründer, S. (2017). Glioblastoma cancer stem cell lines express functional acid sensing ion channels ASIC1a and ASIC3. *Scientific Reports*, 7(1), 13674.

Tzeng, S., & Green, J. (2013). Therapeutic nanomedicine for brain cancer. *Therapeutic Delivery*, *4*(6), 687–704.

Urbańska, K., Sokołowska, J., Szmidt, M., & Sysa, P. (2014). Glioblastoma multiforme—An overview. *Contemporary Oncology*, *18*(5), 307–312.

van Tellingen, O., Yetkin-Arik, B., de Gooijer, M., Wesseling, P., Wurdinger, T., & de Vries, H. (2015). Overcoming the blood-brain tumor barrier for effective glioblastoma treatment. *Drug Resistance Updates*, *19*, 1–12.

Veeriah, S., Brennan, C., Meng, S., Singh, B., Fagin, J. A., Solit, D. B., et al. (2009). The tyrosine phosphatase PTPRD is a tumor suppressor that is frequently inactivated and mutated in glioblastoma and other human cancers. *Proceedings of the National Academy of Sciences of the United States of America*, *106*(23), 9435–9440.

Verhaak, R. G., Hoadley, K. A., Purdom, E., Wang, V., Qi, Y., Wilkerson, M. D., et al. (2010). Integrated genomic analysis identifies clinically relevant subtypes of glioblastoma characterized by abnormalities in PDGFRA, IDH1, EGFR, and NF1. *Cancer Cell*, *17*(1), 98–110.

Vredenburgh, J., Desjardins, A., Herndon, J., 2nd, Dowell, J., Reardon, D., Quinn, J., et al. (2007). Phase II trial of bevacizumab and irinotecan in recurrent malignant glioma. *Clinical Cancer Research*, *13*(4), 1253–1259.

Wakimoto, H., Kesari, S., Farrell, C. J., Curry, W. T., Zaupa, C., Aghi, M., et al. (2009). Human glioblastoma-derived cancer stem cells: Establishment of invasive glioma models and treatment with oncolytic herpes simplex virus vectors. *Cancer Research*, *69*(8), 3472–3481.

Wang, X., Chen, J. X., Liu, J. P., You, C., Liu, Y. H., & Mao, Q. (2014). Gain of function of mutant TP53 in glioblastoma: Prognosis and response to temozolomide. *Annals of Surgical Oncology*, *21*(4), 1337–1344.

Wang, Y., Fei, D., Vanderlaan, M., & Song, A. (2004). Biological activity of bevacizumab, a humanized anti-VEGF antibody in vitro. *Angiogenesis*, 7(4), 335–345.

Weller, M., van den Bent, M., Hopkins, K., Tonn, J. C., Stupp, R., Falini, A., et al. (2014). EANO guideline for the diagnosis and treatment of anaplastic gliomas and glioblastoma. *The Lancet Oncology*, *15*(9), e395–e403.

Westphal, M., Heese, O., Steinbach, J., Schnell, O., Schakert, G., Mehdorn, M., et al. (2015). A randomised, open label phase III trial with nimotuzumab, an anti-epidermal growth factor receptor monoclonal antibody in the treatment of newly diagnosed adult glioblastoma. *European Journal of Cancer*, *51*(4), 522–532.

Wong, E. T., Hess, K. R., Gleason, M. J., Jaeckle, K. A., Kyritsis, A. P., Prados, M. D., et al. (1999). Outcomes and prognostic factors in recurrent glioma patients enrolled onto phase II clinical trials. *Journal of Clinical Oncology*, *17*(8), 2572–2578.

Xu, Y.-Y., Gao, P., Sun, Y., & Duan, Y.-R. (2015). Development of targeted therapies in treatment of glioblastoma. *Cancer Biology & Medicine*, *12*(3), 223–237.

Yang, Q., Guo, C., & Chen, Z. (2015). Profile of nimotuzumab in the treatment of high-grade glioma. *OncoTargets and Therapy*, *8*, 819–825.

Young, G. S. (2007). Advanced MRI of adult brain tumors. *Neurologic Clinics*, *25*(4), 947–973.

Zhu, Y., Guignard, F., Zhao, D., Liu, L., Burns, D. K., Mason, R. P., et al. (2005). Early inactivation of p53 tumor suppressor gene cooperating with NF1 loss induces malignant astrocytoma. *Cancer Cell*, *8*(2), 119–130.

CHAPTER THREE

Dynamical Behavior of Somatostatin-14 and Its Cyclic Analogues as Analyzed in Bulk and on Plasmonic Silver Nanoparticles

Belén Hernández*[,†], Yves-Marie Coïc[‡], Eduardo López-Tobar[§], Santiago Sanchez-Cortes[§], Bruno Baron[¶], Fernando Pflüger[†], Sergei G. Kruglik[||], Régis Cohen[#], Mahmoud Ghomi*[,†,1]

*Laboratoire Matrice Extracellulaire et Dynamique Cellulaire (MEDyC), UMR 7369, Université de Reims, Faculté des Sciences, Reims Cedex 2, France
[†]Groupe de Biophysique Moléculaire, Sorbonne Paris Cité, Université Paris 13, UFR Santé-Médecine-Biologie Humaine, Bobigny, France
[‡]Institut Pasteur, Unité de Chimie des Biomolécules, UMR 3523, Paris Cedex 15, France
[§]Instituto de Estructura de la Materia, IEM-CSIC, Serrano 121, Madrid, Spain
[¶]Institut Pasteur, Plate-Forme de Biophysique de Macromolécules et de leurs Interactions, Paris Cedex 15, France
[||]Laboratoire Jean-Perrin, Sorbonne Université, UPMC Univ. Paris 06, CNRS UMR 8237, Paris, France
[#]Service d'Endocrinologie, Centre Hospitalier de Saint-Denis, Saint-Denis, France
[1]Corresponding author: e-mail address: mahmoud.ghomi@univ-paris13.fr

Contents

Advances in Protein Chemistry and Structural Biology, Volume 112
ISSN 1876-1623
https://doi.org/10.1016/bs.apcsb.2018.01.002

Abstract

Primarily known as the inhibitor of growth hormone release, the role of somatostatin in many other inhibiting activities upon binding to its five G-protein-coupled receptors has been elucidated. Because of the short half-life of somatostatin, a number of synthetic analogues were elaborated for this peptide hormone. Herein, after recalling the main somatostatin therapeutic interests, we present the dynamical behavior of somatostatin-14 and its two currently used synthetic cyclic analogues, octreotide and pasireotide. Physical techniques, such as fluorescence, UV–visible absorption, circular dichroism, Raman spectroscopy, surface-enhanced Raman spectroscopy, and transmission electron microscopy, were jointly used in order to get information on the solution structural features, as well as on the anchoring sites of the three peptides on silver colloids. While somatostatin-14 adopts a rather unordered chain within the submillimolar concentration range, its cyclic analogues were revealed to be ordered, i.e., stabilized either in a type-II′ β-turn (octreotide) or in a face-to-face γ-turn/type-I β-turn (pasireotide) structure. Nevertheless, a progressive structuring trend was observed in somatostatin-14 upon increasing concentration to the millimolar range. Because of their cationic character, the three peptides have revealed their capability to bind onto negatively charged silver nanoparticles. The high affinity of the peptides toward metallic particles seems to be extremely promising for the elaboration of somatostatin-based functionalized plasmonic nanoparticles that can be used in diagnosis, drug delivery, and therapy.

1. INTRODUCTION

1.1 Therapeutic Interest of Somatostatin and its Analogues

Since its identification in 1973 by Brazeau and Guillemin, somatostatin revealed numerous actions in physiology and pathologies (Brazeau et al., 1973). Many somatostatin analogues (SSAs) have been developed to improve the health status with a high-level evidence-based medicine for patients suffering from benign or malignant neuroendocrine-tumors (NETs) (Caplin et al., 2014; Rinke et al., 2009). Nevertheless, among SSAs, octreotide and lanreotide remain the pivotal therapeutic molecules for controlling hormone-related symptoms of functioning NETs and tumor growth (Modlin, Pavel, Kidd, & Gustafsson, 2010). More recently, another SSA, referred to as pasireotide, having a high affinity to its receptors, extended the therapeutic use to other diseases, i.e., Cushing's disease (McKeage, 2013). Furthermore, in the field of nuclear medicine, the local administration with a high level of radioactivity to treat advanced stages of NETs is rendered possible through the coupling of SSA to an adequate radioisotope (Brabander et al., 2017).

1.1.1 Somatostatin in Physiology

Initially discovered as an inhibitor of growth hormone release, somatostatin is now known to inhibit a variety of gastrointestinal processes. It also affects rates of neurotransmission in the central nervous system. Somatostatin is a peptide hormone, having two active forms with 14 and 28 residues, abbreviated as SST-14 and SST-28, respectively. This hormone is expressed throughout the body and inhibits the release of numerous secondary hormones by binding to high-affinity G-protein-coupled somatostatin receptors, which comprises five distinct subtypes (SSTRi, $i=1, \ldots, 5$). SSTRs express in various regions, such as brain, adrenals, pancreas, and gastrointestinal tract.

1.1.2 SSAs Use in Pathologies

Because of its short half-life (about 2 min) in plasma, the clinical use of somatostatin was limited. SSAs combined with slow-release component were shown to have a much longer half-life that can reach 1 month or more. As somatostatin is able to reduce the portal flow, it has been used for managing bleeding of esophageal varices in patients with liver cirrhosis. It may also suppress pepsin secretion and, hence, alleviate upper gastrointestinal bleeding. In gastroenterology, it is also frequently used to dry intestinal secretions and accelerate the healing of digestive fistulas. In endocrinology, SSAs are indicated to control or decrease numerous hormonal secretions: growth hormone in acromegaly, insulin in insulinomas, ACTH in Cushing's diseases, serotonin in carcinoid syndromes, glucagon, etc.

Antiproliferative action of these components is not only used in the previous endocrine diseases but represents the obligatory passage in the therapeutic courses of the NETs (Mohamed et al., 2014). NETs emerge from neuroendocrine cells of the bronchopulmonary or gastrointestinal system. Their evolutions are slow and indolent with a frequent clinical revelation at metastatic stage. Ninety percent of gastroenteropancreatic (GEP)-NETs express SSTRs on cell surface and may benefit from SSAs-based treatments. Each year an estimated 8000 individuals are newly diagnosed with NET in the United States. When locally advanced or metastatic (70% at diagnosis), with nonfeasible surgery, medical treatments and SSAs should be chosen according to tumor features (Yao et al., 2008).

Two randomized studies evaluated the SSAs efficacy in NETs: CLARINET study (lanreotide) and PROMID trial (octreotide) (Caplin et al., 2014; Rinke et al., 2009). Both proved, vs placebo, a reduction of progression-free survival in well-differentiated metastatic NETs. There was no difference for overall survival. Meta-analysis confirms that SSAs have an antiproliferative effect in advanced GEP-NETs, reducing disease progression risk by 41%, with a good safety (Merola, Panzuto, & Delle Fave, 2017).

1.1.3 Radiolabeled SSAs in Imaging and Therapy

The overexpression of SSTRs is a characteristic of NETs and can be used to localize the primary tumor or its metastases by imaging with the radiolabeled SSAs (Table 1). The majority of tumors expressed SSTR2, and also SSTR1, SSTR3, and SSTR5, and a minority expressed SSTR4 (Reubi, Waser, Schaer, & Laissue, 2001). SSTR imaging is necessary for staging, therapy planning, and follow-up. Imaging with radiolabeled SSAs has high specificity, low antigenicity, rapid clearance, and good tissue distribution. Initially Octreoscan™ was the only radiopharmaceutical binding to SSTR2. Because of some limits in precision, other components labeled with Gallium 68 combined with PET scan technology were used. In the PET diagnostics, there are three routine SSA tracers labeled with ^{68}Ga: DOTA-TATE, DOTA-TOC, and DOTA-NOC. All three tracers bind specifically with SSTRs. ^{68}Ga-labeled SSTRs have high sensitivity (82%–97%) and specificity (80%–92%) in the detection of small primary tumors or metastases of NETs (Barrio et al., 2017; Wang et al., 2013).

Table 1 Overview of Radiolabeled Somatostatin Analogues (SSAs) and Their Uses in Medicine

Theranostic Molecule	Somatostatin Analogue (SSA)
Target	SSTRs, especially SSTR2
Planar imaging	SSA labeled with indium-III
SPECT or PET	[^{68}Ga]Ga-DOTA-TATE
	[^{68}Ga]Ga-DOTA-TOC
	[^{68}Ga]Ga-DOTA-NOC
Therapeutic agent	[^{177}Lu]Ga-DOTA-TATE
	[^{177}Lu]Ga-DOTA-TOC
	[^{90}Y]Ga-DOTA-TOC
	[^{90}Y]Ga-DOTA-TATE
Indication	NETs

High level of radioactivity delivered locally can be used to destroy tumoral cells. Peptide receptor radionuclide therapy (PRRT) is a systemic therapy in patients with advanced metastatic NET. PRRT requires a good tumor uptake in the SSTR imaging. For therapeutic purposes, the peptides DOTA-TATE and DOTA-TOC can be labeled with either ^{90}Y or ^{177}Lu. PRRT has yielded very promising results in selected patients (Table 1). It produces good tumor response rates with limited and reversible side effects (Brabander et al., 2017).

1.1.4 SSA Perspectives in Humans

New imaging technology, modification of SSA structure, and binding to other components are the ways of future development of these analogues. Technology of imaging has been improved in recent years by the capacity of detection and better localization of signal in body by multibarret-computerized tomography and magnetic resonance imaging. After static imaging, it is possible to perform functional imaging with an improvement in diagnostic performance of PET–CT imaging of NETs. New radiolabeled ligands and SSAs are in development and could be, in the future, the keys of a more personalized localization and therapy of these tumors (Chatalic, Kwekkeboom, & de Jong, 2015). A functionalized nanoparticle (NP) incorporated with SSAs could modify the targeting, through a preferential binding to cancer cells (Shekhter Zahavi et al., 2017).

1.2 Background Structural Data of SST-14, Octreotide, and Pasireotide

Since the first years following the discovery of SST-14, attempts have rapidly grown up to elucidate the structure–activity relationship of this peptide hormone. It was shown that the presence of the Trp^8-Lys^9 pair of central residues (Fig. 1A) plays a key role in the interaction of SST-14 with SSTRs (Weckbecker et al., 2003). Upon this consideration, it was first thought that SST-14 might have a β-hairpin structure with Trp^8 as the top turn residue. Strikingly, despite the absence of any evident β-sheet marker, the circular dichroism (CD) signal observed in the 100–160 μM concentration range was assigned to a rigid β-turn (Holladay & Puett, 1976; Holladay, Rivier, & Puett, 1977). Later, on the basis of nuclear magnetic resonance (NMR) data obtained in the millimolar (mM) concentration range, it has been concluded that SST-14 cannot adopt a unique rigid structure, but rather fluctuates among several rapidly interconverting conformers, forming somehow a wave-like motion between Cys^3 and Cys^{14} residues

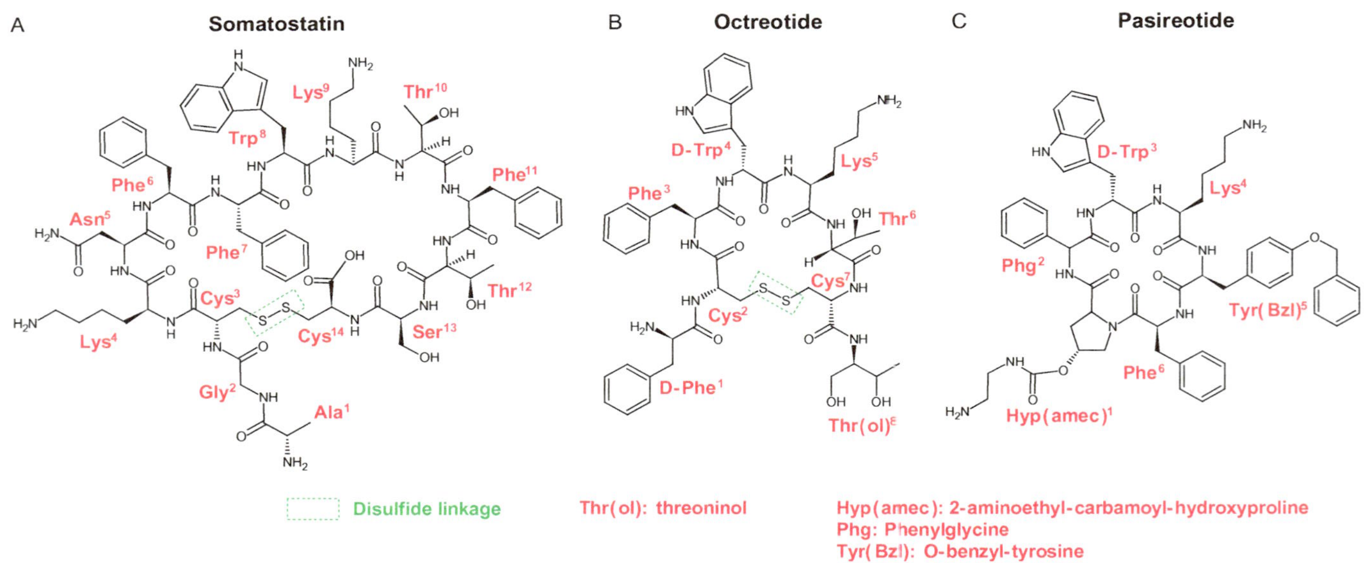

Fig. 1 Chemical composition and amino acid numbering of somatostatin-14 (A), octreotide (B), and pasireotide (C).

(Hallenga et al., 1979, 1980; Knappenberg et al., 1982), In addition, the severe NMR line broadening observed above 15 mM was correlated to the SST-14 aggregation in solution (Hallenga et al., 1980).

The chemical composition of octreotide is closely related to that of SST-14 (Fig. 1B). It should be emphasized that the biological activity of a cyclic SST-14 was shown to be considerably enhanced by the L–D substitution of the central Trp residue (Brown, Rivier, & Vale, 1976; Vale, Grant, Amoss, Blackwell, & Guillemin, 1972). Two D-stereoisomers (D-Phe^1 and D-Trp^4), as well as a C^{ter}-threoninol, Thr^8(ol), were introduced in this SSA. Octreotide is the only SSA for which the crystal structure is available (Pohl et al., 1995). It was shown that three molecules (referred to as I, II, and III) surrounded by water were present in the crystal asymmetric unit. All three molecules have a type-II′ β-turn structure formed around the central D-Trp^4-Lys^5 pair. It should be stressed that β-hairpins are the most occurring structural elements in proteins, allowing connection between helices and β-sheets. A β-turn is composed of four residues, generally referred to as i, $i+1$, $i+2$, and $i+3$, and the folding through a β-turn is made possible by the special values given to the backbone torsion angles (ϕ and ψ) of the two middle $i+1$, $i+2$ residues. On the other hand, the so-called reverse β-turns (named type-I′ and -II′) were recognized to be the most frequent ones (Das, Naganagowda, Karle, & Balaram, 2001; Sibanda & Thornton, 1985). Particularly, a type-II′ β-turn is generally formed with a backbone ϕ_{i+1} value close to +60°, limiting the choice of the amino acid type placed at the $i+1$ position (Gunasekaran, Ramakrishnan, & Balaram, 1997; Kaur & Raghava, 2004). In octreotide, three possible reasons might explain the formation of a type-II′ β-turn: (i) the presence of the D-Trp^4 residue at the turn $i+2$ position, (ii) the shortness of the cyclic part (6 residues instead of 12 in SST-14), (iii) and the presence of a disulfide bridge closing the β-hairpin. Following the analysis of the octreotide crystal structure, ^1H NMR data collected from this peptide in DMSO-d_6 also revealed the presence of two major conformers stabilized by a type-II′ β-turn structure (Melacini, Zhu, & Goodman, 1997).

In contrast to SST-14 and octreotide, the peptide bonds are responsible for maintaining the pasireotide cyclic structure (Fig. 1C). As a consequence, the definition of the N^{ter}–C^{ter} course along the pasireotide backbone is not evident. In this report, we have adopted the same amino acid numbering as that suggested previously (Interlandi, 2009). As in the case of octreotide, pasireotide contains a D-Trp^3-Lys^4 neighboring pair. The presence of three synthetic amino acids, i.e., Hyp(amec)^1, Phg^2, and Tyr(Bzl)^5, is to be remarked in this SSA. Very limited structural data were available for pasireotide. Only a set of NMR-based molecular dynamics (MD) data had conducted to the conclusion that octreotide and pasireotide might have similar structural features, fluctuating between a flat and a saddle-like backbone configuration (Interlandi, 2009; Lewis et al., 2004).

1.3 Plasmonic NPs Used in Biology and Medicine

The main interest of NPs in material science is related to their exceptionally large surface/volume (s/v) ratio. To give a typical example, spherical objects provide an s/v ratio as: $6/D$, where D is their size (diameter). At the nanometric scale, this ratio varies in the decreasing order: 1.2, 0.12, and $0.04\,\mathrm{nm}^{-1}$, for the increasing D values: 5, 50, and 150 nm, respectively. For comparison, the s/v ratio of a ping-pong ball ($D \sim 4$ cm) is as low as $\sim 1.5 \times 10^{-7}\,\mathrm{nm}^{-1}$.

The NPs' remarkable physicochemical properties, different from those encountered in the bulk, correspond to their catalytic activity, magnetism, thermal and electric conductivity, melting point, optical absorption, and scattering.

The use of different types of NPs has gained a continuously increasing attention in biological and medical applications. For instance, we can mention: (i) the organic NPs made of NIPAM/BAM copolymer used for accelerating the aggregation of human β2-microglobulin protein (Linse et al., 2007); (ii) the inorganic NPs, such as those elaborated either from CdTe employed for preventing the fibrillation (Yoo et al., 2011), or those made by uniform maghemite γ-Fe_2O_3 for accelerating/inhibiting the fibrillation of amyloid β (Aβ) peptides involved in Alzheimer's disease (Skaat, Shafir, & Margel, 2011); and (iii) the metallic gold NPs that have shown a large spectrum of action including the inhibition of the aggregation of short size Aβ peptides (Ma, Wei, & Yang, 2013), the acceleration of fibrillation of cysteine-free peptides (Nergiz, Slocik, Naik, & Singamaneni, 2013), as well as the retardation of insulin fibrillogenesis process (Hsieh, Chang, & Cho, 2013).

Focusing particularly on plasmonic NPs, made currently of noble metals (typically Au, Ag, Pt, and Cu) characterized by a high reflectivity, their remarkable optical properties have received a great attention. This type of NPs exhibits strong UV–visible absorption bands that are not present in bulk metal, proving their solution brilliant colors. When a metallic NP is exposed to an electromagnetic wave, the electrons at the surface of the metal (surface plasmon or SP) oscillate at the same frequency as the incident wave. Then the oscillating electrons reradiate electromagnetic wave at the same frequency. Particularly, upon irradiation with a wavelength (λ) that matches the principal plasmon absorption (λ_p), giving rise to the so-called surface plasmonic resonance (SPR), the NPs can absorb and scatter the light beam out of their physical cross sections (Aslan, Lakowicz, & Geddes, 2005; Hutter & Fendler, 2004). It has been evidenced that the absorption maxima of the SP band depends on the metal type, as well as on the size and shape of the nanostructures. Furthermore, the electric field of the scattered light is highly enhanced when the size of NPs is comparable to or smaller than the wavelength of the light exciting the surface plasmons. This phenomenon, referred to as localized surface plasmon resonance (LSPR), is greatly sensitive to the refractive index. As a consequence, the change in refractive index by the adsorption of biological molecules on the plasmonic NPs surface results in a large redshift of λ_p.

The elaboration of colloidal plasmonic NPs, particularly those made of Au and Ag, was described during the last years (Agnihotri, Mukherji, & Mukherji, 2014; Bastús, Comenge, & Puntes, 2011; Liz-Marzán, 2004; Sanchez-Cortes, Garcia-Ramos, Morcillo, & Tinti, 1995; Shenashen, El-Safty, & Elshehy, 2013), along with the synthesis protocols that give rise to a variety of nanoshapes, such as nanospheres, nanowires, nanorods, nanocubes, nanoflowers, dendrites, nanoprisms, nanoplates, and nanodisks. In parallel, the influence of the shape and size of gold nanoparticles (AuNPs) in exploring biological systems has been reported (Lee & El-Sayed, 2006; Li, Jing, Zhang, & Long, 2012; Paige Hall et al., 2008). As far as the pharmaceutical and medical applications of plasmonic NPs are concerned, one can remark an intensive research performed in drug delivery (Abdellatif et al., 2016), cancer and photothermal therapy (Huang, El-Sayed, Qian, & El-Sayed, 2006; Jain, El-Sayed, & El-Sayed, 2007; Qian, Cao, & Long, 2016), and other diseases (Karimi et al., 2016; Kharlamov et al., 2015).

In the field of biomolecular spectroscopy, benefit was taken from the LSPR effect for analyzing the molecules adsorbed on the surface of plasmonic NPs. Especially, in the two spectroscopic methods, i.e., surface-enhanced fluorescence (SEF) and surface-enhanced Raman scattering (SERS), plasmonic NPs were employed in order to obtain electronic and vibrational information on the biomolecules in the vicinity of plasmonic NPs (Sevilla et al., 2012). Because of a considerable amplification of both incident and scattered light beams, SERS permits detection of very low traces of biological molecules in solution; it can also be used as a tool for probing the binding sites of the adsorbed species by means of the vibrational markers (Hernández et al., 2016; López-Tobar et al., 2015).

2. RESULTS AND DISCUSSION

2.1 Bulk Structural Dynamics of SST-14 and Its Analogues

2.1.1 Rotational Freedom of the Trp Residue Evidenced by Fluorescence Data

The normalized emission spectra of SST-14 (Fig. 2A), octreotide (Fig. 2B), and pasireotide (Fig. 2C), obtained by excitation at 290 nm, reveal the fluorescence signal of the unique Trp residue involved in these peptides. A large band is observed at ~350 nm, barely affected by increasing the solution ionic strength (Fig. 2). Nevertheless, in low dielectric constant solvents, such as water/methanol mixture or methanol, the Trp fluorescence of SST-14 undergoes a considerable decrease (between 25% and 50%) along with a 10–15 nm downshift (Hernández et al., 2014). The same behavior was observed in octreotide and pasireotide (Hernández et al., 2016). A similar effect had also been revealed on SST-14 interacting with lipids (Beschiaschvilli & Seeling, 1991; Bhattacharyya & Basak, 1995). The values of anisotropy ratio, $\langle r \rangle$, as derived from the linearly polarized emission spectra of SST-14, octreotide, and pasireotide, are reported in Table 2. Being close to zero, these values are consistent with a high rotational freedom of the Trp residue.

2.1.2 CD Structural Fingerprints

In contrast to their fluorescence spectra, the CD spectra of the three peptides present completely different shapes. The SST-14 dichroic spectrum (Fig. 3A) is characterized by a deep minimum at ~203 nm, remaining unchanged in the 100–500 μM concentration range (Hernández, Carelli, Coïc, De Conninck, & Ghomi, 2009). This signal also remains unchanged by increasing ionic strength (Fig. 3A) or by dissolving the peptide in methanol (Hernández et al., 2009). The whole set of observed data shows that in the submillimolar concentration range, SST-14 adopts a rather unordered chain. We recall that a random chain is characterized by a deep negative signal at ~198 nm (Perczel & Fasman, 1992). In a previous study, it is shown that an open-chain SST-14 (in which Cys^3 and Cys^{14} are replaced by Ser residues) provides the same spectral shape in water as the cyclic SST-14, leading us to conclude that the disulfide linkage has no essential effect on the peptide hormone folding. However, the open-chain SST-14 was shown to adopt a helical conformation in methanol (Hernández et al., 2009).

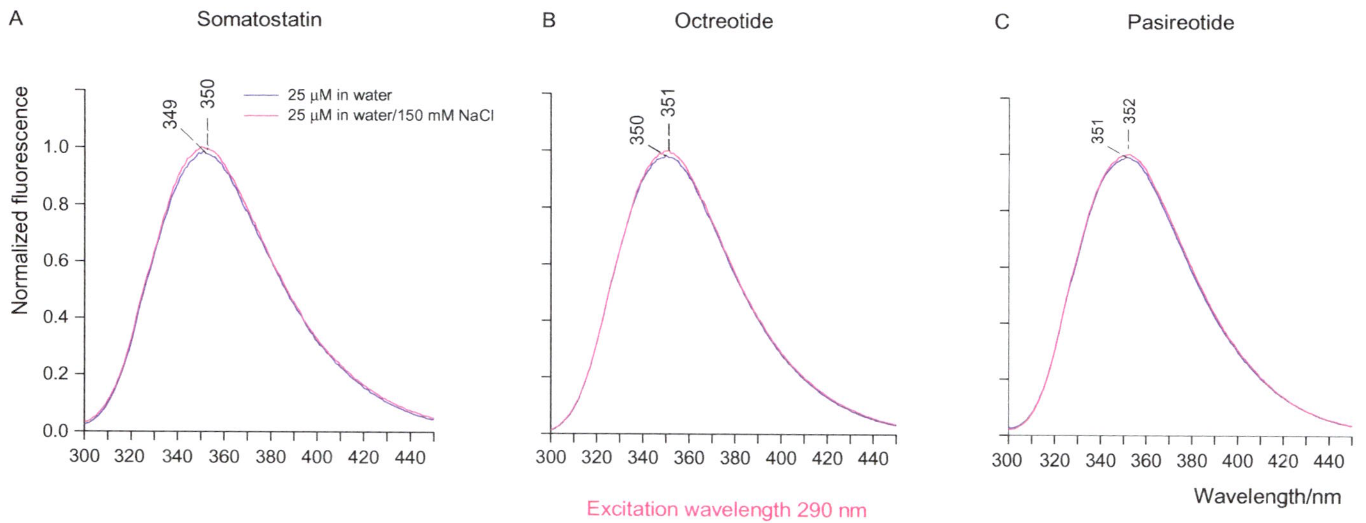

Fig. 2 Normalized fluorescence spectra of somatostatin-14 (A), octreotide (B), and pasireotide (C) obtained by excitation at 290 nm. See Table 2 for anisotropy values.

Table 2 Fluorescence Anisotropy Values, ⟨r⟩ Measured in Aqueous Solution[a,b]

	Pure Water	**Water/150 mM NaCl**
Somatostatin-14	0.022	0.022
Octreotide	0.001	0.001
Pasireotide	0.016	0.001

[a]Each value corresponds to the average of 60 measurements.
[b]Identical anisotropy values were obtained for the three excitation wavelengths at 275, 290, and 295 nm.

The CD spectral shape of octreotide (Fig. 3B) is mainly composed of two negative bands at ~200 and ~217 nm, followed by a large positive band between 240 and 270 nm. In contrast to SST-14, a concentration effect can be observed in the 25–250 μM range, as characterized by a progressive wavelength upshift of the double minima, and a change in its intensity ratio, i.e., $[\Phi_2]/[\Phi_1]$, where $[\Phi_2]$ and $[\Phi_1]$ refer to the ellipticities of the higher and lower wavelength components of the doublet, respectively. CD and NMR-based structural studies on 10 gramicidin S cyclic peptides, all adopting a type-II′ β-turn, revealed that the $[\Phi_2]/[\Phi_1]$ ratio can be taken as an indicator of their hairpin stability (Gibbs, Bjorndahi, Hodges, & Wishart, 2002). Precisely, for a highly stable (compact) β-turn, the $[\Phi_2]/[\Phi_1]$ ratio is close to unity, whereas the decrease of this ratio is consistent with the β-turn instability. In octreotide (Fig. 3B) the $[\Phi_2]/[\Phi_1]$ ratio varies from ~0.4 to ~0.8 upon increasing concentration, showing that the intermolecular interactions lead to a progressive increase of the turn stability.

The pasireotide CD signal (Fig. 3C) does not share any common feature with those observed in SST-14 and octreotide. A very deep minima around 195 nm is followed at its high wavelength side by two weak bands peaking at ~203 and ~213 nm, and a large positive band at ~230 nm. No concentration dependence in this particular signal was observed in the 50–100 μM range (Fig. 3C). As a consequence, one can definitely rule out the previous assumption, upon which pasireotide and octreotide might possess similar structural features (Interlandi, 2009). Moreover, the pasireotide CD spectrum does not resemble any other characteristic β-turn, such as those previously assigned to the type-I, -I′, -II, -II′, and -VIII (Fuchs et al., 2006; Gibbs et al., 2002; Hernández, Coïc, Gouyette, & Ghomi, 2012; Hernández, Coïc, et al., 2014; Kelly & Price, 2000; Mahalakshimi, Shanmugam, Polavarapu, & Balaram, 2005; Perczel & Fasman, 1992).

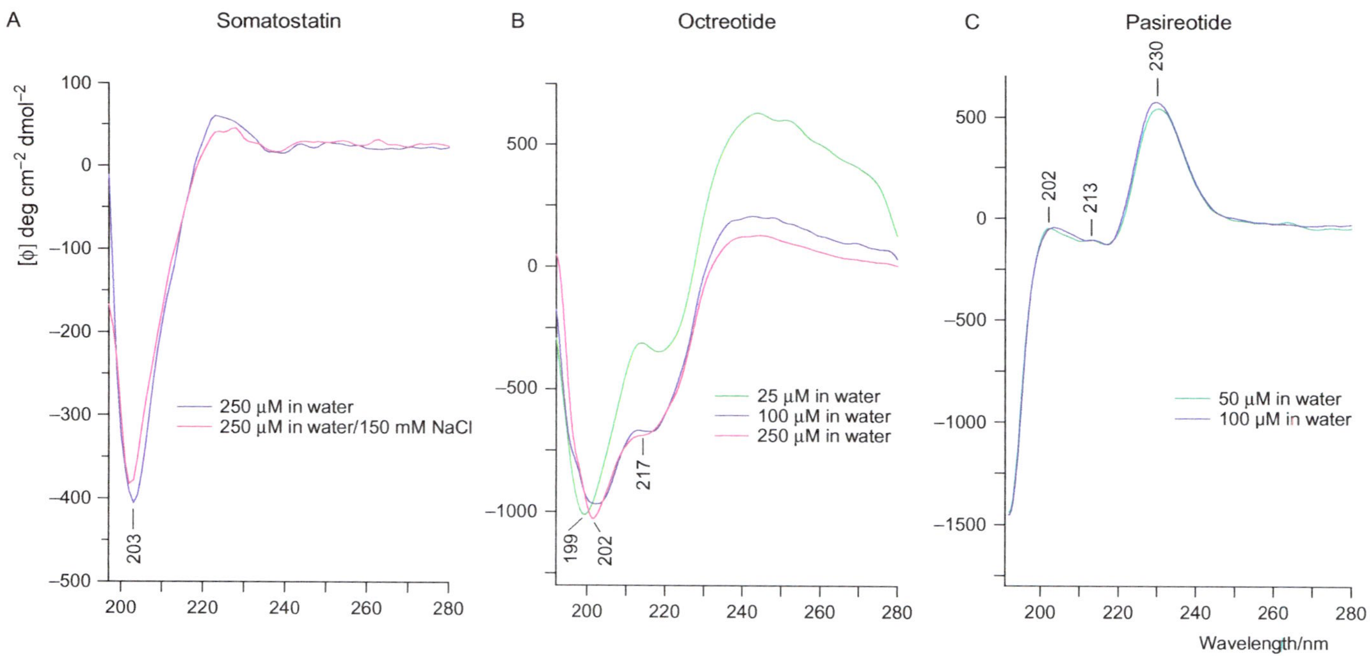

Fig. 3 Circular dichroism spectra of somatostatin-14 (A), octreotide (B), and pasireotide (C) analyzed as a function of concentration in the submillimolar range.

Briefly, while a type-I β-turn provides a CD signal with double minima at ~208/222 nm, resembling that observed in an α-helix, a type-II β-turn is characterized by a positive band located in the 190–210 nm region (Perczel & Fasman, 1992). As far as the reverse β-turn types (i.e., of type-I′ and -II′) are concerned, the CD and NMR spectra recorded in the octapeptides, Boc-Leu-Val-Val-D-Pro-Xxx-Leu-Val-Val-OMe, were consistent with a β-turn formed at their central D-Pro-Xxx. Interestingly, the presence of D-Ala or L-Ala at the position Xxx favors either a stable type-I′ or a type-II′ β-turn, respectively. CD signatures of these peptides observed in methanol and TFE could permit assignment of a large negative band around ~215 nm to the type-I′ β-turn, whereas the type-II′ β-turn was characterized by two partially resolved negative bands with the same intensities, around ~204 and ~218 nm. Finally, the type-VIII β-turn was associated to a CD fingerprint similar to that of type-II′. One can also presume that the particular structure of pasireotide is initiated by the presence of a proline ring included in its $Hyp(amec)^1$ residue. This hypothesis can be no more retained, because the cyclic hexapeptide cyclo-(Phe-D-Pro-Gly-Arg-Gly-Asp), giving a CD fingerprint similar to that of octreotide, presents a structural interconversion between the type-II′ and -I′ β-turn (Bour et al., 2008).

The impossibility of assigning a β-turn structural model to pasireotide led us to think about the presence of a more compact fold in this peptide, i.e., a γ-turn. It should be remarked that there exist two γ-turn types, referred to as classical and reverse, one being the mirror image of the other. Both of them correspond to a tight fold appearing on three residues (named i, $i+1$, and $i+2$), and presenting special values for the backbone torsion angles ϕ_{i+1} and ψ_{i+1} (Milner-White, Ross, Ismail, Belhadj-Mostefa, & Poet, 1988). Previously, the combined use of CD, FT-IR absorption, NMR, and classical MD calculations had revealed the formation of a γ-turn in the tripeptide Ala-Phe-Ala (Motta, Reches, Pappalardo, Andretti, & Gazit, 2005). On the basis of a series of experimental and theoretical structural data derived from different tripeptides (unpublished data), we could finally suggest a reasonable model for pasireotide composed by two face-to-face folds: a reverse γ-turn located on the three residues—Phg^2-D-Trp^3-Lys^4—and a type-I β-turn on the four residues—$Tyr(Bzl)^5$-Phe^6-$Hyp(amec)^1$-Phg^2 (Hernández et al., 2016). Fig. 4C displays the suggested structural model for pasireotide, as compared with those suggested for SST-14 (Fig. 4A) and octreotide (Fig. 4B).

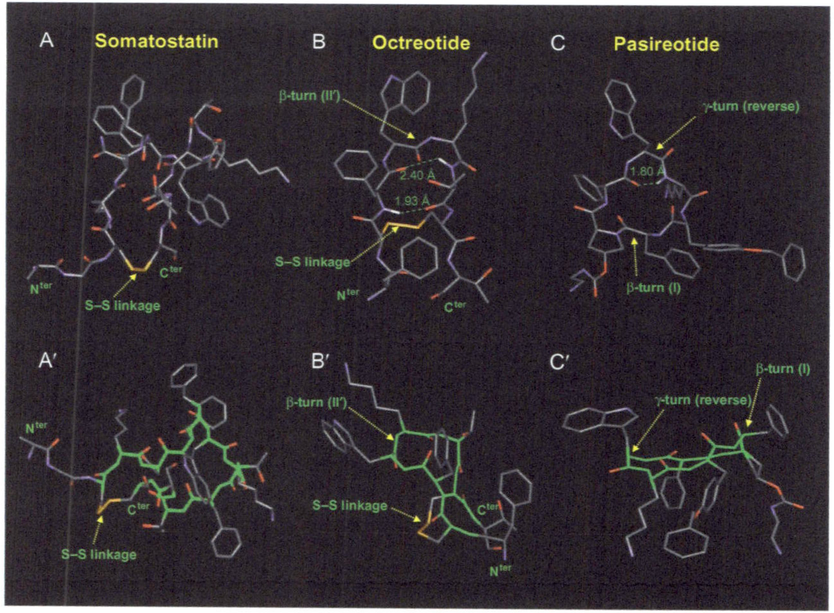

Fig. 4 Graphical representation of the structural models of somatostatin-14 (A and A′), octreotide (B and B′), and pasireotide (C and C′). Two views are selected for each peptide. Hydrogen atoms are removed, except for the molecular groups contributing to intramolecular H-bonds. In A′, B′, and C′ views, the closed part of each peptide is in *green*. Hydrogen bonds are displayed as broken *green lines*, of which the length (in Å) is also reported. Turns and their types, as well as disulfide linkages, are also indicated. *Sources: SST-14: NMR study of SST-14 in 5% D-mannitol. https://www.rcsb.org/pdb/explore.do?structureId=2mi1. The displayed molecule is one of the 10 registered ones, PDB ID: 2MI1. Octreotide: NMR study of the backbone conformational equilibria of sandostatin, minimized average β-sheet structure, PDB ID: 1SOC, http://www.rcsb.org/pdb/explore.do?structureId=1soc. See Hernández, B., López-Tobar, E., Sanchez-Cortes, S., Coïc, Y. M., Baron, B., Chenal, A., et al. (2016). From bulk to plasmonic nanoparticle surfaces: The behaviour of two potent therapeutic peptides, octreotide and pasireotide,* Physical Chemistry Chemical Physics, 18, *24437–24450 for conformational angles and Cartesian coordinates. Pasireotide: DFT-based theoretical calculations. See Hernández, B., López-Tobar, E., Sanchez-Cortes, S., Coïc, Y. M., Baron, B., Chenal, A., et al. (2016). From bulk to plasmonic nanoparticle surfaces: The behaviour of two potent therapeutic peptides, octreotide and pasireotide,* Physical Chemistry Chemical Physics, 18, *24437–24450 for conformational angles and Cartesian coordinates.*

2.1.3 Raman Scattering Structural Fingerprints

In the two last sections, we have emphasized the structural dynamics of the three peptides in the submillimolar concentration range by means of fluorescence and CD data. Here, we follow the structural analysis within the millimolar interval using Raman data. Generally recorded in the middle wavenumber spectral region (1800–400 cm^{-1}), Raman spectra present the advantage of providing a low water contribution (Fig. 5). The solvent background can thus be accurately removed from the peptide Raman spectrum (Fig. 6), enabling an accurate analysis of the molecular vibrational modes.

Owing to their strong modulation of electronic polarizability, the aromatic rings provide strong Raman bands. All three peptides contain a Trp residue, and one (octreotide and pasireotide) or three (SST-14) Phe residues. It should be added that L- and D-stereoisomers provide the same vibrational markers. A previous investigation on the Raman spectra of free amino acids and short peptides has allowed classification and assignment of the most characteristic Trp and Phe markers (Hernández et al., 2010; Hernández, Pflüger, Kruglik, & Ghomi, 2013). One can notice six Phe characteristic bands, referred to as F1, ..., F6, and eight Trp bands, named W1, ..., W8 (Table 3). Being all localized in Trp and Phe aromatic rings, the

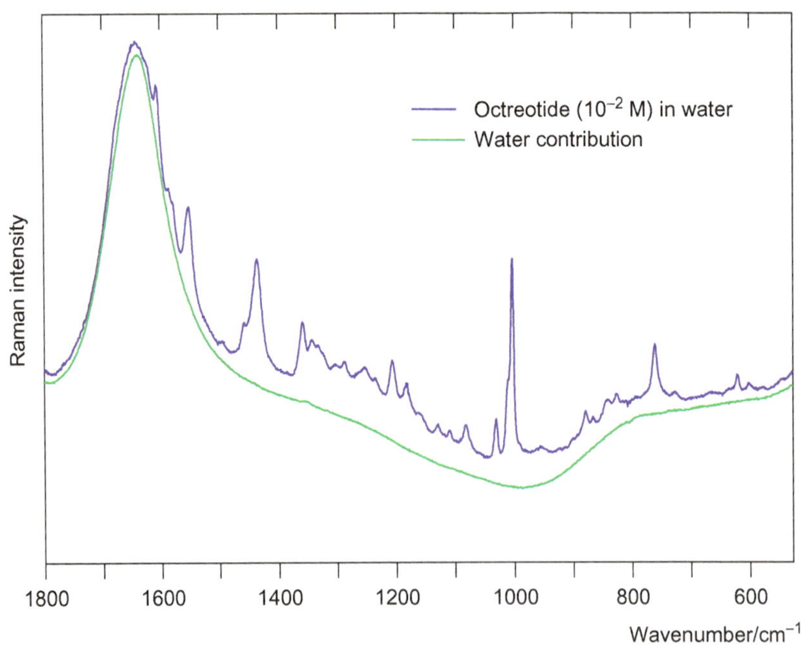

Fig. 5 Raman spectra obtained from the solution sample of octreotide (10^{-2} M) and from the solvent (water).

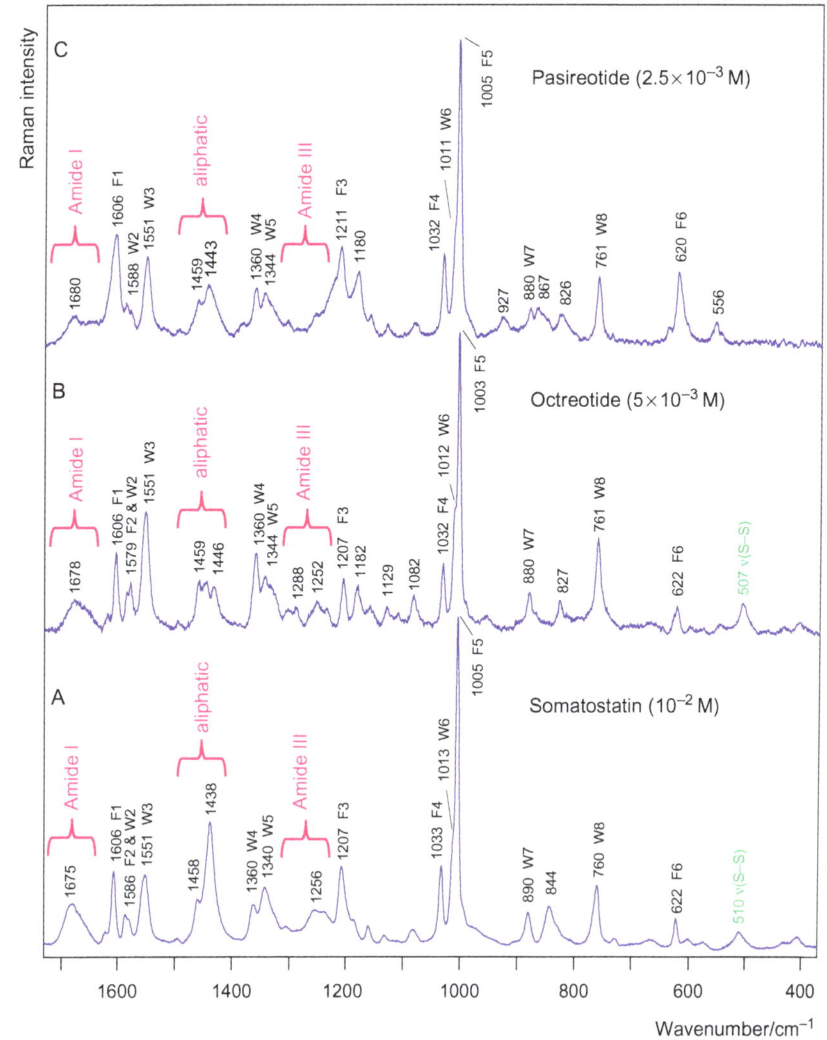

Fig. 6 Comparison of the Raman spectra obtained from the solution samples of somatostatin-14 (A), octreotide (B), and pasireotide (C). Sample concentrations are reported in parentheses. F1–F6 and W1–W8 refer to the characteristic Raman lines of Phe and Trp, respectively. See also Table 3. Amide I and amide III regions, as well as the disulfide marker are also indicated. Sample concentration is reported in parentheses.

wavenumbers of the Trp and Phe markers observed in peptides remain very close to those collected from free amino acids (Table 3). In all cases, the strongest Raman band is F5 (Phe residue) at $\sim$1004 cm^{-1} (Fig. 6), whereas W6 (Trp residue) at $\sim$1012 cm^{-1} is observed as a shoulder at the high wavenumber side of the F5 marker. Despite their aromatic character, the

Table 3 Characteristic Vibrational Modes of Tryptophan and Phenylalanine Observed in Free Amino Acids, Somatostatin-14, Octreotide, and Pasireotide[a]

Free Amino Acid[b]	SST-14@Bulk[c]	SST-14@Ag[d]	Octreotide@Bulk[c]	Octreotide@Ag[d]	Pasireotide@Bulk[c]	Pasireotide@Ag[d]
1621 (W1)						
1606 (F1)	1606 (F1)	1604 (F1)	1606 (F1)	1606 (F1)	1606 (F1)	1608 (F1)
1586 (F2)	1586 (F2 and W2)	1581 (F2 and W2)		1581 (F2 and W2)	1588 (W2)	1580 (F2 and W2, sh)
1579 (W2)			1579 (F2 and W2)			
1552 (W3)	1551 (W3)	1551 (W3)	1551 (W3)	1556 (W3)	1551 (W3)	1553 (W3)
1361 (W4)	1360 (W4)	1362 (W4)	1360 (W4)	1360 (W4)	1360 (W4)	1358 (W4)
1343 (W5)	1340 (W5)	1340 (W5)	1344 (W5)	1344 (W5)	1344 (W5)	1344 (W5)
1207 (F3)	1207 (F3)	1206 (F3)	1207 (F3)	1207 (F3)	1211 (F3)	1210 (F3)
1032 (F4)	1033 (F4)	1031 (F4)	1032 (F4)	1032 (F4)	1032 (F4)	1031 (F4)
1012 (W6)	1013 (W6, sh)		1012 (W6, sh)		1012 (W6, sh)	
1004 (F5)	1005 (F5)	1003 (F5)	1003 (F5)	1005 (F5)	1005 (F5)	1004 (F5)
880 (W7)	890 (W7)	878 (W7)	880 (W7)	880 (W7)	880 (W7)	879 (W7)
758 (W8)	760 (W8)	758 (W8)	761 (W8)	761 (W8)	761 (W8)	761 (W8)
622 (F6)	622 (F6)	622 (F6)	622 (F6)	623 (F6)	622 (F6)	623 (F6)

[a]Observed wavenumbers are in cm^{-1}. In parentheses assignments to tryptophan (W) and phenylalanine (F). *sh*, shoulder.
[b]Vibrational modes of the solution samples containing free amino acids are taken from Hernández, Pflüger, Adenier, Kruglik, and Ghomi (2010).
[c]See Fig. 6 for the bulk Raman spectra of somatostatin, octreotide, and pasireotide.
[d]See Fig. 11 for the SERS spectra of somatostatin, octreotide, and pasireotide adsorbed on silver nanoparticles.

Phg^2 and $Tyr(Bzl)^5$ residues involved in pasireotide do not give rise to resolved and strong Raman markers (Fig. 6C). This might be due to the structural similarity of Phg, as well as the benzyl moiety of $Tyr(Bzl)^5$ with Phe, leading presumably to an overlap of their characteristic markers. In a similar manner, the barely resolved medium intensity bands observed in the 860–840 cm^{-1} region of pasireotide could be assigned to the phenol ring involved in $Tyr(Bzl)^5$ residue (Fig. 6C) (Hernández, Coïc, Pflüger, Kruglik, & Ghomi, 2015).

Raman spectra can provide valuable structural information through their amide I (1700–1640 cm^{-1}) and amide III (1305–1230 cm^{-1}) regions. It should be emphasized that amide I vibrations result from the backbone C=O bond-stretch motion, more or less coupled to its adjacent N—H angular bending. Amide III vibrations basically arise from the backbone N—H angular bending. As a consequence, upon labile hydrogen deuteration (Raman spectra observed in heavy water), amide I vibrations present a ~10 cm^{-1} downshift, whereas amide III markers completely vanish. Both amide (I and III) regions are shown to be conformation sensitive, i.e., specific amide markers can be assigned to α-helix, β-strand/β-sheet, turns, and unordered segments of a peptide chain. Based on a series of analyses made on peptides and proteins, we could propose rational band decomposition in order to achieve the estimation on the populations corresponding to different secondary structural elements (Boukhalfa-Heniche et al., 2004; Guiffo-Soh, Hernández, Coïc, Boukhalfa-Heniche, & Ghomi, 2007; Guiffo-Soh et al., 2008; Hernández, Boukhalfa-Heniche, Seksek, Coïc, & Ghomi, 2006; Hernández et al., 2009; Hernández, Coïc, et al., 2014; Hernández, Coïc, Gouyette, et al., 2012; Hernández, Coïc, Kruglik, et al., 2012; Hernández et al., 2016). Here, four amide I components and five amide III components have been used (Table 4; Fig. 7).

Table 4 Components Used in the Decomposition of Amide I and Amide III Regions, and Their Assignments[a]

Amide I	Amide III	
1690 ± 5	1260 ± 5	Random chain (R)
1660 ± 5	1235 ± 5	β-Strand (S)
1675 ± 5	1305 ± 5	Turn (T)
1650 ± 5	1285 ± 5	Turn (T)
	1250 ± 5	Turn (T)

[a]The band width of each component was kept between 15 and 25 cm^{-1}.

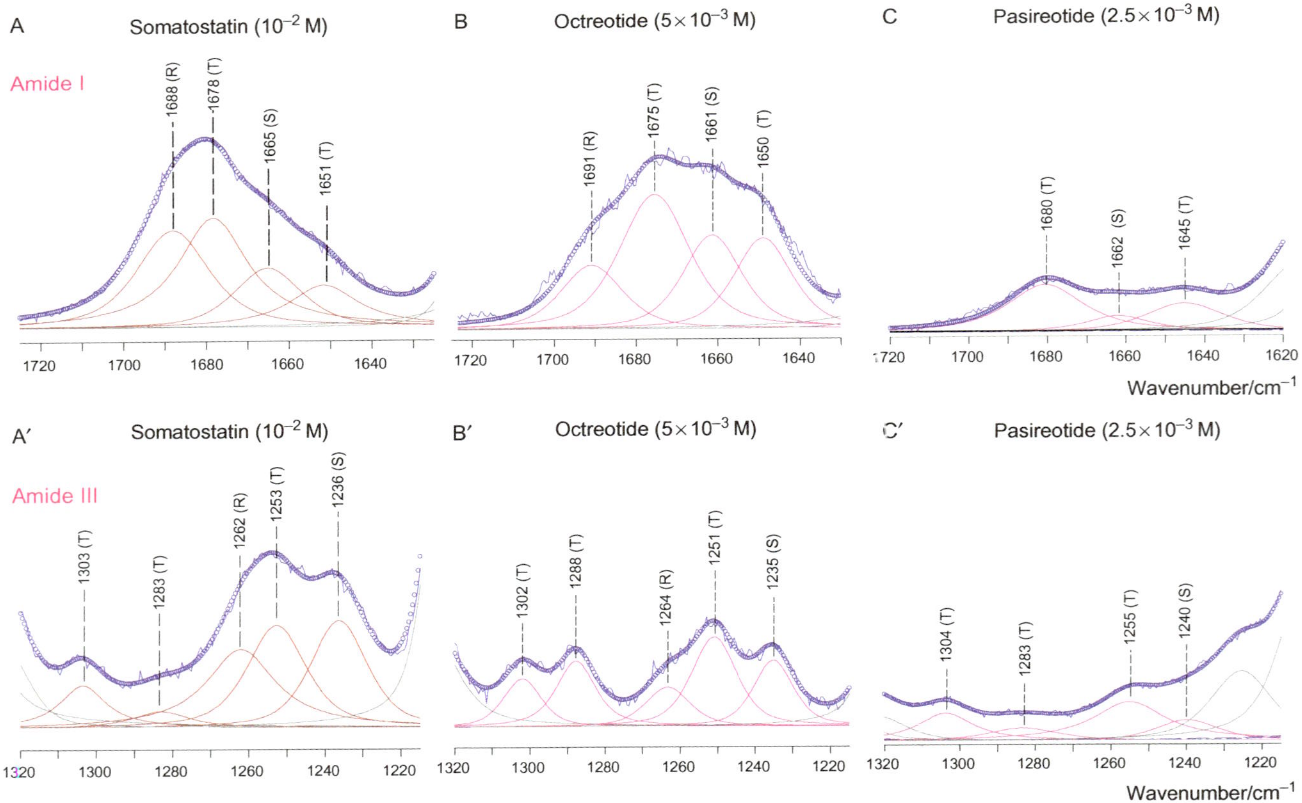

Fig. 7 Band decomposition of the amide I (A, B, and C) and amide III (A′, B′, and C′) of the Raman spectra obtained from the solution samples of SST-14 (A and A′), octreotide (B and B′), and pasireotide (C and C′). In front of each component wavenumber, its assignment to random (R), β-strand (S), and turn (T) is reported. *Circles* correspond to the sum of the components used in the band decomposition.

The histogram presented in Fig. 8 clearly shows that upon increasing concentration the weight (population) of unordered chain decreases in favor of the progressive increase of the β-strand population. In contrast, the turn population manifests a quite constant stability. This effect is more pronounced in SST-14, because of the large size of its turn. It is worth mentioning that pasireotide revealed no detectable unordered population, because of the absence of the amide I component at ~1690 cm^{-1}, and the amide III component at ~1265 cm^{-1} in band decomposition. In pasireotide, the turn population is as high as ~84%; the remaining ~16% population corresponds to β-strand, confirming the rigid structural model proposed for this compound (Fig. 4C and C′).

The concentration-induced structural dynamics of disulfide-containing peptides, such as SST-14 and octreotide, can also be followed through the Raman markers appearing in the 550–500 cm^{-1} spectral region (Fig. 9),

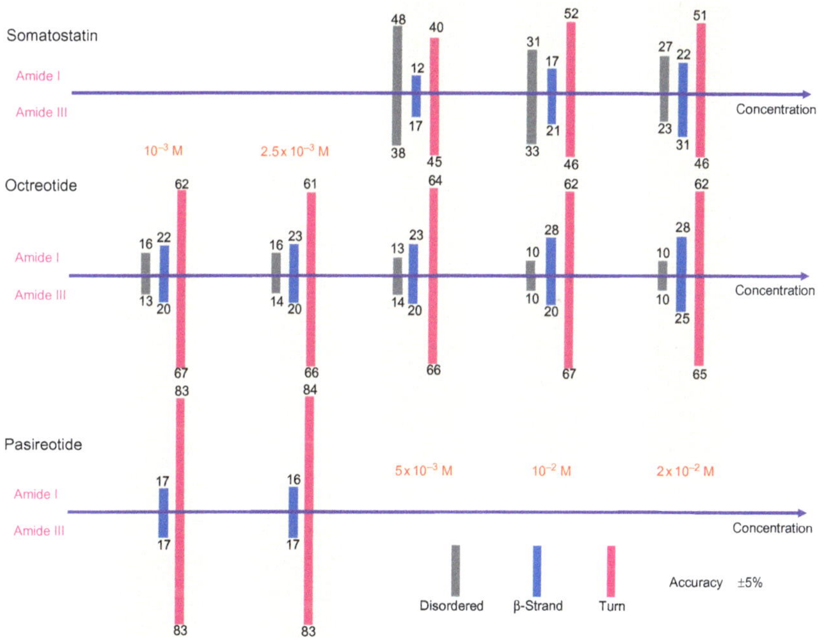

Fig. 8 Histogram representing the normalized weight (population) of different secondary structural elements (disordered, β-strand and turn) obtained by band decomposition in the amide I and amide III regions. From *top* to *bottom*, on each axis are displayed the results corresponding to one of the three analyzed peptides. Data corresponding to the amide I region are indicated above, and those related to the amide III region are reported beneath the same axis. Increasing concentration from *left* to *right*.

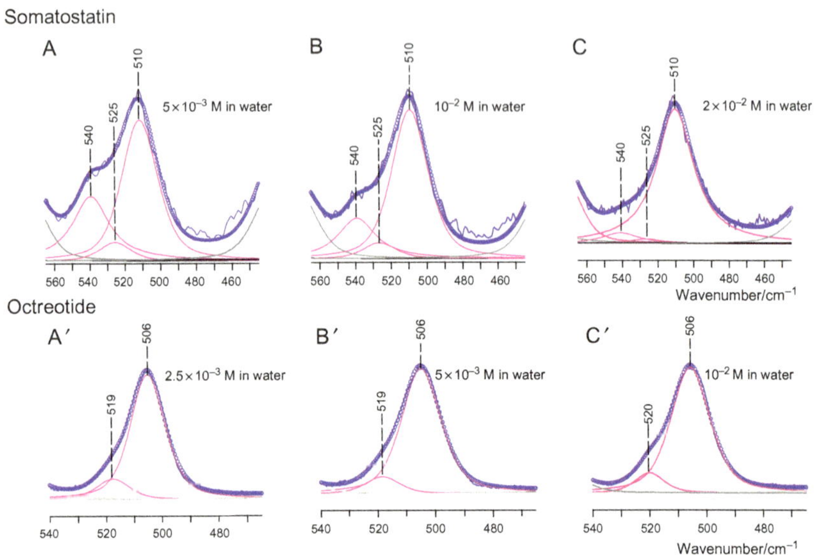

Fig. 9 Band decomposition of the disulfide bond-stretch region observed in the bulk Raman spectra of SST-14 (A, B, and C) and octreotide (A′, B′, and C′). *Circles* correspond to the sum of the components used in decomposition. Increasing concentration is from *left* to *right*. Sample concentration is reported.

which reveals the S—S bond-stretch motion within the Cys-Cys (cystine) moiety formed by the oxidative reaction of the Cys thiol groups. The conformational sensitivity of the S—S bond vibrational motions was evidenced since the earliest Raman studies on simple molecular models (Sugeta, Go, & Miyazawa, 1972; Van Wart & Scheraga, 1976, 1986). Based on these studies, it had been postulated that the Raman marker assigned to the S—S bond-stretch vibration is dependent on the values of the three torsion angles defined around the three successive bonds along the cystine —C—S—S—C— motif. As a consequence, the three frequently observed bands at ~510, ~525, and ~540 cm^{-1} were assigned to *ggg*, *ggt*, and *tgt* orientations (where *g*, *gauche* and *t*, *trans*) of the three mentioned torsion angles. It is worth emphasizing that the structural rigidity, brought about by the concentration increase, can affect the conformational flexibility of the turn and its closing disulfide bridge. While the Raman spectrum of SST-14 at the lowest concentration (5 mM) reveals three components at 510, 525, and 540 cm^{-1} (Fig. 9A), those obtained at higher concentrations (10 and 20 mM) show a considerable decrease of the components at 525 and 540 cm^{-1} (Fig. 9B and C). In other words, the concentration increase permits stabilization of the disulfide bridge in a conformation assignable to the 510 cm^{-1} marker. In octreotide, no remarkable change could be observed in the disulfide region. As shown in Fig. 9A′–C′, the components at 506 and 519 cm^{-1} remain unchanged in the concentration range between 2.5 and 20 mM, revealing the structural stability of its type-II′ β-turn. Note that in octreotide, no component close to 540 cm^{-1} could be found in the mentioned concentration range.

A recent analysis devoted to the disulfide linkage Raman markers has shown that the above-mentioned traditional interpretation based on "three Raman markers vs three conformations" appears to be insufficient for achieving a reliable relationship between the multicomponent Raman bands and the disulfide bridge conformational flexibility (Hernández et al., 2014). A new assumption has been suggested, inspired from a former statistical analysis of the protein data bank (Hogg, 2013; Schmidt, Ho, & Hogg, 2006; Schmidt & Hogg, 2007). This analysis has pointed out the fact that an accurate treatment of a disulfide bridge conformation needs the full consideration of the signs associated to the five torsion angles, referred to as χ_1, χ_2, χ_3, χ_2', χ_1', defined around the five successive bonds in the —Cα—Cβ—S—S—Cβ′—Cα′— segment of cystine. For instance, the ± sign attributed to χ_3 (defined around the S—S bond) determines the handedness (right or left) of the disulfide bridge; the signs of the three middle angles χ_2, χ_3, and χ_2' are related to the disulfide spatial shape, referred to as Staple-, Hook-, and Spiral-like disulfides; and finally the ± signs assigned to the two extreme angles, i.e., χ_1 and χ_1', are responsible for the orientation of the disulfide linkage with respect to peptide backbone(s). Based on this methodology, density functional theory calculations revealed that the S—S bond-stretch wavenumber may cover a large spectral range from 469 to 553 cm^{-1} (Hernández, Pflüger, et al., 2014). Specifically, in SST-14 and octreotide, one can assign the components at ~506 and 510 cm^{-1} (Fig. 9) to ±LHStaple and −LHHook conformers, respectively; the component at ~525 cm^{-1} to +RHHook-type conformer; and finally the component at ~540 cm^{-1} to ±LHSpiral conformer. In these notations, the left side ±, −, and + signs refer to χ_1 and χ_1', while RH and LH designate right- and left-handed disulfide bonds, respectively.

2.2 Adsorption of SST-14 and Its Analogues Onto Silver Nanoparticles

Because of the silver chemical reactivity, the major drawback encountered in the adsorption of SST-14 and octreotide on silver nanoparticles (AgNPs) was the breakdown of their disulfide bond. Thus, a special attention was focused on the preparation protocol of silver colloids, especially on the choice of reductant. In a series of investigations, the integrity of the disulfide linkage on AgNPs was checked by means of SERS data (Hernández et al., 2016; López-Tobar et al., 2015; López-Tobar, Hernández, Ghomi, & Sanchez-Cortes, 2013).

2.2.1 *Aggregation of AgNPs Provoked by the Adsorption of SST-14 and Its Analogues*

Silver colloids present a sharp plasmon resonance band peaking at ~400 nm (Fig. 10A). To induce aggregation of AgNPs, an ionic salt (KNO_3, NaCl, etc.) is generally added to the colloidal solution sample. In fact, this operation leads to screening negative charges at the surface of colloids by the positively

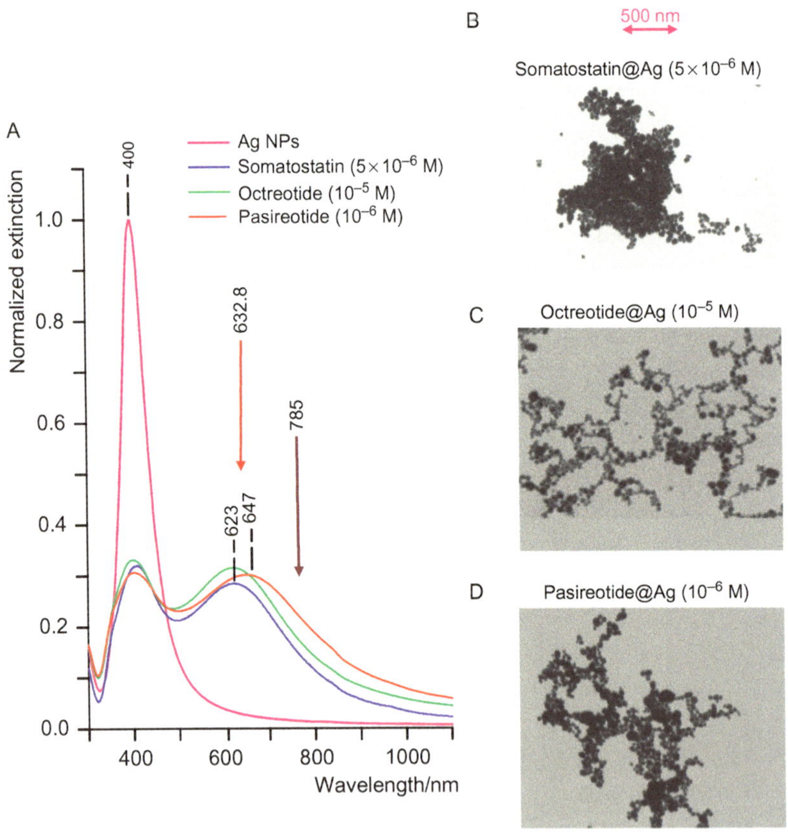

Fig. 10 Normalized extinction spectra of silver colloids alone, and in the presence of SST-14, octreotide, and pasireotide (A). TEM images taken from the same samples containing AgNPs in the presence of SST-14 (B), octreotide (C), and pasireotide (D) are also reported. Sample concentration is reported in *parentheses*.

charged counterions. As a consequence, the repulsive electrostatic interaction between NPs is reduced, facilitating their aggregation. In the case of SST-14 and its analogues, no aggregating agent was needed, because they are all of cationic character. SST-14 can bear up to three positive charges (N^{ter}, Lys^4, and Lys^9). Likewise, octreotide (N^{ter} and Lys^5) and pasireotide (Lys^4 and $Hyp(amec)^1$) may contain two positive charges.

Upon interaction with peptides, the plasmon resonance of AgNPs shows a large redshift (~230 nm), making appear a new large band between 620 and 640 nm (Fig. 10A). As shown, the optimal plasmon band shift is obtained by a low peptide concentration corresponding to the 10^{-5}–10^{-6} M range (Fig. 10A). Transmission electron microscopy (TEM) images confirm the aggregation of Ag colloids interacting with the three cationic peptides (Fig. 10B–D).

2.2.2 SERS Induced by SST-14 and Its Analogues Adsorbed on AgNPs

The amplification of the Raman signal in SERS experiments is greatly due to the formation of the so-called hot spots, i.e., narrow interparticle junctions formed upon AgNP aggregation (Fig. 10B–D). Within the hot spots, both incident and scattered radiations are considerably amplified (Chen, Ding, Luo, Loo, & Yao, 2015; Guerrini, Izquierdo-Lorenzo, Garcia-Ramos, Domingo, & Sanchez-Cortes, 2009), provided that the incident wavelength falls into the plasmon band of aggregated NPs. Fig. 10A shows the position of the two excitation wavelengths, i.e., $\lambda_L = 632.8$ and 785 nm, being in resonance with the large plasmon band appearing upon interaction with peptides.

Fig. 11 displays the SERS spectra of SST-14 at 5×10^{-6} M, octreotide at 10^{-5} M, and pasireotide at 10^{-6} M, adsorbed on Ag particles. The very good signal/noise ratio of SERS spectra allows us to follow the changes occurring in Raman spectra in going from bulk (Fig. 6) to AgNP surface (Fig. 11). A striking similarity is found between the SERS and bulk Raman spectra. For instance, the wavenumbers, as well as the relative intensities of the characteristic Trp and Phe characteristic bands, remain very close to those observed in bulk (Fig. 6; Table 3). It should be emphasized that in SERS spectra, the amide I and amide III regions cannot be compared to those observed in bulk, because only the signal from the backbone segments that are close to metal surface is enhanced. As a consequence, these spectral regions cannot provide reliable information on the whole backbone structural features. The presence of the low-intensity bands located at 498 cm^{-1} in SST-14 (Fig. 11C) and at 506 cm^{-1} in octreotide (Fig. 11B) proves the integrity of the disulfide linkage upon adsorption.

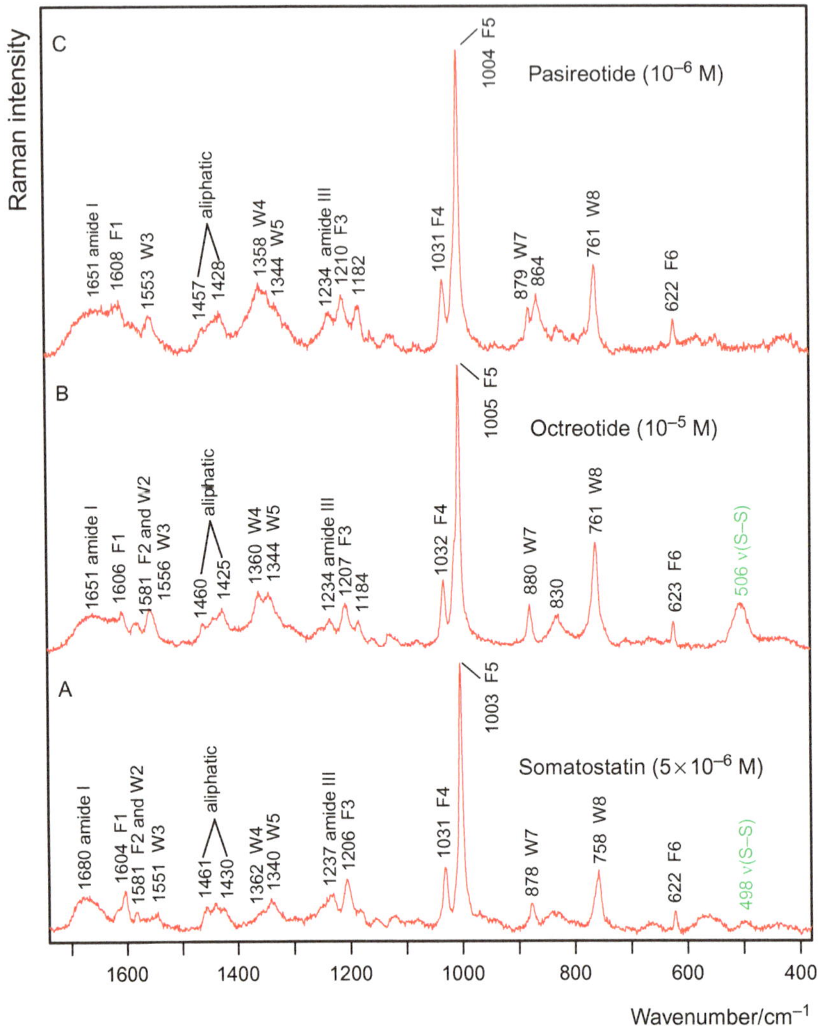

Fig. 11 Comparison of the surface-enhanced Raman spectra obtained from the solution samples of somatostatin-14 (A), octreotide (B), and pasireotide (C). F1–F6 and W1–W8 refer to the characteristic Raman lines of Phe and Trp, respectively. See also Table 3. Amide I and amide III regions as well as the disulfide marker are indicated. Sample concentration is reported in parentheses.

2.2.3 Anchoring Sites of SST-14 and Its Analogues on AgNPs

On the basis of the information provided by the SERS data in the middle wavenumber region (Fig. 11), we can confirm that the aromatic (Phe and Trp) residues cannot be considered as the potential binding site of

SST-14 and its cyclic analogues to AgNPs. The low wavenumber spectral region (<300 cm^{-1}) is known for its capability of providing complementary information on the binding mechanism through the vibrational motions involving metal atoms (Cañamares, Garcia-Ramos, Gómez-Varga, Domingo, & Sanchez-Cortes, 2005; Garrell, Shaw, & Krimm, 1983). Thanks to the presence of Cl^- ions at the surface of silver colloids, a middle intensity SERS marker is observed at ~245 cm^{-1}, ascribed to Ag-Cl^- bond-stretch motion. Fig. 12A clearly shows the presence of a unique band in the case of AgNPs aggregated by addition of KNO_3. This band also appears when the silver particles are aggregated by means of peptides (Fig. 12B–D). Strikingly, in the presence of peptides, a new component appears at ~220 nm. This observation leads us to deduce that there should be two types of Ag-Cl^- vibrations, one of them unperturbed and the other perturbed by the interaction with peptides adsorbed on AgNPs. Surely, this perturbation should be of electrostatic character, presumably between Cl^- anion and one of the cationic sites ($NH_3{}^+$) of the adsorbed peptide.

Based on the provided evidences, three possible adsorption schemes can be suggested for SST-14, i.e., through its Lys^4 (Fig. 13A), Lys^9 (Fig. 13B), or N^{ter} group (Fig. 13C). Likewise, in octreotide the adsorption can be made possible by the intermediate of either Lys^5 (Fig. 14A) or N^{ter} group (Fig. 14B). Finally, in pasireotide, Lys^4 (Fig. 14C) and $Hyp(amec)^1$ (Fig. 14D) can contribute to stabilize the peptide/NP interactions.

3. CONCLUDING REMARKS

3.1 Bulk Structural Dynamics

Because of their particular chemical structure, peptide chains are able to self-assemble upon increasing concentration. Interpeptide interactions are generally stabilized by the formation of interbackbone —C=O...H—N— hydrogen bonds, giving rise to parallel or antiparallel β-sheet structures. Peptide self-assemblies can be further stabilized by other interaction types, such as aliphatic–aliphatic, aliphatic–aromatic, aromatic–aromatic, polar, and electrostatic interactions. In this framework, the structuring role of surrounding water molecules should also be taken into consideration (Otting, Liepinsh, & Wüthrich, 1991; Zhong, Pal, & Zewail, 2011).

As far as SST-14 and its analogues are considered, at low concentrations, i.e., in the submillimolar range, fluorescence anisotropy measurements were consistent with a high rotational freedom of their

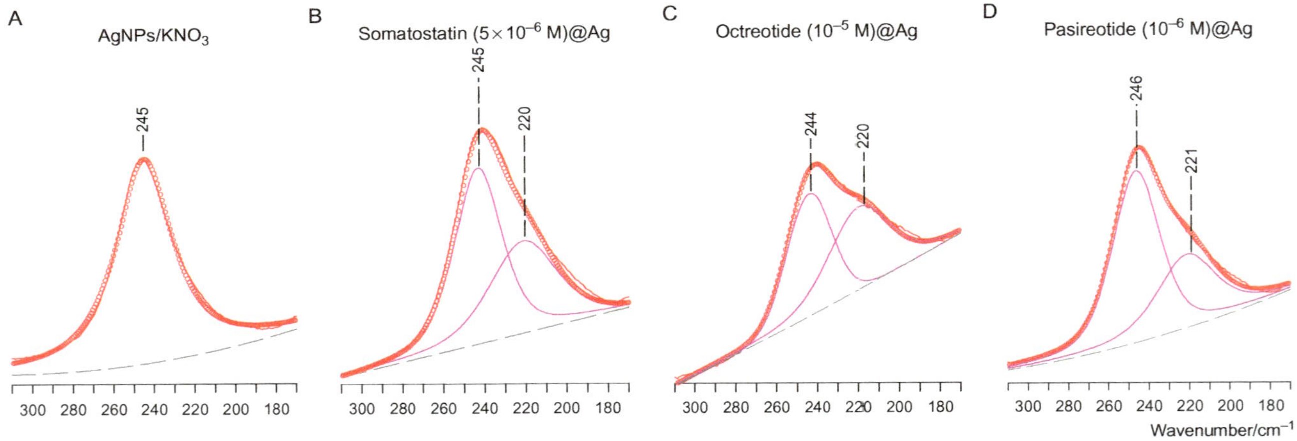

Fig. 12 Band decomposition of the low wavenumber (<300 cm^{-1}) region observed in the surface-enhanced Raman spectra of AgNPs aggregated upon addition of KNO_3 (A), SST-14 (B), octreotide (C), and pasireotide (D). *Circles* correspond to the sum of the components used in decomposition. Sample concentration is reported in *parentheses*.

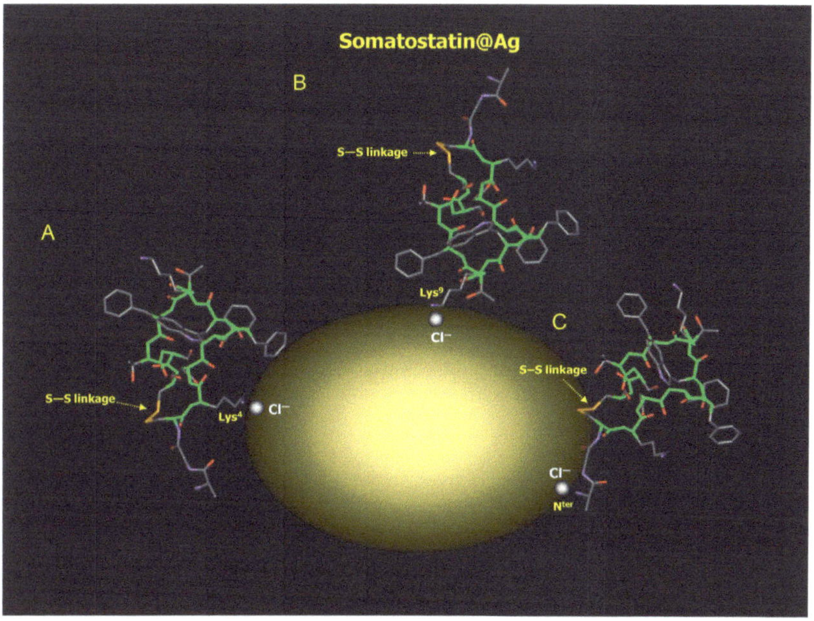

Fig. 13 Binding schemes of SST-14 on silver nanoparticles. The peptide hormone is stabilized on AgNPs by means of the electrostatic interactions involving the negatively charged chloride ions on the plasmonic surface and the positively charged amino group borne by Lys^4 residue (A), Lys^9 residue (B), or N^{ter} group (C).

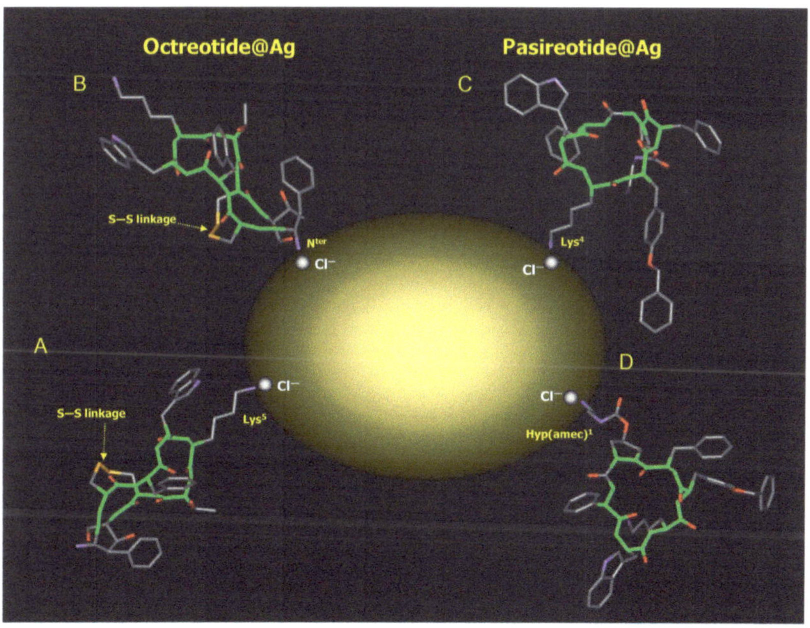

Fig. 14 Binding schemes of octreotide (A and B) and pasireotide (C and D) on silver nanoparticles. Octreotide is stabilized on AgNPs by means of the electrostatic interactions involving the negatively charged chloride ions on the plasmonic surface and the positively charged amino group borne by either Lys^5 residue (A) or N^{ter} group (B). Pasireotide is stabilized on AgNPs through the positively charged amino group borne by either Lys^5 residue (C) or $Hyp(amec)^1$ residue (B).

unique Trp residue. This feature might certainly facilitate the binding of these peptides to SSTRs, in which Trp-Lys adjacent residues should play a key role (Weckbecker et al., 2003). In this range of concentration, SST-14 remains unordered, whereas octreotide and pasireotide adopt stable cyclic structures. As a consequence, the longer and more flexible chain of SST-14 confers this peptide hormone the propensity of binding quite equivalently to its five receptors. In contrast, octreotide, having a shorter turn and more rigid structure, was shown to have a pronounced affinity for SSTR2, and a moderate one for both SSTR3 and SSTR5 (Pawlikowski & Meleń-Mucha, 2004; Weckbecker et al., 2003). Concerning pasireotide, it targets strikingly four out of five SSTRs, with the highest affinity for SSTR5 (Bruns, Lewis, Briner, Meno-Tetang, & Weckbecker, 2002).

Upon increasing the concentration up to the millimolar range, SST-14 becomes progressively structured, i.e., the rate of unordered chain decreases in favor of H-bonded β-strands. At 20 mM, only ~20% of total structural elements correspond to unordered chains. The unordered-to-ordered structural trend is less pronounced in octreotide. Above 10 mM, the unordered chain population of this peptide remains around 10%. In contrast, no indication on unordered structure could be evidenced in pasireotide. A highly rigid cyclic structure was evidenced in the whole micromolar–millimolar concentration interval.

In relation with the fact that peptide/protein hormones are stored as amyloids within secretory granules, dynamical behavior of SST-14 at higher concentrations has also received a considerable attention (Maji et al., 2009; van Grondelle et al., 2007). Recent studies highlighted the spontaneous formation of nanofibrils by gradual SST-14 aggregation in the 15–60 mM concentration range. Nanofibrils are thought to be stabilized by the interaction of side-by-side molecules, forming interpeptide antiparallel H-bonded strands. Above 60 mM, nanofibrils give progressively rise to extended nanofibers upon formation of parallel interchain H-bonds (van Grondelle et al., 2013). Although the disulfide linkage was shown to have no evident effect on the structural dynamics of the peptide hormone at low (submillimolar) concentrations (Hernández et al., 2009), its presence was revealed essential on the nature and morphology of amyloids, as well as on their aggregation/release kinetics (Anoop et al., 2014). No high concentration-induced aggregation analysis has been reported up to now on octreotide. Evidently, such type of exploration cannot be envisaged in pasireotide because of its low water solubility above 2.5 mM.

3.2 Toward Somatostatin-Based Functionalization of Plasmonic NPs

We have shown that the three cationic peptides (SST-14 and its two cyclic analogues) manifest a high affinity for binding spontaneously to the surface of hydroxylamine-reduced AgNPs. This binding does not cause the breakdown of the disulfide bridge in SST-14 and octreotide. The peptide/NP binding process is rendered possible by electrostatic interactions, occurring between the positively charged amino groups of the peptide and the chloride anions covering the colloidal surface. Beyond a considerable plasmon resonance shift, an important amplification of the SERS intensity, ranging between 10^4 and 10^5, was provoked by peptide adsorption, allowing detection of trace amounts (10^{-5}–10^{-6} M) of the peptides in aqueous solution.

During the recent years, it has been noticed that a better addressability of cellular targets and tumor cells can be achieved by the use of functionalized NPs. Limiting our scope to plasmonic NPs, and especially to their functionalization either with somatostatin, or with its analogues, we can mention here some in vitro experiments: (i) The cellular uptake by means of somatostatin-coated AuNPs on HCC-1806, HELA, and U-87 cell lines has been revealed to be significantly higher than that achieved by nonfunctionalized particles (Abdellatif et al., 2016). (ii) The temperature increase induced by irradiation at 532 nm (i.e., into the plasmonic resonance band) of octreotide-coated AuNPs during 5 ns leads to a notable decrease of HELA cells viability (Mendoza-Nava et al., 2013). (iii) Radiolabeled AuNPs may function simultaneously as radiotherapy and thermal ablation systems. A multifunctional protocol was thus tested as follows: Lys^3-bombesin, a peptide having a high affinity for GRP-R receptor (overexpressed in prostate cancer), and a cell penetrating peptide, HIV Tat(49–57) reaching DNA, were conjugated to $^{99m}Tc/^{177}Lu$-labeled AuNPs and tested on PC3 cancer cells. Internalization, followed by irradiation at 532 nm, resulted in a significant decrease of PC3 viability down to 1.3% (Jiménez-Mancilla et al., 2013).

Nevertheless, the application of somatostatin-based functionalized plasmonic NPs for in vivo purposes needs further attention, especially for avoiding the formation of the so-called protein corona around these particles when they are found in biological fluids (Casals, Pfaller, Duschl, Oostingh, & Puntes, 2010; Walkey et al., 2014). To minimize this effect a number of polymer-coated NPs were elaborated (Khan, Gupta, & Nandi, 2013; Pelaz et al., 2015). Recently, the silver-decorated gold nanorods have been successfully used to deliver specific targeted chemotherapy against both MCF7 and PC3 cell lines. In these experiments, doxorubicin, a commonly used chemotherapy, and an antiepithelial cell adhesion molecule antibodies were covalently bound to thiolated polyethylene glycol (PEG)-coated AuNR\Ag. The resultant system was used to deliver chemotherapy drugs to cancer cells (Nima et al., 2017).

4. MATERIAL AND METHODS

4.1 Peptides

Somatostatin-14 (SST-14, Fig. 1A) was purchased from Sigma-Aldrich (Lyon). Octreotide (Fig. 1B) was synthesized by solid-phase peptide synthesis at the Institut Pasteur (Paris, France), following the Fmoc/*t*Bu methodology. The details of this chemical synthesis were reported previously (Hernández, Coïc, Kruglik, et al., 2012). Pasireotide (Fig. 1C) was a gift from Novartis Pharma (Basel). Purity of the three peptides was checked by RP-HPLC. Respective masses of the peptides were determined by mass spectroscopy as follows: SST-14 monoisotopic $[M+H]^+$ 1637.72 d, observed 1637.76 d (Hernández et al., 2009); octreotide $[M+H]^+$ 1019.45 d, observed 1019.46 d (Hernández, Coïc, Kruglik, et al., 2012); pasireotide monoisotopic $[M+H]^+$ 1047.51 d, observed 1047.73 d (Hernández et al., 2016).

4.2 Sample Preparation

4.2.1 Solution Samples of Peptides

Lyophilized powder of peptides was dissolved in pure water from a Millipore filtration system (Guyancourt). Sodium chloride purchased from Merck (Fontenay-sous-Bois) was added to the solution samples, if necessary. 5 mM stock solutions of SST-14 and octreotide were prepared, i.e., at 8.2 mg/mL (SST-14) and 5.5 mg/mL (octreotide), and further diluted to reach the final concentrations needed for CD, fluorescence, and SERS experiments. In pasireotide, taking into account its water solubility, the maximum concentration giving rise to a limpid solution was 2.5 mM (2.6 mg/mL). This sample was the most concentrated one for recording the bulk Raman spectra; it served as stock solution for other experiments needing lower concentrations.

4.2.2 AgNPs Used as Substrates for Peptide Adsorption

AgNP suspension was elaborated by adding 300 mL of sodium hydroxide (Sigma-Aldrich, Lyon) solution (1 M) to 90 mL of a 6×10^{-2} M hydroxylamine hydrochloride (Sigma-Aldrich, Lyon) solution. Then 10 mL of a 1.11×10^{-3} silver nitrate (Sigma-Aldrich, Lyon) aqueous solution was added dropwise to the mixture under vigorous stirring. The solution containing silver colloids turned to brown. Using this protocol, polydisperse colloids are generally obtained, as evidenced by the TEM images in Fig. 10. AgNPs were aged 1 week at 4°C before use. Upon addition the solution containing a cationic peptide to colloidal solution, the color turns to dark gray, revealing AgNPs aggregation. UV–visible and Raman spectra were recorded just after sample preparation.

4.3 Experimental Setups

4.3.1 Fluorescence

Room temperature fluorescence spectra were recorded on a PTI Quanta-Master QM4CW spectrofluorometer (Lawrenceville, NJ) at 25°C using a quartz cell with 10mm path length. Bandwidths were set to 1 and 5nm for excitation and emission, respectively. Values corresponding to anisotropy measurements were 5 and 15nm, respectively. Fluorescence anisotropy, $\langle r \rangle$, was estimated by the relation: $\langle r \rangle = \frac{I_{VV} - GI_{VH}}{I_{VV} + 2GI_{VH}}$, where the G factor is: $G = \frac{I_{HV}}{I_{HH}}$, and I denotes the polarized intensity corresponding to the vertical (V) and horizontal (H) polarization states (with respect to the plane including the incident and fluorescence beams, also referred to as scattering plane) of the excitation (first subscript) and emission (second subscript) radiations (Lackowicz, 2006).

4.3.2 Circular Dichroism

Room temperature CD spectra were analyzed on a JASCO J-810 spectrophotometer (Lisses) within the 190–300nm spectral region. A path length of 1mm and a spectral resolution of 0.2nm were selected. Each spectrum corresponding to an average of five scans was recorded with a speed of 100nm per min. The measured ellipticity for each sample, referred to as $[\phi]_{obs}$, was further normalized to obtain the so-called mean residue ellipticity, $[\phi]$, by using the expression: $[\Phi] = \frac{[\Phi]_{obs}}{ncl}$, where n, c, and l are the number of residues in the peptide, the molar concentration, and the optical path length of the sample, respectively (Lazo & Downing, 2001). The normalized ellipticity was expressed in deg cm^2 dmol^{-1}.

4.3.3 Ultraviolet–Visible Absorption

UV–visible spectra of silver colloids, alone or in the presence of peptides, were recorded at room temperature on a Shimadzu 3600 UV–visible absorption spectrometer (Duisburg) by means of quartz cells having an optical path of 1 cm.

4.3.4 Transmission Electron Microscopy

TEM images were taken at room temperature using a JEOL JEM-2010 electron microscope (Croissy) with an acceleration voltage of 200 kV. The samples were prepared by depositing 10 μL of the suspension containing either AgNPs or their complexes with the peptides, on carbon-coated Cu grids (ref. G400-Cu).

4.3.5 Bulk Raman Scattering

Room temperature Stokes Raman spectra were analyzed at right angle on a Jobin-Yvon T64000 spectrometer (Longjumeau) at single spectrograph configuration, 1200 grooves/mm holographic grating, and a holographic notch filter. Raman data corresponding to 1200 s acquisition time for each spectrum were collected on a liquid nitrogen cooled CCD detection system (Spectrum One, Jobin-Yvon). The effective slit width was set to 5 cm^{-1}. Solution samples were excited by the 488 nm line of an Ar^+ laser, Spectra Physics (Evry), with 200 mW power at the sample.

4.3.6 Surface-Enhanced Raman Scattering

SERS data were collected on a Renishaw Raman inVia spectrometer (Champs-sur-Marne) equipped with an electrically cooled CCD camera. Samples were excited by means of the 632.8 or 785 nm line of a diode laser with an output power of 18 mW at the sample, and 1200 grooves per nm holographic grating. The spectral resolution was 2 cm^{-1}, with a total acquisition time of 10 s for each SERS measurement.

4.3.7 Spectroscopic Data Posttreatment

Buffer subtraction and smoothing of the observed spectra were performed using the GRAMS/AI Z.00 package (Thermo Galactic, Waltham, MA). The analysis of some composite bands was performed by curve fitting using pseudo-Voigt (Gaussian + Lorentzian) functions, with the Lorentzian contribution kept equal to or greater than 50%. Final presentation of Raman spectra was done by means of SigmaPlot package 6.10 (SPSS Inc., Chicago, IL).

ACKNOWLEDGMENTS

The authors would like to thank Novartis Pharma (Basel, Switzerland) to have generously provided the lyophilized sample of pasireotide. This work was supported by the Spanish *Ministerio de Economía y Competitividad* (Project FIS 2014-52212-R). Quantum mechanical calculations were granted access to the HPC resources of CINES/IDRIS under the allocation 2017-A0010805065 made by GENCI (*Grand Equipement National de Calcul Intensif*).

REFERENCES

Abdellatif, A. A., Zayed, G., El-Bakry, A., Zaky, A., Saleem, I. Y., & Tawfeek, H. M. (2016). Novel gold nanoparticles coated with somatostatin as a potential delivery system for targeting somatostatin receptors. *Drug Development and Industrial Pharmacy*, *42*, 1782–1791.

Agnihotri, S., Mukherji, S., & Mukherji, S. (2014). Size-controlled silver nanoparticles synthesized over the range 5–100 nm using the same protocol and their antibacterial efficacy. *RSC Advances*, *4*, 3974–3983.

Anoop, A., Ranganathan, S., Dhaked, B. D., Jha, N. N., Pratihar, S., Ghosh, S., et al. (2014). Elucidating the role of disulfide bond on amyloid formation and fibril reversibility of somatostatin-14. Relevance to its storage and secretion. *The Journal of Biological Chemistry*, *289*, 16884–16903.

Aslan, K., Lakowicz, J. R., & Geddes, C. D. (2005). Plasmon light scattering in biology and medicine: New sensing approaches, visions and perspectives. *Current Opinion in Chemical Biology*, *9*, 538–544.

Barrio, M., Czernin, J., Fanti, S., Ambrosini, V., Binse, I., Du, L., et al. (2017). The impact of somatostatin receptor-directed PET/CT on the management of patients with neuroendocrine tumor: A systematic review and meta-analysis. *Journal of Nuclear Medicine*, *58*, 756–761.

Bastús, N. G., Comenge, J., & Puntes, V. (2011). Kinetically controlled seeded growth synthesis of citrate-reduced nanoparticles of up to 200 nm: Size focusing versus Ostwald ripening. *Langmuir*, *27*, 11098–11105.

Beschiaschvilli, G., & Seeling, J. (1991). Peptide binding to lipid membranes. Spectroscopic studies on the insertion of a cyclic somatostatin analog into phospholipid bilayers. *Biochimica et Biophysica Acta*, *1061*, 78–84.

Bhattacharyya, K., & Basak, S. (1995). Somatostatin in a water-restricted environment: Fluorescence and circular dichroism study in AOT reverse micelles. *Photochemistry and Photobiology*, *62*, 17–23.

Boukhalfa-Heniche, F. Z., Hernández, B., Gaillard, S., Coïc, Y. M., Huynh-Dinh, T., Lecouvey, M., et al. (2004). Complex formation and vectorization of a phosphorothioate oligonucleotide with an amphipathic leucine- and lysine-rich peptide: Study at molecular and cellular levels. *Biopolymers*, *73*, 727–734.

Bour, P., Kim, J., Kapitan, J., Hammer, R. P., Huang, R., Wu, L., et al. (2008). Vibrational circular dichroism and IR spectral analysis as a test of theoretical conformational modelling for a cyclic hexapeptide. *Chirality*, *20*, 1104–1119.

Brabander, T., van der Zwan, W. A., Teunissen, J. J., Kam, B. L., Feelders, R. A., de Herder, W. W., et al. (2017). Long-term efficacy, survival and safety of [^{177}Lu-DOTA^{0}, Tyr^{3}]octreotate in patients with gastroenteropancreatic and bronchial neuroendocrine tumors. *Clinical Cancer Research*, *23*, 4617–4624.

Brazeau, P., Vale, W., Burgus, R., Ling, N., Butcher, M., Rivier, J., et al. (1973). Hypothalamic polypeptide that inhibits the secretion of immunoreactive pituitary growth hormone. *Science*, *179*, 77–79.

Brown, M., Rivier, J., & Vale, W. (1976). Biological activity of somatostatin and somatostatin analogs on inhibition of arginine-induced insulin and glucagon release in the rat. *Endocrinology*, *98*, 336–343.

Bruns, C., Lewis, I., Briner, U., Meno-Tetang, G., & Weckbecker, G. (2002). SOM230: A novel Somatostatin peptidomimetic with broad somatotropin release inhibiting factor (SRIF) receptor binding and a unique antisecretory profile. *European Journal of Endocrinology*, *146*, 707–716.

Cañamares, M. V., Garcia-Ramos, J. V., Gómez-Varga, J. D., Domingo, C., & Sanchez-Cortes, S. (2005). Comparative study of the morphology, aggregation, adherence to glass, and surface-enhanced Raman scattering activity of silver nanoparticles prepared by chemical reduction of Ag^+ using citrate and hydroxylamine. *Langmuir*, *21*, 8546–8553.

Caplin, M. E., Pavel, M., Cwikla, J. B., Phan, A. T., Raderer, M., Sedláčková, E., et al. (2014). Lanreotide in metastatic enteropancreatic neuroendocrine tumors. *The New England Journal of Medicine*, *371*, 224–233.

Casals, E., Pfaller, T., Duschl, A., Oostingh, G. J., & Puntes, V. (2010). Time evolution of the nanoparticle protein corona. *ACS Nano*, *4*, 3623–3632.

Chatalic, K. L., Kwekkeboom, D. J., & de Jong, M. (2015). Radiopeptides for imaging and therapy: A radiant future. *Journal of Nuclear Medicine*, *56*, 1809–1812.

Chen, J., Ding, W., Luo, Z., Loo, B. H., & Yao, J. (2015). Probing single molecules and molecular aggregates: Raman spectroscopic advances. *Journal of Raman Spectroscopy*, *47*, 623–635.

Das, C., Naganagowda, G. A., Karle, I. L., & Balaram, P. (2001). Designed beta-hairpin peptides with defined tight turn stereochemistry. *Biopolymers*, *58*, 335–346.

Fuchs, P. F., Bonvin, A. M., Bochicchio, B., Pepe, A., Alix, A. J., & Tamburro, A. M. (2006). Kinetics and thermodynamics of type VIII beta-turn formation: A CD, NMR, and microsecond explicit molecular dynamics study of the GDNP tetrapeptide. *Biophysical Journal*, *90*, 2745–2759.

Garrell, R. L., Shaw, K. D., & Krimm, S. (1983). Surface enhanced Raman spectroscopy of halide ions on colloidal silver: Morphology and coverage. *Surface Science*, *124*, 613–624.

Gibbs, A. C., Bjorndahi, T. C., Hodges, R. S., & Wishart, D. S. (2002). Probing the structural determinants of type II' beta-turn formation in peptides and proteins. *Journal of the American Chemical Society*, *124*, 1203–1213.

Guerrini, L., Izquierdo-Lorenzo, I., Garcia-Ramos, J. V., Domingo, C., & Sanchez-Cortes, S. (2009). Self-assembly of α,ω-aliphatic diamines on Ag nanoparticles as an effective localized surface plasmon nanosensor based in interparticle hot spots. *Physical Chemistry Chemical Physics*, *11*, 7363–7371.

Guiffo-Soh, G., Hernández, B., Coïc, Y. M., Boukhalfa-Heniche, F. Z., Fadda, G., & Ghomi, M. (2008). Vibrational analysis of amino acids and short peptides in hydrated media. 3. Successive KL repeats induce highly stable beta-strands capable of forming non-H-bonded aggregates. *The Journal of Physical Chemistry B*, *112*, 1282–1289.

Guiffo-Soh, G., Hernández, B., Coïc, Y. M., Boukhalfa-Heniche, F. Z., & Ghomi, M. (2007). Vibrational analysis of amino acids and short peptides in hydrated media. II. Role of KLLL repeats to induce helical conformations in minimalist LK-peptides. *The Journal of Physical Chemistry B*, *111*, 12563–12572.

Gunasekaran, K., Ramakrishnan, C., & Balaram, P. (1997). Beta-hairpins in proteins revisited: Lessons for de novo design. *Protein Engineering*, *10*, 1131–1141.

Hallenga, K., van Binst, G., Knappenberg, M., Brison, J., Michel, A., & Dirkx, J. (1979). The conformational properties of some fragments of the peptide hormone somatostatin. *Biochimica et Biophysica Acta*, *577*, 82–101.

Hallenga, K., van Binst, G., Scarso, A., Michel, A., Knappenberg, M., Dremier, C., et al. (1980). The conformational properties of the peptide hormone somatostatin (III). Assignment and analysis of the ^1H and ^{13}C high resolution NMR spectra of somatostatin in aqueous solution. *FEBS Letters*, *119*, 47–52.

Hernández, B., Boukhalfa-Heniche, F. Z., Seksek, O., Coïc, Y. M., & Ghomi, M. (2006). Secondary conformation of short lysine- and leucine-rich peptides assessed by optical spectroscopies: Effect of chain length, concentration, solvent, and time. *Biopolymers*, *81*, 8–19.

Hernández, B., Carelli, C., Coïc, Y. M., De Conninck, J., & Ghomi, M. (2009). Vibrational analysis of amino acids and short peptides in aqueous media. V. The effect of the disulfide bridge on the structural features of the peptide hormone somatostatin-14. *The Journal of Physical Chemistry B*, *113*, 12796–12803.

Hernández, B., Coïc, Y. M., Baron, B., Kruglik, S. G., Pflüger, F., Cohen, R., et al. (2014). Low concentration structural dynamics of lanreotide and somatostatin-14. *Biopolymers*, *101*, 1019–1028.
Hernández, B., Coïc, Y. M., Gouyette, C., & Ghomi, M. (2012). Probing the interactions of oligodeoxynucleotides with a cationic peptide by Raman scattering. In M. Ghomi (Ed.), *Applications of Raman spectroscopy to biology. From basic studies to disease diagnosis* (pp. 58–71). Amsterdam, The Netherlands: IOS Press.
Hernández, B., Coïc, Y. M., Kruglik, S. G., Carelli, C., Cohen, R., & Ghomi, M. (2012). Octreotide used for probing the type-II′ β-turn CD and Raman markers. *The Journal of Physical Chemistry B*, *116*, 9337–9345.
Hernández, B., Coïc, Y. M., Pflüger, F., Kruglik, S. G., & Ghomi, M. (2015). All characteristic Raman markers of tyrosine and tyrosinate originate from phenol ring fundamental vibrations. *Journal of Raman Spectroscopy*, *47*, 210–220.
Hernández, B., López-Tobar, E., Sanchez-Cortes, S., Coïc, Y. M., Baron, B., Chenal, A., et al. (2016). From bulk to plasmonic nanoparticle surfaces: The behaviour of two potent therapeutic peptides, octreotide and pasireotide. *Physical Chemistry Chemical Physics*, *18*, 24437–24450.
Hernández, B., Pflüger, F., Adenier, A., Kruglik, S. G., & Ghomi, M. (2010). Vibrational analysis of amino acids and short peptides in hydrated media. VIII. Amino acids with aromatic side chains: L-phenylalanine, L-tyrosine, and L-tryptophan. *The Journal of Physical Chemistry B*, *114*, 15319–15330.
Hernández, B., Pflüger, F., Kruglik, S. G., & Ghomi, M. (2013). Characteristic Raman lines of phenylalanine analyzed by a multiconformational analysis. *Journal of Raman Spectroscopy*, *44*, 827–833.
Hernández, B., Pflüger, F., López-Tobar, E., Kruglik, S. G., Garcia-Ramos, J. V., Sanchez-Cortes, S., et al. (2014). Disulfide linkage Raman markers: A reconsideration attempt. *Journal of Raman Spectroscopy*, *45*, 657–664.
Hogg, P. J. (2013). Targeting allosteric disulphide bonds in cancer. *Nature Reviews. Cancer*, *13*, 425–431.
Holladay, L. A., & Puett, D. (1976). Somatostatin conformation: Evidence for a stable intramolecular structure from circular dichroism, diffusion, and sedimentation equilibrium. *Proceedings of the National Academy of Sciences of the United States of America*, *73*, 1199–1202.
Holladay, L. A., Rivier, J., & Puett, D. (1977). Conformational studies on somatostatin and analogues. *Biochemistry*, *16*, 4895–4900.
Hsieh, H., Chang, C. W., & Cho, H. H. (2013). Gold nanoparticles as amyloid fibrillogenesis inhibitors. *Colloids and Surfaces, B: Biointerfaces*, *112*, 525–529.
Huang, X., El-Sayed, I. H., Qian, W., & El-Sayed, M. A. (2006). Cancer cell imaging and photothermal therapy in the near-infrared region by using gold nanorods. *Journal of the American Chemical Society*, *128*, 2115–2120.
Hutter, E., & Fendler, J. H. (2004). Exploitation of localized surface plasmon resonance. *Advanced Materials*, *16*, 1685–1706.
Interlandi, G. (2009). Backbone conformations and side chain flexibility of two somatostatin mimics investigated by molecular dynamics simulations. *Proteins*, *75*, 659–670.
Jain, P. K., El-Sayed, I. H., & El-Sayed, M. A. (2007). Au nanoparticles target cancer. *Nano Today*, *2*, 18–29.
Jiménez-Mancilla, N., Ferro-Flores, G., Santos-Cuevas, C., Ocampo-García, B., Luna-Gutiérrez, M., Azorin-Vega, E., et al. (2013). Multifunctional targeted therapy system based on ^{99m}Tc/^{177}Lu-labeled gold nanoparticles-Tat(49–57)-Lys^{3}-bombesin internalized in nuclei of prostate cancer cells. *Journal of Labelled Compounds and Radiopharmaceuticals*, *56*, 663–671.
Karimi, M., Zare, H., Bakhshian Nik, A., Yazdani, N., Hamrang, M., Mohamed, E., et al. (2016). Nanotechnology in diagnosis and treatment of coronary artery disease. *Nanomedicine*, *11*, 513–530.

Kaur, H., & Raghava, G. P. S. (2004). A neural network method for prediction of β-turn types in proteins using evolutionary information. *Bioinformatics*, *20*, 2751–2758.

Kelly, S. M., & Price, N. C. (2000). The use of circular dichroism in the investigation of protein structure and function. *Current Protein & Peptide Science*, *1*, 349–384.

Khan, S., Gupta, A., & Nandi, C. K. (2013). Controlling the fate of protein corona by tuning surface properties of nanoparticles. *Journal of Physical Chemistry Letters*, *4*, 3747–3752.

Kharlamov, A. N., Tyurnina, A. E., Veselova, V. S., Kovtun, O. P., Shur, V. Y., & Gabinsky, J. L. (2015). Silica-gold nanoparticles for atheroprotective management of plaques: Results of the NANOM-FIM trial. *Nanoscale*, *7*, 8003–8015.

Knappenberg, M., Michel, A., Scarso, A., Brison, J., Zanen, J., Hallenga, K., et al. (1982). The conformational properties of somatostatin. IV. The conformers contributing to the conformation equilibrium of somatostatin in aqueous solution as found by semi-empirical energy calculations and high-resolution NMR experiments. *Biochimica et Biophysica Acta*, *18*, 229–246.

Lackowicz, J. R., Principles of fluorescence spectroscopy (2006). (3rd ed.). New York: Springer.

Lazo, N. D., & Downing, D. T. (2001). Effects of Na_2SO_4 on hydrophobic and electrostatic interactions between amphipathic α-helices. *The Journal of Peptide Research*, *58*, 457–463.

Lee, K. S., & El-Sayed, M. A. (2006). Gold and silver nanoparticles in sensing and imaging: Sensitivity of plasmon response to size, shape, and metal composition. *The Journal of Physical Chemistry*. *B*, *110*, 19220–19225.

Lewis, L., Albert, R., Bauer, W., Chandramouli, N., Pless, J., Oberer, L., et al. (2004). The superior therapeutic properties of SOM230 originate from unique structural elements. *CHIMIA*, *58*, 222–227.

Li, Y., Jing, C., Zhang, L., & Long, Y. T. (2012). Resonance scattering particles as biological nanosensors *in vitro* and *in vivo*. *Chemical Society Reviews*, *41*, 632–642.

Linse, S., Cabaleiro-Lago, C., Xue, W. F., Lynch, I., Lindman, S., Thulin, E., et al. (2007). Nucleation of protein fibrillation by nanoparticles. *Proceedings of the National Academy of Sciences of the United States of America*, *104*, 8691–8696.

Liz-Marzán, L. M. (2004). Nanometals formation and color. *Materials Today*, *7*, 26–31.

López-Tobar, E., Hernández, B., Ghomi, M., Gómez, J., Chenal, A., Garcia Ramos, J. V., et al. (2015). Anchoring sites of fibrillogenic peptide hormone somatostatin-14 on plasmonic nanoparticles. *Journal of Physical Chemistry C*, *119*, 8273–8279.

López-Tobar, E., Hernández, B., Ghomi, M., & Sanchez-Cortes, S. (2013). Stability of the disulfide bond in cystine adsorbed on silver and gold nanoparticles as evidenced by SERS data. *Journal of Physical Chemistry C*, *117*, 1531–1537.

Ma, Q., Wei, G., & Yang, X. (2013). Influence of Au nanoparticles on the aggregation of amyloid-β-(25 – 35) peptides. *Nanoscale*, *5*, 10397–10403.

Mahalakshimi, R., Shanmugam, G., Polavarapu, P. L., & Balaram, P. (2005). Circular dichroism of designed peptide helices and β-hairpins: Analysis of Trp- and Tyr-rich peptides. *Chembiochem*, *6*, 2152–2158.

Maji, S. K., Perrin, M. H., Sawaya, M. R., Jessberger, S., Vadodaria, K., Rissman, R. A., et al. (2009). Functional amyloids as natural storage of peptide hormones in pituitary secretory granules. *Science*, *325*, 328–332.

McKeage, K. (2013). Pasireotide: A review of its use in Cushing's disease. *Drugs*, *73*, 563–574.

Melacini, G., Zhu, Q., & Goodman, M. (1997). Multiconformational NMR analysis of sandostatin (octreotide): Equilibrium between beta-sheet and partially helical structures. *Biochemistry*, *36*, 1233–1241.

Mendoza-Nava, H., Ferro-Flores, G., Ocampo-García, B., Serment-Guerrero, J., Santos-Cuevas, C., Jiménez-Mancilla, N., et al. (2013). Laser heating of gold nanospheres functionalized with octreotide: In vitro effect on HeLa cell viability. *Photomedicine and Laser Surgery*, *31*, 17–22.

Merola, E., Panzuto, F., & Delle Fave, G. (2017). Antiproliferative effect of somatostatin analogs in advanced gastro-entero-pancreatic neuroendocrine tumors: A systematic review and meta-analysis. *Oncotarget*, *8*, 46624–46634.

Milner-White, E. J., Ross, B. M., Ismail, R., Belhadj-Mostefa, K., & Poet, R. (1988). One type of gamma-turn, rather than the other gives rise to chain-reversal in proteins. *Journal of Molecular Biology*, *204*, 777–782.

Modlin, I. M., Pavel, M., Kidd, M., & Gustafsson, B. I. (2010). Somatostatin analogues in the treatment of gastroenteropancreatic neuroendocrine (carcinoid) tumours. *Alimentary Pharmacology & Therapeutics*, *31*, 169–188.

Mohamed, A., Blanchard, M. P., Albertelli, M., Barbieri, F., Brue, T., Niccoli, P., et al. (2014). Pasireotide and octreotide antiproliferative effects and sst2 trafficking in human pancreatic neuroendocrine tumor cultures. *Endocrine-Related Cancer*, *21*, 691–704.

Motta, A., Reches, M. R., Pappalardo, L., Andretti, G., & Gazit, E. (2005). The preferred conformation of tripeptide Ala-Phe-Ala in water is an inverse γ-turn: Implications for protein folding and drug design. *Biochemistry*, *44*, 14170–14178.

Nergiz, S. Z., Slocik, J. M., Naik, R. R., & Singamaneni, S. (2013). Surface defect sites facilitate fibrillation: An insight into adsorption of gold-binding peptides on Au (111). *Physical Chemistry Chemical Physics*, *15*, 11629–11633.

Nima, Z. A., Alwbari, A. M., Dantuluri, V., Hamzah, R. N., Sra, N., Motwani, P., et al. (2017). Targeting nano drug delivery to cancer cells using tunable, multi-layer, silver-decorated gold nanorods. *Journal of Applied Toxicology*, *37*, 1370–1378.

Otting, G., Liepinsh, E., & Wüthrich, K. (1991). Protein hydration in aqueous solution. *Science*, *254*, 974–980.

Paige Hall, W., Anker, J. N., Lin, Y., Modica, J., Mrksich, M., & van Duyne, R. P. (2008). A calcium-modulated plasmon switch. *Journal of the American Chemical Society*, *130*, 5836–5837.

Pawlikowski, M., & Meleń-Mucha, G. (2004). Somatostatin analogs—From new molecules to new applications. *Current Opinion in Pharmacology*, *4*, 608–613.

Pelaz, B., del Pino, P., Maffre, P., Hartmann, R., Gallego, M., Rivera-Fernández, S., et al. (2015). Surface functionalization of nanoparticles with polyethyleneglycol: Effects on protein adsorption and cellular uptake. *ACS Nano*, *9*, 6996–7008.

Perczel, A., & Fasman, G. D. (1992). Quantitative analysis of cyclic beta-turn models. *Protein Science*, *1*, 378–395.

Pohl, E., Heine, A., Sheldrick, G. M., Dauter, Z., Wilson, K. S., Kallen, J., et al. (1995). Structure of octreotide, a somatostatin analogue. *Acta Crystallographica*, *D51*, 48–59.

Qian, R., Cao, Y., & Long, Y. T. (2016). Dual-targeting nanovesicles for in situ intracellular imaging of and discrimination between wild-type and mutant p53. *Angewandte Chemie International Edition*, *55*, 719–723.

Reubi, J., Waser, B., Schaer, J. C., & Laissue, J. A. (2001). Somatostatin receptor sst1–sst5 expression in normal and neoplastic human tissues using receptor autoradiography with subtype-selective ligands. *European Journal of Nuclear Medicine*, *28*, 836–846.

Rinke, A., Muller, H. H., Schade-Brittinger, C., Klose, K. J., Barth, P., Wied, M., et al. (2009). Placebo controlled, double-blind, prospective, randomized study on the effect of octreotide LAR in the control of tumor growth in patients with metastatic neuroendocrine midgut tumors: A report from the PROMID Study Group. *Journal of Clinical Oncology*, *27*, 4656–4663.

Sanchez-Cortes, S., Garcia-Ramos, J. V., Morcillo, G., & Tinti, A. (1995). Morphological study of silver colloids employed in surface- enhanced Raman spectroscopy: Activation when exciting in visible and near-infrared regions. *Journal of Colloid and Interface Science, 175*, 358–368.
Schmidt, B., Ho, L., & Hogg, P. J. (2006). Allosteric disulfide bonds. *Biochemistry, 45*, 7429–7433.
Schmidt, B., & Hogg, P. J. (2007). Search for allosteric disulfide bonds in NMR structures. *BMC Structural Biology*, 7, 49.
Sevilla, P., Lajos, G., de Llanos, R., Sanchez-Cortes, S., Jancura, D., Miškovský, P., et al. (2012). Surface-enhanced fluorescence and Raman spectroscopy (SEF and SERS) of anthraquinone anti tumoral drugs and their complex with biomolecules on Ag nanoparticles. In M. Ghomi (Ed.), *Applications of Raman spectroscopy to biology. From basic studies to disease diagnosis* (pp. 31–57). Amsterdam, The Netherlands: IOS Press.
Shekhter Zahavi, T., Oron-Herman, M., Kostenich, G., Rub, E., Salitra, Y., Buzhansky, L., et al. (2017). Molecular engineering of somatostatin analogue with minimal dipeptide motif induces the formation of functional nanoparticles. *ChemNanoMat, 3*, 27–32.
Shenashen, M. A., El-Safty, S. A., & Elshehy, E. A. (2013). Synthesis, morphological, and properties of silver nanoparticles in potential applications. *Particle and Particle Systems Characterization, 31*, 293–316.
Sibanda, B. L., & Thornton, J. M. (1985). β-Hairpin families in globular proteins. *Nature, 316*, 170–174.
Skaat, H., Shafir, G., & Margel, S. (2011). Acceleration of amyloid-β fibril formation by peptide-conjugated fluorescent-maghemite nanoparticles. *Journal of Nanoparticle Research, 13*, 3521–3534.
Sugeta, H., Go, A., & Miyazawa, T. (1972). S-S and C-S stretching vibrations and molecular conformations of dialkyl disulfides and cystine. *Chemistry Letters, 1*, 83–86.
Vale, W., Grant, G., Amoss, M., Blackwell, R., & Guillemin, R. (1972). Culture of enzymatically dispersed pituitary cells: Functional validation of a method. *Endocrinology, 91*, 562–572.
van Grondelle, W., Lecomte, S., Lopez-Iglesias, C., Manero, J. M., Cherif-Cheikh, R., Paternostre, M., et al. (2013). Lamination and spherulite-like compaction of a hormone's native amyloid-like nanofibrils: Spectroscopic insights into key interactions. *Faraday Discussions, 166*, 163–180.
van Grondelle, W., Lopez-Iglesias, C., Coll, E., Artzner, F., Paternostre, M., Lacombe, F., et al. (2007). Spontaneous fibrillation of the native neuropeptide hormone Somatostatin-14. *Journal of Structural Biology, 160*, 211–213.
Van Wart, H. E., & Scheraga, H. A. (1976). Raman spectra of strained disulfides. Effect of rotation about sulfur-sulfur bonds on sulfur-sulfur stretching frequencies. *The Journal of Physical Chemistry, 80*, 1823–1832.
Van Wart, H. E., & Scheraga, H. A. (1986). Agreement with the disulfide stretching frequency-conformation correlation of Sugeta, Go, and Miyazawa. *Proceedings of the National Academy of Sciences of the United States of America, 83*, 3064–3067.
Walkey, C. D., Olsen, J. B., Song, F., Liu, R., Guo, H., Olsen, D. W. H., et al. (2014). Protein corona fingerprinting predicts the cellular interaction of gold and silver nanoparticles. *ACS Nano, 8*, 2439–2455.
Wang, L., Tang, K., Zhang, Q., Li, H., Wen, Z., Zhang, H., et al. (2013). Somatostatin receptor-based molecular imaging and therapy for neuroendocrine tumors. *BioMed Research International, 2013*, 102819.
Weckbecker, G., Lewis, I., Albert, R., Schmid, H. A., Hoyer, D., & Bruns, C. (2003). Opportunities in somatostatin research: Biological, chemical and therapeutic aspects. *Nature Reviews. Drug Discovery, 2*, 999–1017.

Yao, J. C., Hassan, M., Phan, A., Dagohoy, C., Leary, C., Mares, J. E., et al. (2008). One hundred years after "carcinoid": Epidemiology of and prognostic factors for neuroendocrine tumors in 35,825 cases in the United States. *Journal of Clinical Oncology*, *26*, 3063–3072.

Yoo, S. I., Yang, M., Brender, J. R., Subramanian, V., Sun, K., Joo, N. E., et al. (2011). Inhibition of amyloid peptide fibrillation by inorganic nanoparticle: Functional similarities with proteins. *Angewandte Chemie*, *50*, 5110–5115.

Zhong, D., Pal, S. K., & Zewail, A. H. (2011). Biological water: A critique. *Chemical Physics Letters*, *503*, 1–11.

CHAPTER FOUR

Transglutaminase and Sialyltransferase Enzymatic Approaches for Polymer Conjugation to Proteins

Katia Maso*, Antonella Grigoletto*, Gianfranco Pasut*,†,1
*University of Padua, Padua, Italy
†Veneto Institute of Oncology IOV—IRCCS, Padua, Italy
[1]Corresponding author: e-mail address: gianfranco.pasut@unipd.it

Contents

Abstract

Proteins hold a central role in medicine and biology, also confirmed by the several therapeutic applications based on biologic drugs. Such therapies are of great relevance thanks to high potency and safety of proteins. Nevertheless, many proteins as therapeutics might present issues like fast kidney clearance, rapid enzymatic degradation, or immunogenicity. Such defects implicate frequent administrations or administrations at high doses of the therapeutics, thus yielding or exacerbating potential side effects. A successful technology for improving the clinical profiles of proteins is the conjugation of polymers to the protein surface. The design of a protein–polymer conjugate presents critical aspects that determine the efficacy and safety of the final product. The control over stoichiometry and conjugation site is a strict criterion on which researchers have been intensively focused during the years, in order to obtain homogeneous and batch-to-batch reproducible products. An innovative site-specific conjugation strategy relies on the use of enzymes as tools to mediate polymer conjugation. Enzymatic approaches are attractive because they allow site-selective polymer conjugation at specific protein amino acids. In these reactions, the polymer is a substrate analog that replaces the native substrate. Furthermore, enzymes can count other advantages such as high yields

Advances in Protein Chemistry and Structural Biology, Volume 112
ISSN 1876-1623
https://doi.org/10.1016/bs.apcsb.2018.01.003

of conversion and physiological conditions of reaction. This chapter provides a meaningful description of protein–polymer conjugation through transglutaminase-mediated and sialyltransferase-mediated enzymatic strategies, reporting the mechanism of action and some relevant examples.

1. POLYMER CONJUGATION TO PROTEINS

During the past decades, peptides and proteins have resulted in attractive targets for biological and pharmaceutical research and many of them, produced in large quantities by recombinant DNA techniques, have become therapies of great importance even for disease that cannot be treated with conventional low-molecular-weight drugs (Leader, Baca, & Golan, 2008; Usmani et al., 2017). Among the several benefits, these entities are well tolerated, safe, and characterized by high selectivity and potency. Despite their therapeutic relevance, their use in therapy is often restricted due to considerable limitations and/or poor performance. Peptides and proteins typically might have a rapid body clearance, immunogenicity, physical and chemical instabilities, and might be sensitive to enzymatic degradation (Petrak & Goddard, 1989). In particular, a short plasma half-life, common for proteins having a molecular weight below the kidney excretion threshold, is a significant issue because is reflected in frequent administrations of therapeutics or in administrations at high doses to reach the desired therapeutic effect, but on the other side aggravating side effects (Leonard et al., 1997). Several tools for improving the clinical properties of proteins can be exploited, such as genetic engineering (Platis & Labrou, 2008), advanced protein formulation, and delivery (Fosgerau & Hoffmann, 2015; Jorgensen & Nielsen, 2009). Drug delivery systems (DDSs) are particularly promising in improving the in vivo efficacy of active molecules as they can prolong the in vivo half-life and protect the drug from enzymatic degradation, thus enhancing the in vivo bioactivity. Various drug delivery approaches have been proposed and investigated, including incorporation into micro/nanoparticles, liposomes, or hydrogels and chemical modification with small molecules or polymers. Among others, conjugation of polymers to proteins surface has been a successful technology, which has produced several molecules available in the market (Duncan & Vicent, 2013; Grigoletto, Maso, et al., 2016; Pasut, 2014). Either natural or synthetic polymers have been tested for this purpose, but poly(ethylene glycol) (PEG) has proven so far to be the gold standard, thanks to its unique combination of properties, ranging from biocompatibility to easy chemistry of activation. PEG is, therefore, widely used for the delivery of therapeutic proteins, peptides, small drugs, and oligonucleotides, as well as for surface modification of other DDSs such as liposomes, nanoparticles, and micelles (Turecek, Bossard, Schoetens, & Ivens, 2016). PEGylated proteins generally preserve unchanged secondary and tertiary structures; however, a reduction in biological activity of the conjugate is a common drawback due to the steric hindrance of polymer chains during the interaction between the protein and its ligand (receptor or substrate). Despite the reduced in vitro activity, PEGylated proteins can achieve a better in vivo performance, compared to their unmodified proteins, thanks to the prolonged pharmacokinetic profiles (Ozer & Chilkoti, 2017).

Nevertheless, besides PEG, various natural polysaccharides and synthetic polymers have been also investigated as candidates for protein modification, including hydroxyl-ethyl starch (Besheer, Hertel, Kressler, Mäder, & Pietzsch, 2009), polysialic acid (PSA) (Greco et al., 2013; Gregoriadis, Jain, Papaioannou, & Laing, 2005), hyaluronic acid (Mero et al., 2015, 2014; Mero, Pasqualin, Campisi, Renier, & Pasut, 2013), dextrin (Hardwicke et al., 2008), polyoxazoline (Mero et al., 2008; Viegas et al., 2011). When developing a protein–polymer conjugate, two key points must be considered and have a great impact on the final conjugate stability and efficacy: (I) the conjugation site and (II) the number of polymer chains conjugated per protein unit. When multiple sites in a protein are coupled to polymer chains, the final product is a heterogeneous mixture formed by different isomers, as in the case of PEG-asparaginase and PEG-adenosine deaminase and PEGylated-interferon-α presently in clinical use (Pasut & Veronese, 2012). Product characterization and batch-to-batch reproducibility have been limiting issues for the final drug safety since different isomers can display different pharmacokinetics, pharmacodynamics, and potential side effects. Over the years, new therapeutics have been required to meet more and more strict criteria in terms of homogeneity and reproducibility, in order to improve the quality and get a better control over the manufacturing process. These needs have led researchers to direct their efforts on the development of new site-specific conjugation technologies. Therefore, in the second generation of PEGylated proteins, a number of different chemical and enzymatic coupling strategies have been explored to achieve control over stoichiometry and conjugation site. Chemical examples of site-specific PEGylation are: N-terminal PEGylation (Neulasta®), thiol PEGylation (Cimzia®), bridging PEGylation, and insertion of unnatural amino acids to exploit bioorthogonal click chemistries (Balan et al., 2007; Brocchini et al., 2008; Chapman, 2002; Cornish, Hahn, & Schultz, 1996; Deiters, Cropp, Summerer, Mukherji, & Schultz, 2004; Goel & Stephens, 2010; Kinstler, Molineux, Treuheit, Ladd, & Gegg, 2002; Molineux, 2004; Pasut & Veronese, 2012; Schumacher et al., 2011; Shaunak et al., 2006).

Thanks to the advancement in biotechnology and in genetic engineering, enzymes have emerged as valuable biochemical tools for the posttranslational modification of proteins. Enzymatic methods have attracted much more attention because they allow site-selective modification at specific amino acid substrates in the target protein. In the field of protein–polymer conjugation, enzymes catalyze the reaction between the protein of interest and a polymer that acts like a substrate analog and contains a proper bioorthogonal functional group. Promiscuous enzymes can tolerate substrate analogs with specific chemical groups that are recognized by the catalytic site. Usually, enzymatic techniques count of higher yields without side reactions, with the further advantage that the reactions generally occur under mild conditions (aqueous buffers, physiological pH, and temperature), preserving protein structure and conformation.

In this chapter, transglutaminase-mediated and sialyltransferase-mediated enzymatic approaches for protein–polymer conjugation will be described, reporting the most meaningful examples from literature.

2. ENZYMES AS TOOLS FOR POLYMER CONJUGATION TO PROTEINS

2.1 Transglutaminase

Transglutaminases (TGases) or protein-glutamine γ-glutamyltransferases are a family of enzymes present in several organisms including mammals, invertebrates, plants, and microorganisms (Lorand & Conrad, 1984; Mariniello & Porta, 2005). Human TGases are distributed in most tissues and body fluids like the liver, the epidermis, the prostate, the hair follicles, and the blood. They are involved in a variety of physiological functions such as extracellular matrix remodeling, cell differentiation, blood coagulation, tumor growth, and apoptosis (Griffin, Casadio, & Bergamini, 2002). TGases catalyze cross-linking reaction between proteins by transfer the acyl moiety of the γ-carboxamide group of a protein-bound glutamine residue (acyl donor) to a variety of linear primary amines (acyl acceptors), in particular the ε-amino group of a protein-bound lysine residue (Folk & Finlayson, 1977), thus forming an ε-(γ-glutamyl)lysine isopeptide bond. Usually only one or few glutamine residues, among all those present in a globular protein, are substrates of TGases, while this specificity is lower for acyl acceptors (Lorand et al., 1979), although in the case of protein-bound lysines there is still a certain degree of specificity (Grigoletto, Mero, Yoshioka, Schiavon, & Pasut, 2017). The specificity for the recognition of certain glutamine residues respect all the other present in a protein is characterized by a structure prerequisite. Indeed, glutamine residues that are substrates for the TGase have to be contained in flexible or disordered portions of the protein (Spolaore et al., 2012), allowing the interaction with the catalytic site of the enzyme. TGases are very useful and versatile tools in biochemical experiments of protein modification, because they can be used to tag a glutamine group of a protein with several chemical entities (including chemical moieties, fluorescent probes, other proteins, oligosaccharides, or polymers), which comprise an amino group (Ikura, Yoshikawa, Sasaki, & Chiba, 1981). One of the most used microbial TGase (mTGase) derives from a variant of *Streptomyces mobaraense*. mTGase has a molecular weight of 37.9kDa and the crystal structure is available (Kashiwagi et al., 2002). Contrarily to human TGase, the microbial one is calcium independent and has a higher reaction rate, characteristics that have favored its use as a biochemical tool and for industrial purposes, for example, in the food industry. mTGase has a compact domain, and, in the catalytic site, the key residue for the activity is the Cys64, positioned at the bottom of a deep cleft. The thiol group of Cys64 reacts with the γ-carboxamide group of a Gln residue of the substrate forming a reactive thioacyl moiety that subsequently reacts with an acyl acceptor (amino group), forming a new isopeptide amide bond. mTGase can incorporate various chemical entities and/or functional groups at protein-bound glutamine or also lysine residues (Grigoletto, Mero, Yoshioka, et al., 2017; Spolaore et al., 2016), on the basis of the features of the substrate used (i.e., an amino-containing molecule to label Gln residues in a protein or a Gln-containing molecule to label Lys residues in a protein) (Fig. 1). Such flexibility has placed mTGase at the basis for the development of a variety of strategies and approaches for proteins modification, waking the interests from basic studies to many biotechnological applications (Yokoyama, Nio, & Kikuchi, 2004).

In the field of protein PEGylation, through a mTGase-mediated reaction, an amino-PEG can be used as an acyl acceptor to covalently attach the polymer to a protein of therapeutic interest. In literature, there are many studies of therapeutic proteins and peptides PEGylated by the mTGase approach. For example, in the case of apomyoglobin (apoMb), only Gln91 has been successfully modified with PEG, despite the presence of

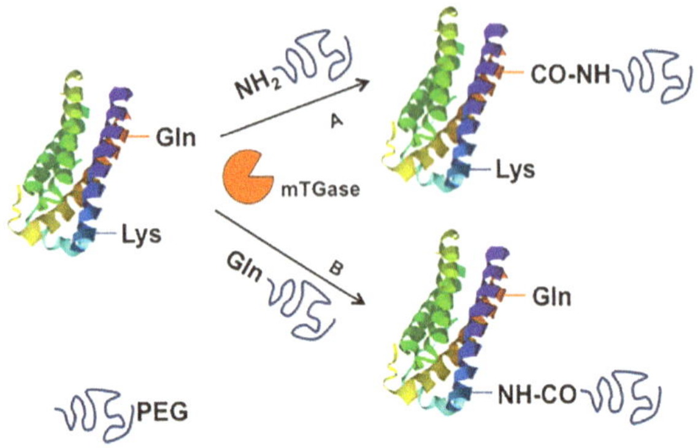

Fig. 1 mTGase-mediated PEGylation of a protein in the presence of an NH_2-PEG (A) or a Gln-containing PEG (B). mTGase can incorporate PEG chains at a Gln protein residue, when an amino-containing PEG is used and at a Lys protein residue in the case when a Gln-containing PEG is used.

6 Gln residues, while hGH, that presents up to 13 Gln residues, was preferentially PEGylated only at Gln40 and Gln141 (Fontana, Spolaore, Mero, & Veronese, 2008). A study directed to examine the selectivity of mTGase has revealed that an important requirement for the acyl donor substrate of mTGase is not only the surface exposure of the Gln residues, but especially the highly conformational flexibility of the peptide sequence containing such glutamine substrate (Fontana et al., 2008). The fact that mTGase-mediated PEGylation of apoMb and hIL-2 led to the obtainment of homogeneous monoPEGylated derivatives is due to the presence of a single glutamine inserted in a very flexible peptide backbone for such proteins. Differently, for hGH, the advantage of the selectivity was lost due to the presence of two Gln residues suitable as acyl donor substrates. Interestingly, it has been demonstrated that by tuning the flexibility of the protein backbone it is possible to control the number of mTGase glutamine substrates. The specificity of mTGase was increased by performing the catalysis in water/organic cosolvents reaction buffers, conditions that induced an increase in the secondary structure of the protein substrates (Grigoletto, Mero, Maso, & Pasut, 2017; Grigoletto, Mero, Zanusso, Schiavon, & Pasut, 2016; Mero, Schiavon, Veronese, & Pasut, 2011). Pasut's lab has successfully tested this approach with hGH and salmon calcitonin (sCT), both having two glutamine substrates of mTGase in physiological buffers (Mero et al., 2011). When the catalysis was conducted in buffers containing up to 50% (v/v) of ethanol, only a single conjugate isomer was obtained instead of

a complex mixture of two positional monoconjugate isomers and the biconjugate (Da Silva Freitas, Mero, & Pasut, 2013). The organic solvent reduced the flexibility in some regions of the target protein by increasing the amount of secondary structure, thus abolishing the main prerequisite of backbone flexibility for an acyl donor. Furthermore, the authors have also demonstrated that mTGase has a great versatility by performing its activity even in a solution containing organic cosolvent.

As reported earlier, mTGase can incorporate PEG chains at the side chain of Gln when an amino-containing PEG is used but, recently, also Lys residues were investigated as possible sites of PEGylation mediated by mTGase. Pasut's lab applied such approach with two proteins: interferon-α (IFN-α) and granulocyte colony-stimulating factor (G-CSF). In the work of IFN-α, two homogeneous monoconjugate isomers were formed with good yields, one at the level of Gln101 and the other at Lys164 (Spolaore et al., 2016). For the conjugation to Gln101 a PEG 20-kDa–NH_2 was used, while to tag the Lys164 residue a PEG derivative containing the *N*-benzyloxycarbonyl-L-glutaminylglycine (ZQG) was used. Among the 12 Gln and 10 Lys residues of IFN-α, only Gln101 and Lys164 were located in highly flexible protein regions and then available for mTGase-mediated derivatization. Both conjugates have maintained the characteristic secondary structure of unmodified IFN-α, retained a good in vitro antiviral activity and shown a prolonged pharmacokinetic profile with respect to native IFN-α. This work has demonstrated that also PEGylation of a pharmaceutical protein at one Lys residue is achievable by using mTGase, which still confirmed its high selectivity. In the case of G-CSF, a mono-PEGylated form of the protein at Lys41 was obtained by mTGase in the presence of PEG-ZQG (Mero et al., 2016). PEG-K41-G-CSF was compared, in term of biophysical and pharmacokinetic studies, to the isomer PEG-Q135-G-CSF, obtained through mTGase conjugation of PEG-NH_2 to Gln135 of G-CSF. The study demonstrated that the isomers have similar in vivo behaviors. Overall these findings have highlighted a further potentiality of mTGase in protein derivatization offering the possibility to target either lysines or glutamines, in both cases with a high site selectivity (Fig. 1).

An alternative method for the obtainment of homogeneous monoconjugates with mTGase was recently proposed by Grigoletto and coworkers by the selective immobilization of the enzyme on solid beads of an inert polysaccharide resin (Grigoletto, Mero, Yoshioka, et al., 2017). The immobilization was selectively directed to the N-terminal residue of mTGase, in order to avoid cross-linking and to minimize the loss of mTGase activity because the catalytic site is far from the N-terminus. mTGase immobilization allowed both the simply enzyme removal from the reaction mixture and the improvement of its stability. The immobilized mTGase (iTGase) was characterized in terms of kinetic parameters, showing an expected reduction of substrate affinity that was counterbalanced by the increase of enzyme stability at different pH values and high temperatures. Second, iTGase was tested in a protein PEGylation study in order to evaluate if the enzyme would maintain the substrate specificity. Two proteins, alpha-lactoalbumin (α-LA) and G-CSF, were PEGylated with iTGase or the free enzyme. In the case of α-LA, when the conjugation was carried out using a PEG-NH_2 and free mTGase, a mixture of mono- and biconjugates was obtained but, interestingly, when the same reaction was performed with iTGase only a monoconjugate derivative was formed (Fig. 2). In the case of G-CSF, iTGase preserved its specificity to form a single monoconjugate derivative but, advantageously, it avoided the undesired formation of the deamidated Q135E G-CSF, which took place when free mTGase was used thus reducing the yield of the conjugation reaction.

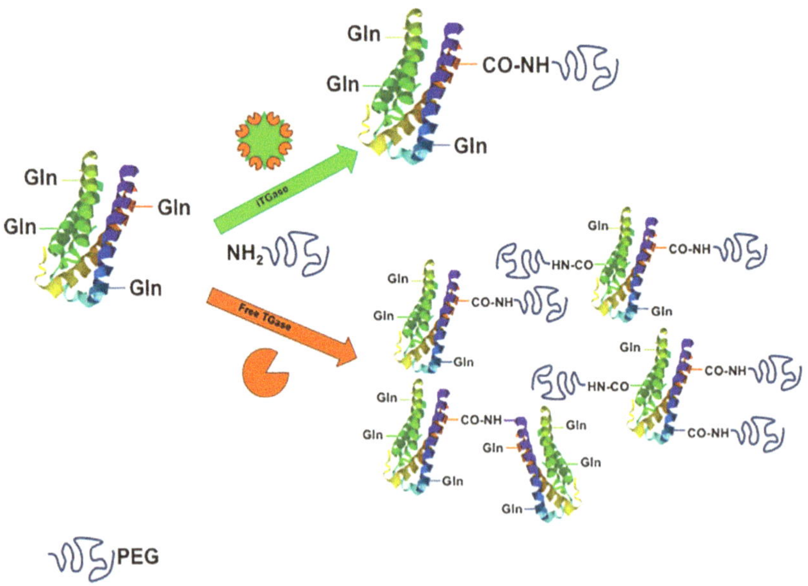

Fig. 2 PEGylation of α-LA in the presence of NH_2-PEG and iTGase (A) or free mTGase (B). PEGylation of α-LA with free mTGase produces a mixture of mono-, bi- and triconjugates and cross-linked proteins, while iTGase yields only a monoconjugate derivative.

It is worth to note that although most of the studies have been based on the use of PEG as conjugating polymer, the applicability of mTGase is not limited to this polymer, but it has been also investigated with other polymers such poly(2-ethyl 2-oxazoline) (Mero, Fang, Pasut, Veronese, & Viegas, 2012).

mTGase has been exploited also for other applications in the field of protein modification. Wakabayashi and collaborators have developed a mTGase enzymatic method to synthesize a one-dimensional protein assembly on a synthetic polymer for biosensing purposes (Wakabayashi, Yahiro, Hayashi, Goto, & Kamiya, 2017). A hybrid entity, composed of multiple proteins and one polymer chain, was constructed with the aim to gain a multivalent effect, as required in biosensing techniques. The challenge was to design a scaffold with a precise control over the conjugation sites, differently from what could be achieved with synthetic random methods. First, 1-hydroxyethyl methacrylate monomer was designed with a triethylene glycol linker and modified with the ZQG dipeptide. After polymerization with acrylamide, the resulted polymer chain was grafted with multiple mTGase substrates with a mole fraction of 4 mol%. A fusion protein composed of an antibody Fc-binding domain (PG) and an alkaline phosphatase enzyme (IPP) was selected to propose a possible application of sensing biomolecules in immunological technology. The IPP–PG fusion protein was genetically modified at the C-terminus, by inserting K-tags substrates of mTGase, to be assembled with the polymer described earlier. Finally, through mTGase-mediated conjugation, the hybrid construct was obtained with an average of 2.1 IPP–PG per polymer chain. The site-specific conjugation allowed the maintenance of the protein functions. The conjugate was used as secondary antibody for the detection of the antigen ovalbumin (OVA) in ELISA and Western blotting, compared to the free K-tag–IPP–PG. In the former case, with the conjugate, the detected signal was increased of about 2.1-fold respect to the detection with free fusion protein, while, for the blotting, a minor amount of OVA was visualized using the IPP–PG–ZQG polymer conjugate. Overall, although the detection limit was not significantly changed, the conjugate system has shown an improved amplification of the signal thanks to the protein clustering effect.

Concluding, all these studies have demonstrated that mTGase can be an important tool for the site-specific modification and polymer conjugation of proteins, offering different opportunities based on its high selectivity and the possibility to precisely tag Gln or Lys residues.

2.2 Sialyltransferases

Glycosylation is one of the major co- or posttranslational modifications of proteins, playing an important role in their stability, folding, and molecular recognition (Rich & Withers, 2009). It is estimated that more than 50% of human proteins are glycosylated. Glycosylation occurs predominantly at the amide nitrogen of asparagine (*N*-glycosylation) and at the hydroxyl group of serine or threonine (*O*-glycosylation) and is performed by a family of glycosyltransferases that shows specificity for both the donor nucleotide-activated sugar and the glycan acceptor site (Paulson & Colley, 1989).

Based on the type of sugar undergoing the transfer, these enzymes are grouped into subfamilies, such as sialyltransferases and galactosyltransferases, which have virtually no sequence homology. Multiple glycosyltransferases (usually 100 or more) are necessary to produce the final carbohydrate chain made of different monosaccharide units. Sialyltransferases are transmembrane glycoproteins located in the Golgi apparatus. They are composed of a short cytoplasmic tail at the N-terminus, a 16–20 amino acids transmembrane domain and a highly variable stem region (20–200 residues) followed by a large C-terminus that is responsible for the catalytic activity (Paulson & Colley, 1989) and faces the luminal side. Despite this general structure, human sialyltransferase families share only few regions of homology that are limited to three consensus sequences called sialylmotifs L (large), S (small), and VS (very small) (Harduin-Lepers et al., 2001). L and S motifs are involved in the binding of the substrates: L motif interacts with the sugar donor, while S motif interacts with the donor and the acceptor. Both these motifs contain a conserved cysteine residue that plays an important role in the catalytic activity of the enzyme. VS motif is located in the C-terminal region and comprehends two conserved residues, a glutamine separated by four amino acids from a histidine. Also, VS motif is believed to be involved in the catalysis. Dimension variations between sialyltransferases are attributed to differences in the length of the stem region that is very flexible because it lacks a rigorous secondary structure organization (Audry et al., 2011). When stem region undergoes proteolytic cleavage, the membrane-bound sialyltransferase is converted in a soluble active form that is secreted in body fluids. Bacterial sialyltransferases are usually promiscuous and accept more substrates compared to the eukaryotic enzymes; moreover, some of them are multifunctional, a feature that helps pathogenic bacteria during host invasion.

Chemoenzymatic remodeling of certain sugar residues, using enzymes involved in protein glycosylation, can be exploited to obtain site-selective

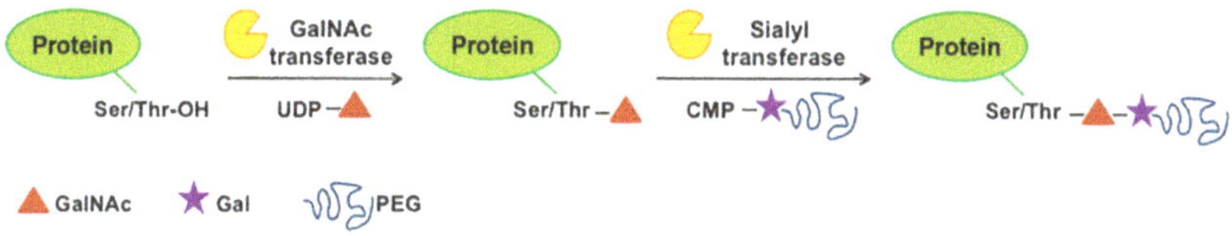

Fig. 3 Enzyme-mediated glycoPEGylation of a nonglycosylated recombinant protein produced in *E. coli*. In the first reaction, a Ser or Thr residue serves as an acceptor for the attachment of GalNAc by GalNAc transferase, creating an *O*-glycosylation site. In the second reaction, a sialyltransferase transfers a PEGylated sialic acid to the new introduced *O*-glycan.

bioconjugation. Following this pathway, in 2004 Neose Technologies patented a new enzymatic PEGylation strategy, termed GlycoPEGylation, which consists in the covalent conjugation of a PEGylated nucleotide sugar to the carbohydrate moiety of a protein (DeFrees et al., 2004; Fig. 3). This strategy relies on the ability of sialyltransferases to attach a modified sialic acid PEGylated at the 5′-amino position in the cytidine monophosphate (CMP) nucleotide (CMP-SA-5-NHCOCH$_2$NHPEG) to an *O*-glycan acceptor site. Since therapeutic proteins produced in *Escherichia coli* lack the glycosylation pattern, an additional step is required to enzymatically introduce the *O*-glycosylation at serine or threonine residues. For this step, it was used GalNAc transferase to selectively link the *N*-acetylgalactosamine (GalNAc) moiety into proteins such as G-CSF and recombinant interferon-alpha2b (IFN-α 2b) expressed in bacteria culture (DeFrees et al., 2006). UDP-GalNAc-polypeptide *N*-acetylgalactosaminyltransferases (ppGaNTases) are a family of up to 20 isoforms of glycosyltransferases that transfer GalNAc from the sugar donor UDP-GalNAc to serine and threonine residues (Ten Hagen, Fritz, & Tabak, 2003). Each isoform of this wide family possesses unique substrate specificities and achieves site-directed *O*-glycosylation of proteins. In their human native form, both G-CSF and IFN-α2b are glycosylated: G-CSF at Thr133 and IFN-α2b at Thr106. Site-specific GalNAc glycosylation of these Thr residues was performed by GalNAc-T2 isoform, then sialyltransferase (ST6GalNAc-I) was used to tether a CMP-sialic acid-PEG 20kDa derivative to the new *O*-glycosylated site (Fig. 3). GlycoPEGylation occurred exclusively to the glycosylated residues of G-CSF and IFN-α2b, introducing a single PEG chain and giving homogenous products with a good preservation of protein bioactivity and the desired prolongation of protein circulation time in blood.

The glycoPEGylated G-CSF, called lipegfilgrastim, has been then developed by Teva Biopharmaceutics under the brand name of Lonquez® and was approved by the European Medicines Agency in July 2013 for reduction in the duration of neutropenia and the incidence of febrile neutropenia in patients treated with cytotoxic chemotherapy for malignancy, with the exception of chronic myeloid leukemia and myelodysplastic syndromes (Guariglia et al., 2016). Neose Technology reported also the glycoPEGylation of granulocyte/macrophage colony-stimulating factor (GM-CSF) that, as native protein, presents multiple sites of glycosylation: Ser5, Ser7, Ser9, and/or Thr10. In this case, GalNAc-T2 incorporated two GalNAc residues, with traces of one and three residues. Ser7 and Ser9 were found to be the major sites. They could not be both glycoPEGylated by ST6GalNAc-I probably because of their close proximity. To overcome such issue, an additional sugar residue was added as "spacer" to each GalNAc residue in order to reduce the steric hindrance of PEG. Core1 GalT enzyme added a galactose to 3-*O*-position of GalNAc and the ST3Gal-1 enzyme was then employed to transfer the sialic acid-PEG derivative to the 3-*O*-position of galactose. This approach based on galactose spacers allowed the obtainment of conjugate isomer mixture composed by at least 75% of di-PEGylated GM-CSF.

Another example of glycoPEGylation of a native glycoprotein was reported by Østergaard and coworkers in which a sialic acid-branched PEG 40 kDa (PEG_2 40 kDa) derivative was conjugated to *N*-glycans of factor IX (FIX), using a three-step procedure (Østergaard et al., 2011). First, sialic acid residues of FIX were removed by a sialidase in order to expose galactosyl moieties. Then, FIX was enzymatically conjugated to CMP-sialic acid-PEG_2 40 kDa by incubation in the presence of sialyltransferase ST3Gal-III that tethered the polymer to terminal galactosyl units of the glycans. This reaction is time dependent and the number of PEG moieties added can be controlled by adjusting the incubation time (Giorgi, Agusti, & De Lederkremer, 2014). Finally, the unreacted galactoses were capped with sialic acid, by exploiting the enzymatic reaction above reported but using the substrate CMP-sialic acid, to prevent hepatic elimination through the asialoglycoprotein receptor. FIX is vitamin K dependent and participates in the intrinsic coagulation pathway where is converted from the zymogen to the activated form by tissue factor (TF)–factor VIIa (FVIIa) complex or factor XIa (FXIa) (Di Scipio, Kurachi, & Davie, 1978). Physiologic activation of FIX (MW 57 kDa) consists in the limited proteolysis at Arg145 and Arg180 liberating a 35-amino acid activation peptide (11 kDa) that carries the 2 *N*-linked glycans. The activated protein is composed of a light chain (MW 18 kDa) and a heavy chain (MW 25 kDa) that are held together by a disulfide bond. It is important that PEGylation does not impair the activation process of FIX. Interestingly, the activation of glycoPEGylated FIX followed the same pathway of FIX with the release of the native activated FIX (FIXa), which retained its enzymatic activity, whereas PEG was removed together with the activation peptide.

Another meaningful example of glycoPEGylation of a coagulation factor is N8-GP, a new recombinant factor VIII (rFVIII) site specifically conjugated to PEG at the *O*-glycan on B domain of rFVIII (Stennicke et al., 2013). Turoctocog alfa rFVIII has been developed by Novo Nordisk (Bagsværd, Denmark) in CHO cells and has a truncated B domain of 21 amino acids (Thim et al., 2010). N8-GP was obtained by glycoPEGylation in which the terminal sialic acid on the *O*-glycan is replaced by a sialic acid attached to a PEG_2 40kDa. When N8-GP is activated by thrombin, the truncated B domain is removed, together with the polymer, releasing the active FVIII (FVIIIa) that has the same primary structure of the native protein (Fig. 4). Turoctocog alfa contains also four *N*-glycosylation sites, potentially exploitable for selective glycoPEGylation; however, *O*-glycosylation site was selected for the reason stated above that this site is removed during FVIII activation. In the first step of turoctocog alfa modification, all sialic acids were removed from the glycans through a sialidase, then, ST3Gal-1 sialic acid transferase was used to selectively tether a PEG-modified sialic acid onto the *O*-glycan (Stennicke et al., 2013). It was not necessary to remove the sialidase from the reaction solution before performing polymer

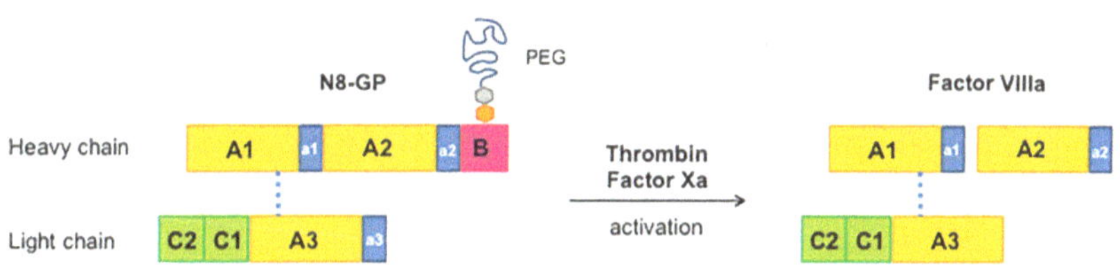

Fig. 4 Factor VIII circulates in blood as a noncovalent heterodimer composed of a heavy chain (A1, A2, and B domains) and a light chain (A3, C1, and C2 domains) that are kept together by a divalent metal bridge (*dotted line*). The lower case "a" designates short segments connecting the domains: a1 and a2 follow A1 and A2 domains, respectively, while a3 precedes A3 domain. After proteolytic cleavage mediated by thrombin or factor Xa, B domain is released and a conformational change occurs in the C2 domain. Activated FVIII is a heterotrimer in which A1 and A3 domains retain the metal ion bond and the A2 domain is associated to the A1 domain through electrostatic interactions. In this figure, a schematic representation of N8-GP before and after activation is reported. N8 (turoctocog) is a recombinant form of factor VIII having a reduced B domain, 21 amino acids. In N8-GP, a 40kDa PEG is linked to the *O*-glycan of B domain. After activation, N8-GP has the same primary structure of native FVIII.

conjugation since the enzyme does not release the polymer from the *O*-glycan. Finally, the remaining nonsialylated glycans were capped by adding an excess of CMP-sialic acid and the ST3Gal-III enzyme. In vivo studies have shown that N8-GP has an increased time of circulation in blood and a prolonged duration of the effect (Agersø et al., 2012; Tiede et al., 2013).

Interestingly, Wakarchuk and coworkers demonstrated the application of bacterial sialyltransferases to generate polysialylated proteins, transferring PSA chains to natural *N*-linked glycans (Lindhout et al., 2011). Polysialylation of therapeutics was first proposed by Gregoriadis in 1993 as an alternative to PEG to improve drugs pharmacokinetic and reduce proteins immunogenicity (Gregoriadis, Fernandes, Mital, & McCormack, 2000; Gregoriadis, McCormack, Wang, & Lifely, 1993). PSAs are highly hydrophilic, biodegradable polymers composed of *N*-acetylneuraminic acid (sialic acid) units arranged in a linear structure, and are naturally exploited by several bacteria during infection to achieve protection against the host immune system (Gregoriadis et al., 2005). A variety of chemistries have been used to generate polysialylated proteins that were shown to preserve their biological activity and present reduced immunogenicity (Fernandes & Gregoriadis, 1996, 1997, 2001; Gregoriadis et al., 2000; Jain et al., 2003). Enzymatic polysialylation, described by Wakarchuk and coworkers in 2011, provides a useful tool for site-selective protein conjugation. PSA conjugation to three glycoproteins with different *N*-linked glycan compositions (the human therapeutic proteins alpha-1-antitrypsin (A1AT), factor IX, and bovine fetuin) was performed by using a bacterial polysialyltransferase from *Neisseria meningitides*. To exploit its function, this enzyme requires a specific motif constituted by at least two sialic acid residues on protein glycans, which in these cases were introduced incubating the proteins with Cst-II, a bifunctional α2,3/α2,8 sialyltransferase from *Campylobacter jejuni*. Polysialylation reaction could be controlled adjusting the concentration of the acceptor protein: at low concentrations, a long PSA chains (more than 70 sialic acids residues) were attached to proteins, while at high concentrations the length of PSA chains were shorter.

In summary glycotransferases site specifically tether polymers at glycosylation sites giving well-defined products with enhanced pharmacokinetic properties determined by the amount and size of conjugated polymers. Perturbation of protein structure is limited by the fact that polymer attachment happens at a naturally modified site. The approval of glycoPEGylated recombinant G-CSF in the European market has proven that this strategy can offer an attractive and functional way to create clinically relevant therapeutics.

3. DISCUSSION AND OUTLOOK

After almost 50 years of continuous developments, polymer conjugation to proteins has clearly proved the validity of its initial promises, especially in the prolongation of pharmacokinetic and reduction of immunogenicity; advantages that have generated many clinically approved biologics. The driving force for the recent advances in the field has been dictated by the need to foster new site-selective conjugation methods. Among the several perfect suits approaches, like, for example, the introduction of unnatural amino acids in the protein sequence (Wang, Brock, Herberich, & Schultz, 2001), the strategy to use enzymes as tools to mediate polymer conjugation to proteins is particularly relevant. Enzymes have been exploited for their selectivity, and furthermore, they work in physiological conditions, thus better preserving the target protein. Enzymatic polymer conjugation can be designed on the basis of specific requirements, and, usually, the strategy follows two directions: (i) to mimic the Nature's posttranslational modification, as above shown for sialyltransferases that tether a polymer at the natural glycosylation sites of proteins, thus copying the glycosylation process, or (ii) to achieve polymer conjugation to specific amino acids that cannot be selectively modified by chemical methods, as in the case of TGases that are able to label protein's glutamines, which are considered unreactive sites.

Enzymes have brought a new wave of innovation in polymer conjugation to proteins, and their contributions will generate a new class of well-defined products with the first candidates already entered in the clinical practice.

ACKNOWLEDGMENTS

The authors were supported by Italian Ministry of Health research programme "Ricerca Finalizzata" (GR-2011-02351128).

REFERENCES

Agersø, H., Stennicke, H. R., Pelzer, H., Olsen, E. N., Merricks, E. P., Defriess, N. A., et al. (2012). Pharmacokinetics and pharmacodynamics of turoctocog alfa and N8-GP in haemophilia A dogs. *Haemophilia*, *18*(6), 941–947. https://doi.org/10.1111/j.1365-2516.2012.02896.x.

Audry, M., Jeanneau, C., Imberty, A., Harduin-Lepers, A., Delannoy, P., & Breton, C. (2011). Current trends in the structure-activity relationships of sialyltransferases. *Glycobiology*, *21*(6), 716–726. https://doi.org/10.1093/glycob/cwq189.

Balan, S., Choi, J. W., Godwin, A., Teo, I., Laborde, C. M., Heidelberger, S., et al. (2007). Site-specific PEGylation of protein bisulfide bonds using a three-carbon bridge. *Bioconjugate Chemistry*, *18*(1), 61–76. https://doi.org/10.1021/bc0601471.

Besheer, A., Hertel, T. C., Kressler, J., Mäder, K., & Pietzsch, M. (2009). Enzymatically catalyzed HES conjugation using microbial transglutaminase: Proof of feasibility. *Journal of Pharmaceutical Sciences*, *98*(11), 4420–4428. https://doi.org/10.1002/jps.21675.

Brocchini, S., Godwin, A., Balan, S., Choi, J. W., Zloh, M., & Shaunak, S. (2008). Disulfide bridge based PEGylation of proteins. *Advanced Drug Delivery Reviews*, *60*(1), 3–12. https://doi.org/10.1016/j.addr.2007.06.014.

Chapman, A. P. (2002). PEGylated antibodies and antibody fragments for improved therapy: A review. *Advanced Drug Delivery Reviews*, *54*(4), 531–545. https://doi.org/10.1016/S0169-409X(02)00026-1.

Cornish, V. W., Hahn, K. M., & Schultz, P. G. (1996). Site-specific protein modification using a ketone handle. *Journal of the American Chemical Society*, *118*(34), 8150–8151. https://doi.org/10.1021/ja961216x.

Da Silva Freitas, D., Mero, A., & Pasut, G. (2013). Chemical and enzymatic site specific pegylation of hGH. *Bioconjugate Chemistry*, *24*(3), 456–463. https://doi.org/10.1021/bc300594y.

DeFrees, S., Wang, Z. G., Xing, R., Scott, A. E., Wang, J., Zopf, D., et al. (2006). GlycoPEGylation of recombinant therapeutic proteins produced in Escherichia coli. *Glycobiology*, *16*(9), 833–843. https://doi.org/10.1093/glycob/cwl004.

DeFrees, S., Zopf, D., Bayer, R., Bowe, C., Hakes, D., & Chen, X. (2004). *Glycopegylation methods and proteins/peptides produced by the methods*. US20040132640.

Deiters, A., Cropp, T. A., Summerer, D., Mukherji, M., & Schultz, P. G. (2004). Site-specific PEGylation of proteins containing unnatural amino acids. *Bioorganic and Medicinal Chemistry Letters*, *14*(23), 5743–5745. https://doi.org/10.1016/j.bmcl.2004.09.059.

Di Scipio, R. G., Kurachi, K., & Davie, E. W. (1978). Activation of human factor IX (Christmas factor). *The Journal of Clinical Investigation*, *61*(6), 1528–1538. https://doi.org/10.1172/JCI109073.

Duncan, R., & Vicent, M. J. (2013). Polymer therapeutics-prospects for 21st century: The end of the beginning. *Advanced Drug Delivery Reviews*, *65*(1), 60–70. https://doi.org/10.1016/j.addr.2012.08.012.

Fernandes, A. I., & Gregoriadis, G. (1996). Synthesis, characterization and properties of sialylated catalase. *Biochimica et Biophysica Acta—Protein Structure and Molecular Enzymology*, *1293*(1), 90–96. https://doi.org/10.1016/0167-4838(95)00227-8.

Fernandes, A. I., & Gregoriadis, G. (1997). Polysialylated asparaginase: Preparation, activity and pharmacokinetics. *Biochimica et Biophysica Acta—Protein Structure and Molecular Enzymology*, *1341*(1), 26–34. https://doi.org/10.1016/S0167-4838(97)00056-3.

Fernandes, A. I., & Gregoriadis, G. (2001). The effect of polysialylation on the immunogenicity and antigenicity of asparaginase: Implication in its pharmacokinetics. *International Journal of Pharmaceutics*, *217*(1–2), 215–224. https://doi.org/10.1016/S0378-5173(01)00603-2.

Folk, J. E., & Finlayson, J. S. (1977). The ε-(γ-glutamyl)lysine crosslink and the catalytic role of transglutaminases. *Advances in Protein Chemistry*, *31*(C), 1–133. https://doi.org/10.1016/S0065-3233(08)60217-X.

Fontana, A., Spolaore, B., Mero, A., & Veronese, F. M. (2008). Site-specific modification and PEGylation of pharmaceutical proteins mediated by transglutaminase. *Advanced Drug Delivery Reviews*, *60*(1), 13–28. https://doi.org/10.1016/j.addr.2007.06.015.

Fosgerau, K., & Hoffmann, T. (2015). Peptide therapeutics: Current status and future directions. *Drug Discovery Today*, *20*(1), 122–128. https://doi.org/10.1016/j.drudis.2014.10.003.

Giorgi, M. E., Agusti, R., & De Lederkremer, R. M. (2014). Carbohydrate PEGylation, an approach to improve pharmacological potency. *Beilstein Journal of Organic Chemistry, 10*, 1433–1444. https://doi.org/10.3762/bjoc.10.147.

Goel, N., & Stephens, S. (2010). Certolizumab pegol. *MAbs, 2*(2), 137–147. https://doi.org/10.4161/mabs.2.2.11271.

Greco, F., Arif, I., Botting, R., Fante, C., Quintieri, L., Clementi, C., et al. (2013). Polysialic acid as a drug carrier: Evaluation of a new polysialic acid–epirubicin conjugate and its comparison against established drug carriers. *Polymer Chemistry, 4*(5), 1600. https://doi.org/10.1039/c2py20876h.

Gregoriadis, G., Fernandes, A., Mital, M., & McCormack, B. (2000). Polysialic acids: Potential in improving the stability and pharmacokinetics of proteins and other therapeutics. *Cellular and Molecular Life Sciences, 57*(13–14), 1964–1969. https://doi.org/10.1007/PL00000676.

Gregoriadis, G., Jain, S., Papaioannou, I., & Laing, P. (2005). Improving the therapeutic efficacy of peptides and proteins: A role for polysialic acids. *International Journal of Pharmaceutics, 300*(1–2), 125–130. https://doi.org/10.1016/j.ijpharm.2005.06.007.

Gregoriadis, G., McCormack, B., Wang, Z., & Lifely, R. (1993). Polysialic acids: Potential in drug delivery. *FEBS Letters, 315*(3), 271–276. https://doi.org/10.1016/0014-5793(93)81177-2.

Griffin, M., Casadio, R., & Bergamini, C. M. (2002). Transglutaminases: Nature's biological glues. *Biochemical Journal, 368*(2), 377–396. https://doi.org/10.1042/bj20021234.

Grigoletto, A., Maso, K., Mero, A., Rosato, A., Schiavon, O., & Pasut, G. (2016). Drug and protein delivery by polymer conjugation. *Journal of Drug Delivery Science and Technology, 32*, 132–141. https://doi.org/10.1016/j.jddst.2015.08.006.

Grigoletto, A., Mero, A., Maso, K., & Pasut, G. (2017). *Transgultaminase-mediated nanoarmoring of enzymes by PEGylation. In C. V. Kumar (Ed.), Methods in enzymology* (pp. 317–346). Oxford: Elsevier. https://doi.org/10.1016/BS.MIE.2017.01.002.

Grigoletto, A., Mero, A., Yoshioka, H., Schiavon, O., & Pasut, G. (2017). Covalent immobilisation of transglutaminase: Stability and applications in protein PEGylation. *Journal of Drug Targeting, 25*(9–10), 856–864. https://doi.org/10.1080/1061186X.2017.1363211.

Grigoletto, A., Mero, A., Zanusso, I., Schiavon, O., & Pasut, G. (2016). Chemical and enzymatic site specific PEGylation of hGH: The stability and in vivo activity of PEG-N-terminal-hGH and PEG-Gln141-hGH conjugates. *Macromolecular Bioscience, 16*(1), 50–56. https://doi.org/10.1002/mabi.201500282.

Guariglia, R., Martorelli, M. C., Lerose, R., Telesca, D., Milella, M. R., & Musto, P. (2016). Lipegfilgrastim in the management of chemotherapy-induced neutropenia of cancer patients. *Biologics, 10*, 1–8. https://doi.org/10.2147/BTT.S58597.

Harduin-Lepers, A., Vallejo-Ruiz, V., Krzewinski-Recchi, M. A., Samyn-Petit, B., Julien, S., & Delannoy, P. (2001). The human sialyltransferase family. *Biochimie, 83*(8), 727–737. https://doi.org/10.1016/S0300-9084(01)01301-3.

Hardwicke, J., Ferguson, E. L., Moseley, R., Stephens, P., Thomas, D. W., & Duncan, R. (2008). Dextrin-rhEGF conjugates as bioresponsive nanomedicines for wound repair. *Journal of Controlled Release, 130*(3), 275–283. https://doi.org/10.1016/j.jconrel.2008.07.023.

Ikura, K., Yoshikawa, M., Sasaki, R., & Chiba, H. (1981). Incorporation of amino acids into food proteins by transglutaminase. *Agricultural and Biological Chemistry, 45*(11), 2587–2592. https://doi.org/10.1080/00021369.1981.10864924.

Jain, S., Hreczuk-Hirst, D. H., McCormack, B., Mital, M., Epenetos, A., Laing, P., et al. (2003). Polysialylated insulin: Synthesis, characterization and biological activity in vivo. *Biochimica et Biophysica Acta, 1622*(1), 42–49. https://doi.org/10.1016/S0304-4165(03)00116-8.

Jorgensen, L., & Nielsen, H. M. (2009). *Delivery technologies for biopharmaceuticals: Peptides, proteins, nucleic acids and vaccines*. West Sussex: John Wiley & Sons, Ltd. https://doi.org/10.1002/9780470688397.

Kashiwagi, T., Yokoyama, K., Ishikawa, K., Ono, K., Ejima, D., Matsui, H., et al. (2002). Crystal structure of microbial transglutaminase from Streptoverticillium mobaraense. *Journal of Biological Chemistry*, *277*(46), 44252–44260. https://doi.org/10.1074/jbc.M203933200.

Kinstler, O., Molineux, G., Treuheit, M., Ladd, D., & Gegg, C. (2002). Mono-N-terminal poly(ethylene glycol)-protein conjugates. *Advanced Drug Delivery Reviews*, *54*(4), 477–485. https://doi.org/10.1016/S0169-409X(02)00023-6.

Leader, B., Baca, Q. J., & Golan, D. E. (2008). Protein therapeutics: A summary and pharmacological classification. *Nature Reviews Drug Discovery*, 7(1), 21–39. https://doi.org/10.1038/nrd2399.

Leonard, J. P., Sherman, M. L., Fisher, G. L., Buchanan, L. J., Larsen, G., Atkins, M. B., et al. (1997). Effects of single-dose interleukin-12 exposure on interleukin-12-associated toxicity and interferon-gamma production. *Blood*, *90*(7), 2541–2548. https://doi.org/10.1182/blood-2015-01-625004.

Lindhout, T., Iqbal, U., Willis, L. M., Reid, A. N., Li, J., Liu, X., et al. (2011). Site-specific enzymatic polysialylation of therapeutic proteins using bacterial enzymes. *Proceedings of the National Academy of Sciences of the United States of America*, *108*(18), 7397–7402. https://doi.org/10.1073/pnas.1019266108.

Lorand, L., & Conrad, S. M. (1984). Transglutaminases. *Molecular and Cellular Biochemistry*, *58*(1–2), 9–35. https://doi.org/10.1007/bf00240602.

Lorand, L., Parameswaran, K. N., Stenberg, P., Tong, Y. S., Velasco, P. T., Jonsson, N., et al. (1979). Specificity of guinea pig liver transglutaminase for amine substrates. *Biochemistry*, *18*(9), 1756–1765. https://doi.org/10.1021/bi00576a019.

Mariniello, L., & Porta, R. (2005). Transglutaminases as biotechnological tools. *Progress in Experimental Tumor Research*, *38*, 174–191. https://doi.org/10.1159/000084240.

Mero, A., Campisi, M., Caputo, M., Cuppari, C., Rosato, A., Schiavon, O., et al. (2015). Hyaluronic acid as a protein polymeric carrier: An overview and a report on human growth hormone. *Current Drug Targets*, *16*(13), 1503–1511. https://doi.org/10.2174/1389450116666150107151906.

Mero, A., Campisi, M., Favero, M., Barbera, C., Secchieri, C., Dayer, J. M., et al. (2014). A hyaluronic acid-salmon calcitonin conjugate for the local treatment of osteoarthritis: Chondro-protective effect in a rabbit model of early OA. *Journal of Controlled Release*, *187*, 30–38. https://doi.org/10.1016/j.jconrel.2014.05.008.

Mero, A., Fang, Z., Pasut, G., Veronese, F. M., & Viegas, T. X. (2012). Selective conjugation of poly(2-ethyl 2-oxazoline) to granulocyte colony stimulating factor. *Journal of Controlled Release*, *159*(3), 353–361. https://doi.org/10.1016/j.jconrel.2012.02.025.

Mero, A., Grigoletto, A., Maso, K., Yoshioka, H., Rosato, A., & Pasut, G. (2016). Site-selective enzymatic chemistry for polymer conjugation to protein lysine residues: PEGylation of G-CSF at lysine-41. *Polymer Chemistry*, 7(42), 6545–6553. https://doi.org/10.1039/C6PY01616B.

Mero, A., Pasqualin, M., Campisi, M., Renier, D., & Pasut, G. (2013). Conjugation of hyaluronan to proteins. *Carbohydrate Polymers*, *92*(2), 2163–2170. https://doi.org/10.1016/j.carbpol.2012.11.090.

Mero, A., Pasut, G., Via, L. D., Fijten, M. W. M., Schubert, U. S., Hoogenboom, R., et al. (2008). Synthesis and characterization of poly(2-ethyl 2-oxazoline)-conjugates with proteins and drugs: Suitable alternatives to PEG-conjugates? *Journal of Controlled Release*, *125*(2), 87–95. https://doi.org/10.1016/j.jconrel.2007.10.010.

Mero, A., Schiavon, M., Veronese, F. M., & Pasut, G. (2011). A new method to increase selectivity of transglutaminase mediated PEGylation of salmon calcitonin and human growth hormone. *Journal of Controlled Release*, *154*(1), 27–34. https://doi.org/10.1016/j.jconrel.2011.04.024.

Molineux, G. (2004). The design and development of pegfilgrastim (PEG-rmetHuG-CSF, Neulasta). *Current Pharmaceutical Design*, *10*(11), 1235–1244. https://doi.org/10.2174/1381612043452613.

Østergaard, H., Bjelke, J. R., Hansen, L., Petersen, L. C., Pedersen, A. A., Elm, T., et al. (2011). Prolonged half-life and preserved enzymatic properties of factor IX selectively PEGylated on native N-glycans in the activation peptide. *Blood*, *118*(8), 2333–2341. https://doi.org/10.1182/blood-2011-02-336172.

Ozer, I., & Chilkoti, A. (2017). Site-specific and stoichiometric stealth polymer conjugates of therapeutic peptides and proteins. *Bioconjugate Chemistry*, *28*(3), 713–723. https://doi.org/10.1021/acs.bioconjchem.6b00652.

Pasut, G. (2014). Polymers for protein conjugation. *Polymer*, *6*(1), 160–170. https://doi.org/10.3390/polym6010160.

Pasut, G., & Veronese, F. M. (2012). State of the art in PEGylation: The great versatility achieved after forty years of research. *Journal of Controlled Release*, *161*(2), 461–472. https://doi.org/10.1016/j.jconrel.2011.10.037.

Paulson, J. C., & Colley, K. J. (1989). Glycosyltransferases. Structure, localization, and control of cell type-specific glycosylation. *Journal of Biological Chemistry*, *264*(30), 17615–17618.

Petrak, K., & Goddard, P. (1989). Transport of macromolecules across the capillary walls. *Advanced Drug Delivery Reviews*, *3*(2), 191–214. https://doi.org/10.1016/0169-409X(89)90010-0.

Platis, D., & Labrou, N. E. (2008). Chemical and genetic engineering strategies to improve the potency of pharmaceutical proteins and enzymes. *Current Medicinal Chemistry*, *15*(19), 1940–1955. https://doi.org/10.2174/092986708785132924.

Rich, J. R., & Withers, S. G. (2009). Emerging methods for the production of homogeneous human glycoproteins. *Nature Chemical Biology*, *5*(4), 206–215. https://doi.org/10.1038/nchembio.148.

Schumacher, F. F., Nobles, M., Ryan, C. P., Smith, M. E. B., Tinker, A., Caddick, S., et al. (2011). In situ maleimide bridging of disulfides and a new approach to protein PEGylation. *Bioconjugate Chemistry*, *22*(2), 132–136. https://doi.org/10.1021/bc1004685.

Shaunak, S., Godwin, A., Choi, J. W., Balan, S., Pedone, E., Vijayarangam, D., et al. (2006). Site-specific PEGylation of native disulfide bonds in therapeutic proteins. *Nature Chemical Biology*, *2*(6), 312–313. https://doi.org/10.1038/nchembio786.

Spolaore, B., Raboni, S., Ramos Molina, A., Satwekar, A., Damiano, N., & Fontana, A. (2012). Local unfolding is required for the site-specific protein modification by transglutaminase. *Biochemistry*, *51*(43), 8679–8689. https://doi.org/10.1021/bi301005z.

Spolaore, B., Raboni, S., Satwekar, A. A., Grigoletto, A., Mero, A., Montagner, I. M., et al. (2016). Site-specific transglutaminase-mediated conjugation of interferon α-2b at glutamine or lysine residues. *Bioconjugate Chemistry*, *27*(11), 2695–2706. https://doi.org/10.1021/acs.bioconjchem.6b00468.

Stennicke, H. R., Kjalke, M., Karpf, D. M., Balling, K. W., Johansen, P. B., Elm, T., et al. (2013). A novel B-domain O-glycoPEGylated FVIII (N8-GP) demonstrates full efficacy and prolonged effect in hemophilic mice models. *Blood*, *121*(11), 2108–2116. https://doi.org/10.1182/blood-2012-01-407494.

Ten Hagen, K. G., Fritz, T. A., & Tabak, L. A. (2003). All in the family: The UDP-GalNAc: polypeptide N-acetylgalactosaminyltransferases. *Glycobiology*, *13*(1), 1R–16R. https://doi.org/10.1093/glycob/cwg007.

Thim, L., Vandahl, B., Karlsson, J., Klausen, N. K., Pedersen, J., Krogh, T. N., et al. (2010). Purification and characterization of a new recombinant factor VIII (N8). *Haemophilia, 16*(2), 349–359. https://doi.org/10.1111/j.1365-2516.2009.02135.x.

Tiede, A., Brand, B., Fischer, R., Kavakli, K., Lentz, S. R., Matsushita, T., et al. (2013). Enhancing the pharmacokinetic properties of recombinant factor VIII: First-in-human trial of glycoPEGylated recombinant factor VIII in patients with hemophilia A. *Journal of Thrombosis and Haemostasis, 11*(4), 670–678. https://doi.org/10.1111/jth.12161.

Turecek, P. L., Bossard, M. J., Schoetens, F., & Ivens, I. A. (2016). PEGylation of biopharmaceuticals: A review of chemistry and nonclinical safety information of approved drugs. *Journal of Pharmaceutical Sciences, 105*(2), 460–475. https://doi.org/10.1016/j.xphs.2015.11.015.

Usmani, S. S., Bedi, G., Samuel, J. S., Singh, S., Kalra, S., Kumar, P., et al. (2017). THPdb: Database of FDA-approved peptide and protein therapeutics. *PLoS One, 12*(7), e0181748. https://doi.org/10.1371/journal.pone.0181748.

Viegas, T. X., Bentley, M. D., Harris, J. M., Fang, Z., Yoon, K., Dizman, B., et al. (2011). Polyoxazoline: Chemistry, properties, and applications in drug delivery. *Bioconjugate Chemistry, 22*(5), 976–986. https://doi.org/10.1021/bc200049d.

Wakabayashi, R., Yahiro, K., Hayashi, K., Goto, M., & Kamiya, N. (2017). Protein-grafted polymers prepared through a site-specific conjugation by microbial transglutaminase for an immunosorbent assay. *Biomacromolecules, 18*(2), 422–430. https://doi.org/10.1021/acs.biomac.6b01538.

Wang, L., Brock, A., Herberich, B., & Schultz, P. G. (2001). Expanding the genetic code of Escherichia coli. *Science, 292*(5516), 498–500. https://doi.org/10.1126/science.1060077.

Yokoyama, K., Nio, N., & Kikuchi, Y. (2004). Properties and applications of microbial transglutaminase. *Applied Microbiology and Biotechnology, 64*(4), 447–454. https://doi.org/10.1007/s00253-003-1539-5.

CHAPTER FIVE

Chimeric Small Antibody Fragments as Strategy to Deliver Therapeutic Payloads

Sandra Aguiar*, Joana Dias*, Ana M. Manuel†, Roberto Russo†, Pedro M.P. Gois†, Frederico A. da Silva*, Joao Goncalves†,1

*Centro de Investigação Interdisciplinar em Sanidade Animal (CIISA), Faculdade de Medicina Veterinária, Universidade de Lisboa, Avenida da Universidade Técnica, Lisboa, Portugal
†iMed.ULisboa—Research Institute for Medicines, Faculty of Pharmacy, Universidade de Lisboa, Lisbon, Portugal
[1]Corresponding author: e-mail address: joao.goncalves@ff.ul.pt

Contents

Abstract

Antibody–drug conjugates (ADCs) represent an innovative class of biopharmaceuticals, which aim at achieving a site-specific delivery of cytotoxic agents to the target cell. The use of ADCs represents a promising strategy to overcome the disadvantages of conventional pharmacotherapy of cancer or neurological diseases, based on cytotoxic or immunomodulatory agents. ADCs consist of monoclonal antibodies attached to biologically active drugs by means of cleavable chemical linkers. Advances in technologies for the coupling of antibodies to cytotoxic drugs promise to deliver greater control of drug pharmacokinetic properties and to significantly improve pharmacodelivery applications, minimizing exposure of healthy tissue.

The clinical success of brentuximab vedotin and trastuzumab emtansine has led to an extensive expansion of the clinical ADC pipeline. Although the concept of an ADC seems simple, designing a successful ADC is complex and requires careful selection of

Advances in Protein Chemistry and Structural Biology, Volume 112
ISSN 1876-1623
https://doi.org/10.1016/bs.apcsb.2018.03.002

the receptor antigen, antibody, linker, and payload. In this review, we explore insights in the antibody and antigen requirements needed for optimal payload delivery and support the development of novel and improved ADCs for the treatment of cancer and neurological diseases.

1. RECOMBINANT ANTIBODIES AND MARKET SIZE

The discovery of hybridoma technology, described by Köhler and Milstein (1975), and the resulting ability to develop monoclonal antibodies (mAbs), initiated a new era in antibody research and their clinical development. Yet, despite representing a major breakthrough in antibody-based therapy, early studies demonstrated that murine mAbs presented properties that could limit their use for human applications. One of the most important drawbacks was the high immunogenicity of these mAbs in a clinical context leading to the generation of antimouse antibodies and limiting their therapeutic benefit. Moreover, murine mAbs presented short serum half-lives and inability to trigger human effector function (Aires da Silva, Corte-Real, & Goncalves, 2008; Hwang & Foote, 2005). These shortcomings led to the manipulation of mAbs by recombinant DNA technology into more human variants, decreasing immunogenicity while increasing efficacy and facilitating the development of antibody-based therapies (Aires da Silva et al., 2008; Jones, Dear, Foote, Neuberger, & Winter, 1986). As a result, over the last three decades a total of 74 antibody-based molecules have been approved for a wide range of indications, such as autoimmune diseases, cancer, and infectious diseases, among others (Fig. 1). Moreover, an equal number of antibody candidates have reached late-phase clinical trials, clearly demonstrating the potential of these biological molecules for human therapeutics (Strohl, 2017). In 2017, five blockbuster antibodies reached the top 10 best-selling innovative drugs (Table 1) (Strohl, 2017). In 2016, the global antibodies market was estimated to be valued at USD 105 billion and is projected to reach USD 341 billion by 2026 end (Future Market Insight, 2016).

Most marketed antibodies are conventional full-length IgG molecules. Nevertheless, the versatility of the antibody molecule and the capacity to engineering it into novel antibody scaffolds, such as bispecific antibodies and smaller antibody fragments, allow to design novel constructs for a range of applications proven to be far more appealing for human therapeutics (Aires da Silva et al., 2008). As a result, in 2017, four small recombinant antibody formats had already been approved for clinical applications (Table 2),

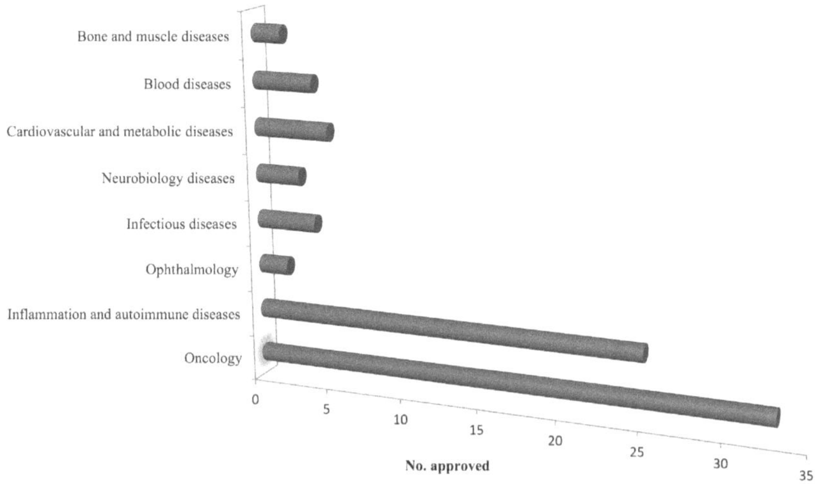

Fig. 1 Approved monoclonal antibodies and Fc fusion proteins for therapeutical use. *Based on Strohl, W. R. (2017). Current progress in innovative engineered antibodies.* Protein & Cell, *1–35. https://doi.org/10.1007/s13238-017-0457-8.*

Table 1 Best-Selling Antibody Drugs in 2017 (Strohl, 2017)

Drug (Generic Name)	Application	Company	2016 Sales (USD)
Humira (adalimumab)	Autoimmune diseases	AbbVie	16 Bn
Rituxan (rituximab)	Cancer	Roche + Biogen	8.5 Bn
Remicade (infliximab)	Autoimmune diseases	Johnson & Johnson and Merck	7.8 Bn
Avastin (bevacizumab)	Cancer	Roche	6.7 Bn
Herceptin (trastuzumab)	Cancer	Roche	6.7 Bn

either as neutralizing agents or as radiolabeled products for imaging, while many others are already in clinical trials.

In addition to its use as a therapeutic agent on their own, the antibody molecule can also be devised as drug transport shuttle. To take advantage of this property, several antibody scaffolds are being engineered to transport cytotoxic drugs, radiotherapeutics molecules, or immunocytokines greatly increasing the applicability of these biodrugs. These features substantially increase the applications of antibodies and lead to the next generation of antibody-based molecules such as ADCs, bispecific and bifunctional antibodies, and cell-surface-expressed antibodies (CARS). Within this context,

Table 2 Antibody Fragments Approved (Strohl, 2017)

Name	Type	Company	Application	Year of Approval (the United States)
Abciximab	Fab fragment	Centocor (now J&J)/Lilly	Coronary diseases	1994
Ranibizumab	Fab fragment	Genentech Novartis	Ophthalmic diseases	2006
Certolizumab pegol	Fab fragment	UCB/Schwartz	Rheumatoid arthritis and Crohn's disease	2009
Idarucizumab	Fab fragment	Boehringer Ingelheim	Drug reversal	2015

this chapter provides an update of this fast-moving market mainly focusing in the development of antibody fragments as a potent emerging class of therapeutics for drug delivery.

2. STRUCTURE AND ANTIBODY FUNCTION

In mammals' immune system, the antibody format presents two identical heavy chains (H) and light chains (L) in a Y-shaped structure (Fig. 2). The L chains belong to kappa (κ) or lambda (λ) subtypes and H chains to α, δ, ε, γ, or μ isotypes. The isotype is determined by sequence and length of heavy-chain constant domains, distributing antibodies among five classes: IgA, IgD, IgE, IgG, and IgM, respectively. Because of its predominance in human serum, importance for immune response, and excellent specificity characteristics, IgG constitutes the dominant format in immunotherapy (Schroeder & Cavacini, 2010; Turner, 1998; Williams & Barclay, 1988).

In IgG class, each heavy chain is organized into three constant domains (CH1, CH2, and CH3) and one variable domain (VH), whereas the light chain is constituted only by a constant (C_L) and a variable domain (VL). Variable domains are positioned into antibody N-terminus and together with the C_L and CH1 regions constitute the fragment antigen binding (Fab) (Fig. 2). This fragment is linked by a flexible sequence (hinge) to the CH2 and CH3 domains, components of the fragment crystallizable (Fc), where carbohydrates motifs are attached. Antibody integrity is maintained by several interdomain disulfide bonds along with the highly conserved intradomain ones. The variable domains mediate antibody specificity and

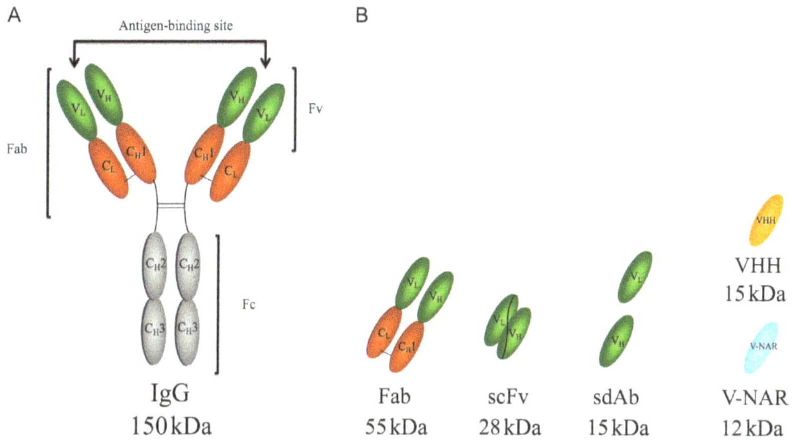

Fig. 2 Schematic representation structure of a conventional IgG antibody (A) and antibody fragments (B) of biotechnological and clinical interest. (A) IgG antibodies comprise a pair of identical heavy and light chains linked by disulfide bonds. Light chains contain one constant domain (CL) and one variable domain (VL), while heavy chains contain three constant domains (CH1, CH2, and CH3) and one variable domain (VH). The variable domains of both the heavy and light chains are responsible for the antigen-binding site of the molecule. The Fc constant region recruits effector functions of the immune system. Constant light (C_L) and heavy (C_H) chain domains are represented in *orange* and *gray*. Variable light (VL) and heavy (VH) chain domains are represented in *green*. (B) The engineering of antibody fragments that can be generated from an intact conventional IgG: antigen-binding fragment (Fab), single-chain Fv fragment (scFv), heavy and light single-domain antibodies (VL and VH sdAbs), natural camelid variable domain (VHH), and shark variable domains (V-NAR).

affinity toward antigen mainly through three hypervariable loops, named complementary-determining regions (CDRs; H1, H2, H3, L1, L2, and L3), with the six CDRs of each antibody "arm" setting up the antigen-binding site. CDR loops are supported by four relatively conserved β-sheet strands, the framework sequences. The Fc region is the major responsible for antibody effector functions through complement and gamma Fc receptor (FcγR) binding. On the other hand, this IgG portion prolongs antibody half-life through a recycling mechanism dependent on neonatal Fc receptor binding (FcRn) (Schroeder & Cavacini, 2010; Turner, 1998; Williams & Barclay, 1988).

In antibody-based therapies, the goal is to eliminate or neutralize the pathogenic organism or the disease target. For this purpose, mAbs achieve their therapeutic effect through various mechanisms. In some cases, antibody binding can directly and effectively block the activity of many pathogens. In other scenarios, effective treatment requires a more general immune response, and antibodies must boost effector functions such as antibody-dependent cellular cytotoxicity (ADCC) and/or complement-dependent cytotoxicity (CDC). In ADCC responses, antibodies bind to antigens on target cells and the antibody Fc domains engage Fc receptors (FcγRs) on the surface of effector cells, such as macrophages and natural killer cells. These cells in turn trigger phagocytosis or lyse the targeted cell. In CDC, antibodies kill the targeted cells by triggering the complement cascade at the cell surface. These effector functions contribute to the therapeutic efficacy of several antibodies in clinical settings, in which the destruction of target cells is desired, as for the removal of tumor cells or virally infected cells.

The major challenge in cancer treatment is that tumors often develop resistance to antibody therapy. Downstream signaling pathways can be mutated (KRAS/BRAF), which limits the antibody's capacity to inhibit growth factor receptor signaling (Van Cutsem et al., 2011). Tumors can overexpress complement inhibitory receptors such as CD46, CD55, and CD59, thereby blocking CDC (Jurianz, Maslak, Garcia-Schüler, Fishelson, & Kirschfink, 1999). Overexpression of certain HLA molecules (i.e., HLA-E and -G) that inhibit NK-cell-mediated ADCC has been described for various tumors (Levy et al., 2008; Lin et al., 2007), and the tumor microenvironment can be infiltrated by T-regulatory cells and myeloid-derived suppressor cells that suppress antitumor immune response (Greten, Manns, & Korangy, 2011). One approach to overcome such limitations in efficacy is the conjugation of cytotoxic compounds to mAbs. These ADCs combine the tumor specificity, pharmacokinetics, and biodistribution properties of antibodies with the potent cell-killing activity of small molecules. This concept was already postulated in the early 20th century by Paul Ehrlich who reasoned that if a compound could be made to selectively target a disease-causing organism, then a toxin for that organism could be delivered along with the agent of selectivity (Ehrlich, 1913). Hence, a "magic bullet" would be created that only killed the targeted organism. An antibody would be extremely suitable for this purpose as antibodies can selectively bind to tumor antigens while maintaining a long half-life in circulation (~21 days for IgG1).

The first generation of ADCs was conjugated with clinically approved chemotherapeutic agents such as vinblastine, mitomycin, methotrexate, and doxorubicin (Perez et al., 2014). Although promising, these ADCs showed limited clinical success, which was mostly attributed to the low potency of the conjugated drug. The second generation of ADCs made use of more potent payloads, namely, calicheamicin, auristatin, and maytansine analogs. Besides, pharmacokinetics of the conjugation linker was optimized and fully human mAbs were used to solve immunogenicity problems observed with murine mAbs. In 2000 the first ADC, gemtuzumab ozogamicin (Mylotarg®), was clinically approved for use in refractory acute myeloid leukemia (AML) (Sievers et al., 2001). Here, a CD33 antibody was conjugated to calicheamicin, a drug that specifically binds to DNA and generates single- and double-strand DNA breaks. Gemtuzumab ozogamicin received accelerated approval for the treatment of patients with relapse AML. Unfortunately, 10 years later, gemtuzumab ozogamicin was withdrawn from the US market due to lack of clinical benefit (Petersdorf et al., 2009). A confirmatory phase III trial showed no improvement in clinical benefit for patients who received standard chemotherapy plus gemtuzumab ozogamicin, but instead a greater number of deaths occurred in the group of patients who received gemtuzumab ozogamicin compared with those receiving chemotherapy alone (Petersdorf et al., 2009). Several factors have been identified that have limited the clinical efficacy of gemtuzumab ozogamicin, including poor stability of the acid-labile conjugation linker, heterogeneous drug loading (approximately 50% of the CD33 antibodies are unconjugated), and sensitivity to multidrug resistance pumps that are often overexpressed in AML. More recently, two novel ADCs were approved by the FDA for the treatment of Hodgkin lymphoma and anaplastic large-cell lymphoma, brentuximab vedotin (Adcetris®), and HER2-positive breast cancer, trastuzumab emtansine (Kadcyla®) (Burris et al., 2011; Younes et al., 2010). These ADCs showed improved linker stability and pharmacokinetics. Their clinical success has led to an impressive expansion of the clinical ADC pipeline.

3. ANTIBODY FRAGMENTS AND SINGLE-DOMAIN ANTIBODIES

Most marketed antibodies are comprised of a full-length IgG molecule that provides for long half-life and effector functions. However, there are a range of therapeutic applications in which other antibody formats may be more desirable. For instance, in some conditions a long antibody serum half-life results in poor contrast in imaging applications, and inappropriate activation of Fc receptor-expressing cells may lead to massive cytokine release and associated toxic effects. In addition, due to high molecular weight (~150 kDa), IgG antibodies are known to diffuse poorly into solid tumors and clear slowly from the body. Therefore, to avoid Fc-associated effects in some clinical settings and to address the size limitations of IgGs, smaller antibody molecules such as the antigen-binding fragment (Fab) or the variable fragment (Fv) are becoming more attractive therapeutic agents (Hudson, 1998; Presta, 2003).

Cleavage of the IgG molecule with papain or recombinant technology can be used to yield Fab and Fv fragments (Fig. 2) (Bird et al., 1988; Hudson, 1999; Hudson & Souriau, 2003; Skerra & Plückthun, 1988). Both these fragments are stable, small molecular entities that exhibit most of the same properties as when they are part of the whole antibody molecule. The smaller Fv fragment (fragment variable) is composed of the V_H and V_L regions only. The recombinant version of the Fv fragment is termed the single-chain variable fragment (scFv) and the two variable regions are joined artificially by a flexible peptide linker, to produce scFv with improved antibody folding and stability. Linkers of different compositions may result in different levels of oligomerization and may also alter the interface of the V_H and V_L regions (Fig. 2). When the linker is short (typically five amino acids), two scFv molecules can associate and form a stable noncovalent dimer, called diabody (Db, ~50 kDa) (Holliger & Hudson, 2005).

In addition to molecular mass, also binding affinity and avidity may have a direct impact on the targeting performance of antibody products. The affinity constant describes the interaction of an antibody with its antigen. It is determined by the kinetic "on-rate" and "off-rate" constants, according to the equation $K_a = K_{on}/K_{off}$. A high "functional affinity" (also called avidity) can be obtained when two or more binding sites in an antibody molecule interact with repetitive antigen sites on the target. In some cases, a high binding affinity can be detrimental for therapeutic antibodies that target solid tumors (Adams et al., 2001). Antibodies with an extremely high affinity (above $10^9\ M^{-1}$) bind irreversibly to the first antigen encountered at the periphery of the tumor, hindering the ability of antibodies to penetrate in the tumor mass. In general, high-affinity fragments do not penetrate deeper in the tumor mass, unless all antigen molecules are saturated in the periphery, an effect called "binding-site barrier" (Adams et al., 2001). By contrast, medium high-affinity fragments can penetrate deeply into the tumor. Affinity constants in the 10^8–$10^9\ M^{-1}$ range often give acceptable tumor accumulation. These values can be obtained either by engineering the antibody-binding site or by increasing the avidity, developing multivalent antibodies.

While the scFv possesses only one binding site, diabody molecules have the same avidity as the parental IgG but are small enough to be excreted via the renal clearance pathway. The fusion of a scFv fragment with the dimerization domain CH3 of IgG or CH4 of IgE gives rise to a bivalent small immunoprotein (SIP, ~75 kDa) or "minibody," a stable homodimer stabilized by a disulfide bond at its C-terminus (Chames, Van Regenmortel, Weiss, & Baty, 2009). Antibodies in SIP format have a high binding avidity and are mainly excreted via the liver. Small antibody fragments (e.g., Fab, diabody, and scFv) can rapidly extravasate and diffuse into solid tumors, but they exhibit short serum half-life, resulting in a rapid renal elimination before a substantial tumor uptake can occur. The rapid clearance from the blood circulation is also due to the lack of effector functions and consequent inability of interaction with the salvage receptor FcRn. In contrast, larger antibodies (e.g., IgG) are eliminated via the hepatobiliary route and show a prolonged residence in circulation but display relatively poor diffusion from the vasculature to the tumor mass due to their size.

Another promising alternative to conventional IgGs are single-domain antibodies (sdAbs). sdAbs are the smallest functional antigen-binding fragments of an antibody that can be isolated from conventional IgGs (e.g., from human, murine, or rabbit) or obtained from naturally occurring antibodies devoid of light chain that were discovered in two types of organisms, the camelids (camels and llamas) and cartilaginous fish (wobbegong and nurse shark) (Aires da Silva et al., 2004; Greenberg et al., 1995; Hamers-Casterman et al., 1993; Holliger & Hudson, 2005; Holt, Herring, Jespers, Woolven, & Tomlinson, 2003; Muyldermans, 2001). Due to their small size, sdAbs show improved tissue penetration and are able to reach targets not easily accessible by conventional antibodies, such as enzyme active sites or canyons in receptors molecules. Moreover, in the same way as Fab and scFv, sdAbs lack the Fc domain of a full IgG antibody, presenting a low nonspecific uptake in tissues that highly express Fc receptors. Additional important advantages include their high stability, low immunogenicity, and lower manufacturing cost. Furthermore, sdAbs can be easily attached to other proteins, peptides, small molecules, or nanoparticles by simple molecular biology or chemical procedures (Holliger & Hudson, 2005). Taking into consideration all these advantages and properties, Fab, scFv, and sdAb antibody scaffolds are definitely a promising alternative to conventional antibodies as potential drug delivery vectors. In the next section, a brief overview of the most effective bioconjugation methods will be presented with a focus on their use to modify antibody fragments.

4. CONJUGATION METHODS FOR ANTIBODIES

In recent years, significant advances have been made in the chemical functionalization of antibodies exploring innovative bioconjugation techniques. These methods target selectively the specific reactivity of amino acid side chains and for that reason can also be applied to different proteins, among which antibody fragments. Briefly, the most current bioconjugation methods are:

(a) Chemical modification of proteins at natural amino acids

Chemical modification of proteins is a challenging task, as these complex macromolecules require specific conditions to maintain their proper folding and function. As a result, the investigation in this field has evolved around the development of reactions that occur with fast kinetics at physiological conditions, without the need of additives that could damage proteins (Means & Feeney, 1990). In order to obtain a reaction with the aforementioned characteristics, the research in this field has focused on exploiting amino acids that have unique side-chain reactivity, such as lysine, cysteine, and, to a lesser extent, serine. These residues present nucleophilic side chains that can react selectively with various electrophiles (Shannon & Weerapana, 2015), making them privileged targets for protein modification. Even though there are many protein modification reactions available nowadays (de Graaf, Kooijman, Hennink, & Mastrobattista, 2009; Stephanopoulos & Francis, 2011; Yang, Li, & Chen, 2014), this section will focus on the modification of lysine and cysteine in the context of antibody functionalization.

- Lysine

Given the nucleophilicity of its amine side chain, lysine has been a preferential amino acid residue to be exploited for bioconjugation purposes. The overall rationale for lysine modification relies on the nucleophilic attack of the amino group side chain of lysine onto an electrophilic group, most commonly activated esters, to form an amide bond.

One of the most commonly used reagents for lysine conjugation are NHS-activated esters (Anderson, Zimmerman, & Callahan, 1964), which are easy to install and react quickly with the lysine ε-amino group under slightly alkaline conditions. This technique was used for the production of gemtuzumab ozogamicin (Hamann et al., 2002) (Wyeth) and trastuzumab emtansine (Lewis Phillips et al., 2008) (Immunogen), two of the first ADCs to reach the market (Fig. 3).

Although very efficient, lysine modification typically suffers from a lack of selectivity that severely limits the performance of the overall construct

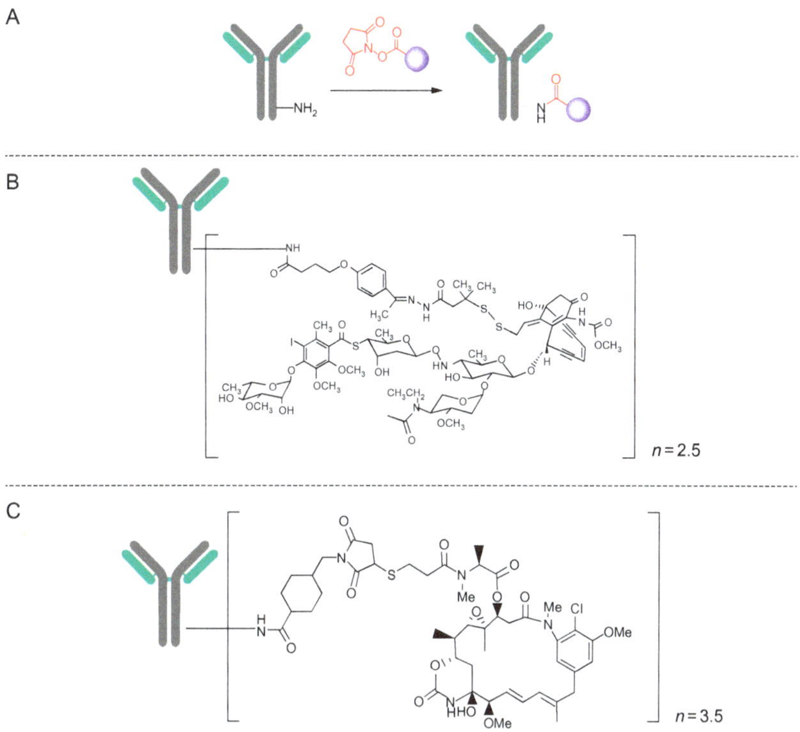

Fig. 3 (A) Functionalization of a lysine ε-amino group (*black*) with a payload (*blue*) activated with an NHS-ester (*red*). (B) Gemtuzumab ozogamicin. (C) Trastuzumab emtansine. *Panel (B): Adapted from Hamann, P. R., Hinman, L. M., Hollander, I., Beyer, C. F., Lindh, D., Holcomb, R., et al. (2002). Gemtuzumab ozogamicin, a potent and selective anti-CD33 antibody–calicheamicin conjugate for treatment of acute myeloid leukemia.* Bioconjugate Chemistry, 13*(1), 47–58. https://doi.org/10.1021/bc010021y; Panel (C): Adapted from Lewis Phillips, G. D., Li, G., Dugger, D. L., Crocker, L. M., Parsons, K. L., Mai, E., et al. (2008). Targeting HER2-positive breast cancer with trastuzumab-DM1, an antibody-cytotoxic drug conjugate.* Cancer Research, 68*(22), 9280–9290. https://doi.org/10.1158/0008-5472.CAN-08-1776.*

yields. The fundamental problem with lysine modification derives from the fact that protein surfaces are rich in lysine residues that can react with the modifying agent, generating a mixture of heterogeneously modified products. This lack of selectivity is a serious issue in the generation of ADCs, as it leads to products with different drug-to-antibody ratios (DARs) and batch-to-batch heterogenicity.

Heterogeneous ADCs with batch to batch DAR variability reflect only part of the problem, as two ADCs with the same DAR number can be different entities with the same amount of drug loading, but on different positions of the antibody. A study from Wang, Amphlett, Blättler, Lambert, and Zhang (2005) analyzed a population of ADCs with DARs from 0 to 6, generated from the modification of an IgG containing 86 lysine residues, and found that each batch could contain over 4.5 million unique species. Batch-to-batch heterogeneity of ADCs hinders the development of bioconjugates with therapeutic utility, as it is linked to pharmacokinetic problems, narrow therapeutic window, aggregation, and increased incidence of side effects (Hamblett et al., 2004; Junutula et al., 2008).

It is not surprising, given the negative effects of heterogeneous protein modification, that the efforts in modern bioconjugation chemistry have been directed to the development of reactions that are not only selective, but more importantly, site-specific, allowing for a better control of the product homogeneity (Agarwal & Bertozzi, 2015; Akkapeddi et al., 2016; Flemming, 2014).

- Cysteine

The sulfhydryl side chain of cysteines is another potential site for bioconjugation, thanks to its unique reactivity toward a particular set of electrophiles. Unlike lysines, cysteine residues are quite scarce on the surface of proteins (Liu & May, 2012); this allows for a superior control over the distribution of the modification. Cysteine modification is also not susceptible to unwanted modification of the N-terminus of a protein, since the moieties that react with cysteines are usually not attacked by amino groups. The thiol side chain of cysteine is able to react with a variety of Michael acceptors (Cal, Bernardes, & Gois, 2014), among which maleimide moieties are the most commonly utilized, given the ease of their preparation and the fast reaction kinetics in water (Fig. 4A) (Baldwin & Kiick, 2011). The first protocol for antibody-cysteine modification with maleimides involved the reduction of all the interchain disulfide bridges, followed by partial reoxidation to restore the bridging on all except a pair of cysteine residues that were then functionalized (Sun et al., 2005).

This approach was used for the production of the FDA-approved ADC brentuximab vedotin (Seattle Genetics and Millennium Pharmaceuticals/Takeda Oncology) (Senter & Sievers, 2012), shown in Fig. 4B. This modification technique afforded good results in terms of DAR and batch-to-batch homogeneity, although the difficulty in controlling the reduction and oxidation steps resulted in a nonnegligible variance in drug loadings. The insertion of a cysteine residue by site-directed mutagenesis on the antibody structure allows to bypass the problem entirely, as it can be functionalized without needing preliminary reduction steps, thus leaving

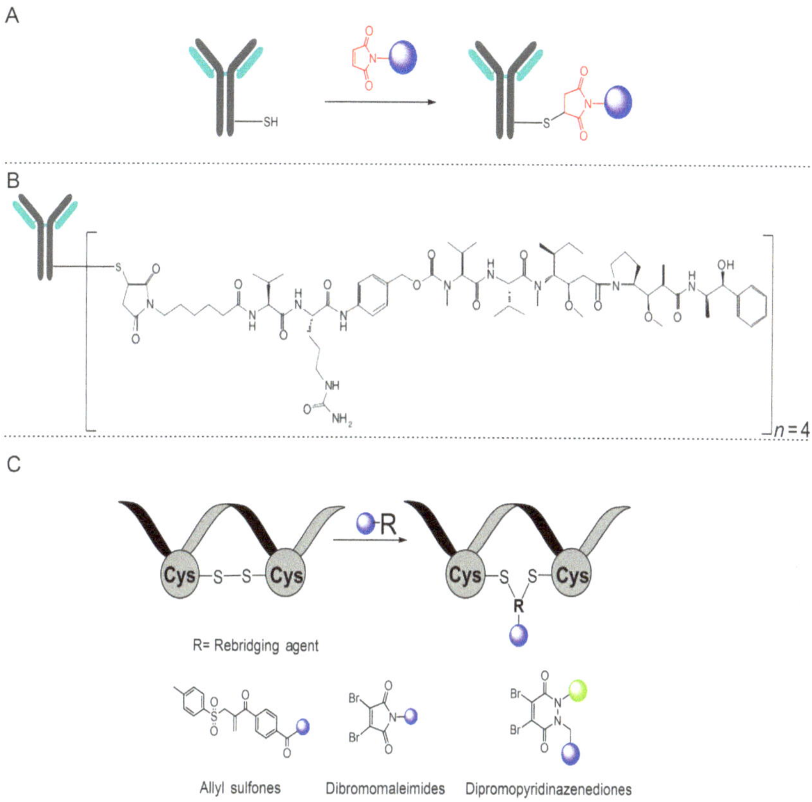

Fig. 4 (A) Modification of a cysteine residue (*black*) with a payload (*blue*) bearing a maleimide moiety (*red*). (B) Brentuximab vedotin. (C) Schematic representation of disulfide rebridging reactions and examples of disulfide rebridging reagents.

the interchain disulfide bonds intact. ADCs generated with this methodology, called THIOMABs, showed improved efficacy and safety, which derive from the excellent control exerted on the bioconjugation step (Junutula et al., 2008; Pillow et al., 2014).

Another avenue for cysteine modification, called disulfide rebridging, consists in the insertion of a small molecule into a disulfide bridge, conferring high stability toward reductive conditions, while maintaining intact the tertiary structure of the protein (Fig. 4C) (Kuan, Wang, & Weil, 2016). This methodology does not require reduction/oxidation steps nor site-directed mutagenesis and has been shown to produce ADCs with high and homogeneous DARs. There are several moieties that have been exploited as rebridging agents, all with good results in terms of stability and homogeneity, including vinylsulfones (Shaunak et al., 2006), dibromomaleimides

(Smith et al., 2010), and dibromopyridiazinediones (Maruani et al., 2015). Dibromopyridiazinediones, in particular, are very interesting because they allow the insertion of two different functionalities with one modification.

5. CHEMICAL MODIFICATION OF ANTIBODY FRAGMENTS

The chemical modification of antibody fragments has been carried out by applying the same techniques already established for full antibodies and proteins in general, allowing for the exploration of a wide range of functionalities that can be added on these structures. This section will compile the advancements in chemical functionalization of antibody fragments, grouped by the type of modification that was performed and the function installed.

(a) Pegylation

One of the most common modifications performed on antibody fragments is the attachment of one or more polyethylene glycol (PEG) chains to the peptidic structure. The purpose of this modification is to increase the circulation half-life of these therapeutics, given their susceptibility to a faster renal clearance due to their smaller size compared to full IgGs. Indeed, various reports have shown that functionalizing Fabs with a 40 kDa PEG chain gives them a similar blood half-life to full IgGs (Chapman et al., 1999; Wei et al., 2005).

Cysteine modification was one of the first avenues to be explored for PEG functionalization (Kitamura et al., 1991). Regrettably, classic cysteine–maleimide chemistry did not prove to be sufficiently selective for this purpose, as this approach failed to discriminate between interchain and hinge cysteine residues, where the former consists of the undesirable target, ultimately resulting in the loss of function. One of the chemical challenges in cysteine pegylation lies in the need of reducing agents in order to activate the cysteine residues. The mild protocols that only reduce the hinge solvent-accessible cysteines are thiol-based (glutathione, mercaptoethanol, cysteine) and can form adducts with the cysteine that is intended to be activated, blocking it and reducing dramatically the pegylation efficiency. On the other hand, though nonthiol-based activators (such as TCEP) cannot form these adducts, their stronger reductive character results in the unwanted reduction of the interchain disulfide bonds, exposing them to pegylation.

A promising solution to this problem is the use of PEG-bis-sulfone reagents that act as rebridging agents allowing for homogeneous and efficient pegylation. These reagents rapidly lose a molecule of sulfinic acid, yielding allyl sulfones in situ that functionalize disulfide bridges previously reduced with DTT (Khalili, Godwin, Choi, Lever, & Brocchini, 2012). Another example of the use of rebridging agents for the homogeneous pegylation of Fabs includes the use of dibromomaleimides that show fast and complete modification in equimolar conditions, resulting in a clean protocol that requires minimal to no purification of the final constructs (Jones et al., 2012).

Humphreys et al. (2007) approached the problem of selective pegylation by producing engineered Fabs in which interchain cysteines were substituted with serine residues, allowing for selective functionalization of the hinge cysteine residues with remarkable efficiency (88%–94%, murine vs human). Notably, these engineered Fabs showed similar stability and in vivo half-lives compared to the native ones.

Another interesting approach was used by Hofer et al. that engineered a C-terminal selenocysteine-engineered Fab derived from rituximab, which allowed a very specific modification on the selenocysteine side chain (Hofer, Skeffington, Chapman, & Rader, 2009). The chemoselectivity of this technique arises from the fact that, even though they show similar reactivity in Michael additions, selenocysteine and cysteine have distinct pK_a values, the former being more acidic than the latter. Thus, this apparently innocuous difference was exploited to perform the conjugation reaction at pH 5, at which cysteine is completely protonated and consequently less nucleophilic, while Sec is 50% deprotonated and therefore retains the ability to act as a nucleophile toward maleimides. This technique led to a selective modification in mildly acidic and reducing conditions, allowing the development of pegylated constructs, as well as conjugates with fluorescent probes or biotin for imaging and proteomics applications.

Antibodies can also be modified with biodegradable polymers (e.g., PASylation®, a PEG mimetic based on proline, alanine, and serine) or fused to human serum albumin (HSA), without affecting the binding and activity of the protein (Kontermann, 2011). HSA is able to prevent the renal elimination of the antibody fragment because this protein is actively recycled by its interaction with FcRn, leading to a substantial increase in circulation time.

(b) Fab–drug conjugates

Even though the majority of therapeutically relevant antibody fragments exert a standalone effect, there is also growing interest in using these biomolecules as targeting agents for the delivery of therapeutic and imaging agents. One of the first examples of Fab–drug conjugates comes from Cao's group (Fig. 5) (Wang et al., 2007) that identified a human Fab that targets epidermal growth factor receptor (EGFR) and is able to enter the cell membrane

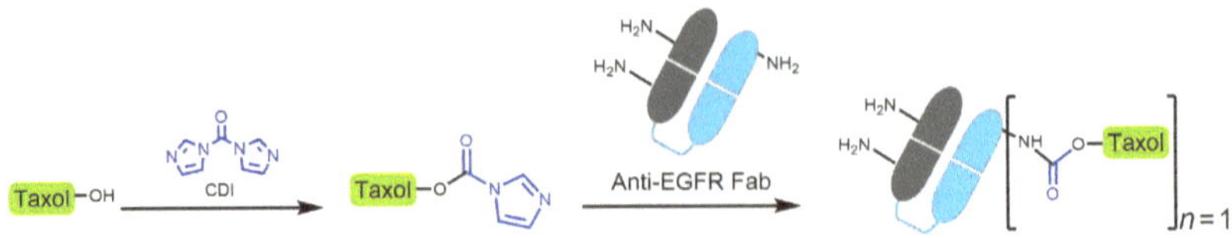

Fig. 5 Taxol (*green*) activation with CDI (*blue*) and subsequent conjugation with the lysine of anti-EGFR Fab.

through receptor binding. This Fab, identified by phage-display, was used to produce a drug conjugate with Taxol, a potent cytotoxic drug, exploiting lysine modification with a method that is conceptually similar to what has been described previously, but experimentally different. In this case, Taxol was activated with *N,N′*-carbonyldiimidazole (CDI) to form an activated ester, which subsequently reacted with the Fab, yielding a Fab–drug conjugate with a methodology akin to NHS ester chemistry. The resulting construct is probably far from homogeneous, although DAR was determined by mass spectrometry to be around 1. Nonetheless, the conjugate showed very good activity and selectivity for inducing apoptosis in EGFR-overexpressing cells.

Another example, coming from the R&D group of Seattle Genetics®, involves the use of an engineered anti-CD30 diabody for the construction of conjugates with monomethyl auristatin E or F (Kim et al., 2008). Harnessing the power of protein engineering, the selected diabody was produced in order to present four solvent-accessible cysteine residues, which allowed for modification using maleimide chemistry. The resulting constructs presented remarkable efficacy since they showed similar potency to the related conjugates derived from the full IgG, higher clearance rate, and overall better pharmacokinetics. Disulfide rebridging techniques have also been used in this context, showing good results both in terms of modification efficiency and homogeneity. Godwin et al. successfully used allyl sulfones rebridging reagents to attach mono methyl auristatin A to a Fab derived from trastuzumab digestion, yielding a highly homogeneous conjugate with a DAR value of 1 (Badescu et al., 2014). The use of dibromomaleimides in this context has been reported by Caddick and Boyle that demonstrated an efficient and regioselective modification of a Fab derived from trastuzumab with Zn-based photosensitizers for photodynamic therapy (Bryden et al., 2014).

One of Pfizer® research group has also reported the development of a new solid-phase technique for the site-selective dual modification of antibodies, also applicable to antibody fragments (Fig. 6) (Puthenveetil et al., 2016).

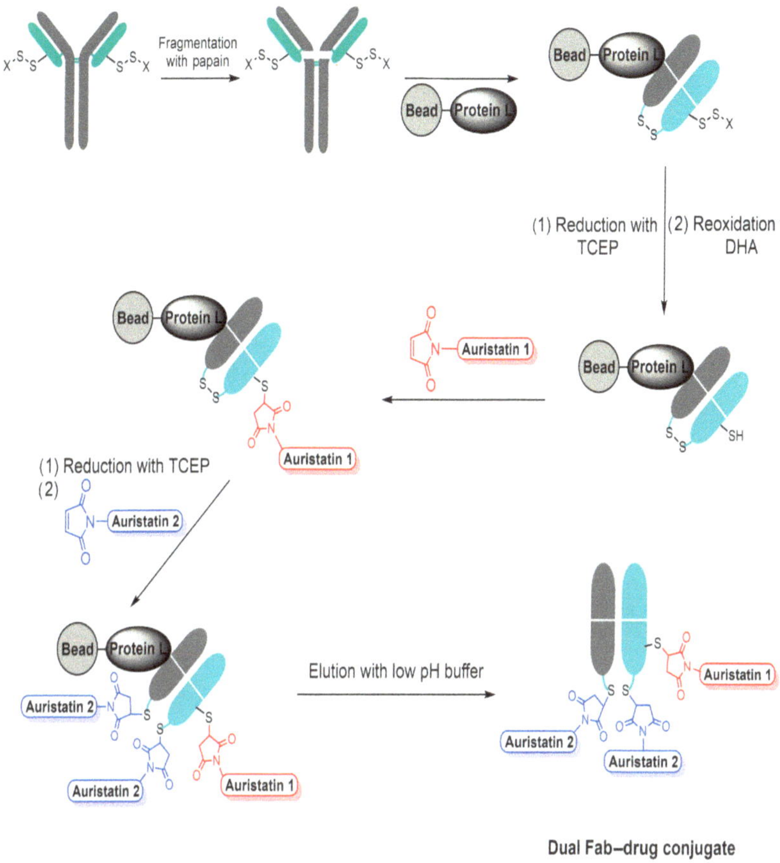

Fig. 6 Solid-phase production of disubstituted Fab–drug conjugates.

This system relies in the use of protein L-modified beads and various reduction/oxidation steps of the disulfide bridges and allowed to site-specific attachment of two different auristatin derivatives on the same Fab, with remarkable selectivity and homogeneity.

In 2012, Dario Neri group described the synthesis of diabody–drug conjugates using an engineered F8 antibody mutant, bearing a C-terminal cysteine and an aldehyde derivative of Cematodin as the drug of choice (Fig. 7) (Casi, Huguenin-Dezot, Zuberbühler, Scheuermann, & Neri, 2012). The insertion of C-terminal cysteine in the diabody was exploited for the formation of a thiazolidine linkage with the aldehyde moiety of the drug, resulting in homogeneous conjugates with a DAR value of 2. The traceless nature of thiazolidine ligation makes it very appealing in these applications, because, after the drug release, the antibody fragment is left unmodified and able to

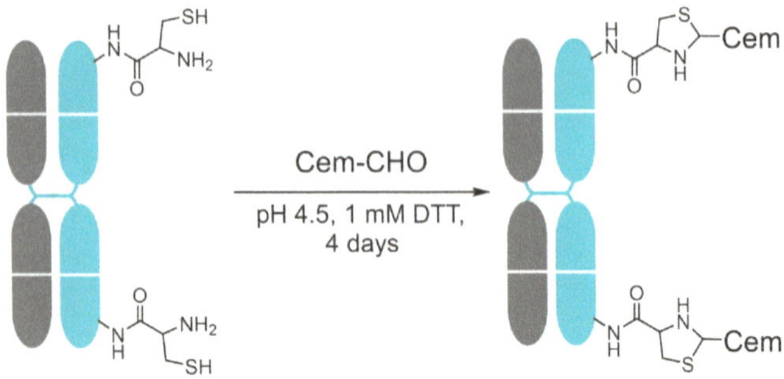

Fig. 7 Thiazolidine formation between C-terminal cysteines of an engineered diabody and an aldehyde derivative of cematodine (Cem-CHO).

exert its biological activity. The formation of the thiazolidine is however rather slow, as it requires 4 days at pH 4.5 to proceed.

(c) Nanoparticle–antibody fragment conjugates

Nanoparticles are excellent platforms for biomedical applications, given their ability to deliver therapeutic entities at a higher concentration than traditional methods (Annapragada, 2015). Nanoparticles are also tunable, functionalizable, and biocompatible, and they can be composed of a wide variety of materials, making them an attractive option for the development of therapeutics. For a more in-depth overview the reader can refer to a recent review from Richards, Maruani, and Chudasama (2016).

Even though they carry all the qualities mentioned earlier, nanoparticles do not possess intrinsic targeting capacities to selectively accumulate in the desired area. The main way to achieve selectivity with nontargeted nanoparticles has been through passive targeting, such as exploiting the "enhanced permeability and retention" (EPR) effect of tumor vasculature (Baetke, Lammers, & Kiessling, 2015).

Given the recent successes in the use of antibodies as targeting ligands, they were quickly adopted to functionalize the surface of nanoparticles, yielding targeting particles for a variety of applications.

From a chemical point of view, the biggest challenge in building antibody–nanoparticle conjugates is minimizing the impact that the conjugation will have on the structure of both partners. Although some positive results emerged, antibodies failed to fully deliver the expected efficacy as targeting moieties for nanoparticles. In fact, antibody-decorated nanoparticles showed immunogenicity (Qiao et al., 2016), poor stability (Wang et al., 2008), and rapid elimination (Kontermann, 2006), a series of drawbacks that can be traced back to the random orientation of full antibodies on the surface of nanoparticles.

The surface of nanoparticles can be functionalized with many techniques, involving both covalent ligation and noncovalent interactions. For the purpose of generating nanoparticle–antibody conjugates, covalent linkage is preferred as it confers better stability (Arruebo, Valladares, & González-Fernández, 2009) and allows to select the orientation of the antibody. Among the plethora of available techniques for the functionalization of nanoparticles with proteins, cysteine chemistry is one of the most widely used (Arruebo et al., 2009; Oh, Park, Joo, & Lee, 2015; Sapsford et al., 2013).

When using cysteine chemistry with full IgGs, this is often the case, as they contain several cysteine residues that can be anchored to the nanoparticle. When full antibodies are attached to a nanoparticle in such a fashion, the result is a nonoriented arrangement of proteins on the particle surface. Moreover, given their large size, full antibodies fail to attach on the surface of nanoparticles in large number, resulting in poor loadings. These issues can be avoided by implementing antibody fragments instead of full IgGs. Fabs, given their smaller size, can be attached selectively to the surface of nanoparticles via the hinge cysteine residue, yielding constructs with grater surface loading and a precise orientation of the ligands on the nanoparticle surface, yielding nanoparticles that outperform the ones decorated with full IgGs (Fig. 8) (Cheng & Allen, 2008; Greene et al., 2017; Manjappa, Goel, Gude, & Ramachandra Murthy, 2014; Sapra, Moase, Ma, & Allen, 2004). Given the many advantages that antibody fragments bring as targeting ligands for nanoparticles, it is not surprising to see a wide range of applications being developed with these constructs.

Conjugation of an antibody fragment on the surface of liposomes yields particles called immunoliposomes, which have been extensively studied as

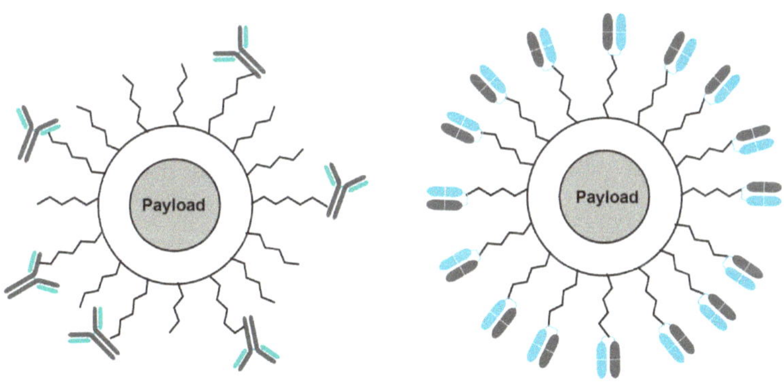

Fig. 8 Schematic representation of the advantages in orientation and loading of Fabs on nanoparticles surface (*right*) vs full antibodies (*left*).

nanoparticles for drug delivery (Cheng & Allen, 2010; Kontermann, 2006). In particular, doxorubicin-loaded immunoliposomes (Cheng & Allen, 2008; Sapra et al., 2004) and doxorubicin-loaded liposomes are already approved as a treatment for various cancers (Caelyx®, Myocet®, Doxil®). Aside from immunoliposomes, there are also examples of applications using polymeric micelles (Ahn et al., 2015), PEG–PLA polymersomes (Xiangbao et al., 2014), and iron oxide nanoparticles (Quarta et al., 2015). The functionalization of magnetically active nanoparticles, such as superparamagnetic properties of iron oxide nanoparticles (SPIONs), with antibody fragments also yielded promising results in the area of magnetic field therapy and theranostics (Cui et al., 2009; Yang et al., 2009). Lastly, immunoliposomes have also been used as carriers for targeted gene therapy, with two formulations already in clinical trials (van der Meel, Vehmeijer, Kok, Storm, & van Gaal, 2013).

(d) Use of antibody fragments for imaging

Targeted nanoparticles have shown great utility also as imaging agents in various applications such as MRI (Yang et al., 2009), PET (Xing, Zhao, Shi, Conti, & Chen, 2014), SPECT (Xing, Zhao, Conti, & Chen, 2014), and optical imaging (Rüger et al., 2014). Nanoparticles in general possess a high loading capability, which culminates in imaging agents with high signal/noise ratios. This in addition to the targeting properties of the antibody moiety makes them very interesting vehicles for contrast agents as their fast clearance kinetics have a positive effect in reducing the signal/noise ratio.

Chatalic and his collaborators have reported the modification of an anti–prostate-specific membrane antigen nanobody via maleimide chemistry on an engineered C-terminal cysteine. This functionalization was carried out with a maleimide-diethylenetriaminepentaacetic acid moiety, capable of chelating ^{111}Indium for SPECT/CT purposes. The resulting construct showed good biodistribution and fast clearance, as proven by the good quality of the SPECT/CT images acquired just a few hours after the injection in animal models (Chatalic et al., 2015).

Molecular imaging with fluorescent probes does not require expensive equipment nor the use of protections, but it is limited by the low penetration of light into tissue, making it an excellent technique for the visualization of superficial tumors. In this framework, CAIX-specific nanobodies, targeting hypoxic tissues, have been modified at the C-terminal engineered cysteine with IRDye800CW-Maleimide. The resulting constructs have been tested in mouse models and both the in vivo and the ex vivo imaging experiments showed good results in terms of biodistribution and signal/noise ratio (van Brussel et al., 2016). A similar approach was described by Oliveira et al. that conjugated the near-infrared fluorophore IRDye800CW to the anti-EGFR nanobody 7D12, resulting in a platform that allowed the visualization of tumors as early as 30 min after the infusion (Oliveira et al., 2012).

The use of dual-affinity antibody fragments for pretargeting has also been employed for imaging purposes. For example, engineered diabodies specific for carcinoembryonic antigen and a ^{67}Ga-modified hapten have shown good results in PET imaging of human colon tumor xenografts in athymic mice (Griffiths et al., 2004).

Another factor to keep in mind is that some particles, such as SPIONs (Alric et al., 2016), gold nanoparticles, and quantum dots (Xu, Liu, Brown, Christian, & Hornby, 2012), present intrinsic properties that can be exploited for imaging purposes, without the need to load the particle with an imaging agent.

6. ANTIBODY FRAGMENTS FOR ONCOLOGY AND CNS APPLICATIONS

As mentioned earlier, IgGs are the most abundant immunoglobulins in the blood and the main format for therapeutic antibody. Yet, IgG antibodies are known to diffuse poorly into solid tumors and unable to transpose complex membranes such as the blood–brain barrier (BBB). As such, smaller antibody molecules (Fab, scFv, sdAb) are becoming more attractive therapeutic agents (Aires da Silva et al., 2008). One of the main advantages of these antibody fragments is that they might reach targets not easily recognized by conventional IgGs. The small size of antibody fragments and single domains improves their ability to penetrate tumors, cells, and complex barriers such as the BBB and leads to rapid clearance from the circulation through the kidney. In some therapeutic applications, the rapid clearance is beneficial. Yet, if the increase of circulation is needed, the half-life of the antibody can be increased by linking molecules such as PEG (Aires da Silva et al., 2008). In addition, the reduced size of these antibody fragments enables the production using biological systems such as bacteria or yeast. As such, antibody fragments can be easily engineered and expressed making them suitable for creating bifunctional proteins or bispecific antibodies. These characteristics are highly desirable when designing antibody-based delivery vectors and may provide several advantages over currently used IgG-base technologies (Abulrob, Sprong, Van Bergen en Henegouwen, & Stanimirovic, 2005). A brief overview of some examples of antibody fragments for oncology and central nervous system (CNS) disease applications will be presented later (Fig. 9).

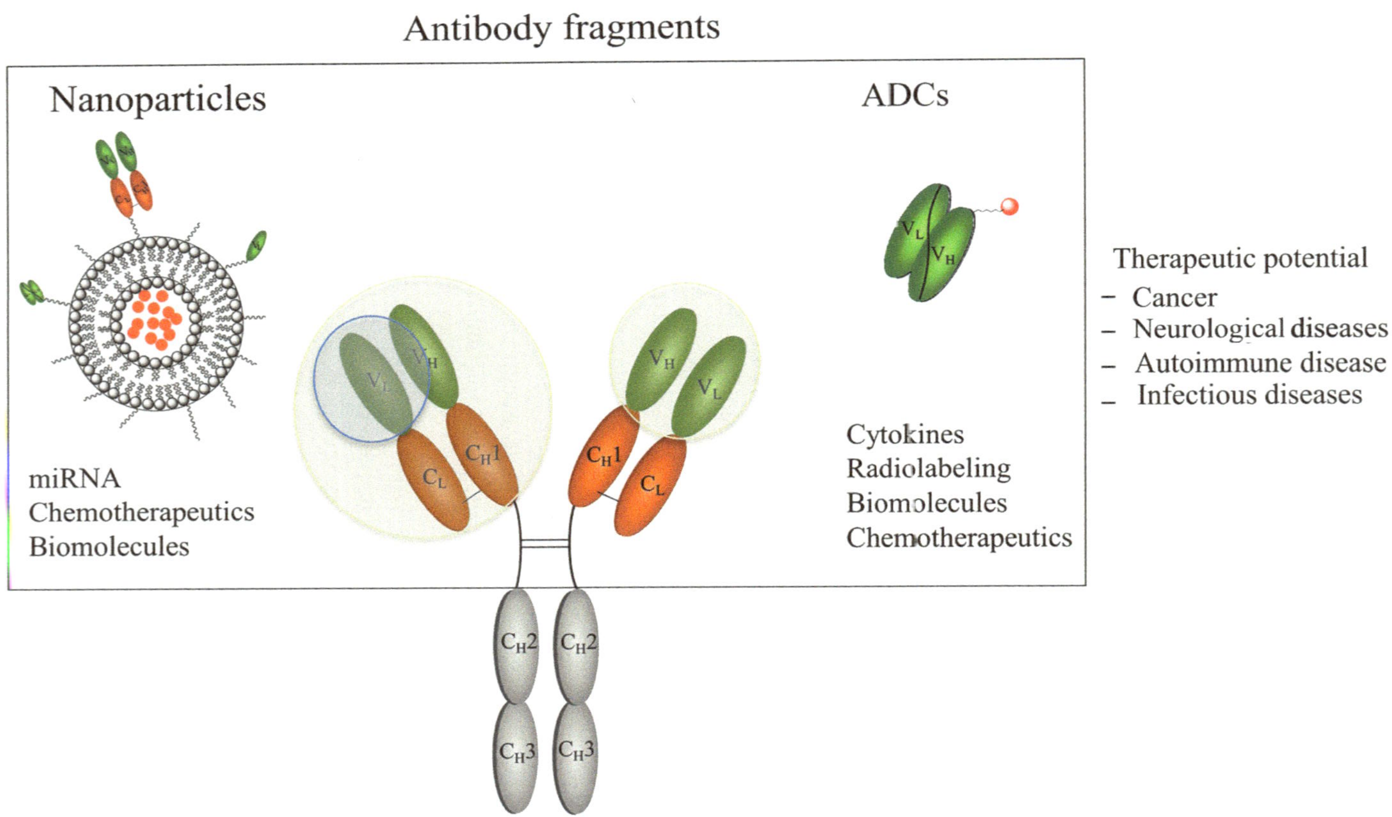

Fig. 9 Different strategies to modify antibody fragments and therapeutic applications that can be envisioned from these engineering antibodies.

6.1 Oncology

The ability to engage the immune system to induce killing of opsonized cells via Fc-mediated mechanisms led to the notion that mAbs might have great potential for the treatment of cancer. Numerous mAbs for the treatment of cancer have been developed in the past two decades, some of which have revolutionized treatment of cancers such as non-Hodgkin lymphoma (rituximab) and breast cancer (trastuzumab). In addition to Fc-mediated effector functions, therapeutic antibodies can exert antitumor activity through a number of different mechanisms, for example, inhibition of growth factor receptor signaling and induction of receptor downmodulation (e.g., zalutumumab). Furthermore, therapeutic antibodies may interact with the tumor microenvironment, for example, by inhibiting regulatory interactions between tumor cells and the adaptive immune system (e.g., ipilimumab, PD-1, PD-L1). Although the generation of tumor-targeting antibodies has generally been very successful, only a limited number of antibodies have been clinically effective (Nelson, 2010). Currently, 17 mAbs have been approved for the treatment of cancer by the FDA and 15 by the European Medicines Agency (EMA) in addition to a comparable number of antibodies for the treatment of inflammatory, cardiovascular, infectious, and other diseases.

The possibility to produce various engineered antibody fragments with dual specificity for a tumor marker and for a hapten has provided new tools for pretargeting applications in radiotherapy. These biomolecules are administered in the first step of the treatment, in order to decorate the surface of the tumor cells, while exposing the hapten-recognizing portion. In a second phase of the treatment, the hapten bearing the radionucleotide is infused in the patient and a portion of this entity will be recognized and bound tightly by the antibody fragment, while the rest will be rapidly eliminated from the system. This approach enables to direct the localization of the radioactive nucleus on the tumor tissue, avoiding the exposure of healthy tissues and achieving promising therapeutic effects (Sharkey & Goldenberg, 2005; Zhang et al., 2002).

Small immunoproteins may be the ideal format for cancer therapy since they exhibit higher total tumor uptake, better tumor-to-blood ratio, and faster clearance than intact IgG or Fab (Chames et al., 2009). Moreover, a short serum half-life may be desirable in cases where the risk of toxicity in healthy tissues increases with prolonged exposure. For example, when antibodies carry radionuclides, rapid tumor uptake and fast blood clearance can limit the bone marrow toxicity typically associated with the use of full IgGs in radioimmunotherapy (Tijink et al., 2009).

The value of antibody fragments as targeting moieties for cancer therapy has been also explored in the context of gene therapy, as it allows to deliver polyplexes inside MUC1-overexpressing tumor cells. To this end, an MUC1-specific nanobody has been modified at an engineered C-terminal cysteine through maleimide chemistry, attaching a branched PEG-polyethylenimine moiety. This functionalization allowed for the condensation of plasmids coding a transcriptionally targeted truncated-Bid (tBid) killer gene and their delivery to the cancer cell, where they are activated under the control of the cancer-specific MUC1 promoter (Sadeqzadeh et al., 2011).

Historically, cancer pharmacotherapy has typically relied on the use of cytotoxic agents. The use of these drugs was based on the idea that these agents would preferentially kill rapidly dividing cancer cells, while sparing normal cells. These chemotherapeutic agents included antifolates and nucleoside analogs (methotrexate, thioguanine, 5-fluorouracil, and 6-mercaptopurine), DNA-interacting agents (cisplatin), antitumor antibiotics (anthracyclines and actinomycin D), and microtubule polymerization inhibitors/promoters (vinca alkaloids and taxanes) (DeVita & Chu, 2008). Despite the large number of anticancer drugs that received marketing approval by the end of the 1980s, their efficacy continued to be reduced by the lack of tumor selectivity, resulting in the damage of healthy tissues in rapid proliferation, including bone marrow and gastrointestinal tract. Most anticancer drugs had to be used near the maximum tolerated dose (MTD) to achieve a clinically significant therapeutic effect but could only be used for a limited time period due to uptake and cumulative damage in normal tissues (Krall, Scheuermann, & Neri, 2013; van der Veldt et al., 2010, 2013). Multidrug therapy (i.e., the concomitant use of different cancer drugs with different mechanisms of action and non-overlapping toxicities) sometimes improved antitumor activity (Frei, 1972); however, systemic toxicity has remained a major drawback and cures can be achieved only in a small set of metastatic cancer types. Conventional cytotoxic agents display an insufficient therapeutic index (MTD/minimum efficacious dose), resulting in a narrow therapeutic window (Chari, Miller, & Widdison, 2014). The rapid clearance of small molecules and the increased interstitial pressure in solid tumors further decrease the antitumor activity of chemotherapy. As such, an ideal anticancer strategy would decrease the minimum effective dose and increase the MTD. Active research in the area of natural products has led to the discovery of extremely potent cytotoxic compounds, which could be considered for ligand-based pharmacodelivery applications. These agents include potent tubulin agents, such as maytansinoids, dolastatins, and cryptophycins (all strong inhibitors of tubulin polymerization with IC_{50} in the picomolar range), but also duocarmycins (highly cytotoxic minor groove binders and DNA-alkylating agents). Although these drugs failed to demonstrate clinical benefits on their own, demonstrating that the increase of the potency by itself does not increase the therapeutic index of chemotherapy, they represent attractive payloads for the preparation of ADCs.

ADCs are designed to take advantage of the specificity, pharmacokinetic, and biodistribution properties of an antigen-targeting antibody and the potent cell-killing activity of small molecules (Sievers & Senter, 2013). ADCs can improve the therapeutic index of anticancer drugs by modulating their distribution in the body, while at the same time enhancing the anti-tumor activity of the unmodified antibodies (Wu & Senter, 2005). Although the concept of ADCs is simple, over 20 years of research have so far only resulted in two clinically approved products: Adcetris® (for the treatment of Hodgkin lymphoma and anaplastic large cell lymphoma) and Kadcyla® (for the treatment of breast cancer) (Senter & Sievers, 2012; Verma et al., 2012). While this recent clinical success further stimulated research in the field of ADC, with more than 20 products being currently in clinical testing, many aspects of ADC design and development are still not fully understood. This includes the identification of ideal targets, antibody format, as well as linker-payload combination.

So far, no antibody fragment has been approved for cancer therapy, even though antibody fragments used for the pharmacodelivery of cytokines are showing promising results for the treatment of cancer and chronic inflammation. L19-IL2, a tumor-targeting immunocytokine consisting of the human scFv antibody fragment L19 (specific to the alternatively-spliced EDB domain of fibronectin, a well-characterized marker of tumor neovasculature) linked to the human IL2, has demonstrated strong therapeutic activity in animal cancer models and has been studied in phase II clinical trials (Eigentler et al., 2011; Schwager, Hemmerle, Aebischer, & Neri, 2013).

6.2 CNS Diseases

A new wave of innovative antibody-based therapies promises significant breakthroughs in various ranges of diseases. Yet, in the field of CNS, drug discovery and development researchers are experiencing difficulties in developing CNS drugs that can complete clinical trials and reach regulatory approval. In fact, as reviewed by Strohl (2017) not a single CNS antibody-based molecule has been approved by FDA (Strohl, 2017).

This is particularly disturbing for major unmet needs in the CNS area such as Alzheimer's, Parkinson's, stroke, and brain cancers.

In fact, treatment of CNS diseases has been a major drug development challenge. Despite intensive research on the structure of the brain and the advances of medical and nanotechnology, many neurological diseases remain undertreated by effective therapies and this is not due to the lack of good candidate drugs. A major bottleneck in successful development of CNS drugs is the discovery and design of drugs that can cross the BBB (Chen & Liu, 2012).

The BBB is a complex endothelial barrier that separates the brain from the circulating blood. It is a highly regulated and efficient barrier composed of a continuous layer of microvascular endothelial cells firmly sealed with tight junctions that prevent paracellular route transport of molecules (Abbott, Patabendige, Dolman, Yusof, & Begley, 2010). It maintains the homeostasis of the brain and protects it against undesirable molecules and pathogens. Consequently, only small molecules (<400 Da) with appropriated lipophilicity, molecular weight (MW), and charge can freely diffuse across the BBB. Higher MW substances essential to the brain can only transpose by either transport proteins, specific receptor-mediated transcytosis (RMT), or adsorptive transcytosis (AMT). As a result, the majority of small molecules (MW > 500 Da), proteins, and peptides do not freely cross the BBB which results in 100% of large protein drugs and >98% of small molecule drugs unable to reach its CNS target, representing a major obstacle for the accumulation of the majority biologically active molecules in the brain (Neves, Aires-da-Silva, Corte-Real, & Castanho, 2016; Pardridge, 2012).

A promising approach to circumvent the BBB impenetrability is to develop CNS drug delivery strategies that take advantage of BBB transport systems. Due to the specificity of the interaction, RMT has been strategically used for brain drug delivery. In the RMT, specific and highly expressed receptors, such as insulin receptor (IR), transferrin receptor (TfR), and LDL receptor, when bound to its ligand, internalize the ligand–receptor complex. Following the release of the ligand at the abluminal (brain) side of the brain capillary endothelium, the receptor recycles back to the luminal membrane, allowing the delivery of macromolecules important for brain function (Farrington et al., 2014; Gabathuler, 2010).

To take advantage of this transport system, mAbs targeting endogenous BBB receptors are being reengineered as "Trojan Horses," or nanoscaled molecular platforms, to cross the BBB in a controlled manner and carry therapeutic compounds without damaging the cells (Lajoie & Shusta, 2015; Pardridge, 2016; Pardridge & Boado, 2012). mAbs are particularly appealing brain shuttles because not only can they be designed to specifically target BBB receptors that undergo RMT, but also be easily conjugated with other molecules or nanoparticles with therapeutic features as previously mentioned.

The most well-characterized receptors that have been used to transport therapeutic molecules across the BBB are the IR and TfR (Pardridge, 2016). Several studies have already validated the use of anti-human IR IgG antibodies to transpose the BBB (Boado, Zhang, Zhang, & Pardridge, 2007; Pardridge, 2012; Pardridge, Kang, Buciak, & Yang, 1995). Moreover, fusion of this IR antibody to different lysosomal enzymes (Boado, Hui, Lu, & Pardridge, 2012; Boado, Ka-Wai Hui, Zhiqiang Lu, & Pardridge, 2014; Boado, Lu, Hui, & Pardridge, 2014; Boado & Pardridge, 2017; Pardridge, 2012), peptide radiopharmaceuticals (Wu, Yang, & Pardridge, 1997), and DNA-encapsulated pegylated liposomes (Zhang, Schlachetzki, & Pardridge, 2003) validated the use of receptor-specific antibodies for brain transport and demonstrated their potential applications in terms of therapeutics and diagnostic of neurological diseases.

Similarly, to insulin, antibodies toward TfR have also been widely validated as effective BBB drug delivery vectors. Applications include amyloid-β-targeting liposomes, decorated with an anti-TfR antibody (Salvati et al., 2013); development of a bivalent antibody with affinity to TfR and BACE (beta-secretase-1) (Yu et al., 2011); reengineering of bispecific antibody with two arms each binding to TfR and APP (amyloid-beta, Aβ) (Niewoehner et al., 2014); and TfR antibodies conjugated to pegylated liposomes for the brain delivery of nonviral plasmid DNA (Xia, Boado, Zhang, Chu, & Pardridge, 2008). Nevertheless, despite the validation of antibodies toward insulin and transferrin receptors as potential brain shuttles, these mAbs are mainly conventional IgG molecules meaning that the nonspecific tissue uptake by highly expressed Fc receptors is also high, leading to a decrease of brain targeting.

A promising alternative to these conventional BBB mAbs is the engineering of antibody fragments such as Fab, scFv, and sdAbs. As previously mentioned fragmentation of antibodies results in modified physiochemical features of these therapeutic molecules particularly interesting for BBB targeting, such as reaching targets not easily accessible by conventional antibodies. More importantly, these fragments can be easily attached to other proteins, peptides, small molecules, or nanoparticles by simple molecular biology or chemical procedures, which makes them ideal to function as brain-targeting drug delivery vectors (Abulrob et al., 2005; Aires da Silva et al., 2008; Nelson, 2010). In line with this, significant focus is being given for the development of antibody fragments with brain-targeting of RMT receptors for drug delivery.

Two candidate brain-targeting sdAbs, FC5 and FC44, have been obtained by phage display library of camelid single-domain antibodies (VHHs) and demonstrated significant accumulation in brain of mice and rapid elimination via kidneys and liver (Muruganandam, Tanha, Narang, & Stanimirovic, 2002). FC5 was further characterized to engage active RMT (Abulrob et al., 2005). To establish FC5 as a BBB delivery vector, this sdAb was complexed with human Fc domain and the opioid peptide dalargin (Dal) in mono- or bivalent fusions and was shown to markedly enhance brain exposure, as well as pharmacological potency in the CNS. This suggested FC5 as a potential BBB carrier to increase concentrations of biotherapeutics in the CNS (Farrington et al., 2014). FC5 was further fused in an scFv format to the N-terminus of an IgG against glutamate metabotropic receptor (mGluR1) to shuttle it across the BBB, achieving pharmacological activity with a 10-fold enrichment of the antibody in the brain parenchyma compared with a synthetic mGluR1 inhibitor (Webster et al., 2016).

Similarly, Jones and collaborators performed a study to identify novel antibody fragments for brain targeting. In their approach, a nonimmune human scFv library was panned against a BBB model comprised of rat endothelial cells in order to discover new BBB-targeting molecules (Jones, Stutz, Zhou, Marks, & Shusta, 2014). Three scFvs were isolated (scFv38, scFv29, and scFv15) from the screening, with two having shown to preferentially bind to the rat brain microvasculature in vivo (scFv15 or scFv38) (Jones et al., 2014), although the ability to transpose the BBB and function as a drug delivery vector remains to be demonstrated.

In another study, Niewoehner et al. explored if using monovalent binding to the TfR, instead of the conventional IgG, would increase brain exposure. To achieve this, authors engineered two different types of brain shuttle constructs with a single-chain (sc)Fab fragment of an anti-TfR monoclonal antibody (mAb) fused to either one or both C-terminal ends of the heavy chain of an anti-Aβ mAb (mAb31), which binds to β-amyloid plaques. The authors observed that the binding properties toward the TfR were preserved in the two brain shuttle constructs. However, the sFab-binding affinity was slightly reduced (five- to eight-fold) when compared to the dFab construct and was dependent on TfR-coating density. Their in vivo experiments demonstrated significant uptake of both shuttle constructs. Nevertheless, while dFab remained within the BECs, sFab efficiently entered the CNS compartment reaching the amyloid plaques which possibly indicates the existence of different internalization pathways (Niewoehner et al., 2014). Curiously, this study contrast with the FC5 llama single-domain antibody in which the dimer construct appeared to be more efficiently transcytosed than the monomer (Farrington et al., 2014).

Despite the need to identify RMT receptors other than insulin and transferrin, these receptors are still being quite used to understand the biology of antibody BBB transport, validate new antibody shuttles, and improve platforms for brain penetration. Kim and collaborators have developed a systemic nanodelivery platform for tumor targeting in glioblastoma multiforme (GBM), an aggressive and lethal form of brain tumor in adults. This consisted of a p53 immunoliposome nanocomplex (designated as p53-scL) for systemic delivery of molecular medicines where the WTp53 gene therapy was encapsulated within a cationic liposome decorated with an anti-TfR scFv, designed to target intracranial tumors. According to the authors, the nanocomplex was able to cross the BBB and efficiently deliver its payload which resulted in apoptosis induction in intracranial GBM xenografts (Kim et al., 2014).

In conclusion, while the BBB still constitutes the major obstacle for neurological diseases pharmaceuticals, antibody-based therapies hold a great promise for CNS diseases. Apart from the therapeutic application of antibodies by themselves, these molecules, and antibody fragments, are exceptionally appealing to be used as drug delivery shuttles for the BBB. The ability of delivering brain-specific therapies allows drug targeting to the disease and reduces systemic distribution of drugs, which leads to decreased toxicity and side effects. Yet, there is still much to be learned about BBB transport and improving antibodies for transcytotic delivery to the CNS. Furthermore, currently exploited RMT targets (IR and TfR) for the delivery of large molecules across the BBB are broadly expressed in tissues and implicated in metabolically critical functions (Stanimirovic, Kemmerich, Haqqani, & Farrington, 2014). Therefore, there is an urgent need to develop more effective transmigrating antibodies as well as to discover novel BBB endogenous receptors, which could provide a more selectively delivery into the brain.

7. CONCLUSIONS

The development of ADCs ushered a new era of investigation at the interface between chemistry, immunology, and biotechnology that yielded impressive results and brought many innovative therapeutics to the clinic. The initial enthusiasm for these new technologies has been toned down by the discovery of various limitations to the use of full antibodies, some of which were addressed, while others are intrinsic to the use of such large biomolecules. In the early 2000s, in order to alleviate some of the disadvantages of the use of IgGs as targeting agents, there was a surge in interest in the use of small molecules for this task (Adams, Smothers, Srinivasan, & Hoos, 2015; Krall et al., 2013). As we look at the promising results of full antibodies and small molecules as targeting agents, we can see that many of the drawbacks associated with both can be traced back to pharmacokinetic issues that are, in the end, dependent on the size of these entities. In this framework, the development of antibody fragments can be the answer to obtaining the best of both worlds and has already shown to outperform both IgGs and small molecules in various applications. As the field of antibody fragments grows, new and interesting structures are developed, expanding the immunological toolbox at our disposal with structures that can show unique and valuable properties (Holliger & Hudson, 2005).

As the fast-paced development and discovery of new engineered antibody fragments give us new entities to work with, it will be crucial to combine all these advances with smart approaches for their chemical functionalization in order to fully explore their potential. With this mindset, the development of strong interdisciplinary investigation will be highly beneficial as the synergistic convergence of chemistry, biology, and biotechnology can bring these biomolecules to solve many unmet needs in both research and the clinic.

ACKNOWLEDGMENTS

The authors thank Portuguese Funding scheme (Fundação para a Ciência e Tecnologia, FCT) for financial support (Grants SFRH/BPD/100522/2014, IF/01010/2013, BBB/BIO/0508/2014, SFRH/BD/90514/ 2012, PTDC/QEQMED/5512/2014, PTDC/QEQ-QOR/1434/2014, EU Marie Sklodowska-Curie ITN ProteinConjugates, and SAICTPAC/0019/2015). This project has received funding from European Structural & Investment Funds through the COMPETE Programme and from National Funds through FCT—*Fundação para a Ciência e a Tecnologia* under the Programme grant SAICTPAC/0019/2015.

REFERENCES

Abbott, N. J., Patabendige, A. A. K., Dolman, D. E. M., Yusof, S. R., & Begley, D. J. (2010). Structure and function of the blood-brain barrier. *Neurobiology of Disease*, *37*(1), 13–25. https://doi.org/10.1016/j.nbd.2009.07.030.

Abulrob, A., Sprong, H., Van Bergen en Henegouwen, P., & Stanimirovic, D. (2005). The blood-brain barrier transmigrating single domain antibody: Mechanisms of transport and antigenic epitopes in human brain endothelial cells. *Journal of Neurochemistry*, *95*(4), 1201–1214. https://doi.org/10.1111/j.1471-4159.2005.03463.x.

Adams, G. P., Schier, R., McCall, A. M., Simmons, H. H., Horak, E. M., Alpaugh, R. K., et al. (2001). High affinity restricts the localization and tumor penetration of single-chain fv antibody molecules. *Cancer Research*, *61*(12), 4750–4755.

Adams, J. L., Smothers, J., Srinivasan, R., & Hoos, A. (2015). Big opportunities for small molecules in immuno-oncology. *Nature Reviews Drug Discovery*, *14*(9), 603–622. https://doi.org/10.1038/nrd4596.

Agarwal, P., & Bertozzi, C. R. (2015). Site-specific antibody–drug conjugates: The nexus of bioorthogonal chemistry, protein engineering, and drug development. *Bioconjugate Chemistry*, *26*(2), 176–192. https://doi.org/10.1021/bc5004982.

Ahn, J., Miura, Y., Yamada, N., Chida, T., Liu, X., Kim, A., et al. (2015). Antibody fragment-conjugated polymeric micelles incorporating platinum drugs for targeted therapy of pancreatic cancer. *Biomaterials*, *39*, 23–30. https://doi.org/10.1016/j.biomaterials.2014.10.069.

Aires da Silva, F., Corte-Real, S., & Goncalves, J. (2008). Recombinant antibodies as therapeutic agents: Pathways for modeling new biodrugs. *BioDrugs: Clinical Immunotherapeutics, Biopharmaceuticals and Gene Therapy*, *22*(5), 301–314.

Aires da Silva, F., Santa-Marta, M., Freitas-Vieira, A., Mascarenhas, P., Barahona, I., Moniz-Pereira, J., et al. (2004). Camelized rabbit-derived VH single-domain intrabodies against Vif strongly neutralize HIV-1 infectivity. *Journal of Molecular Biology*, *340*(3), 525–542. https://doi.org/10.1016/j.jmb.2004.04.062.

Akkapeddi, P., Azizi, S.-A., Freedy, A. M., Cal, P. M. S. D., Gois, P. M. P., & Bernardes, G. J. L. (2016). Construction of homogeneous antibody–drug conjugates using site-selective protein chemistry. *Chemical Science*, 7(5), 2954–2963. https://doi.org/10.1039/C6SC00170J.

Alric, C., Aubrey, N., Allard-Vannier, É., di Tommaso, A., Blondy, T., Dimier-Poisson, I., et al. (2016). Covalent conjugation of cysteine-engineered scFv to PEGylated magnetic nanoprobes for immunotargeting of breast cancer cells. *RSC Advances*, *6*(43), 37099–37109. https://doi.org/10.1039/C6RA06076E.

Anderson, G. W., Zimmerman, J. E., & Callahan, F. M. (1964). The use of esters of N-hydroxysuccinimide in peptide synthesis. *Journal of the American Chemical Society*, *86*(9), 1839–1842. https://doi.org/10.1021/ja01063a037.

Annapragada, A. (2015). Advances in nanoparticle imaging technology for vascular pathologies. *Annual Review of Medicine*, *66*(1), 177–193. https://doi.org/10.1146/annurev-med-042513-125551.

Arruebo, M., Valladares, M., & González-Fernández, Á. (2009). *Antibody-conjugated nanoparticles for biomedical applications. Journal of Nanomaterials, 2009.* (Article ID 439389). https://doi.org/10.1155/2009/439389.

Badescu, G., Bryant, P., Bird, M., Henseleit, K., Swierkosz, J., Parekh, V., et al. (2014). Bridging disulfides for stable and defined antibody drug conjugates. *Bioconjugate Chemistry*, *25*(6), 1124–1136. https://doi.org/10.1021/bc500148x.

Baetke, S. C., Lammers, T., & Kiessling, F. (2015). Applications of nanoparticles for diagnosis and therapy of cancer. *The British Journal of Radiology*, *88*(1054), 20150207. https://doi.org/10.1259/bjr.20150207.

Baldwin, A. D., & Kiick, K. L. (2011). Tunable degradation of maleimide–thiol adducts in reducing environments. *Bioconjugate Chemistry*, *22*(10), 1946–1953. https://doi.org/10.1021/bc200148v.

Bird, R. E., Hardman, K. D., Jacobson, J. W., Johnson, S., Kaufman, B. M., Lee, S. M., et al. (1988). Single-chain antigen-binding proteins. *Science (New York, N.Y.)*, *242*(4877), 423–426.

Boado, R. J., Hui, E. K.-W., Lu, J. Z., & Pardridge, W. M. (2012). Glycemic control and chronic dosing of rhesus monkeys with a fusion protein of iduronidase and a monoclonal antibody against the human insulin receptor. *Drug Metabolism and Disposition: The Biological Fate of Chemicals*, *40*(10), 2021–2025. https://doi.org/10.1124/dmd.112.046375.

Boado, R. J., Ka-Wai Hui, E., Zhiqiang Lu, J., & Pardridge, W. M. (2014). Insulin receptor antibody-iduronate 2-sulfatase fusion protein: Pharmacokinetics, anti-drug antibody, and safety pharmacology in rhesus monkeys. *Biotechnology and Bioengineering*, *111*(11), 2317–2325. https://doi.org/10.1002/bit.25289.

Boado, R. J., Lu, J. Z., Hui, E. K.-W., & Pardridge, W. M. (2014). Insulin receptor antibody-sulfamidase fusion protein penetrates the primate blood-brain barrier and reduces glycosoaminoglycans in Sanfilippo type A cells. *Molecular Pharmaceutics*, *11*(8), 2928–2934. https://doi.org/10.1021/mp500258p.

Boado, R. J., & Pardridge, W. M. (2017). Brain and organ uptake in the rhesus monkey in vivo of recombinant iduronidase compared to an insulin receptor antibody-iduronidase fusion protein. *Molecular Pharmaceutics*, *14*(4), 1271–1277. https://doi.org/10.1021/acs.molpharmaceut.6b01166.

Boado, R. J., Zhang, Y., Zhang, Y., & Pardridge, W. M. (2007). Humanization of anti-human insulin receptor antibody for drug targeting across the human blood-brain barrier. *Biotechnology and Bioengineering*, *96*(2), 381–391. https://doi.org/10.1002/bit.21120.

Bryden, F., Maruani, A., Savoie, H., Chudasama, V., Smith, M. E. B., Caddick, S., et al. (2014). Regioselective and stoichiometrically controlled conjugation of photodynamic sensitizers to a HER2 targeting antibody fragment. *Bioconjugate Chemistry*, *25*(3), 611–617. https://doi.org/10.1021/bc5000324.

Burris, H. A., Rugo, H. S., Vukelja, S. J., Vogel, C. L., Borson, R. A., Limentani, S., et al. (2011). Phase II study of the antibody drug conjugate trastuzumab-DM1 for the treatment of human epidermal growth factor receptor 2 (HER2)-positive breast cancer after prior HER2-directed therapy. *Journal of Clinical Oncology: Official Journal of the American Society of Clinical Oncology*, *29*(4), 398–405. https://doi.org/10.1200/JCO.2010.29.5865.

Cal, P. M. S. D., Bernardes, G. J. L., & Gois, P. M. P. (2014). Cysteine-selective reactions for antibody conjugation. *Angewandte Chemie (International Edition in English)*, *53*(40), 10585–10587. https://doi.org/10.1002/anie.201405702.

Casi, G., Huguenin-Dezot, N., Zuberbühler, K., Scheuermann, J., & Neri, D. (2012). Site-specific traceless coupling of potent cytotoxic drugs to recombinant antibodies for pharmacodelivery. *Journal of the American Chemical Society*, *134*(13), 5887–5892. https://doi.org/10.1021/ja211589m.

Chames, P., Van Regenmortel, M., Weiss, E., & Baty, D. (2009). Therapeutic antibodies: Successes, limitations and hopes for the future. *British Journal of Pharmacology*, *157*(2), 220–233. https://doi.org/10.1111/j.1476-5381.2009.00190.x.

Chapman, A. P., Antoniw, P., Spitali, M., West, S., Stephens, S., & King, D. J. (1999). Therapeutic antibody fragments with prolonged in vivo half-lives. *Nature Biotechnology*, *17*(8), 780–783. https://doi.org/10.1038/11717.

Chari, R. V. J., Miller, M. L., & Widdison, W. C. (2014). Antibody-drug conjugates: An emerging concept in cancer therapy. *Angewandte Chemie (International Edition in English)*, *53*(15), 3796–3827. https://doi.org/10.1002/anie.201307628.

Chatalic, K. L. S., Veldhoven-Zweistra, J., Bolkestein, M., Hoeben, S., Koning, G. A., Boerman, O. C., et al. (2015). A novel ^{111}In-labeled anti-prostate-specific membrane antigen nanobody for targeted SPECT/CT imaging of prostate cancer. *Journal of Nuclear Medicine: Official Publication, Society of Nuclear Medicine, 56*(7), 1094–1099. https://doi.org/10.2967/jnumed.115.156729.

Chen, Y., & Liu, L. (2012). Modern methods for delivery of drugs across the blood–brain barrier. *Advanced Drug Delivery Reviews, 64*(7), 640–665. https://doi.org/10.1016/j.addr.2011.11.010.

Cheng, W. W., & Allen, T. M. (2008). Targeted delivery of anti-CD19 liposomal doxorubicin in B-cell lymphoma: A comparison of whole monoclonal antibody, Fab' fragments and single chain Fv. *Journal of Controlled Release: Official Journal of the Controlled Release Society, 126*(1), 50–58. https://doi.org/10.1016/j.jconrel.2007.11.005.

Cheng, W. W., & Allen, T. M. (2010). The use of single chain Fv as targeting agents for immunoliposomes: An update on immunoliposomal drugs for cancer treatment. *Expert Opinion on Drug Delivery*, 7(4), 461–478. https://doi.org/10.1517/17425240903579963.

Cui, D., Han, Y., Li, Z., Song, H., Wang, K., He, R., et al. (2009). Fluorescent magnetic nanoprobes for in vivo targeted imaging and hyperthermia therapy of prostate cancer. *Nano Biomedicine and Engineering, 1*(1), 61–74. https://doi.org/10.5101/nbe.v1i1.p61-74.

de Graaf, A. J., Kooijman, M., Hennink, W. E., & Mastrobattista, E. (2009). Nonnatural amino acids for site-specific protein conjugation. *Bioconjugate Chemistry, 20*(7), 1281–1295. https://doi.org/10.1021/bc800294a.

DeVita, V. T., & Chu, E. (2008). A history of cancer chemotherapy. *Cancer Research, 68*(21), 8643–8653. https://doi.org/10.1158/0008-5472.CAN-07-6611.

Ehrlich, P. (1913). Address in pathology, on chemiotherapy: Delivered before the seventeenth international congress of medicine. *British Medical Journal, 2*(2746), 353–359.

Eigentler, T. K., Weide, B., de Braud, F., Spitaleri, G., Romanini, A., Pflugfelder, A., et al. (2011). A dose-escalation and signal-generating study of the immunocytokine L19-IL2 in combination with dacarbazine for the therapy of patients with metastatic melanoma. *Clinical Cancer Research: An Official Journal of the American Association for Cancer Research, 17*(24), 7732–7742. https://doi.org/10.1158/1078-0432.CCR-11-1203.

Farrington, G. K., Caram-Salas, N., Haqqani, A. S., Brunette, E., Eldredge, J., Pepinsky, B., et al. (2014). A novel platform for engineering blood-brain barrier-crossing bispecific biologics. *The FASEB Journal, 28*(11), 4764–4778. https://doi.org/10.1096/fj.14-253369.

Flemming, A. (2014). Antibody engineering: Fine-tuning antibody–drug conjugates. *Nature Reviews Drug Discovery, 13*, 178. https://doi.org/10.1038/nrd4266.

Frei, E. (1972). Combination cancer therapy: Presidential address. *Cancer Research, 32*(12), 2593–2607.

Future Market Insight. (2016). Antibodies Market: Launch of Innovative Drugs and Easy Availability of Biosimilar Is Expected to Boost the Demand: Global Industry Analysis and Opportunity Assessment 2016–2026. Retrieved from https://www.futuremarketinsights.com/reports/antibodies-market.

Gabathuler, R. (2010). Approaches to transport therapeutic drugs across the blood-brain barrier to treat brain diseases. *Neurobiology of Disease, 37*(1), 48–57. https://doi.org/10.1016/j.nbd.2009.07.028.

Greenberg, A. S., Avila, D., Hughes, M., Hughes, A., McKinney, E. C., & Flajnik, M. F. (1995). A new antigen receptor gene family that undergoes rearrangement and extensive somatic diversification in sharks. *Nature, 374*(6518), 168–173. https://doi.org/10.1038/374168a0.

Greene, M. K., Richards, D. A., Nogueira, J. C. F., Campbell, K., Smyth, P., Fernández, M., et al. (2017). Forming next-generation antibody–nanoparticle conjugates through the oriented installation of non-engineered antibody fragments. *Chemical Science, 9*(1), 79–87. https://doi.org/10.1039/C7SC02747H.

Greten, T. F., Manns, M. P., & Korangy, F. (2011). Myeloid derived suppressor cells in human diseases. *International Immunopharmacology, 11*(7), 802–807. https://doi.org/10.1016/j.intimp.2011.01.003.

Griffiths, G. L., Chang, C.-H., McBride, W. J., Rossi, E. A., Sheerin, A., Tejada, G. R., et al. (2004). Reagents and methods for PET using bispecific antibody pretargeting and 68Ga-radiolabeled bivalent hapten-peptide-chelate conjugates. *Journal of Nuclear Medicine: Official Publication, Society of Nuclear Medicine, 45*(1), 30–39.

Hamann, P. R., Hinman, L. M., Hollander, I., Beyer, C. F., Lindh, D., Holcomb, R., et al. (2002). Gemtuzumab ozogamicin, a potent and selective anti-CD33 antibody–calicheamicin conjugate for treatment of acute myeloid leukemia. *Bioconjugate Chemistry, 13*(1), 47–58. https://doi.org/10.1021/bc010021y.

Hamblett, K. J., Senter, P. D., Chace, D. F., Sun, M. M. C., Lenox, J., Cerveny, C. G., et al. (2004). Effects of drug loading on the antitumor activity of a monoclonal antibody drug conjugate. *Clinical Cancer Research, 10*(20), 7063–7070. https://doi.org/10.1158/1078-0432.CCR-04-0789.

Hamers-Casterman, C., Atarhouch, T., Muyldermans, S., Robinson, G., Hamers, C., Songa, E. B., et al. (1993). Naturally occurring antibodies devoid of light chains. *Nature, 363*(6428), 446–448. https://doi.org/10.1038/363446a0.

Hofer, T., Skeffington, L. R., Chapman, C. M., & Rader, C. (2009). Molecularly defined antibody conjugation through a selenocysteine interface. *Biochemistry, 48*(50), 12047–12057. https://doi.org/10.1021/bi901744t.

Holliger, P., & Hudson, P. J. (2005). Engineered antibody fragments and the rise of single domains. *Nature Biotechnology, 23*(9), 1126–1136. https://doi.org/10.1038/nbt1142.

Holt, L. J., Herring, C., Jespers, L. S., Woolven, B. P., & Tomlinson, I. M. (2003). Domain antibodies: Proteins for therapy. *Trends in Biotechnology, 21*(11), 484–490. https://doi.org/10.1016/j.tibtech.2003.08.007.

Hudson, P. J. (1998). Recombinant antibody fragments. *Current Opinion in Biotechnology, 9*(4), 395–402.

Hudson, P. J. (1999). Recombinant antibody constructs in cancer therapy. *Current Opinion in Immunology, 11*(5), 548–557.

Hudson, P. J., & Souriau, C. (2003). Engineered antibodies. *Nature Medicine, 9*(1), 129–134. https://doi.org/10.1038/nm0103-129.

Humphreys, D. P., Heywood, S. P., Henry, A., Ait-Lhadj, L., Antoniw, P., Palframan, R., et al. (2007). Alternative antibody Fab′ fragment PEGylation strategies: Combination of strong reducing agents, disruption of the interchain disulphide bond and disulphide engineering. *Protein Engineering, Design and Selection, 20*(5), 227–234. https://doi.org/10.1093/protein/gzm015.

Hwang, W. Y. K., & Foote, J. (2005). Immunogenicity of engineered antibodies. *Methods (San Diego, Calif.), 36*(1), 3–10. https://doi.org/10.1016/j.ymeth.2005.01.001.

Jones, P. T., Dear, P. H., Foote, J., Neuberger, M. S., & Winter, G. (1986). Replacing the complementarity-determining regions in a human antibody with those from a mouse. *Nature, 321*(6069), 522–525. https://doi.org/10.1038/321522a0.

Jones, M. W., Strickland, R. A., Schumacher, F. F., Caddick, S., Baker, J. R., Gibson, M. I., et al. (2012). Polymeric dibromomaleimides as extremely efficient disulfide bridging bioconjugation and pegylation agents. *Journal of the American Chemical Society, 134*(3), 1847–1852. https://doi.org/10.1021/ja210335f.

Jones, A. R., Stutz, C. C., Zhou, Y., Marks, J. D., & Shusta, E. V. (2014). Identifying blood-brain-barrier selective single-chain antibody fragments. *Biotechnology Journal, 9*(5), 664–674. https://doi.org/10.1002/biot.201300550.

Junutula, J. R., Raab, H., Clark, S., Bhakta, S., Leipold, D. D., Weir, S., et al. (2008). Site-specific conjugation of a cytotoxic drug to an antibody improves the therapeutic index. *Nature Biotechnology, 26*(8), 925–932. https://doi.org/10.1038/nbt.1480.

Jurianz, K., Maslak, S., Garcia-Schüler, H., Fishelson, Z., & Kirschfink, M. (1999). Neutralization of complement regulatory proteins augments lysis of breast carcinoma cells targeted with rhumAb anti-HER2. *Immunopharmacology*, *42*(1–3), 209–218.

Khalili, H., Godwin, A., Choi, J., Lever, R., & Brocchini, S. (2012). Comparative binding of disulfide-bridged PEG-Fabs. *Bioconjugate Chemistry*, *23*(11), 2262–2277. https://doi.org/10.1021/bc300372r.

Kim, K. M., McDonagh, C. F., Westendorf, L., Brown, L. L., Sussman, D., Feist, T., et al. (2008). Anti-CD30 diabody-drug conjugates with potent antitumor activity. *Molecular Cancer Therapeutics*, 7(8), 2486–2497. https://doi.org/10.1158/1535-7163.MCT-08-0388.

Kim, S.-S., Rait, A., Kim, E., Pirollo, K. F., Nishida, M., Farkas, N., et al. (2014). A nanoparticle carrying the p53 gene targets tumors including cancer stem cells, sensitizes glioblastoma to chemotherapy and improves survival. *ACS Nano*, *8*(6), 5494–5514. https://doi.org/10.1021/nn5014484.

Kitamura, K., Takahashi, T., Yamaguchi, T., Noguchi, A., Noguchi, A., Takashina, K., et al. (1991). Chemical engineering of the monoclonal antibody A7 by polyethylene glycol for targeting cancer chemotherapy. *Cancer Research*, *51*(16), 4310–4315.

Köhler, G., & Milstein, C. (1975). Continuous cultures of fused cells secreting antibody of predefined specificity. *Nature*, *256*(5517), 495–497.

Kontermann, R. E. (2006). Immunoliposomes for cancer therapy. *Current Opinion in Molecular Therapeutics*, *8*(1), 39–45.

Kontermann, R. E. (2011). Strategies for extended serum half-life of protein therapeutics. *Current Opinion in Biotechnology*, *22*(6), 868–876. https://doi.org/10.1016/j.copbio.2011.06.012.

Krall, N., Scheuermann, J., & Neri, D. (2013). Small targeted cytotoxics: Current state and promises from DNA-encoded chemical libraries. *Angewandte Chemie International Edition*, *52*(5), 1384–1402. https://doi.org/10.1002/anie.201204631.

Kuan, S. L., Wang, T., & Weil, T. (2016). Site-selective disulfide modification of proteins: Expanding diversity beyond the proteome. *Chemistry—A European Journal*, *22*(48), 17112–17129. https://doi.org/10.1002/chem.201602298.

Lajoie, J. M., & Shusta, E. V. (2015). Targeting receptor-mediated transport for delivery of biologics across the blood-brain barrier. *Annual Review of Pharmacology and Toxicology*, *55*, 613–631. https://doi.org/10.1146/annurev-pharmtox-010814-124852.

Levy, E. M., Bianchini, M., Von Euw, E. M., Barrio, M. M., Bravo, A. I., Furman, D., et al. (2008). Human leukocyte antigen-E protein is overexpressed in primary human colorectal cancer. *International Journal of Oncology*, *32*(3), 633–641.

Lewis Phillips, G. D., Li, G., Dugger, D. L., Crocker, L. M., Parsons, K. L., Mai, E., et al. (2008). Targeting HER2-positive breast cancer with trastuzumab-DM1, an antibody-cytotoxic drug conjugate. *Cancer Research*, *68*(22), 9280–9290. https://doi.org/10.1158/0008-5472.CAN-08-1776.

Lin, A., Yan, W.-H., Xu, H.-H., Gan, M.-F., Cai, J.-F., Zhu, M., et al. (2007). HLA-G expression in human ovarian carcinoma counteracts NK cell function. *Annals of Oncology: Official Journal of the European Society for Medical Oncology*, *18*(11), 1804–1809. https://doi.org/10.1093/annonc/mdm356.

Liu, H., & May, K. (2012). Disulfide bond structures of IgG molecules: Structural variations, chemical modifications and possible impacts to stability and biological function. *MAbs*, *4*(1), 17–23. https://doi.org/10.4161/mabs.4.1.18347.

Manjappa, A. S., Goel, P. N., Gude, R. P., & Ramachandra Murthy, R. S. (2014). Anti-neuropilin 1 antibody Fab' fragment conjugated liposomal docetaxel for active targeting of tumours. *Journal of Drug Targeting*, *22*(8), 698–711. https://doi.org/10.3109/1061186X.2014.910792.

Maruani, A., Smith, M. E. B., Miranda, E., Chester, K. A., Chudasama, V., & Caddick, S. (2015). A plug-and-play approach to antibody-based therapeutics via a chemoselective dual click strategy. *Nature Communications*, *6*, 6645. https://doi.org/10.1038/ncomms7645.

Means, G. E., & Feeney, R. E. (1990). Chemical modifications of proteins: History and applications. *Bioconjugate Chemistry*, *1*(1), 2–12. https://doi.org/10.1021/bc00001a001.

Muruganandam, A., Tanha, J., Narang, S., & Stanimirovic, D. (2002). Selection of phage-displayed llama single-domain antibodies that transmigrate across human blood-brain barrier endothelium. *FASEB Journal: Official Publication of the Federation of American Societies for Experimental Biology*, *16*(2), 240–242. https://doi.org/10.1096/fj.01-0343fje.

Muyldermans, S. (2001). Single domain camel antibodies: Current status. *Journal of Biotechnology*, *74*(4), 277–302.

Nelson, A. L. (2010). Antibody fragments. *MAbs*, *2*(1), 77–83.

Neves, V., Aires-da-Silva, F., Corte-Real, S., & Castanho, M. A. R. B. (2016). Antibody approaches to treat brain diseases. *Trends in Biotechnology*, *34*(1), 36–48. https://doi.org/10.1016/j.tibtech.2015.10.005.

Niewoehner, J., Bohrmann, B., Collin, L., Urich, E., Sade, H., Maier, P., et al. (2014). Increased brain penetration and potency of a therapeutic antibody using a monovalent molecular shuttle. *Neuron*, *81*(1), 49–60. https://doi.org/10.1016/j.neuron.2013.10.061.

Oh, J.-H., Park, D. H., Joo, J. H., & Lee, J.-S. (2015). Recent advances in chemical functionalization of nanoparticles with biomolecules for analytical applications. *Analytical and Bioanalytical Chemistry*, *407*(29), 8627–8645. https://doi.org/10.1007/s00216-015-8981-y.

Oliveira, S., van Dongen, G. A. M. S., Stigter-van Walsum, M., Roovers, R. C., Stam, J. C., Mali, W., et al. (2012). Rapid visualization of human tumor xenografts through optical imaging with a near-infrared fluorescent anti-epidermal growth factor receptor nanobody. *Molecular Imaging*, *11*(1), 33–46.

Pardridge, W. M. (2012). Drug transport across the blood–brain barrier. *Journal of Cerebral Blood Flow & Metabolism*, *32*(11), 1959–1972. https://doi.org/10.1038/jcbfm.2012.126.

Pardridge, W. M. (2016). CSF, blood-brain barrier, and brain drug delivery. *Expert Opinion on Drug Delivery*, *13*(7), 963–975. https://doi.org/10.1517/17425247.2016.1171315.

Pardridge, W. M., & Boado, R. J. (2012). Reengineering biopharmaceuticals for targeted delivery across the blood-brain barrier. *Methods in Enzymology*, *503*, 269–292. https://doi.org/10.1016/B978-0-12-396962-0.00011-2.

Pardridge, W. M., Kang, Y. S., Buciak, J. L., & Yang, J. (1995). Human insulin receptor monoclonal antibody undergoes high affinity binding to human brain capillaries in vitro and rapid transcytosis through the blood-brain barrier in vivo in the primate. *Pharmaceutical Research*, *12*(6), 807–816.

Perez, H. L., Cardarelli, P. M., Deshpande, S., Gangwar, S., Schroeder, G. M., Vite, G. D., et al. (2014). Antibody-drug conjugates: Current status and future directions. *Drug Discovery Today*, *19*(7), 869–881. https://doi.org/10.1016/j.drudis.2013.11.004.

Petersdorf, S., Kopecky, K., Stuart, R. K., Larson, R. A., Nevill, T. J., Stenke, L., et al. (2009). Preliminary results of southwest oncology group study S0106: An international intergroup phase 3 randomized trial comparing the addition of gemtuzumab ozogamicin to standard induction therapy versus standard induction therapy followed by a second randomization to post-consolidation gemtuzumab ozogamicin versus no additional therapy for previously untreated acute myeloid leukemia. *Blood*, *114*(22), 790.

Pillow, T. H., Tien, J., Parsons-Reponte, K. L., Bhakta, S., Li, H., Staben, L. R., et al. (2014). Site-specific trastuzumab maytansinoid antibody–drug conjugates with improved therapeutic activity through linker and antibody engineering. *Journal of Medicinal Chemistry*, *57*(19), 7890–7899. https://doi.org/10.1021/jm500552c.

Presta, L. (2003). Antibody engineering for therapeutics. *Current Opinion in Structural Biology, 13*(4), 519–525.

Puthenveetil, S., Musto, S., Loganzo, F., Tumey, L. N., O'Donnell, C. J., & Graziani, E. (2016). Development of solid-phase site-specific conjugation and its application toward generation of dual labeled antibody and fab drug conjugates. *Bioconjugate Chemistry, 27*(4), 1030–1039. https://doi.org/10.1021/acs.bioconjchem.6b00054.

Qiao, S., Zhao, Y., Geng, S., Li, Y., Hou, X., Liu, Y., et al. (2016). A novel double-targeted nondrug delivery system for targeting cancer stem cells. *International Journal of Nanomedicine, 11*, 6667–6678. https://doi.org/10.2147/IJN.S116230.

Quarta, A., Bernareggi, D., Benigni, F., Luison, E., Nano, G., Nitti, S., et al. (2015). Targeting FR-expressing cells in ovarian cancer with Fab-functionalized nanoparticles: A full study to provide the proof of principle from in vitro to in vivo. *Nanoscale, 7*(6), 2336–2351. https://doi.org/10.1039/C4NR04426F.

Richards, D. A., Maruani, A., & Chudasama, V. (2016). Antibody fragments as nanoparticle targeting ligands: A step in the right direction. *Chemical Science, 8*(1), 63–77. https://doi.org/10.1039/C6SC02403C.

Rüger, R., Tansi, F. L., Rabenhold, M., Steiniger, F., Kontermann, R. E., Fahr, A., et al. (2014). In vivo near-infrared fluorescence imaging of FAP-expressing tumors with activatable FAP-targeted, single-chain Fv-immunoliposomes. *Journal of Controlled Release: Official Journal of the Controlled Release Society, 186*, 1–10. https://doi.org/10.1016/j.jconrel.2014.04.050.

Sadeqzadeh, E., Rahbarizadeh, F., Ahmadvand, D., Rasaee, M. J., Parhamifar, L., & Moghimi, S. M. (2011). Combined MUC1-specific nanobody-tagged PEG-polyethylenimine polyplex targeting and transcriptional targeting of tBid transgene for directed killing of MUC1 over-expressing tumour cells. *Journal of Controlled Release: Official Journal of the Controlled Release Society, 156*(1), 85–91. https://doi.org/10.1016/j.jconrel.2011.06.022.

Salvati, E., Re, F., Sesana, S., Cambianica, I., Sancini, G., Masserini, M., et al. (2013). Liposomes functionalized to overcome the blood-brain barrier and to target amyloid-β peptide: The chemical design affects the permeability across an in vitro model. *International Journal of Nanomedicine, 8*, 1749–1758. https://doi.org/10.2147/IJN.S42783.

Sapra, P., Moase, E. H., Ma, J., & Allen, T. M. (2004). Improved therapeutic responses in a xenograft model of human B lymphoma (Namalwa) for liposomal vincristine versus liposomal doxorubicin targeted via anti-CD19 IgG2a or Fab′ fragments. *Clinical Cancer Research, 10*(3), 1100–1111. https://doi.org/10.1158/1078-0432.CCR-03-0041.

Sapsford, K. E., Algar, W. R., Berti, L., Gemmill, K. B., Casey, B. J., Oh, E., et al. (2013). Functionalizing nanoparticles with biological molecules: Developing chemistries that facilitate nanotechnology. *Chemical Reviews, 113*(3), 1904–2074. https://doi.org/10.1021/cr300143v.

Schroeder, H. W., & Cavacini, L. (2010). Structure and function of immunoglobulins. *The Journal of Allergy and Clinical Immunology, 125*(2 Suppl. 2), S41–S52. https://doi.org/10.1016/j.jaci.2009.09.046.

Schwager, K., Hemmerle, T., Aebischer, D., & Neri, D. (2013). The immunocytokine L19-IL2 eradicates cancer when used in combination with CTLA-4 blockade or with L19-TNF. *The Journal of Investigative Dermatology, 133*(3), 751–758. https://doi.org/10.1038/jid.2012.376.

Senter, P. D., & Sievers, E. L. (2012). The discovery and development of brentuximab vedotin for use in relapsed Hodgkin lymphoma and systemic anaplastic large cell lymphoma. *Nature Biotechnology, 30*(7), 631–637. https://doi.org/10.1038/nbt.2289.

Shannon, D. A., & Weerapana, E. (2015). Covalent protein modification: The current landscape of residue-specific electrophiles. *Current Opinion in Chemical Biology, 24*, 18–26. https://doi.org/10.1016/j.cbpa.2014.10.021.

Sharkey, R. M., & Goldenberg, D. M. (2005). Perspectives on cancer therapy with radiolabeled monoclonal antibodies. *Journal of Nuclear Medicine: Official Publication, Society of Nuclear Medicine*, *46*(Suppl. 1), 27S–115S.

Shaunak, S., Godwin, A., Choi, J.-W., Balan, S., Pedone, E., Vijayarangam, D., et al. (2006). Site-specific PEGylation of native disulfide bonds in therapeutic proteins. *Nature Chemical Biology*, *2*(6), 312–313. https://doi.org/10.1038/nchembio786.

Sievers, E. L., Larson, R. A., Stadtmauer, E. A., Estey, E., Löwenberg, B., Dombret, H., et al. (2001). Efficacy and safety of gemtuzumab ozogamicin in patients with CD33-positive acute myeloid leukemia in first relapse. *Journal of Clinical Oncology: Official Journal of the American Society of Clinical Oncology*, *19*(13), 3244–3254. https://doi.org/10.1200/JCO.2001.19.13.3244.

Sievers, E. L., & Senter, P. D. (2013). Antibody-drug conjugates in cancer therapy. *Annual Review of Medicine*, *64*, 15–29. https://doi.org/10.1146/annurev-med-050311-201823.

Skerra, A., & Plückthun, A. (1988). Assembly of a functional immunoglobulin Fv fragment in *Escherichia coli*. *Science (New York, NY)*, *240*(4855), 1038–1041.

Smith, M. E. B., Schumacher, F. F., Ryan, C. P., Tedaldi, L. M., Papaioannou, D., Waksman, G., et al. (2010). Protein modification, bioconjugation, and disulfide bridging using bromomaleimides. *Journal of the American Chemical Society*, *132*(6), 1960–1965. https://doi.org/10.1021/ja908610s.

Stanimirovic, D., Kemmerich, K., Haqqani, A. S., & Farrington, G. K. (2014). Engineering and pharmacology of blood-brain barrier-permeable bispecific antibodies. *Advances in Pharmacology (San Diego, California)*, *71*, 301–335. https://doi.org/10.1016/bs.apha.2014.06.005.

Stephanopoulos, N., & Francis, M. B. (2011). Choosing an effective protein bioconjugation strategy. *Nature Chemical Biology*, 7(12), 876–884. https://doi.org/10.1038/nchembio.720.

Strohl, W. R. (2017). Current progress in innovative engineered antibodies. *Protein & Cell*, 1–35. https://doi.org/10.1007/s13238-017-0457-8.

Sun, M. M. C., Beam, K. S., Cerveny, C. G., Hamblett, K. J., Blackmore, R. S., Torgov, M. Y., et al. (2005). Reduction – alkylation strategies for the modification of specific monoclonal antibody disulfides. *Bioconjugate Chemistry*, *16*(5), 1282–1290. https://doi.org/10.1021/bc050201y.

Tijink, B. M., Perk, L. R., Budde, M., Stigter-van Walsum, M., Visser, G. W. M., Kloet, R. W., et al. (2009). 124I-L19-SIP for immuno-PET imaging of tumour vasculature and guidance of 131I-L19-SIP radioimmunotherapy. *European Journal of Nuclear Medicine and Molecular Imaging*, *36*(8), 1235–1244. https://doi.org/10.1007/s00259-009-1096-y.

Turner, M. (1998). Antibodies and their receptors. In *Immunology* (5th ed., pp. 71–82). London: Mosby International Ltd.

van Brussel, A. S. A., Adams, A., Oliveira, S., Dorresteijn, B., El Khattabi, M., Vermeulen, J. F., et al. (2016). Hypoxia-targeting fluorescent nanobodies for optical molecular imaging of pre-invasive breast cancer. *Molecular Imaging and Biology: MIB: The Official Publication of the Academy of Molecular Imaging*, *18*(4), 535–544. https://doi.org/10.1007/s11307-015-0909-6.

Van Cutsem, E., Köhne, C.-H., Láng, I., Folprecht, G., Nowacki, M. P., Cascinu, S., et al. (2011). Cetuximab plus irinotecan, fluorouracil, and leucovorin as first-line treatment for metastatic colorectal cancer: Updated analysis of overall survival according to tumor KRAS and BRAF mutation status. *Journal of Clinical Oncology: Official Journal of the American Society of Clinical Oncology*, *29*(15), 2011–2019. https://doi.org/10.1200/JCO.2010.33.5091.

van der Meel, R., Vehmeijer, L. J. C., Kok, R. J., Storm, G., & van Gaal, E. V. B. (2013). Ligand-targeted particulate nanomedicines undergoing clinical evaluation: Current status. *Advanced Drug Delivery Reviews*, *65*(10), 1284–1298. https://doi.org/10.1016/j.addr.2013.08.012.

van der Veldt, A. A. M., Hendrikse, N. H., Smit, E. F., Mooijer, M. P. J., Rijnders, A. Y., Gerritsen, W. R., et al. (2010). Biodistribution and radiation dosimetry of 11C-labelled docetaxel in cancer patients. *European Journal of Nuclear Medicine and Molecular Imaging, 37*(10), 1950–1958. https://doi.org/10.1007/s00259-010-1489-y.

van der Veldt, A. A. M., Lubberink, M., Mathijssen, R. H. J., Loos, W. J., Herder, G. J. M., Greuter, H. N., et al. (2013). Toward prediction of efficacy of chemotherapy: A proof of concept study in lung cancer patients using [^{11}C]docetaxel and positron emission tomography. *Clinical Cancer Research: An Official Journal of the American Association for Cancer Research, 19*(15), 4163–4173. https://doi.org/10.1158/1078-0432.CCR-12-3779.

Verma, S., Miles, D., Gianni, L., Krop, I. E., Welslau, M., Baselga, J., et al. (2012). Trastuzumab emtansine for HER2-positive advanced breast cancer. *New England Journal of Medicine, 367*(19), 1783–1791. https://doi.org/10.1056/NEJMoa1209124.

Wang, L., Amphlett, G., Blättler, W. A., Lambert, J. M., & Zhang, W. (2005). Structural characterization of the maytansinoid–monoclonal antibody immunoconjugate, huN901–DM1, by mass spectrometry. *Protein Science, 14*(9), 2436–2446. https://doi.org/10.1110/ps.051478705.

Wang, Y., Dossey, A. M., Froude, J. W., Lubitz, S., Tzur, D., Semenchenko, V., et al. (2008). PSA fluoroimmunoassays using anti-PSA ScFv and quantum-dot conjugates. *Nanomedicine (London, England), 3*(4), 475–483. https://doi.org/10.2217/17435889.3.4.475.

Wang, X., Zhu, J., Zhao, P., Jiao, Y., Xu, N., Grabinski, T., et al. (2007). In vitro efficacy of immuno-chemotherapy with anti-EGFR human Fab-TAXOL conjugate on A431 epidermoid carcinoma cells. *Cancer Biology & Therapy, 6*(6), 980–987.

Webster, C. I., Caram-Salas, N., Haqqani, A. S., Thom, G., Brown, L., Rennie, K., et al. (2016). Brain penetration, target engagement, and disposition of the blood–brain barrier-crossing bispecific antibody antagonist of metabotropic glutamate receptor type 1. *The FASEB Journal, 30*(5), 1927–1940. https://doi.org/10.1096/fj.201500078.

Wei, S., Lightwood, D., Ladyman, H., Cross, S., Neale, H., Griffiths, M., et al. (2005). Modulation of CSF-1-regulated post-natal development with anti-CSF-1 antibody. *Immunobiology, 210*(2–4), 109–119. https://doi.org/10.1016/j.imbio.2005.05.005.

Williams, A. F., & Barclay, A. N. (1988). The immunoglobulin superfamily—Domains for cell surface recognition. *Annual Review of Immunology, 6*, 381–405. https://doi.org/10.1146/annurev.iy.06.040188.002121.

Wu, A. M., & Senter, P. D. (2005). Arming antibodies: Prospects and challenges for immunoconjugates. *Nature Biotechnology, 23*(9), 1137–1146. https://doi.org/10.1038/nbt1141.

Wu, D., Yang, J., & Pardridge, W. M. (1997). Drug targeting of a peptide radiopharmaceutical through the primate blood-brain barrier in vivo with a monoclonal antibody to the human insulin receptor. *Journal of Clinical Investigation, 100*(7), 1804–1812.

Xia, C.-F., Boado, R. J., Zhang, Y., Chu, C., & Pardridge, W. M. (2008). Intravenous glial-derived neurotrophic factor gene therapy of experimental parkinson's disease with trojan horse liposomes and a tyrosine hydroxylase promoter. *The Journal of Gene Medicine, 10*(3), 306–315. https://doi.org/10.1002/jgm.1152.

Xiangbao, Y., Linquan, W., Mingwen, H., Fan, Z., Kai, W., Xin, Y., et al. (2014). Humanized anti-VEGFR-2 ScFv-As2O3-stealth nanoparticles, an antibody conjugate with potent and selective anti-hepatocellular carcinoma activity. *Biomedicine & Pharmacotherapy = Biomedecine & Pharmacotherapie, 68*(5), 597–602. https://doi.org/10.1016/j.biopha.2014.04.006.

Xing, Y., Zhao, J., Conti, P. S., & Chen, K. (2014). Radiolabeled nanoparticles for multimodality tumor imaging. *Theranostics, 4*(3), 290–306. https://doi.org/10.7150/thno.7341.

Xing, Y., Zhao, J., Shi, X., Conti, P. S., & Chen, K. (2014). Recent development of radiolabeled nanoparticles for PET imaging. *Austin Journal of Nanomedicine & Nanotechnology. 2*(2), 1016. (ISSN:2381–8956).

Xu, W., Liu, L., Brown, N. J., Christian, S., & Hornby, D. (2012). Quantum dot-conjugated anti-GRP78 scFv inhibits cancer growth in mice. *Molecules (Basel, Switzerland)*, *17*(1), 796–808. https://doi.org/10.3390/molecules17010796.

Yang, M., Li, J., & Chen, P. R. (2014). Transition metal-mediated bioorthogonal protein chemistry in living cells. *Chemical Society Reviews*, *43*(18), 6511–6526. https://doi.org/10.1039/C4CS00117F.

Yang, L., Mao, H., Wang, Y. A., Cao, Z., Peng, X., Wang, X., et al. (2009). Single chain epidermal growth factor receptor antibody conjugated nanoparticles for in vivo tumor targeting and imaging. *Small (Weinheim an Der Bergstrasse, Germany)*, *5*(2), 235–243. https://doi.org/10.1002/smll.200800714.

Younes, A., Bartlett, N. L., Leonard, J. P., Kennedy, D. A., Lynch, C. M., Sievers, E. L., et al. (2010). Brentuximab vedotin (SGN-35) for relapsed CD30-positive lymphomas. *The New England Journal of Medicine*, *363*(19), 1812–1821. https://doi.org/10.1056/NEJMoa1002965.

Yu, Y. J., Zhang, Y., Kenrick, M., Hoyte, K., Luk, W., Lu, Y., et al. (2011). Boosting brain uptake of a therapeutic antibody by reducing its affinity for a transcytosis target. *Science Translational Medicine*, *3*(84), 84ra44. https://doi.org/10.1126/scitranslmed.3002230.

Zhang, Y., Schlachetzki, F., & Pardridge, W. M. (2003). Global non-viral gene transfer to the primate brain following intravenous administration. *Molecular Therapy: The Journal of the American Society of Gene Therapy*, 7(1), 11–18.

Zhang, M., Yao, Z., Garmestani, K., Axworthy, D. B., Zhang, Z., Mallett, R. W., et al. (2002). Pretargeting radioimmunotherapy of a murine model of adult T-cell leukemia with the α-emitting radionuclide, bismuth 213. *Blood*, *100*(1), 208–216. https://doi.org/10.1182/blood-2002-01-0107.

FURTHER READING

Nelson, A. L., Dhimolea, E., & Reichert, J. M. (2010). Development trends for human monoclonal antibody therapeutics. *Nature Reviews Drug Discovery*, *9*(10), 767–774. https://doi.org/10.1038/nrd3229.

CHAPTER SIX

Smart Cell-Penetrating Peptide-Based Techniques for Intracellular Delivery of Therapeutic Macromolecules

Yang He*,†, Feng Li‡,1, Yongzhuo Huang*,†,1
*Shanghai Institute of Materia Medica, Chinese Academy of Sciences, Shanghai, China
†University of Chinese Academy of Sciences, Beijing, China
‡Harrison School of Pharmacy, Auburn University, Auburn, AL, United states
[1]Corresponding authors: e-mail address: fzl0023@auburn.edu; yzhuang@simm.ac.cn

Contents

Abstract

Many therapeutic macromolecules must enter cells to take their action. However, their treatment outcomes are often hampered by their poor transportation into target cells. Therefore, efficient intracellular delivery of these macromolecules is critical for improving their therapeutic efficacy. Cell-penetrating peptide (CPP)-based approaches are one of the most efficient methods for intracellular delivery of macromolecular therapeutics. Nevertheless, poor specificity is a significant concern for systemic administrated

Advances in Protein Chemistry and Structural Biology, Volume 112
ISSN 1876-1623
https://doi.org/10.1016/bs.apcsb.2018.01.004

CPP-based delivery systems. This chapter will review recent advances in CPP-mediated macromolecule delivery with a focus on various smart strategies which not only enhance the intracellular delivery but also improve the targeting specificity.

1. INTRODUCTION

Therapeutic macromolecules, such as proteins, peptides, antibodies, vaccines, nucleic acids, polysaccharides, and polyesters, are playing an increasingly crucial role in treating tumors (Wang, Chen, et al., 2014; Wang, Wang, et al., 2014), diabetes (Holman et al., 2007), Alzheimer's diseases (Angelov, Garamus, Drechsler, & Angelova, 2017), etc. From 1986 to 2014, the US Food and Drug Administration (FDA) approved a total of 125 therapeutic biological drugs, including the first-in-class, the advance-in-class, and the addition-in-class. Most of these drugs have been approved after 1996 and more than half of them belong to the first-in-class category (Miller & Lanthier, 2015).

Therapeutic macromolecules have several unique advantages. (1) These drugs have excellent biocompatibility after humanization (Wang, Chen, et al., 2014; Wang, Wang, et al., 2014). (2) Compared with traditional small molecule drugs, macromolecule drugs have better targeting specificity, and thus have less toxic side effects (Takakura & Hashida, 1996). (3) Many macromolecule drugs (e.g., antibody) have good pharmacokinetic properties and showed prolonged circulation time in the blood (Maeda, Sawa, & Konno, 2001). (4) Some macromolecule drugs, such as antibodies, can recognize and bind to specific cell surface receptors with high efficiency (Torchilin & Lukyanov, 2003).

However, the application of many biologics (e.g., nucleic acids, cytoplasmic active proteins), has several formidable delivery challenges, including maintaining stability in human body, overcoming multiple biological barriers, improving cellular uptake, and achieving effective drug release in the cytoplasm (Tan et al., 2014; Torchilin & Lukyanov, 2003). Therefore, efficient intracellular delivery is a very critical issue for their therapeutic efficacy and potential clinical applications.

Cell-penetrating peptides (CPPs), also known as protein transduction domain (PTD), have been effectively used to enhance the delivery of therapeutic macromolecules. CPPs refer to short polypeptides which have strong cell membrane-penetrating abilities. CPPs are usually rich in basic amino acids such as arginine and lysine. These CPPs usually have less than 30 amino acids (Tan et al., 2014). CPPs can enhance the delivery of therapeutic macromolecules into mammalian cells (Mano, Teodosio, Paiva, Simoes, & Pedroso de Lima, 2005), insects cells (Chen et al., 2012), and plant cells (Chang, Chou, Chen, Liu, & Lee, 2007). They have been widely used in various areas including cell biology, gene therapy, and drug delivery.

2. AN OVERVIEW OF CPPs

The discovery of CPPs can be traced back to 1988 when Green and Frankel confirmed that transactivating protein TAT from human immunodeficiency virus (HIV)-1 could be internalized by cells and enter the cytoplasm and nucleus (Frankel & Pabo, 1988; Green & Loewenstein, 1988). In 1994, Barsoum's group found that the TAT protein could efficiently enter cultured cells in vitro and demonstrated the feasibility of utilizing a TAT-mediated method for delivering macromolecules into living cells (Fawell et al., 1994). In 1999, a landmark work by Dowdy's group showed β-galactosidase was efficiently delivered to different tissues of mice without losing biological activity when delivered as a fusion protein, TAT-β-galactosidase. This fusion protein even showed the ability to cross the blood–brain barrier (BBB) (Schwarze & Dowdy, 1999). Barsoum and Dowdy's works laid a foundation for the wide application of CPPs in biomedical field.

Although the cell penetration mechanism of CPPs has not been fully understood yet, it is generally believed that cellular uptake is greatly influenced by the peptide's physicochemical properties, such as charge density and length of the peptide. In addition, CPPs can bind to lipids and membrane-associated proteins on the surface of cell membranes through electrostatic interactions (Nakase, Takeuchi, Tanaka, & Futaki, 2008). There are several other mechanisms: (1) many CPPs can directly penetrate cell membranes. This energy-independent transmembrane model has been widely used to explain the membrane-penetrating process of TAT (Shin, Zhang, Min, Lee, Byun, et al., 2014; Shin, Zhang, Min, Lee, Moon, et al., 2014). Briefly, the interaction between fusion proteins and cell membranes triggers the unfolding of fusion proteins. Then unfolded fusion proteins directly cross cell membranes. Once inside cells, fusion proteins are refolded with the assistance of the molecular chaperone system (Dowdy, 2000). It was recently reported that the original unfolded form of proteins may also penetrate membranes through large swallowing (Kaplan, Wadia, & Dowdy, 2005). (2) Some CPPs (e.g., penetratin) cross cell membranes by forming a transient structure through translocation. Penetratin dimers bind to negatively charged phospholipids present in cell membranes and then form an inverted micelle structure in the lipid bilayer (Derossi et al., 1996). (3) The endocytosis mediated by lipid raft/caveolin (Mayor & Pagano, 2007), clathrin (Vandenbroucke, De Smedt, Demeester, & Sanders, 2007), and macropinocytosis (Futaki, Nakase, Tadokoro, Takeuchi, & Jones, 2007), may also contribute to the cellular uptake of some CPPs. Depending on the types of CPPs and target cells, different mechanisms may be involved. In Table 1, we list representative CPPs derived from various origins and their cell penetration mechanisms.

CPPs have been widely used in the delivering of therapeutics for various diseases, such as tumors (Shin, Zhang, Min, Lee, Byun, et al., 2014; Shin, Zhang, Min, Lee, Moon, et al., 2014), diabetes (He et al., 2013), heart diseases (Bian et al., 2007), brain diseases (Lin, Xie, et al., 2016; Lin, Zhao, et al., 2016), and neurodegenerative disorders (Yang et al., 2013). For example, Dietz et al. constructed a recombinant protein TAT-Bcl-XL by introducing TAT to an antiapoptotic protein Bcl-XL (Dietz, 2002).

Table 1 Representative Cell-Penetrating Peptides and Their Membrane Penetration Mechanisms

CPPs	Origin	Proposed Mechanism
$TAT_{(48-60)}$	HIV-1 transcriptional activator	Direct penetration, pore formation
Penetratin $(pAntp)_{(43-58)}$	Antennapedia *Drosophila melanogaster*	Direct penetration, endocytosis
Polyarginines	Model peptide (chimeric)	Direct penetration, endocytosis
pVEC	Murine vascular endothelial cadherin	Direct penetration, transporter mediated
Pep-1	Chimeric	Direct penetration, pore formation
Transportan	Galanin–mastoparan (chimeric)	Endocytosis, direct penetration
MAP	Model amphipathic peptides (chimeric)	Multiple mechanisms

Adapted from Koren, E., & Torchilin, V. P. (2012). Cell-penetrating peptides: Breaking through to the other side. *Trends in Molecular Medicine, 18*(7), 385–393. https://doi.org/10.1016/j.molmed.2012.04.012 by permission of the Elsevier Ltd.

Intraocular injection of TAT-Bcl-XL efficiently delivered Bcl-XL into the retinal ganglion cell layer and rescued neurons from degeneration. Yang's group linked gelonin (a ribosome-inactivating protein) with low molecular weight protamine (LMWP) or TAT. Both of them showed good in vitro and in vivo tumor inhibitory activity (Park et al., 2005; Shin et al., 2015). CPPs have also been applied for the delivery of nanoparticulates, including liposomes, PLGA nanoparticles (NPs), and albumin NPs. For example, our group linked LMWP to bovine serum albumin (BSA) by a covalent conjugation approach. Albumin NPs prepared with this conjugated albumin could be used to deliver anticancer drugs (e.g., paclitaxel and fenretinide) (Lin, Xie, et al., 2016; Lin, Zhao, et al., 2016). The prepared NP formulation showed improved intracellular delivery of drugs and excellent anticancer activities against malignant brain tumor (Lin, Xie, et al., 2016; Lin, Zhao, et al., 2016).

In addition, CPPs have been successfully used for intracellular delivery of various nucleic acid-based therapeutics including oligonucleotides, nucleic acid peptides (PNAs), small interference RNAs (siRNAs), and plasmid DNAs (van den Berg & Dowdy, 2011). For example, Davidson et al. linked penetratin to the 5′-end of the siRNA sense strand. This conjugate could efficiently deliver siRNA into primary neuronal cells and downregulates the target protein (Davidson et al., 2004). Chiu et al. covalently linked TAT with siRNA to enhance its intracellular delivery (Chiu, Ali, Chu, Cao, & Rana, 2004). Recently, Langer's group showed that human serum albumin (HSA) NPs modified with a CPP, R9, were superior to a commercial transfection reagent and other nanoparticulate systems in terms of cell uptake and transfection efficiency (Mesken, Iltzsche, Mulac, & Langer, 2017). The safety of this HSA-R9/plasmid system was confirmed with several assays including WST-1 assay, LDH release, and caspase-3 activation assay, suggesting a great potential of this CPP-modified gene delivery vector.

Although CPPs have many advantages, there are several obstacles for their biomedical applications. First, the toxicity of CPPs is a significant concern. Although most CPPs only show toxicity at a high concentration, some CPPs (e.g., amphiphilic PaCPPs) have an apparent toxic effect even at a low concentration (Ziegler, 2008). Second, exogenous CPPs have poor in vivo stability. CPPs are easily degraded by enzymes in the circulation system or lose their functions in vivo under physiological and pathological conditions. The degradation of CPPs poses a key impact on their cellular uptake efficiency (Palm, Jayamanne, Kjellander, & Hallbrink, 2007). A potential solution to the stability problem is the use of stable D-form amino acids that are able to resist the enzymatic degradation and thus improve the in vivo stability of CPPs (Schorderet et al., 2005). Moreover, it is also promising to employ the peptide stapling techniques to prepare the stable CPPs (Dietrich et al., 2017). Last but not the least, CPP-based delivery systems also showed poor specificity and caused toxic side effects. Usually, CPPs can penetrate not only target tissues or cells but also normal tissues or cells. The delivery of toxic therapeutic agents into normal tissues or cells would lead to toxic side effects.

An ideal drug delivery system should be able to effectively deliver drugs to target sites, and reduce the amount of drug exposure in other organs. In order to achieve this goal, "smart" CPP-based techniques have been proposed for drug delivery applications. These systems utilize a prodrug like "shielding-and-then-activating" strategy (Huang, Jiang, et al., 2013; Huang, Shao, et al., 2013; Hsieh, Hung, & Fang, 2009). "Shielding" refers to masking CPPs in nontarget sites and reduce the unwanted intracellularization. Common methods for "shielding" include steric hindrance, electrostatic interaction, and carrier encapsulation. "Activating" refers to the reinstallation of cell penetration ability once delivery systems reach the target site and in response to specific physical and chemical stimuli. In this chapter, we will review various "shielding-and-then-activating" strategies and discuss some representative "smart" CPP-based delivery systems (Scheme 1).

3. SMART CPP-BASED, PRODRUG LIKE TECHNIQUES

Prodrug is a common approach for achieving tissue-specific drug delivery and reducing toxic side effects. As a case in point, tumor microenvironment (TME), which is characterized by acidic pH, elevated temperature, and overexpressed proteases, provide a useful target for triggering the activation of prodrugs. For example, Liu et al. developed a protease-activatable prodrug, in which the cytotoxic agent doxorubicin (Dox) was conjugated with a peptide that can be specifically cleaved by a tumor-associated protease, legumain (Liu, Sun, Huang, Janda, & Edgington, 2003). The conjugate was inactive as a prodrug which was activated through the cleavage peptide by legumain in tumors to release the parent drug Dox.

We previously used the term "macromolecular prodrugs" for the "shielding-and-then-activating" strategy in CPP-based delivery (Huang et al., 2010). A key design feature of macromolecular prodrugs is the "off-and-on" switch in response to the physical or chemical stimuli. The specificity

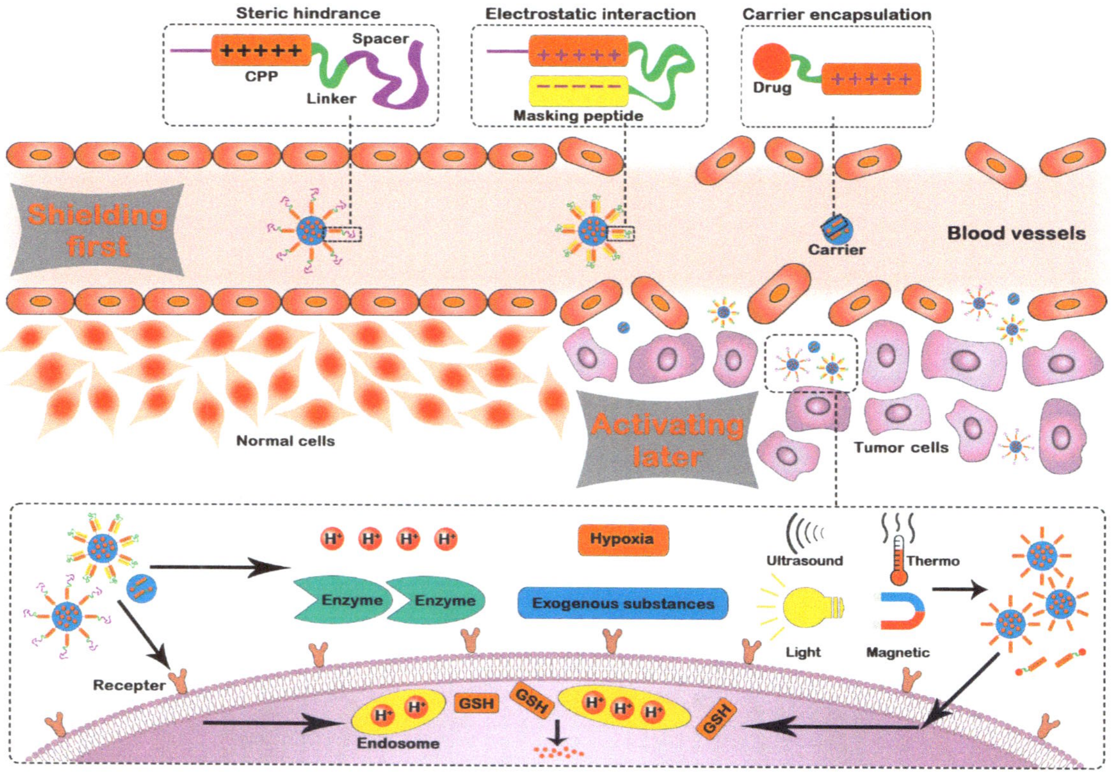

Scheme 1 Illustration of the "shielding-and-then-activating" CPP-based drug delivery strategy.

and efficiency of such switch is a determinant for the success of the prodrug design. Therefore, the selection of a proper "shielding" and "activating" mechanism is essential for this type of delivery systems.

4. CPP-SHIELDING STRATEGY

4.1 Steric Hindrance Methods

Long chain polymers have been used in many studies as a steric hindrance to mask the strong positive charge of CPPs. Polymers such as poly(ethylene glycol) (PEG) (Torchilin, Benson, & Toth, 2008), poly(phosphoester)s (PPEs) (Schottler et al., 2016) could be conjugated to CPP-based systems. Although this strategy can effectively reduce cellular uptake by the normal tissues or cells, PEG can also block or reduce the cellular uptake in target cells as well, thus creating a problem called "PEG dilemma" (Hatakeyama, Akita, & Harashima, 2013). Therefore, PEG should be removed efficiently on target sites and expose the CPP to facilitate cellular penetration. For example, to mask nonspecific uptake, PEGs were linked to lysine residues of CPP–peptide nucleic acids (PNAs) conjugate through azobenzene bonds (Lee, Moroz, Castagner, & Leroux, 2014). The azobenzene bonds could be cleaved by azoreductase secreted by many bacteria existing in the colonic mucosa site, resulting in the detachment of PEG, and the release of the CPP-PNA conjugate (Fig. 1).

4.2 Charge-Masking Methods

There are two commonly used charge-masking strategies for smart CPP-based delivery. One is the antibody targeted triggered electrically modified prodrug type strategy (ATTEMPTS) developed by Yang's group (Ye, Shin, Liang, He, & Yang, 2015). In this method, a CPP–protein conjugate and heparinized antibody forms a prodrug type inactive complex through the electrostatic interaction between CPP and heparin. The antibody can provide targeting ability and prodrug can be activated by giving the exogenous protamine that competitively binds with heparin antibody and thereby release the CPP protein. Another is the hairpin-structured activatable CPP (ACPP) delivery method developed by Tsien's lab (Jiang et al., 2004). The hairpin structure was composed of a cationic CPP (e.g., polyarginie), an anionic sequence (e.g., poly(glutamic acid)), and a cleavable linker (e.g., a protease substrate peptide). CPP can be effectively blocked through the interaction with negatively charged component. Once

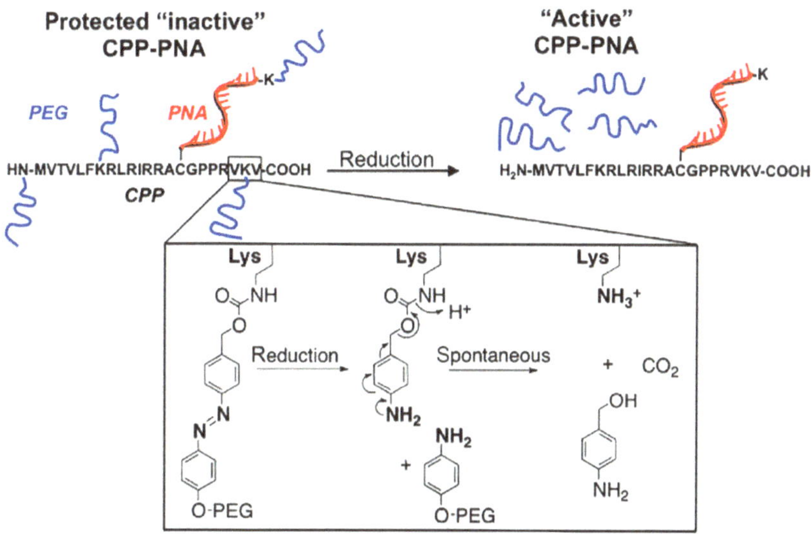

Fig. 1 Scheme of the activatable PEG-masked CPP-PNA. The reduction of azobenzene bonds removes the PEG shield and thus activates CPP-PNA. *Reproduced from Lee, S. H., Moroz, E., Castagner, B., & Leroux, J. C. (2014). Activatable cell penetrating peptide–peptide nucleic acid conjugate via reduction of azobenzene PEG chains.* Journal of the American Chemical Society, 136*(37), 12868–12871. https://doi.org/10.1021/ja507547w. Copyright © 2014, American Chemical Society.*

the substrate peptide linker was cleaved, the CPP-cargo will be released for executing its biological functions.

4.3 Carrier Encapsulation Methods

In addition, the CPP drugs can be encapsulated into a vehicle and released in the target site. For example, siRNA-CPP conjugates were loaded into thermal and magnetic dual-responsive liposomes that could be directed to the tumor under magnetic field (Yang, Xie, et al., 2016; Yang, Yang, et al., 2016). The encapsulated siRNA-CPP would then be released in the tumor site where liposomes were disrupted by the magnetic thermal effect.

5. CPP-ACTIVATING STRATEGY

5.1 TME-Responsive Systems

TME is characterized by specific tumor cell surface markers, various special tumor-associated cells (e.g., immune cells, endothelial cells, and fibroblasts), and other components such as cytokines, enzymes, new blood vessels, and extracellular matrix. Usually, we can take advantage of specific features of TME to realize tumor-targeted drug delivery (Spill, Reynolds, Kamm, & Zaman, 2016; Wu & Dai, 2017). Here, we will discuss smart CPP-based drug delivery systems which are designed according to specific features in TME.

5.1.1 pH-Sensitive Systems

Owing to the sustained chronic proliferation and differentiation of tumor cells, they reconstruct the metabolic process to satisfy the energy needs. Unlike the mitochondrial oxidative phosphorylation of normal cells, the glucose metabolism of tumor cells is restricted to the glycolysis to a large extent, which is the "Weinberg effect" (Hanahan & Weinberg, 2011). Owing to the "Weinberg effect," TME has increased lactic acid content and has an acidic pH value. Various systems were designed to utilize acid pH environment to improve drug delivery into tumors.

5.1.1.1 Acid-Cleavable Linker

Many chemical linkers (e.g., hydrazone bonds (Di Stefano, Lanza, Kratz, Merina, & Fiume, 2004) and amide bonds (Cao et al., 2014; Xu et al., 2007)) can be hydrolyzed under weak acid conditions but remain stable under neutral or weak alkaline conditions. Nanocarriers can be designed based on these acid-cleavable linkers. For example, a liposome was designed which includes a TAT conjugated to the liposome through a short PEG spacer, and a long chain PEG-phosphatidylethanolamine (PEG-PE) with a pH-sensitive hydrazone bond. In addition, PEG-PE was further conjugated with antimyosin antibody 2G4 or nucleosome-specific antibody 2C5 (Sawant et al., 2006; Torchilin, 2008). The extension of PEG-PE long chain helped "masking" TAT while antibody enhanced the drug delivery to the tumor site through active targeting. Under the acidic pH of the TME, the long chain PEG will be detached after the cleavage of hydrazone bonds, and the unmasked TAT will enhance cellular uptake (Fig. 2).

5.1.1.2 Acid-Extendable Linker

Another strategy depends on the application of materials that undergo confirmation change in response to acidic conditions. The imidazole group of histidine (pK_a 6.5) has a slight negative charge under normal physiological condition but is positively charged in the acidic TME (Shi et al., 2015). Poly-histidine was used as a linker to attach TAT to the surface of nanocarriers, which can expose TAT through a "pop-up" mechanism. Briefly, both PEG_{2K}

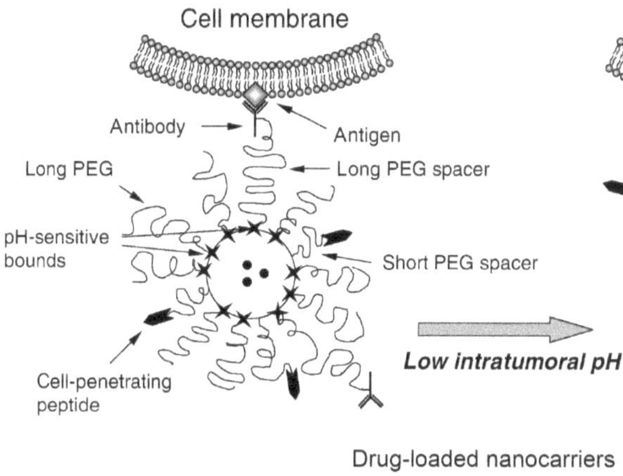

Fig. 2 Scheme of the pH-sensitive nanocarriers. The nanocarrier is modified with CPPs through short chain PEG spacers. CPPs are shielded by long chain PEG-PE. The PEG-PE are decorated at distal termini with a specific antibody. PEG-PE are linked by acidic pH-sensitive hydrazone bonds. The system accumulates in the tumor mainly depending on antibody-mediated active tumor targeting. Under mild acidic pH TME, protective long chain PEG detached from the surface of nanocarriers after the hydrolysis of hydrazone bonds. CPPs are then exposed and facilitate cellular uptake of nanocarriers. *Reproduced from Torchilin, V. P. (2008). Tat peptide-mediated intracellular delivery of pharmaceutical nanocarriers.* Advanced Drug Delivery Reviews, *60(4–5), 548–558. https://doi.org/10.1016/j.addr.2007.10.008 by permission of the Elsevier B.V.*

conjugated TAT and $PEG_{3.4K}$ were immobilized in a diblock copolymer micelle (Lee et al., 2008). Poly-histidine was unionized during the circulation in the blood, where TAT was hidden and attached onto the surface of the nanocarrier through a hydrophobic interaction. Therefore, cellular uptake of nanocarriers by nonmalignant normal tissues or cells with neutral pH was minimized. However, imidazole group of poly-histidine was ionized and become positively charged in response to the acidic TME. Owing to the electrostatic repulsion between TAT and nanocarriers, TAT was exposed when the PEG_{2K}-TAT chain stretched out in a "pop-up" way (Fig. 3).

5.1.1.3 Acid-Activatable CPPs

Acid-activatable CPPs were also developed to avoid nonspecific interaction and off-target effects. Shen's group developed a kind of TAT which did not have cell penetration activities until it is activated in response to acid pH in TME (Jin et al., 2013; Zhang, Gao, & He, 2015). In their studies, TAT containing two lysines and five arginines was selected. Since the guanidine

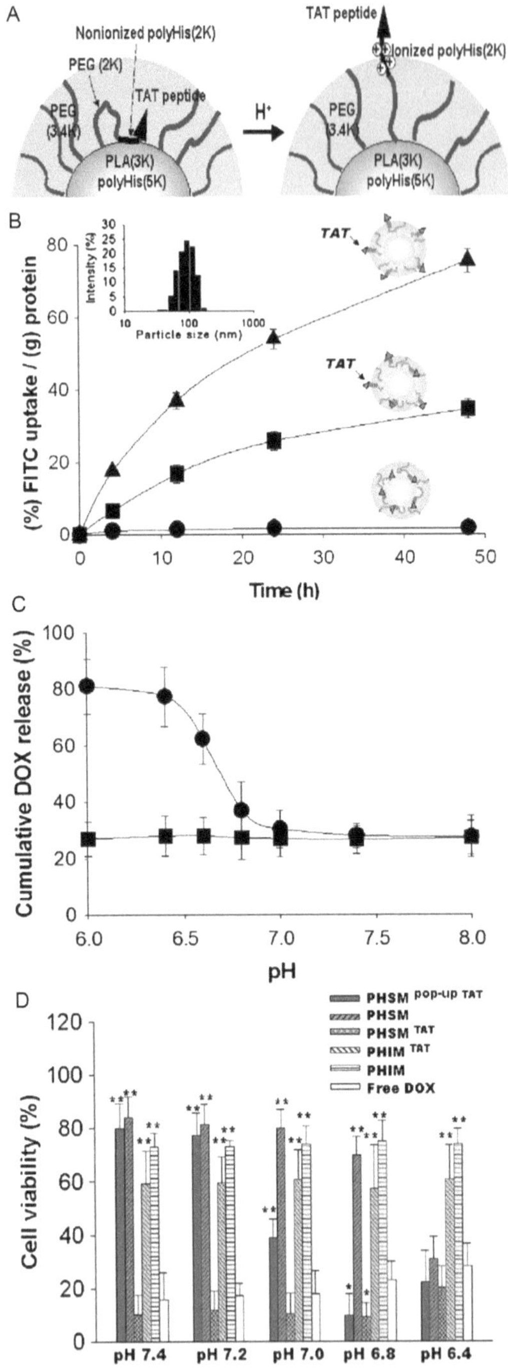

Fig. 3 (A) Scheme of the pH-sensitive "pop-up" diblock copolymer micelle system for tumor targeting. (B) Cellular uptake of PHSM$^{pop\text{-}up\ TAT}$ in cultured MCF-7 cells: pH 7.4 (●), pH 7.0 (■), pH 6.8 (▲). Particle size and size distributions of PHSM$^{pop\text{-}up\ TAT}$ suspensions (in PBS) at pH 7.4 were measured by dynamic light scattering. Each data point represents an average with standard deviation ($n = 3$). (C) Cumulative DOX release from PHSM$^{pop\text{-}up\ TAT}$ (●) with decreasing pH. PHIM^{TAT} (■) was used as a control group ($n = 3$).

group of arginine was not easily amidized, the terminal primary amine of the lysine residue was chosen for amidization treatment to block nonselective cellular uptake. This amidized TAT (aTAT) was used in their study to decorate the corona of poly(ethylene glycol)-*block*-poly(ε-caprolactone) (PEG-PCL) micelles. When modified micelles were exposed to acidic TME, amide bonds linked with the TAT lysine residue were rapidly hydrolyzed, resulting in the activation of TAT, and facilitate cellular uptake of micelles. In addition, TAT may also be activated intracellularly in the lysosome and promote the drug delivery into the cytoplasm through "proton sponge effect" (Cai et al., 2014) (Fig. 4).

5.1.1.4 Acid-Induced Charge Reversal

A large number of studies showed that negatively charged surface of drug delivery systems prolonged their circulation time in the blood and avoided uptake in nontargeting cells (Zhang et al., 2016). Delivery systems modified with CPPs usually have a positive surface charge and thus have rapid clearance and nonspecific cellular uptake. "Charge reversal" strategy is proposed to "mask" the positive charge of CPPs until the delivery systems reach target tissues or cells. An acid-induced charge-reversal nanocarrier was recently developed to promote tumor-specific cellular uptake and facilitate lysosome escape (Fig. 5) (Han et al., 2015). Amphiphilic diblock copolymer poly (L-lysine)-*block*-poly(L-leucine) (PLL-PLLeu) was synthesized to form positively charged micelles. Lysine residues of the poly-lysine block were amidated with 2,3-dimethylmaleic anhydride (DMA) to form β-carboxylic amides. In addition, succinyl chloride amidated TAT was conjugated with the polymer through a click chemistry to form a negatively charged PLLeu-PLL(DMA)-Tat (SA) (PPDTS) micelle. When the system arrived at the tumor site, lysine residues were exposed after the hydrolysis of the β-carboxylic amides. The surface of micelles was converted into positive charge, facilitating micelle internalization into tumor cells. Furthermore, the amide bond

(D) Drug-sensitive MCF-7 cell killing efficacy of $PHSM^{pop\text{-}up\ TAT}$ (DOX 10 μg/ml equivalent) as a function of pH after 48-h incubation. Five control experiments (PHSM, $PHSM^{TAT}$, $PHIM^{TAT}$, PHIM, and free DOX) were conducted ($n = 9$) ($^{*}P< 0.05$, $^{**}P < 0.01$ compared to free DOX). *Adapted with Lee, E. S., Gao, Z., Kim, D., Park, K., Kwon, I. C., & Bae, Y. H. (2008). Super pH-sensitive multifunctional polymeric micelle for tumor pH(e) specific TAT exposure and multidrug resistance.* Journal of Controlled Release, 129*(3), 228–236. https://doi.org/10.1016/j.jconrel.2008.04.024 by permission of the Elsevier B.V.*

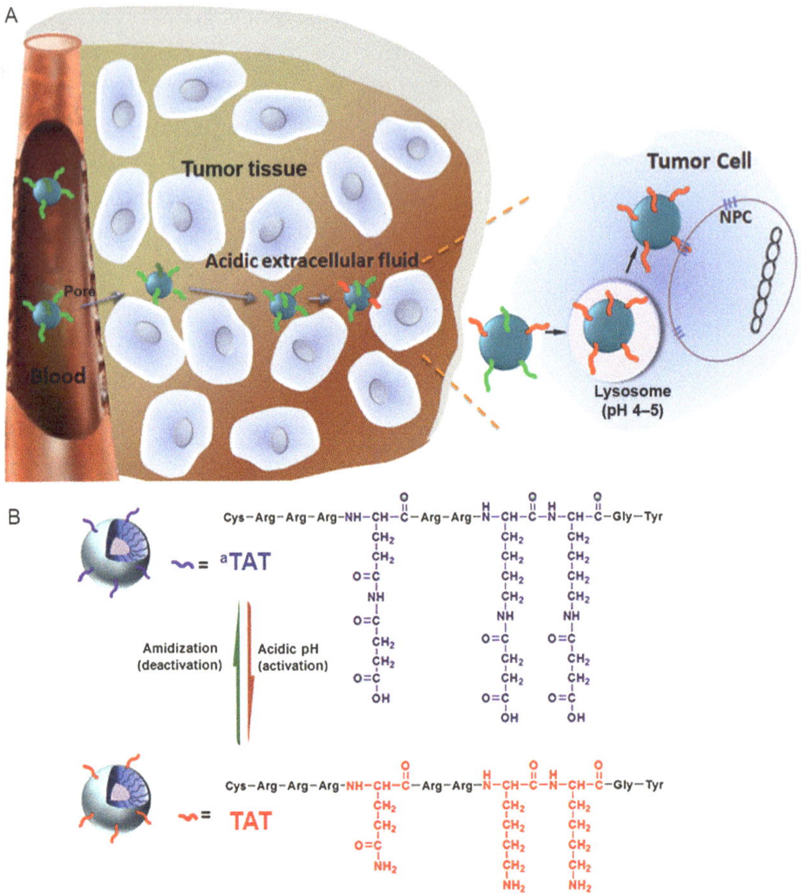

Fig. 4 Scheme of acid-activatable cell-penetrating peptides for in vivo tumor-targeted drug delivery. (A) Nanosized micelles modified by amidated TAT (aTAT) are activated step by step in acidic tumor extracellular fluid and lysosome. (B) The amidation of the lysine residues of TAT and the hydrolysis activation of aTAT under an acidic pH. *Reproduced from Jin, E., Zhang, B., Sun, X., Zhou, Z., Ma, X., Sun, Q., et al. (2013). Acid-active cell-penetrating peptides for in vivo tumor-targeted drug delivery.* Journal of the American Chemical Society, 135*(2), 933–940. https://doi.org/10.1021/ja311180x. Copyright © 2013, American Chemical Society.*

between TAT peptide and succinyl chloride was hydrolyzed at late endosome and lysosome which have an even lower pH (4.0–6.0) (Erazo-Oliveras, Muthukrishnan, Baker, Wang, & Pellois, 2012). This conversion will activate lysosomal escape and promote TAT peptide-mediated nucleus targeting (Han et al., 2015; Nori & Kopecek, 2005).

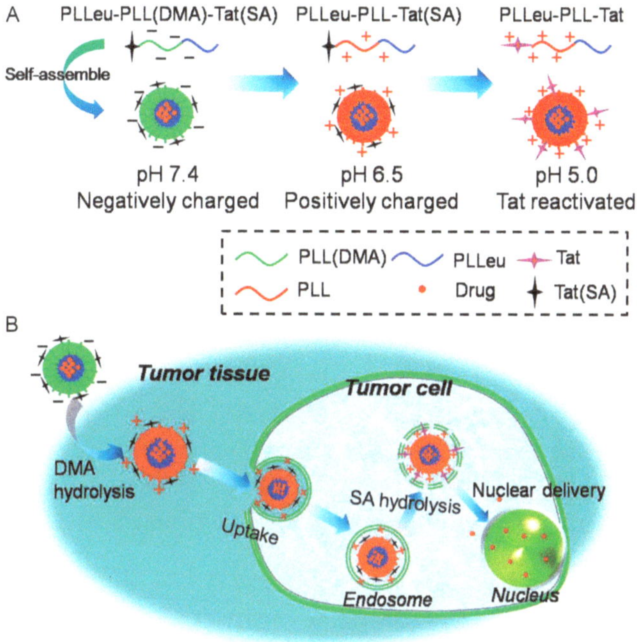

Fig. 5 Scheme of dual acid-induced and charge-reversal nanocarrier. (A) Self-assembled negatively charged micelles stepwise undergo charge reversal in response to mildly acidic tumor tissues (pH ≈ 6.5) and the reactivation of Tat in more acidic endosomes (pH ≈ 5.0). (B) The intracellular and intranuclear delivery of the drug-loaded micelle. *Reproduced from Han, S. S., Li, Z. Y., Zhu, J. Y., Han, K., Zeng, Z. Y., Hong, W., et al. (2015). Dual-pH sensitive charge-reversal polypeptide micelles for tumor-triggered targeting uptake and nuclear drug delivery.* Small, 11*(21), 2543–2554. https://doi.org/10.1002/smll.201402865 with permission from Wiley.*

In another study, the surface of drug-loaded mesoporous silica nanoparticles (MSN) was modified with fluorescein FITC linked TAT peptide. Citraconic anhydride (Cit) was used to modify TAT peptide and mask its positive charge so that the surface of MSN would be negatively charged. Positively charged YSA peptide tagged with a quencher molecule (YSA-BHQ1) was attached to the surface of TAT-FITC modified MSN through the electrostatic attraction. YSA could enhance drug delivery into tumors through Ephrin A2 receptor (EphA2)-mediated active targeting. Meanwhile, the fluorescence of FITC was quenched by BHQ1 when they are close to each other. At the acidic TME, the positive charge of TAT was recovered after the hydrolysis of citraconic anhydride-amino bond and resulted in the dissociation between

TAT-FITC and YSA-BHQ1. This resulted in the activation of nucleus targeting ability of TAT as well as recovery of FITC fluorescence. The system demonstrated a potential as a theranostic agent for simultaneously tumor diagnosis and therapy (Zhao, Zhao, Wang, Fan, & Wu, 2016).

5.1.2 Enzyme-Triggered Systems

The earliest versions ACPPs can be activated by specific enzymes in the tumor site to have their cell penetration properties (He et al., 2015). Delivery systems with ACPPs can increase the enrichment of payloads in the tumor site and improve their uptake by cancer cells (Macewan, Callahan, & Chilkoti, 2010). The activation of ACPPs usually depends on the cleavage of substrate peptide linkers by enzymes such as proteases (Choi, Swierczewska, Lee, & Chen, 2012; Reissmann, 2014), esterases (Lee, Park, Park, & Na, 2013), and hyaluronidase (HAase) (Lee ct al., 2016). In Section 5.1.2.1, we will briefly introduce some examples of these systems which have been used in tumor diagnosis and therapy.

5.1.2.1 Matrix Metalloproteinase (MMP)

When the polycation peptide is connected to the polyanionic peptide segment, the electrostatic interaction between them temporarily masks the cell-penetrating ability of the polycationic peptide. Enzymatic cleavage of the linker between polycationic and polyanionic peptides could interrupt their interaction and thus unmasks the sheltered cell penetration abilities of CPPs. This type of system was first developed by Tsien's group who designed a "hairpin"-shaped targeted delivery system for tumor diagnostic imaging. In their design, a CPP R9 and a polyanionic E8 were linked by a MMP2/9 substrate peptide linker (PLGLAG). R9 and E8 were attracted to each other through electrostatic interactions, which masked the cell penetration ability of R9. The cell penetration ability of R9 was recovered after the cleavage of the linker by MMP2/9 which were often overexpressed in tumor cells (Aguilera, Olson, Timmers, Jiang, & Tsien, 2009; Jiang et al., 2004). Later, many researchers did a series of studies on this type of system (He et al., 2016). For example, Jiang's group constructed a TME dual-trigger activatable CPPs (dtACPPs) nanosystem for the delivery of VEGF siRNA expressing plasmid and doxorubicin (Huang, Jiang, et al., 2013; Huang, Shao, et al., 2013). Recently, our group also successfully utilized this ACPPs strategy to deliver a macromolecular drug, trichosanthin (TCS) (Chen, Zhang, Jin, Li, et al., 2017; Chen, Zhang, Jin, Tang, Wang, et al., 2017; Chen, Zhang, Jin, Tang, Wu, et al., 2017). In our study, TCS-LMWP-MSP was prepared as a recombinant fusion protein. Then, a PEG chain was conjugated to the MSP, a MMPs substrate linker, converting TCS-LMWP-MSP into an inactive prodrug, TCS-LMWP-MSP-PEG. This prodrug reaches tumor through the EPR effect (passive targeting) after i.v. injection. The masking PEG chain was removed after the cleavage of the MSP linker by MMP2 enzyme overexpressed by HT1080 cells. Then, the unmasked LMWP could promote cellular uptake of TCS, which could inactivate the ribosome, inhibit protein synthesis, and result in cell death (Fig. 6). This system also demonstrated the ability to overcome drug resistance when used in combination with paclitaxel liposomes and showed a synergistic antitumor effect (Chen, Zhang, Jin, Li, et al., 2017; Chen, Zhang, Jin, Tang, Wang, et al., 2017; Chen, Zhang, Jin, Tang, Wu, et al., 2017).

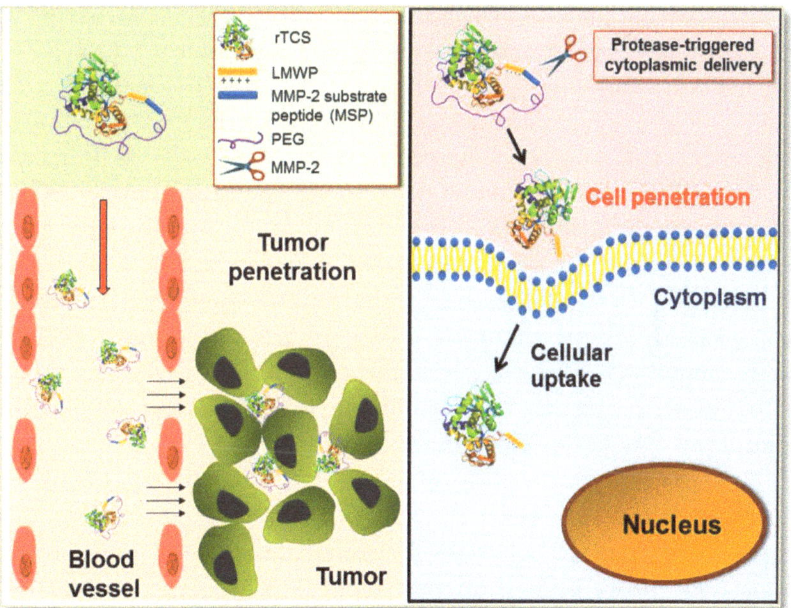

Fig. 6 Scheme of the MMP2-sensitive dtACPPs system for the intracellular delivery of rTCS macromolecule. *Reproduced from Chen, Y., Zhang, M., Jin, H., Tang, Y., Wang, H., Xu, Q., et al. (2017). Intein-mediated site-specific synthesis of tumor-targeting protein delivery system: Turning PEG dilemma into prodrug-like feature.* Biomaterials, 116, *57–68. https://doi.org/10.1016/j.biomaterials.2016.11.033. Copyright © 2017, Elsevier.*

5.1.2.2 Legumain

Legumain is another protease which has high specificity for the hydrolysis of acyl bonds of asparaginyl asparagine (Li et al., 2013). Under normal physiological conditions, legumain is mainly expressed in renal tubules and assists in the renal tubular reabsorption (Miller, Matthews, Reinheckel, Fleming, & Watts, 2011). Legumain is overexpressed in certain tumors. Liu et al. designed a legumain-triggered liposomal drug delivery system (Liu et al., 2014). In their study, an endothelin copolymer alanine–alanine–asparagine (AAN) was conjugated to TAT lysine residue through a legumain cleavable amide bond. The conjugation of AAN led to a significant reduction of transmembrane transport capacity of TAT and thus effectively reduced the nonspecific cell internalization. The modified TAT was attached to the surface of drug-loaded liposomes with PEG_{2000} as a spacer. Legumain presents at the tumor cleaved AAN and activated the membrane-penetrating ability of TAT.

Similarly, our lab developed legumain-activated imaging probe for monitoring the development of inflammatory tumors (Jiang et al., 2014). We designed a hairpin structure composed of poly(arginine) (polyR) and poly(glutamic acid) (polyE) connected by a legumain cleavage linker Pro-Thr-Asn (PTN). The fluorophore Cy5 and quencher QSY21 were conjugated with polyR and polyE, respectively. The system forms a "hairpin" structure and the fluorescence of the Cy5 was quenched through fluorescence resonance energy transfer (FRET) due to its proximity to the quencher QSY21. When tested in SW620 xenograft tumor mouse model, this probe was stable in the normal tissue with the fluorescence of the Cy5 being quenched. At the tumor site, the legumain sensitive bond was cleaved and Cy5 fluorescence was recovered, showing a strong fluorescence signal in contrast to normal tissues. This system provided a powerful tool to monitor tumor development, invasion, and metastasis.

5.1.2.3 Hyaluronidase (HAase)

HAase is overexpressed in various tumors including breast tumor, prostate tumor, and urinary bladder tumor (Stern, 2008). Particularly, it is overexpressed in tumor infiltration area and during the tumor transfer process (Jojovic, Delpech, Prehm, & Schumacher, 2002). HAase could also be utilized to design ACPPs-based drug delivery systems. Jiang et al. modified drug-loaded liposomes with a CPP R6H4 which was masked by a layer of hyaluronic acid (HA) coating (Jiang et al., 2012). The hydrophilic HA coating protected the system from plasma protein adsorption, enhanced its stability, and prolonged its systemic circulation after i.v. injection. The degradation of HA by HAase at the tumor site could unmask R6H4 and enhances the cellular uptake of liposomes.

In addition to the enzymes described above, thrombin, prostate-specific antigen (PSA), and other biological factors overexpressed in specific organs or tissues could also be used to cleave specific bonds and served as "triggers" for different drug delivery systems (He et al., 2015).

5.1.3 Hypoxia

Owing to the rapid proliferation of tumor cells, TME is associated with insufficient supply of oxygen and characterized by a hypoxia condition. Hypoxia-induced factor (HIF) is abundant in the hypoxic TME. HIF-1α contains an oxygen-dependent degradation domain (ODD) which would be degraded by oxygen (Harada et al., 2007). A plasmid was constructed to express cell-penetrating recombinant protein TAT-ODD-p53. This protein was stable in hypoxic tumor tissue and effectively promoted the apoptosis of tumor cells. However, the ODD would be degraded by the ubiquitin-proteasome system in normal tissues by an oxygen-dependent way, and thus avoid toxicity in normal tissues (Lee & Lu, 2014; Yu et al., 2009). In addition, other tumor suppressors, such as caspases, could be used to substitute P53 in the construction of recombinant protein, TAT-ODD drug (Harada, Hiraoka, & Kizaka-Kondoh, 2002).

5.2 Competitive Activation Systems

The mechanism of "competitive activation" is based on the electrostatic attraction and removal between cations and anions. ATTEMPTS delivery system is a good example (Huang et al., 2010; Ye et al., 2015). Generally, the system consists of three parts: (1) drug modified with CPP through a sensitive bond, (2) targeting antibody conjugated with heparin, and (3) exogenous protamine (Huang, Jiang, et al., 2013; Huang, Shao, et al., 2013). Cationic CPP and anionic heparin form complexes through electrostatic interaction. The complex formation masks the cell penetration ability of CPP. The conjugated antibody could enhance drug accumulation in target tissues or organs. Exogenous protamine was injected followed by the administration of drug-CPP/antibody–heparin complex. Owing to the more powerful binding between protamine and heparin, the exogenous protamine would compete for the binding site in the heparin and lead to the disassociate of CPP drug from the complex. The released CPP drug with "unmasked" CPPs would have "restored" cell-penetrating capability (Fig. 7).

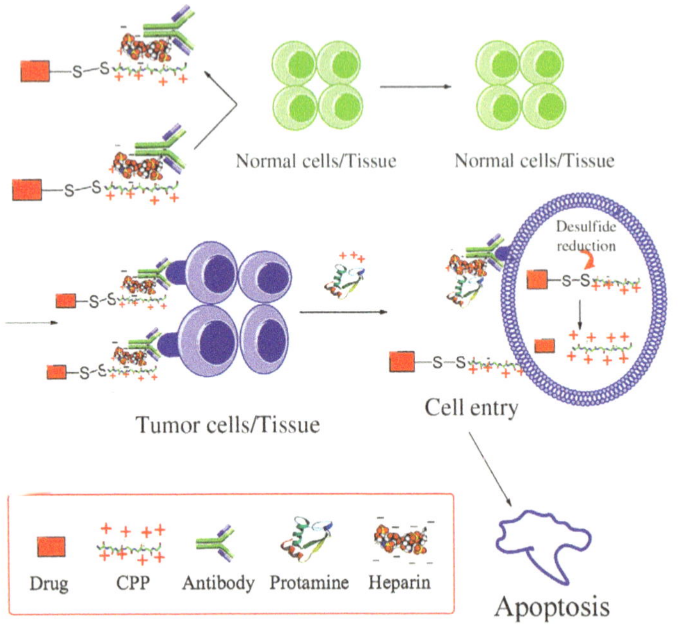

Fig. 7 The principle of the "ATTEMPTS" system. *Reproduced from Huang, Y., Jiang, Y., Wang, H., Wang, J., Shin, M. C., Byun, Y., et al. (2013). Curb challenges of the "Trojan horse" approach: Smart strategies in achieving effective yet safe cell-penetrating peptide-based drug delivery.* Advanced Drug Delivery Reviews, 65*(10), 1299–1315. https://doi.org/10.1016/j.addr.2012.11.007 by permission of the Elsevier B.V.*

The complex formed between heparin and CPP through electrostatic interaction could mask the cell penetration capability of CPP and thus avoid nonspecific cellular uptake in normal tissues or organs. The ATTEMPTs delivery system has been applied to deliver a variety of protein drugs with poor cell permeability, including tissue plasminogen activator (t-PA) (Yang et al., 2005), gelonin (Shin, Zhang, Min, Lee, Byun, et al., 2014; Shin, Zhang, Min, Lee, Moon, et al., 2014), and asparaginase (Kwon et al., 2008). The same strategy was also used in the preparation of polyrotaxane-doxorubicin conjugates (PR-DOX) based supramolecular assemblies (Moon, Kwon, Lee, Park, & Yang, 2007).

5.3 Physical Stimuli-Responsive Systems

Numerous CPP-based drug delivery systems have been designed to respond to physical stimuli such as ultrasound, ultraviolet light, temperature, and magnetic field. The advantage of these systems is that the release of drugs can be triggered through physical stimuli generated externally with a specific instrument.

5.3.1 Ultrasound

Ultrasound has been successfully used to facilitate the delivery of drugs or genes into specific tissues (Vives, Schmidt, & Pelegrin, 2008). Ultrasound can form small pores on the surface of the cell membrane and increase intracellular penetration of extracellular media. This is a process achieved through cavitation effect (Sonoda et al., 2007). Ultrasound-targeted microbubble destruction (UTMD) technology developed based on this principle has been used in the delivery of macromolecule such as genes and proteins. Similarly, ultrasound could also be used to trigger the release of other cargos from microbubbles. Tian's group encapsulated TAT and siRNA against epidermal growth factor receptor (EGFR) in microbubbles which release cargos in response to external ultrasound radiation at the target site. The released TAT can enhance the cellular uptake of siRNAs. This system demonstrated enhanced siRNA delivery efficiency and showed good anticancer effects on a triple-negative breast cancer (TNBC) tumor model (Fig. 8) (Jing et al., 2016).

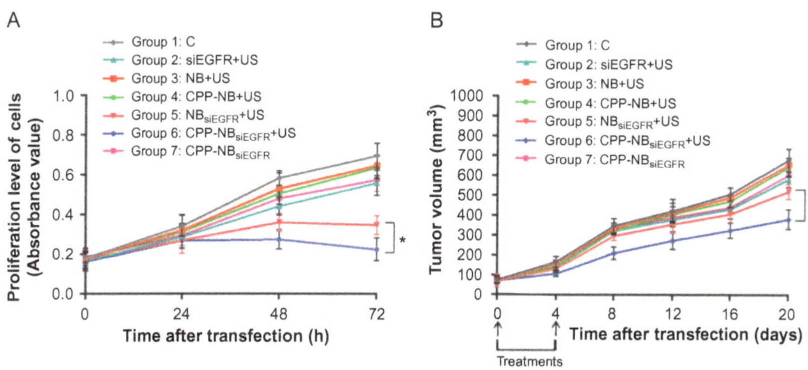

Fig. 8 (A) Cell proliferation curves for tumor cells after transfection with EGFR siRNA. (B) Tumor volume change curves of the subcutaneously transplanted xenograft tumors. *Reproduced from Jing, H., Cheng, W., Li, S., Wu, B., Leng, X., Xu, S., et al. (2016). Novel cell-penetrating peptide-loaded nanobubbles synergized with ultrasound irradiation enhance EGFR siRNA delivery for triple negative breast cancer therapy.* Colloids and Surfaces B: Biointerfaces, 146, *387–395. https://doi.org/10.1016/j.colsurfb.2016.06 by permission of the Elsevier B.V.*

The incorporation of active targeting moiety in the above-mentioned system may further improve the overall therapeutic effect. In another study, ultrasound contrast agent (UCA) was used to coencapsulate TAT and pEGFP-N1 plasmid. Then, mouse antihuman P-selectin monoclonal antibody was conjugated to the surface of UCA to prepare an active targeting system. P-selectin is a cell adhesion molecule which is stored in endothelial cells and blood platelets under normal physiological conditions. When ischemic organ damage occurs, the P-selectin is selectively overexpressed on the surface of endothelial cells of the injured organ. Therefore, P-selectin can serve as a molecular target for ischemic injury. TAT and genes are protected from enzymatic digestion when they are encapsulated in microbubbles. Once the microbubbles reach injury site, the release of TAT and plasmid is triggered through UTMD. Then, TAT can enhance the gene transfer into the target tissue. This technique provides a promising approach for the diagnosis and treatment of ischemic diseases (Ren et al., 2014).

5.3.2 Light

5.3.2.1 Ultraviolet (UV) Light

Light is an ideal external trigger because the modest short-term light exposure is not only easy to control but also harmless to cells (Amamoto et al., 2015; Forman, Dietrich, & Todd Monroe, 2007). A typical photoactivating CPP-based therapeutic macromolecule delivery system is described here. Briefly, the CPP (green tree shape) was attached to the N-(2-hydroxypropyl) methacrylamide(HPMA) copolymer scaffold. The positive charge of the CPP lysine residues was sheltered by a light-sensitive group, 6-nitroveratryl-carbonyl (Novc) (red pentagonal star shape). Thus, CPP was in the "caged" state and lost its cell-penetrating properties in the dark condition. When irradiated with the UV light, the photolabile group Nvoc was removed and the cell penetration capability of CPP was restored (Fig. 9). This system was used to deliver fluorescein FITC (P-(cCPP)-FITC) and proapoptotic drug peptide$_D$ (KLAKLAK)$_2$-(P-(cCPP)-KLAK), respectively. A significant cellular uptake of fluorescence was observed in cells treated with 50 mg/mL of P-(cCPP)-FITC after 10 min of UV light exposure. Cells treated with 40 mM and above (P-(cCPP)-KLAK) showed significant cytotoxic effects after irradiation and incubation for a certain period of time (Shamay, Adar, Ashkenasy, & David, 2011).

Later, Mei's group adapted a similar design in their nanostructured lipid carrier (NLC)-based system (Yang, Yang, Xie, Cai, & Mei, 2014). NLC was

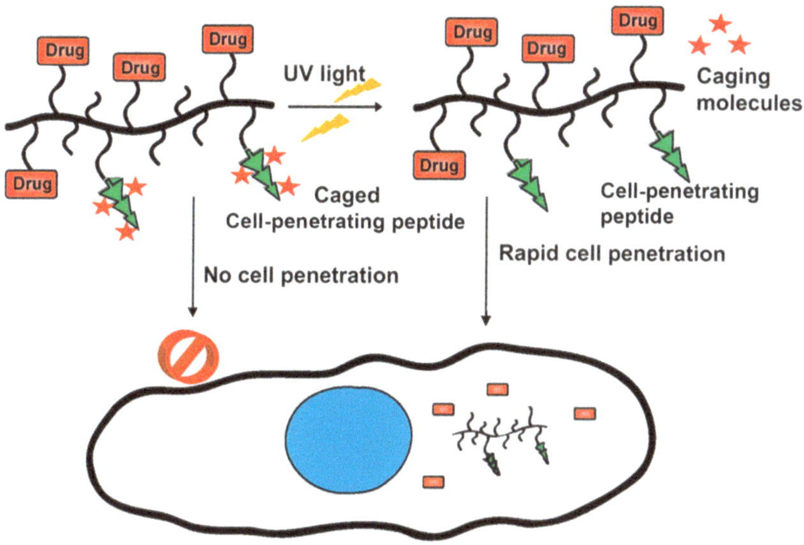

Fig. 9 Scheme of the UV-light activating CPP-based macromolecular drug delivery system. Under dark condition, CPP is in "caged" state and loses its penetrating property. After being irradiated by UV light, the photolabile group Nvoc is deshielded and cell-penetrating activity of CPP is restored. *Reproduced from Shamay, Y., Adar, L., Ashkenasy, G., & David, A. (2011). Light induced drug delivery into cancer cells.* Biomaterials, *32(5), 1377–1386. https://doi.org/10.1016/j.biomaterials.2010.10.029 by permission of Elsevier Ltd.*

selected a delivery carrier. In their study, they attached a DSPE-PEG_{2000} segment to a CPP, penetratin. The UV-sensitive group Nvoc remained unchanged. Then, NLC was modified with DSPE-PEG_{2000}-penetratin. In the dark, the whole system cannot penetrate through the cell membrane. While treated by UV light, the photolabile protective group is released and penetrating ability of CPP is restored.

5.3.2.2 Near-Infrared (NIR) Light

The NIR light can also be applied to a photoactivating drug delivery system. It is superior to UV light for in vivo application due to its better tissue penetration ability. The general design principle of NIR-responsive systems is similar to the UV-responsive counterparts with UV-responsive groups being replaced by NIR-responsive groups. A recent study used NIR two-photon excitation-responsive dye, 1-(bromomethyl)-4,5-dimethoxy-nitrobenzene, which can be cleaved by NIR light (Fig. 10) (Yang et al., 2015). In contrast to the low tissue permeability (often confined to surface tissue) and

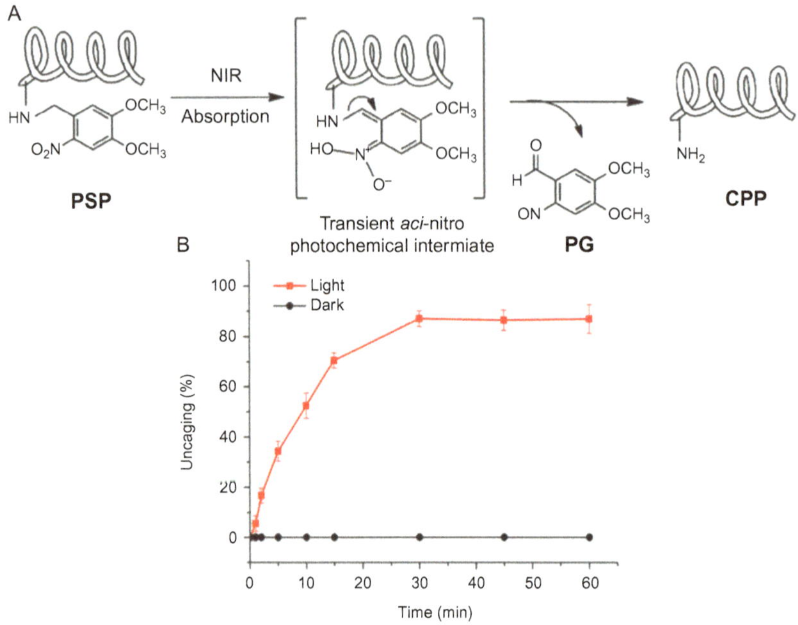

Fig. 10 An example of NIR light activating CPP drug delivery system. (A) The unblocking process of the photosensitive peptide (PSP). (B) The percentage of uncaged CPP as a function of the NIR light irradiation time. *Reproduced from Yang, Y., Yang, Y., Xie, X., Cai, X., Wang, Z., Gong, W., et al. (2015). A near-infrared two-photon-sensitive peptide-mediated liposomal delivery system.* Colloids and Surfaces B: Biointerfaces, 128, *427–438. https://doi.org/10.1016/j.colsurfb.2015.02.041 by permission of Elsevier B.V.*

potential damage to tissues associated with the short-wavelength UV light, NIR with a longer wavelength has a better tissue penetration ability. NIR light in the range of 650–900 nm can effectively penetrate the biological tissue and reach a depth of 10 cm without cause any damage (Weissleder, 2001). Thus, it provides a promising strategy for the diagnosis and treatment applications in deep tissue lesions.

5.3.3 Temperature

In recent years, research on thermosensitive materials is gaining increasing attentions (Riess, 2003). Thermosensitive materials can change their solubility, conformation, and intermolecular interactions in response to temperature change (Zhuang, Gordon, Ventura, Li, & Thayumanavan, 2013). Poly (N-isopropylacrylamide) (PNIPAAm) is a typical thermosensitive polymer that can undergo a phase transition in response to temperature change.

When the temperature is higher than PNIPAAm's lower critical solution temperature (LCST, about 32°C), the material tends to fold into a spherical shape. This material expands outward and interacts with water molecules to form intermolecular hydrogen bonds below the LCST. PNIPAAm can be used to effectively mask the strong positive charge of CPPs (Zhang, Mao, & Gao, 2014). For example, thermally sensitive quantum dots (TSQDs) were constructed by immobilizing CPP and biotin-conjugated PNIPAAm on the surface of quantum dots for the delivery of siRNA or other gene drugs (Kim, Lee, Kim, Jeong, & Park, 2010). When the temperature was at 25°C, CPP was blocked by the extended PNIPAAm chain. When the temperature was adjusted to 37°C, PNIPAAm chain dehydrated and shrank, leading to the unmasking of CPP (Fig. 11).

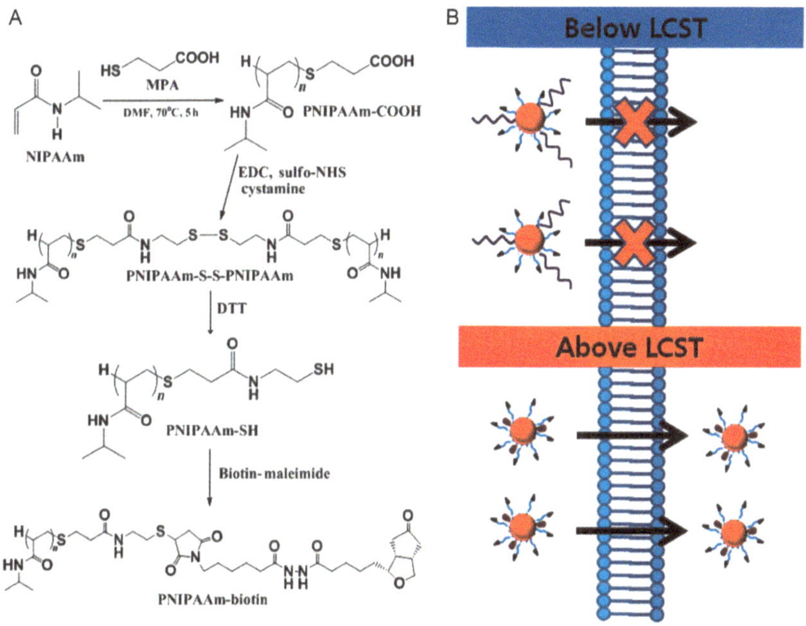

Fig. 11 Thermosensitive CPP-based drug delivery system. Thermosensitive material (e.g., PNIPAAm) expands outward below LCST to block the positive charge of CPP but folds on the surface of the QDs above LCST to expose CPP. (A) Scheme for synthesizing PNIPAAm-biotin. (B) Schematic illustration of TSQDs and their CPP "shielding/deshielding" behavior for modulating cellular uptake in response to temperature. *Adapted from Kim, C., Lee, Y., Kim, J. S., Jeong, J. H., & Park, T. G. (2010). Thermally triggered cellular uptake of quantum dots immobilized with poly(N-isopropylacrylamide) and cell penetrating peptide.* Langmuir, *26(18), 14965–14969. https://doi.org/10.1021/la102632m. Copyright © 2010, American Chemical Society.*

5.3.4 *Magnetic Field*

Magnetic nanoparticles have been widely used in magnetic resonance imaging (MRI) and drug delivery (Berry & Curtis, 2003; Smith et al., 2010). Several studies demonstrated the feasibility of using magnetic nanoparticles in combination with CPPs for enhancing drug delivery. Zhao et al. developed a novel TAT-conjugated nuclear-shell structured magnetic silica-coated Fe_3O_4 nanoparticle and investigated its ability to penetrate the BBB and to enter the nucleus of glioma cells (Zhao, Shang, et al., 2016). Results showed that TAT and magnetic field could synergistically promote the cross of these NPs through BBB and enhance their cellular uptake in human U251 cell nucleus. In another study, Yang et al. encapsulated siRNA-CPP and magnetic Fe_3O_4 into a thermosensitive liposome (Yang, Xie, et al., 2016; Yang, Yang, et al., 2016). The liposome could be enriched in disease lesions under the induction of a magnetic field. Then, an alternating magnetic field produced by a copper coil with alternating current (AC) could induce Fe_3O_4 to generate heat and trigger the release of siRNA-CPP from the thermossensitive liposome (TSL). Once released, CPP can enhance the cellular uptake of conjugated siRNA. Since CPPs were encapsulated in the liposome, nonspecific uptake before thermoinduction was minimized (Lin, Xie, et al., 2016; Lin, Zhao, et al., 2016).

5.4 Combination of Multiple Strategies

As described above, many strategies can be applied to produce "smart" CPPs based systems with minimal nonspecific uptake and enhanced drug targeting. In fact, these strategies can be used simultaneously in a single system to further improve drug delivery efficiency. ATTEMPTS system is a typical example which included a combination of multiple strategies to enhance drug delivery (Huang et al., 2010). In the ATTEMPS system, the antibody was utilized to enhance drug targeting through the antibody/receptor interaction. In addition, the CPP was masked through the electrostatic interaction with negatively charged heparin to avoid nonspecific cellular uptake.

Our laboratory recently developed a dual-targeted nanocomposite drug delivery system for the combination of "active targeting" and "tumor–microenvironment–enzymatic activation" strategies (Chen, Zhang, Jin, Li, et al., 2017; Chen, Zhang, Jin, Tang, Wang, et al., 2017; Chen, Zhang, Jin, Tang, Wu, et al., 2017). In short, lactoferrin was used as a targeting group to selectively bind to low-density lipoprotein-associated protein-1 (LRP-1) overexpressed by the brain capillaries endothelial cell (Liu et al., 2013). Then, rTCS-LMWP complex was exposed after the cleavage of the MMP2 substrate. As a CPP, LMWP promoted the infiltration and penetration of TCS into glioma cells and enhanced the antitumor activity.

As a matter of fact, cancer cells also exhibit many unique features which are different from normal cells, including lower pH (Erazo-Oliveras et al., 2012), abnormal energy metabolism (Hanahan & Weinberg, 2011), overexpressed glutathione (GSH) (Meng, Hennink, & Zhong, 2009), and increased reactive oxygen species (ROS) (Cairns, Harris, & Mak, 2011). By responding to two or more factors, the delivery strategies are often used in combination. For example, a TSL was designed for siRNA delivery based on a combination of receptor targeting, hyperthermia, and redox sensitive (Yang, Xie, et al., 2016; Yang, Yang, et al., 2016). The surface of TSLs was modified with asparagine–glycine–arginine (NGR) group which specifically targeted tumor cells that overexpress CD13 molecules (Yang, Yang, Xie, Cai, Zhang, et al., 2014). siRNA and CPP were linked through a redox-sensitive disulfide bond and then siRNA-S-S-CPP was encapsulated into a TSL. The system is stable and resistant to enzymatic digestion. In addition, the nonspecific cellular uptake is also minimized. The release of siRNA-S-S-CPP was trigger by the external heat simulation applied to the tumor site. Once siRNA-S-S-CPP released from the liposome, CPP promotes its cellular uptake. Cleavage of the disulfide bond by GSH in the cytoplasm caused the release of free siRNA and the silence of the target oncogene (Fig. 12). Properly designed system with multiple strategy combination can enhance drug delivery synergistically.

6. PERSPECTIVE

With many breakthroughs and rapid progress in the biotechnology, there is an increasing number of applications of therapeutic biological macromolecules such as recombinant plasmids, polypeptides, and proteins. However, delivery obstacles such as cell membrane, BBB, other inherent biological defense mechanisms limit the application of macromolecules (Murriel & Dowdy, 2006). CPPs are described as a "Trojan horse" because they can carry macromolecular cargos across various types of biological barriers and achieve efficient intracellular or even intranuclear delivery (Wadia, Stan, & Dowdy, 2004; Zhang, Wu, Li, & Feng, 2015). However, whether the "Trojan horse" will follow the designed route to the desired tissue site

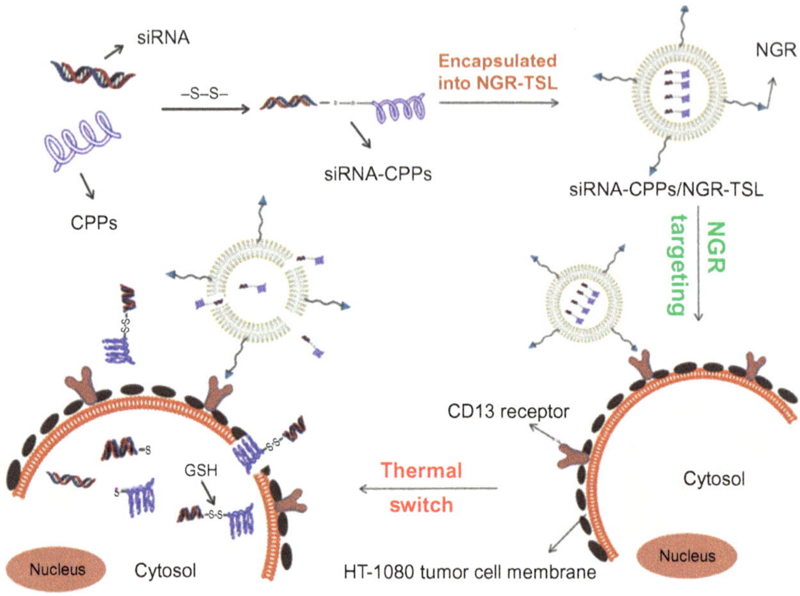

Fig. 12 Use of NGR targeting, thermal switch, and redox sensitive as combination in siRNA-CPPs/NGR-thermosensitive liposome. *Reproduced from Yang, Y., Yang, Y., Xie, X., Xu, X., Xia, X., Wang, H., et al. (2016). Dual stimulus of hyperthermia and intracellular redox environment triggered release of siRNA for tumor-specific therapy.* International Journal of Pharmaceutics, 506*(1–2), 158–173. https://doi.org/10.1016/j.ijpharm.2016.04.035 by permission of Elsevier B.V.*

remains to be confirmed (Shi, Qi, Xiang, & Zhang, 2014; Vives, 2005). In fact, delivery systems with unmodified CPP usually have a wide distribution in vivo, due to lack of cell specificity of unmodified TAT. Therefore, there is a great need for methods which can minimize the nonspecific cellular uptake of CPPs in normal tissues or organs.

Over the past few decades, cell penetration mechanisms of CPPs have been extensively studied. Although the mechanism of direct penetrating through the membrane has been widely accepted, there are still some other possible mechanisms such as endocytosis (Mayor & Pagano, 2007). Recent studies showed that CPPs could bind to tumor surface receptors with some specificity. Syndecan-4 is a heparan sulfate proteoglycan that acts as an endogenous membrane-related receptor. It is involved in the uptake of peptides through clathrin-mediated endocytosis. The specific interaction between CPPs and proteoglycan on the surface of cells plays an important role (Kawaguchi et al., 2016). Besides, CPPs rich in arginine and lysine can bind to the permeability-regulating receptor, neuropilin-1 (NRP1), and enhance the delivery of fusion proteins into tumors (Kadonosono et al., 2015).

In addition, several other membrane proteins have been involved in the internalization of arginine-rich CPPs (Fotinmleczek et al., 2005; Jobin & Alves, 2014).

In this review, we summarized "smart" CPP-based drug delivery strategies to minimize systemic toxicity which occurs after intravenous injection. To avoid nonspecific uptake, a prodrug type delivery system is adapted. Briefly, the CPPs are encapsulated in a delivery carrier or are modified to mask the cell membrane penetration ability during systemic circulation. CPPs would be unmasked in response to a specific endogenous environment condition or an external trigger. Activated or exposed CPPs can facilitate cell penetration and improve therapeutic effect.

Nevertheless, the prodrug design is still imperfect. Besides, the direct CPP-macromolecule conjugates or recombinant fusion expression products are the one-to-one modes. The combination between polymer nanocarriers and CPPs has been advocated more often in order to improve drug delivery capacity. Nanocarriers can also target tumors through the enhanced permeability and retention (EPR) effect (Kurrikoff, Gestin, & Langel, 2016). A receptor-mediated active tumor targeting could also be adapted to further reduce nonspecific tissue distribution and improve the therapeutic efficacy. The design of the multifunctional delivery systems is an appealing strategy. We can possibly combine active targeting strategies with other approaches (Shin, Zhang, Min, Lee, Byun, et al., 2014; Shin, Zhang, Min, Lee, Moon, et al., 2014). Last but not the least, though there are increasing number of publications, the clinical research and development of CPPs remains slow, suggesting the need for further studies, especially in vivo studies.

ACKNOWLEDGMENTS

We thank the National Basic Research Program of China (973 Program 2014CB931900) and NSFC, China (81373357, 81422048, 81673382, and 81521005) for the support. We also thank the support from Shanghai Institute of Materia Medica, CAS (CASIMM0120153023) and from Auburn University Start-up fund.

CONFLICT OF INTEREST

The authors declare no conflict of interest.

REFERENCES

Aguilera, T. A., Olson, E. S., Timmers, M. M., Jiang, T., & Tsien, R. Y. (2009). Systemic in vivo distribution of activatable cell penetrating peptides is superior to that of cell penetrating peptides. *Integrative Biology*, *1*(5–6), 371. https://doi.org/10.1039/b904878b.

Amamoto, T., Hirata, T., Takahashi, H., Kamiya, M., Urano, Y., Santa, T., et al. (2015). Spatiotemporal activation of molecules within cells using silica nanoparticles responsive to blue-green light. *Journal of Materials Chemistry B, 3*(37), 7427–7433. https://doi.org/10.1039/c5tb01165e.

Angelov, B., Garamus, V. M., Drechsler, M., & Angelova, A. (2017). Structural analysis of nanoparticulate carriers for encapsulation of macromolecular drugs. *Journal of Molecular Liquids, 235(*, 83–89. https://doi.org/10.1016/j.molliq.2016.11.064.

Berry, C. C., & Curtis, A. S. G. (2003). Functionalisation of magnetic nanoparticles for applications in biomedicine. *Journal of Physics D Applied Physics, 36*(36), R198–R206 [199]. https://doi.org/10.1088/0022-3727/36/13/203.

Bian, J., Popovic, Z. B., Benejam, C., Kiedrowski, M., Rodriguez, L. L., & Penn, M. S. (2007). Effect of cell-based intercellular delivery of transcription factor GATA4 on ischemic cardiomyopathy. *Circulation Research, 100*(11), 1626–1633. https://doi.org/10.1161/01.RES.0000269778.75877.68.

Cai, X., Zhu, H., Dong, H., Li, Y., Su, J., & Shi, D. (2014). Suppression of VEGF by reversible-PEGylated histidylated polylysine in cancer therapy. *Advanced Healthcare Materials, 3*(11), 1818–1827. https://doi.org/10.1002/adhm.201400063.

Cairns, R. A., Harris, I. S., & Mak, T. W. (2011). Regulation of cancer cell metabolism. *Nature Reviews. Cancer, 11*(2), 85–95. https://doi.org/10.1038/nrc2981.

Cao, J., Su, T., Zhang, L., Liu, R., Wang, G., He, B., et al. (2014). Polymeric micelles with citraconic amide as pH-sensitive bond in backbone for anticancer drug delivery. *International Journal of Pharmaceutics, 471*(1–2), 28–36. https://doi.org/10.1016/j.ijpharm.2014.05.010.

Chang, M., Chou, J. C., Chen, C. P., Liu, B. R., & Lee, H. J. (2007). Noncovalent protein transduction in plant cells by macropinocytosis. *The New Phytologist, 174*(1), 46–56. https://doi.org/10.1111/j.1469-8137.2007.01977.x.

Chen, Y.-J., Liu, B. R., Dai, Y.-H., Lee, C.-Y., Chan, M.-H., Chen, H.-H., et al. (2012). A gene delivery system for insect cells mediated by arginine-rich cell-penetrating peptides. *Gene, 493*(2), 201–210. https://doi.org/10.1016/j.gene.2011.11.060.

Chen, Y. Z., Zhang, M., Jin, H. Y., Li, D. D., Xu, F., Wu, A. H., et al. (2017). Glioma dual-targeting nanohybrid protein toxin constructed by Intein-mediated site-specific ligation for multistage booster delivery. *Theranostics*, 7(14), 3489–3503. https://doi.org/10.7150/thno.20578.

Chen, Y., Zhang, M., Jin, H., Tang, Y., Wang, H., Xu, Q., et al. (2017a). Intein-mediated site-specific synthesis of tumor-targeting protein delivery system: Turning PEG dilemma into prodrug-like feature. *Biomaterials, 116*, 57–68. https://doi.org/10.1016/j.biomaterials.2016.11.033.

Chen, Y., Zhang, M., Jin, H., Tang, Y., Wu, A., Xu, Q., et al. (2017b). Prodrug-like, PEGylated protein toxin trichosanthin for reversal of chemoresistance. *Molecular Pharmaceutics, 14*(5), 1429–1438. https://doi.org/10.1021/acs.molpharmaceut.6b00987.

Chiu, Y. L., Ali, A., Chu, C. Y., Cao, H., & Rana, T. M. (2004). Visualizing a correlation between siRNA localization, cellular uptake, and RNAi in living cells. *Chemistry & Biology, 11*(8), 1165–1175. https://doi.org/10.1016/j.chembiol.2004.06.006.

Choi, K. Y., Swierczewska, M., Lee, S., & Chen, X. (2012). Protease-activated drug development. *Theranostics, 2*(2), 156–178. https://doi.org/10.7150/thno.4068.

Davidson, T. J., Harel, S., Arboleda, V. A., Prunell, G. F., Shelanski, M. L., Greene, L. A., et al. (2004). Highly efficient small interfering RNA delivery to primary mammalian neurons induces microRNA-like effects before mRNA degradation. *The Journal of Neuroscience, 24*(45), 10040–10046. https://doi.org/10.1523/JNEUROSCI.3643-04.2004.

Derossi, D., Calvet, S., Trembleau, A., Brunissen, A., Chassaing, G., & Prochiantz, A. (1996). Cell internalization of the third helix of the antennapedia homeodomain is receptor-independent. *Journal of Biological Chemistry, 271*(30), 18188–18193. https://doi.org/10.1074/jbc.271.30.18188.

Di Stefano, G., Lanza, M., Kratz, F., Merina, L., & Fiume, L. (2004). A novel method for coupling doxorubicin to lactosaminated human albumin by an acid sensitive hydrazone bond: Synthesis, characterization and preliminary biological properties of the conjugate. *European Journal of Pharmaceutical Sciences*, *23*(4–5), 393–397. https://doi.org/10.1016/j.ejps.2004.09.005.

Dietrich, L., Rathmer, B., Ewan, K., Bange, T., Heinrichs, S., Dale, T. C., et al. (2017). Cell permeable stapled peptide inhibitor of Wnt signaling that targets beta-catenin protein–protein interactions. *Cell Chemical Biology*, *24*(8), 958–968. e955. https://doi.org/10.1016/j.chembiol.2017.06.013.

Dietz, G. (2002). Inhibition of neuronal apoptosis in vitro and in vivo using TAT-mediated protein transduction. *Molecular and Cellular Neuroscience*, *21*(1), 29–37. https://doi.org/10.1006/mcne.2002.1165.

Dowdy, S. F. (2000). Protein transduction: Unrestricted delivery into all cells? *Trends in Cell Biology*, *10*(7), 290–295. https://doi.org/10.1016/S0962-8924(00)01771-2.

Erazo-Oliveras, A., Muthukrishnan, N., Baker, R., Wang, T. Y., & Pellois, J. P. (2012). Improving the endosomal escape of cell-penetrating peptides and their cargos: Strategies and challenges. *Pharmaceuticals (Basel)*, *5*(11), 1177–1209. https://doi.org/10.3390/ph5111177.

Fawell, S., Seery, J., Daikh, Y., Moore, C., Chen, L. L., Pepinsky, B., et al. (1994). Tat-mediated delivery of heterologous proteins into cells. *Proceedings of the National Academy of Sciences of the United States of America*, *91*(2), 664–668. https://doi.org/10.1073/pnas.91.2.664.

Forman, J., Dietrich, M., & Todd Monroe, W. (2007). Photobiological and thermal effects of photoactivating UVA light doses on cell cultures. *Photochemical & Photobiological Sciences*, *6*(6), 649–658. https://doi.org/10.1039/b616979a.

Fotinmleczek, M., Welte, S., Mader, O., Duchardt, F., Fischer, R., Hufnagel, H., et al. (2005). Cationic cell-penetrating peptides interfere with TNF signalling by induction of TNF receptor internalization. *Journal of Cell Science*, *118*(15), 3339–3351. https://doi.org/10.1242/jcs.02460.

Frankel, A. D., & Pabo, C. O. (1988). Cellular uptake of the tat protein from human immunodeficiency virus. *Cell*, *55*(6), 1189–1193. https://doi.org/10.1016/0092-8674(88)90263-2.

Futaki, S., Nakase, I., Tadokoro, A., Takeuchi, T., & Jones, A. T. (2007). Arginine-rich peptides and their internalization mechanisms. *Biochemical Society Transactions*, *35*(Pt 4), 784. https://doi.org/10.1042/BST0350784.

Green, M., & Loewenstein, P. M. (1988). Autonomous functional domains of chemically synthesized human immunodeficiency virus tat-activator protein. *Cell*, *55*(6), 1179–1188. https://doi.org/10.1016/0092-8674(88)90262-0.

Han, S. S., Li, Z. Y., Zhu, J. Y., Han, K., Zeng, Z. Y., Hong, W., et al. (2015). Dual-pH sensitive charge-reversal polypeptide micelles for tumor-triggered targeting uptake and nuclear drug delivery. *Small*, *11*(21), 2543–2554. https://doi.org/10.1002/smll.201402865.

Hanahan, D., & Weinberg, R. A. (2011). Hallmarks of cancer: The next generation. *Cell*, *144*(5), 646–674. https://doi.org/10.1016/j.cell.2011.02.013.

Harada, H., Hiraoka, M., & Kizaka-Kondoh, S. (2002). Antitumor effect of TAT-oxygen-dependent degradation-caspase-3 fusion protein specifically stabilized and activated in hypoxic tumor cells. *Cancer Research*, *62*(7), 2013–2018.

Harada, H., Kizaka-Kondoh, S., Li, G., Itasaka, S., Shibuya, K., Inoue, M., et al. (2007). Significance of HIF-1-active cells in angiogenesis and radioresistance. *Oncogene*, *26*(54), 7508–7516. https://doi.org/10.1038/sj.onc.1210556.

Hatakeyama, H., Akita, H., & Harashima, H. (2013). The polyethyleneglycol dilemma: Advantage and disadvantage of PEGylation of liposomes for systemic genes and nucleic acids delivery to tumors. *Biological & Pharmaceutical Bulletin*, *36*(6), 892–899. https://doi.org/10.1248/bpb.b13-00059.

He, F., Cao, L., Zhang, X. J., Xiang, B., Cao, D. Y., & Qi, X. R. (2015). The application of enzyme-sensitive activatable cell-penetrating peptides to targeted delivery system. *Yao xue xue bao = Acta pharmaceutica Sinica*, *50*(2), 141–147.

He, H., Sheng, J., David, A. E., Kwon, Y. M., Zhang, J., Huang, Y., et al. (2013). The use of low molecular weight protamine chemical chimera to enhance monomeric insulin intestinal absorption. *Biomaterials*, *34*(31), 7733–7743. https://doi.org/10.1016/j.biomaterials.2013.06.047.

He, H., Sun, L., Ye, J., Liu, E., Chen, S., Liang, Q., et al. (2016). Enzyme-triggered, cell penetrating peptide-mediated delivery of anti-tumor agents. *Journal of Controlled Release*, *240*, 67–76. https://doi.org/10.1016/j.jconrel.2015.10.040.

Holman, R. R., Thorne, K. I., Farmer, A. J., Davies, M. J., Keenan, J. F., Paul, S., et al. (2007). Addition of biphasic, prandial, or basal insulin to oral therapy in type 2 diabetes. *The New England Journal of Medicine*, *357*(357), 1716–1730. https://doi.org/10.1056/NEJMoa075392.

Hsieh, P. W., Hung, C. F., & Fang, J. Y. (2009). Current prodrug design for drug discovery. *Current Pharmaceutical Design*, *15*(19), 2236 2250. https://doi.org/10.2174/138161209788682523.

Huang, Y., Jiang, Y., Wang, H., Wang, J., Shin, M. C., Byun, Y., et al. (2013). Curb challenges of the "Trojan horse" approach: Smart strategies in achieving effective yet safe cell-penetrating peptide-based drug delivery. *Advanced Drug Delivery Reviews*, *65*(10), 1299–1315. https://doi.org/10.1016/j.addr.2012.11.007.

Huang, Y., Park, Y. S., Wang, J., Moon, C., Kwon, Y. M., Chung, H. S., et al. (2010). ATTEMPTS system: A macromolecular prodrug strategy for cancer drug delivery. *Current Pharmaceutical Design*, *16*(21), 2369–2376. https://doi.org/10.2174/138161210791920441.

Huang, S., Shao, K., Liu, Y., Kuang, Y., Li, J., An, S., et al. (2013). Tumor-targeting and microenvironment-responsive smart nanoparticles for combination therapy of antiangiogenesis and apoptosis. *ACS Nano*, 7(3), 2860. https://doi.org/10.1021/nn400548g.

Jiang, Y., Lu, J., Wang, Y., Zeng, F., Wang, H., Peng, H., et al. (2014). Molecular-dynamics-simulation-driven design of a protease-responsive probe for in-vivo tumor imaging. *Advanced Materials*, *26*(48), 8174–8178. https://doi.org/10.1002/adma.201403547.

Jiang, T., Olson, E. S., Nguyen, Q. T., Roy, M., Jennings, P. A., & Tsien, R. Y. (2004). Tumor imaging by means of proteolytic activation of cell-penetrating peptides. *Proceedings of the National Academy of Sciences*, *101*(51), 17867–17872. https://doi.org/10.1073/pnas.0408191101.

Jiang, T., Zhang, Z., Zhang, Y., Lv, H., Zhou, J., Li, C., et al. (2012). Dual-functional liposomes based on pH-responsive cell-penetrating peptide and hyaluronic acid for tumor-targeted anticancer drug delivery. *Biomaterials*, *33*(36), 9246–9258. https://doi.org/10.1016/j.biomaterials.2012.09.027.

Jin, E., Zhang, B., Sun, X., Zhou, Z., Ma, X., Sun, Q., et al. (2013). Acid-active cell-penetrating peptides for in vivo tumor-targeted drug delivery. *Journal of the American Chemical Society*, *135*(2), 933–940. https://doi.org/10.1021/ja311180x.

Jing, H., Cheng, W., Li, S., Wu, B., Leng, X., Xu, S., et al. (2016). Novel cell-penetrating peptide-loaded nanobubbles synergized with ultrasound irradiation enhance EGFR siRNA delivery for triple negative breast cancer therapy. *Colloids and Surfaces B: Biointerfaces*, *146*, 387–395. https://doi.org/10.1016/j.colsurfb.2016.06.037.

Jobin, M. L., & Alves, I. D. (2014). On the importance of electrostatic interactions between cell penetrating peptides and membranes: A pathway toward tumor cell selectivity? *Biochimie*, *107*(Pt. A), 154–159. https://doi.org/10.1016/j.biochi.2014.07.022.

Jojovic, M., Delpech, B., Prehm, P., & Schumacher, U. (2002). Expression of hyaluronate and hyaluronate synthase in human primary tumours and their metastases in scid mice. *Cancer Letters*, *188*(1–2), 181–189. https://doi.org/10.1016/S0304-3835(02)00037-X.

Kadonosono, T., Yamano, A., Goto, T., Tsubaki, T., Niibori, M., Kuchimaru, T., et al. (2015). Cell penetrating peptides improve tumor delivery of cargos through neuropilin-1-dependent extravasation. *Journal of Controlled Release*, *201*, 14–21. https://doi.org/10.1016/j.jconrel.2015.01.011.

Kaplan, I. M., Wadia, J. S., & Dowdy, S. F. (2005). Cationic TAT peptide transduction domain enters cells by macropinocytosis. *Journal of Controlled Release*, *102*(1), 247–253. https://doi.org/10.1016/j.jconrel.2004.10.018.

Kawaguchi, Y., Takeuchi, T., Kuwata, K., Chiba, J., Hatanaka, Y., Nakase, I., et al. (2016). Syndecan-4 is a receptor for clathrin-mediated endocytosis of arginine-rich cell-penetrating peptides. *Bioconjugate Chemistry*, *27*(4), 1119–1130. https://doi.org/10.1021/acs.bioconjchem.6b00082.

Kim, C., Lee, Y., Kim, J. S., Jeong, J. H., & Park, T. G. (2010). Thermally triggered cellular uptake of quantum dots immobilized with poly(N-isopropylacrylamide) and cell penetrating peptide. *Langmuir*, *26*(18), 14965–14969. https://doi.org/10.1021/la102632m.

Kurrikoff, K., Gestin, M., & Langel, U. (2016). Recent in vivo advances in cell-penetrating peptide-assisted drug delivery. *Expert Opinion on Drug Delivery*, *13*(3), 373–387. https://doi.org/10.1517/17425247.2016.1125879.

Kwon, Y. M., Li, Y. T., Liang, J. F., Park, Y. J., Chang, L. C., & Yang, V. C. (2008). PTD-modified ATTEMPTS system for enhanced asparaginase therapy: A proof-of-concept investigation. *Journal of Controlled Release*, *130*(3), 252–258. https://doi.org/10.1016/j.jconrel.2008.06.017.

Lee, E. S., Gao, Z., Kim, D., Park, K., Kwon, I. C., & Bae, Y. H. (2008). Super pH-sensitive multifunctional polymeric micelle for tumor pH(e) specific TAT exposure and multidrug resistance. *Journal of Controlled Release*, *129*(3), 228–236. https://doi.org/10.1016/j.jconrel.2008.04.024.

Lee, H., Hong, B. J., Lee, J. H., Yeo, S., Jung, H. Y., Chung, J., et al. (2016). Hyaluronate-death receptor 5 antibody conjugates for targeted treatment of liver metastasis. *Biomacromolecules*, *17*(9), 3085–3093. https://doi.org/10.1021/acs.biomac.6b01022.

Lee, J.-H., & Lu, H. (2014). Chimeric p53 as an alternative therapy for hypoxic tumors. *Cancer Biology & Therapy*, *11*(1), 108–110. https://doi.org/10.4161/cbt.11.1.14526.

Lee, S. H., Moroz, E., Castagner, B., & Leroux, J. C. (2014). Activatable cell penetrating peptide–peptide nucleic acid conjugate via reduction of azobenzene PEG chains. *Journal of the American Chemical Society*, *136*(37), 12868–12871. https://doi.org/10.1021/ja507547w.

Lee, C. S., Park, W., Park, S. J., & Na, K. (2013). Endolysosomal environment-responsive photodynamic nanocarrier to enhance cytosolic drug delivery via photosensitizer-mediated membrane disruption. *Biomaterials*, *34*(36), 9227–9236. https://doi.org/10.1016/j.biomaterials.2013.08.037.

Li, N., Liu, Q., Su, Q., Wei, C., Lan, B., Wang, J., et al. (2013). Effects of legumain as a potential prognostic factor on gastric cancers. *Medical Oncology*, *30*(3), 621. https://doi.org/10.1007/s12032-013-0621-9.

Lin, W., Xie, X., Yang, Y., Fu, X., Liu, H., Yang, Y., et al. (2016). Thermosensitive magnetic liposomes with doxorubicin cell-penetrating peptides conjugate for enhanced and targeted cancer therapy. *Drug Delivery*, *23*(9), 3436–3443. https://doi.org/10.1080/10717544.2016.1189983.

Lin, T., Zhao, P., Jiang, Y., Tang, Y., Jin, H., Pan, Z., et al. (2016). Blood–brain-barrier-penetrating albumin nanoparticles for biomimetic drug delivery via albumin-binding protein pathways for antiglioma therapy. *ACS Nano*, *10*(11), 9999–10012. https://doi.org/10.1021/acsnano.6b04268.

Liu, Z., Jiang, M., Kang, T., Miao, D., Gu, G., Song, Q., et al. (2013). Lactoferrin-modified PEG-co-PCL nanoparticles for enhanced brain delivery of NAP peptide following intranasal administration. *Biomaterials*, *34*(15), 3870–3881. https://doi.org/10.1016/j.biomaterials.2013.02.003.

Liu, C., Sun, C., Huang, H., Janda, K., & Edgington, T. (2003). Overexpression of legumain in tumors is significant for invasion/metastasis and a candidate enzymatic target for prodrug therapy. *Cancer Research*, *63*(11), 2957–2964.

Liu, Z., Xiong, M., Gong, J., Zhang, Y., Bai, N., Luo, Y., et al. (2014). Legumain protease-activated TAT-liposome cargo for targeting tumours and their microenvironment. *Nature Communications*, *5*(4280). https://doi.org/10.1038/ncomms5280.

Macewan, S. R., Callahan, D. J., & Chilkoti, A. (2010). Stimulus-responsive macromolecules and nanoparticles for cancer drug delivery. *Nanomedicine*, *5*(5), 793–806. https://doi.org/10.2217/nnm.10.50.

Maeda, H., Sawa, T., & Konno, T. (2001). Mechanism of tumor-targeted delivery of macromolecular drugs, including the EPR effect in solid tumor and clinical overview of the prototype polymeric drug SMANCS. *Journal of Controlled Release*, *74*(1–3), 47–61. https://doi.org/10.1016/S0168-3659(01)00309-1.

Mano, M., Teodosio, C., Paiva, A., Simoes, S., & Pedroso de Lima, M. C. (2005). On the mechanisms of the internalization of S4(13)-PV cell-penetrating peptide. *The Biochemical Journal*, *390*(Pt 2), 603–612. https://doi.org/10.1042/BJ20050577.

Mayor, S., & Pagano, R. E. (2007). Pathways of clathrin-independent endocytosis. *Nature Reviews. Molecular Cell Biology*, *8*(8), 603–612. https://doi.org/10.1038/nrm2216.

Meng, F., Hennink, W. E., & Zhong, Z. (2009). Reduction-sensitive polymers and bioconjugates for biomedical applications. *Biomaterials*, *30*(12), 2180–2198. https://doi.org/10.1016/j.biomaterials.2009.01.026.

Mesken, J., Iltzsche, A., Mulac, D., & Langer, K. (2017). Modifying plasmid-loaded HSA-nanoparticles with cell penetrating peptides—Cellular uptake and enhanced gene delivery. *International Journal of Pharmaceutics*, *522*(1–2), 198–209. https://doi.org/10.1016/j.ijpharm.2017.03.006.

Miller, K. L., & Lanthier, M. (2015). Regulatory watch: Innovation in biologic new molecular entities: 1986–2014. *Nature Reviews. Drug Discovery*, *14*(2), 83. https://doi.org/10.1038/nrd4535.

Miller, G., Matthews, S. P., Reinheckel, T., Fleming, S., & Watts, C. (2011). Asparagine endopeptidase is required for normal kidney physiology and homeostasis. *The FASEB Journal*, *25*(5), 1606–1617. https://doi.org/10.1096/fj.10-172312.

Moon, C., Kwon, Y. M., Lee, W. K., Park, Y. J., & Yang, V. C. (2007). In vitro assessment of a novel polyrotaxane-based drug delivery system integrated with a cell-penetrating peptide. *Journal of Controlled Release*, *124*(1–2), 43–50. https://doi.org/10.1016/j.jconrel.2007.08.029.

Murriel, C. L., & Dowdy, S. F. (2006). Influence of protein transduction domains on intracellular delivery of macromolecules. *Expert Opinion on Drug Delivery*, *3*(6), 739–746. https://doi.org/10.1517/17425247.3.6.739.

Nakase, I., Takeuchi, T., Tanaka, G., & Futaki, S. (2008). Methodological and cellular aspects that govern the internalization mechanisms of arginine-rich cell-penetrating peptides. *Advanced Drug Delivery Reviews*, *60*(4–5), 598–607. https://doi.org/10.1016/j.addr.2007.10.006.

Nori, A., & Kopecek, J. (2005). Intracellular targeting of polymer-bound drugs for cancer chemotherapy. *Advanced Drug Delivery Reviews*, *57*(4), 609–636. https://doi.org/10.1016/j.addr.2004.10.006.

Palm, C., Jayamanne, M., Kjellander, M., & Hallbrink, M. (2007). Peptide degradation is a critical determinant for cell-penetrating peptide uptake. *Biochimica et Biophysica Acta*, *1768*(7), 1769–1776. https://doi.org/10.1016/j.bbamem.2007.03.029.

Park, Y. J., Chang, L. C., Liang, J. F., Moon, C., Chung, C. P., & Yang, V. C. (2005). Non-toxic membrane translocation peptide from protamine, low molecular weight protamine (LMWP), for enhanced intracellular protein delivery: In vitro and in vivo study. *FASEB Journal, 19*(11), 1555–1157. https://doi.org/10.1096/fj.04-2322fje.

Reissmann, S. (2014). Cell penetration: Scope and limitations by the application of cell-penetrating peptides. *Journal of Peptide Science, 20*(10), 760–784. https://doi.org/10.1002/psc.2672.

Ren, J., Zhang, P., Tian, J., Zhou, Z., Liu, X., Wang, D., et al. (2014). A targeted ultrasound contrast agent carrying gene and cell-penetrating peptide: Preparation and gene transfection in vitro. *Colloids and Surfaces B: Biointerfaces, 121*, 362–370. https://doi.org/10.1016/j.colsurfb.2014.06.017.

Riess, G. (2003). Micellization of block copolymers. *Progress in Polymer Science, 28*(7), 1107–1170. https://doi.org/10.1016/s0079-6700(03)00015-7.

Sawant, R. M., Hurley, J. P., Salmaso, S., Kale, A., Tolcheva, E., Levchenko, T. S., et al. (2006). "SMART" drug delivery systems: Double-targeted pH-responsive pharmaceutical nanocarriers. *Bioconjugate Chemistry, 17*(4), 943–949. https://doi.org/10.1021/bc060080h.

Schorderet, D. F., Manzi, V., Canola, K., Bonny, C., Arsenijevic, Y., Munier, F. L., et al. (2005). D-TAT transporter as an ocular peptide delivery system. *Clinical & Experimental Ophthalmology, 33*(6), 628–635. https://doi.org/10.1111/j.1442-9071.2005.01108.x.

Schottler, S., Becker, G., Winzen, S., Steinbach, T., Mohr, K., Landfester, K., et al. (2016). Protein adsorption is required for stealth effect of poly(ethylene glycol)- and poly(phosphoester)-coated nanocarriers. *Nature Nanotechnology, 11*(4), 372–377. https://doi.org/10.1038/nnano.2015.330.

Schwarze, S. R., & Dowdy, S. F. (1999). In vivo protein transduction: Delivery of a biologically active protein into the mouse. *Science, 285*(5433), 1569–1572. https://doi.org/10.1126/science.285.5433.1569.

Shamay, Y., Adar, L., Ashkenasy, G., & David, A. (2011). Light induced drug delivery into cancer cells. *Biomaterials, 32*(5), 1377–1386. https://doi.org/10.1016/j.biomaterials.2010.10.029.

Shi, K., Li, J., Cao, Z., Yang, P., Qiu, Y., Yang, B., et al. (2015). A pH-responsive cell-penetrating peptide-modified liposomes with active recognizing of integrin αvβ3 for the treatment of melanoma. *Journal of Controlled Release, 217*, 138–150. https://doi.org/10.1016/j.jconrel.2015.09.009.

Shi, N. Q., Qi, X. R., Xiang, B., & Zhang, Y. (2014). A survey on "Trojan Horse" peptides: Opportunities, issues and controlled entry to "Troy". *Journal of Controlled Release, 194*, 53–70. https://doi.org/10.1016/j.jconrel.2014.08.014.

Shin, M. C., Zhang, J., Min, K. A., Lee, K., Byun, Y., David, A. E., et al. (2014a). Cell-penetrating peptides: Achievements and challenges in application for cancer treatment. *Journal of Biomedical Materials Research. Part A, 102*(2), 575–587. https://doi.org/10.1002/jbm.a.34859.

Shin, M. C., Zhang, J., Min, K. A., Lee, K., Moon, C., Balthasar, J. P., et al. (2014b). Combination of antibody targeting and PTD-mediated intracellular toxin delivery for colorectal cancer therapy. *Journal of Controlled Release, 194*, 197–210. https://doi.org/10.1016/j.jconrel.2014.08.030.

Shin, M. C., Zhao, J., Zhang, J., Huang, Y., He, H., Wang, M., et al. (2015). Recombinant TAT-gelonin fusion toxin: Synthesis and characterization of heparin/protamine-regulated cell transduction. *Journal of Biomedical Materials Research. Part A, 103*(1), 409–419. https://doi.org/10.1002/jbm.a.35188.

Smith, C. A., de la Fuente, J., Pelaz, B., Furlani, E. P., Mullin, M., & Berry, C. C. (2010). The effect of static magnetic fields and tat peptides on cellular and nuclear uptake of magnetic nanoparticles. *Biomaterials*, *31*(15), 4392–4400. https://doi.org/10.1016/j.biomaterials.2010.01.096.

Sonoda, S., Tachibana, K., Uchino, E., Yamashita, T., Sakoda, K., Sonoda, K. H., et al. (2007). Inhibition of melanoma by ultrasound-microbubble-aided drug delivery suggests membrane permeabilization. *Cancer Biology & Therapy*, *6*(8), 1276–1283. https://doi.org/10.4161/cbt.6.8.4485.

Spill, F., Reynolds, D. S., Kamm, R. D., & Zaman, M. H. (2016). Impact of the physical microenvironment on tumor progression and metastasis. *Current Opinion in Biotechnology*, *40*, 41–48. https://doi.org/10.1016/j.copbio.2016.02.007.

Stern, R. (2008). Hyaluronidases in cancer biology. *Seminars in Cancer Biology*, *18*(4), 275–280. https://doi.org/10.1016/j.semcancer.2008.03.017.

Takakura, Y., & Hashida, M. (1996). Macromolecular carrier systems for targeted drug delivery: Pharmacokinetic considerations on biodistribution. *Pharmaceutical Research*, *13*(6), 820–831. https://doi.org/10.1023/A:1016084508097.

Tan, J., Cheong, H., Park, Y. S., Kim, H., Zhang, M., Moon, C., et al. (2014). Cell-penetrating peptide-mediated topical delivery of biomacromolecular drugs. *Current Pharmaceutical Biotechnology*, *15*(3), 231. https://doi.org/10.2174/1389201015666140617094320.

Torchilin, V. P. (2008). Tat peptide-mediated intracellular delivery of pharmaceutical nanocarriers. *Advanced Drug Delivery Reviews*, *60*(4–5), 548–558. https://doi.org/10.1016/j.addr.2007.10.008.

Torchilin, V. P., Benson, H., & Toth, I. (2008). Cell penetrating peptide-modified pharmaceutical nanocarriers for intracellular drug and gene delivery. *Biopolymers*, *90*(5), 604–610. https://doi.org/10.1002/bip.20989.

Torchilin, V. P., & Lukyanov, A. N. (2003). Peptide and protein drug delivery to and into tumors: Challenges and solutions. *Drug Discovery Today*, *8*(6), 259–266. https://doi.org/10.1016/S1359-6446(03)02623-0.

van den Berg, A., & Dowdy, S. F. (2011). Protein transduction domain delivery of therapeutic macromolecules. *Current Opinion in Biotechnology*, *22*(6), 888–893. https://doi.org/10.1016/j.copbio.2011.03.008.

Vandenbroucke, R. E., De Smedt, S. C., Demeester, J., & Sanders, N. N. (2007). Cellular entry pathway and gene transfer capacity of TAT-modified lipoplexes. *Biochimica et Biophysica Acta*, *1768*(3), 571–579. https://doi.org/10.1016/j.bbamem.2006.11.006.

Vives, E. (2005). Present and future of cell-penetrating peptide mediated delivery systems: "Is the Trojan horse too wild to go only to troy?" *Journal of Controlled Release*, *109*(1–3), 77–85. https://doi.org/10.1016/j.jconrel.2005.09.032.

Vives, E., Schmidt, J., & Pelegrin, A. (2008). Cell-penetrating and cell-targeting peptides in drug delivery. *Biochimica et Biophysica Acta*, *1786*(2), 126–138. https://doi.org/10.1016/j.bbcan.2008.03.001.

Wadia, J. S., Stan, R. V., & Dowdy, S. F. (2004). Transducible TAT-HA fusogenic peptide enhances escape of TAT-fusion proteins after lipid raft macropinocytosis. *Nature Medicine*, *10*(3), 310–315. https://doi.org/10.1038/nm996.

Wang, Z., Chen, Y., Liu, E., Gong, J., Shin, M. C., & Huang, Y. (2014). CPP-mediated protein delivery in a noncovalent form: Proof-of-concept for percutaneous and intranasal delivery. *Protein & Peptide Letters*, *21*(11), 1129–1136. https://doi.org/10.2174/0929866521666140807121903.

Wang, H., Wang, H., Liang, J., Jiang, Y., Guo, Q., Peng, H., et al. (2014). Cell-penetrating apoptotic peptide/p53 DNA nanocomplex as adjuvant therapy for drug-resistant breast cancer. *Molecular Pharmaceutics*, *11*(10), 3352–3360. https://doi.org/10.1021/mp5001058.

Weissleder, R. (2001). A clearer vision for in vivo imaging. *Nature Biotechnology*, *19*(4), 316. https://doi.org/10.1038/86684.

Wu, T., & Dai, Y. (2017). Tumor microenvironment and therapeutic response. *Cancer Letters*, *387*, 61–68. https://doi.org/10.1016/j.canlet.2016.01.043.

Xu, P., Van Kirk, E. A., Zhan, Y., Murdoch, W. J., Radosz, M., & Shen, Y. (2007). Targeted charge-reversal nanoparticles for nuclear drug delivery. *Angewandte Chemie (International Ed. in English)*, *46*(26), 4999–5002. https://doi.org/10.1002/anie.200605254.

Yang, V. C., Naik, S. S., Song, H., Dombkowski, A. A., Crippen, G., & Liang, J. F. (2005). Construction and characterization of a t-PA mutant for use in ATTEMPTS: A drug delivery system for achieving targeted thrombolysis. *Journal of Controlled Release*, *110*(1), 164–176. https://doi.org/10.1016/j.jconrel.2005.09.027.

Yang, Y., Xie, X., Xu, X., Xia, X., Wang, H., Li, L., et al. (2016). Thermal and magnetic dual-responsive liposomes with a cell-penetrating peptide–siRNA conjugate for enhanced and targeted cancer therapy. *Colloids and Surfaces B: Biointerfaces*, *146*, 607–615. https://doi.org/10.1016/j.colsurfb.2016.07.002.

Yang, Y., Yang, Y., Xie, X., Cai, X., & Mei, X. (2014). Preparation and characterization of photo-responsive cell-penetrating peptide-mediated nanostructured lipid carrier. *Journal of Drug Targeting*, *22*(10), 891–900. https://doi.org/10.3109/1061186X.2014.940589.

Yang, Y., Yang, Y., Xie, X., Cai, X., Wang, Z., Gong, W., et al. (2015). A near-infrared two-photon-sensitive peptide-mediated liposomal delivery system. *Colloids and Surfaces B: Biointerfaces*, *128*, 427–438. https://doi.org/10.1016/j.colsurfb.2015.02.041.

Yang, Y., Yang, Y., Xie, X., Cai, X., Zhang, H., Gong, W., et al. (2014). PEGylated liposomes with NGR ligand and heat-activable cell-penetrating peptide–doxorubicin conjugate for tumor-specific therapy. *Biomaterials*, *35*(14), 4368–4381. https://doi.org/10.1016/j.biomaterials.2014.01.076.

Yang, Y., Yang, Y., Xie, X., Xu, X., Xia, X., Wang, H., et al. (2016). Dual stimulus of hyperthermia and intracellular redox environment triggered release of siRNA for tumor-specific therapy. *International Journal of Pharmaceutics*, *506*(1–2), 158–173. https://doi.org/10.1016/j.ijpharm.2016.04.035.

Yang, Z.-Z., Zhang, Y.-Q., Wang, Z.-Z., Wu, K., Lou, J.-N., & Qi, X.-R. (2013). Enhanced brain distribution and pharmacodynamics of rivastigmine by liposomes following intranasal administration. *International Journal of Pharmaceutics*, *452*(1–2), 344–354. https://doi.org/10.1016/j.ijpharm.2013.05.009.

Ye, J., Shin, M. C., Liang, Q., He, H., & Yang, V. C. (2015). 15 years of ATTEMPTS: A macromolecular drug delivery system based on the CPP-mediated intracellular drug delivery and antibody targeting. *Journal of Controlled Release*, *205*, 58–69. https://doi.org/10.1016/j.jconrel.2014.12.002.

Yu, Z., Wu, J., Wu, S., Jia, P., Tong, Y., Wu, X., et al. (2009). A recombinant cell-permeable p53 fusion protein is selectively stabilized under hypoxia and inhibits tumor cell growth. *Cancer Letters*, *279*(1), 101–107. https://doi.org/10.1016/j.canlet.2009.01.030.

Zhang, Q., Gao, H., & He, Q. (2015). Taming cell penetrating peptides: Never too old to teach old dogs new tricks. *Molecular Pharmaceutics*, *12*(9), 3105–3118. https://doi.org/10.1021/acs.molpharmaceut.5b00428.

Zhang, J., Li, M., Yuan, Z., Wu, D., Chen, J. d., & Feng, J. (2016). Stepwise-activable multifunctional peptide-guided prodrug micelles for cancerous cells intracellular drug release. *Journal of Nanoparticle Research*, *18*(10), 299. https://doi.org/10.1007/s11051-016-3616-6.

Zhang, W., Mao, Z., & Gao, C. (2014). Preparation of TAT peptide-modified poly(N-isopropylacrylamide) microgel particles and their cellular uptake, intracellular distribution, and influence on cytoviability in response to temperature change. *Journal of Colloid and Interface Science*, *434*, 122–129. https://doi.org/10.1016/j.jcis.2014.07.031.

Zhang, J., Wu, D., Li, M. F., & Feng, J. (2015). Multifunctional mesoporous silica nanoparticles based on charge-reversal plug-gate nanovalves and acid-decomposable

ZnO quantum dots for intracellular drug delivery. *ACS Applied Materials & Interfaces*, 7(48), 26666–26673. https://doi.org/10.1021/acsami.5b08460.

Zhao, X., Shang, T., Zhang, X., Ye, T., Wang, D., & Rei, L. (2016). Passage of magnetic tat-conjugated $Fe_3O_4@SiO_2$ nanoparticles across in vitro blood–brain barrier. *Nanoscale Research Letters*, *11*(1), 451. https://doi.org/10.1186/s11671-016-1676-2.

Zhao, J., Zhao, F., Wang, X., Fan, X., & Wu, G. (2016). Secondary nuclear targeting of mesoporous silica nano-particles for cancer-specific drug delivery based on charge inversion. *OncoTarget*, 7(43), 70100–70112. http://doi.org/10.18632/oncotarget.12149.

Zhuang, J., Gordon, M. R., Ventura, J., Li, L., & Thayumanavan, S. (2013). Multi-stimuli responsive macromolecules and their assemblies. *Chemical Society Reviews*, *42*(17), 7421–7435. https://doi.org/10.1039/c3cs60094g.

Ziegler, A. (2008). Thermodynamic studies and binding mechanisms of cell-penetrating peptides with lipids and glycosaminoglycans. *Advanced Drug Delivery Reviews*, *60*(4–5), 580–597. https://doi.org/10.1016/j.addr.2007.10.005.

CHAPTER SEVEN

In Silico Tools and Databases for Designing Peptide-Based Vaccine and Drugs

Salman Sadullah Usmani*,†, Rajesh Kumar*,†, Sherry Bhalla*, Vinod Kumar*,†, Gajendra P.S. Raghava*,†,1

*Center for Computational Biology, Indraprastha Institute of Information Technology, New Delhi, India
†Bioinformatics Centre, CSIR-Institute of Microbial Technology, Chandigarh, India
[1]Corresponding author: e-mail address: raghava@imtech.res.in

Contents

Abstract

The prolonged conventional approaches of drug screening and vaccine designing prerequisite patience, vigorous effort, outrageous cost as well as additional manpower. Screening and experimentally validating thousands of molecules for a specific therapeutic property never proved to be an easy task. Similarly, traditional way of vaccination includes administration of either whole or attenuated pathogen, which raises toxicity

Advances in Protein Chemistry and Structural Biology, Volume 112
ISSN 1876-1623
https://doi.org/10.1016/bs.apcsb.2018.01.006

and safety issues. Emergence of sequencing and recombinant DNA technology led to the epitope-based advanced vaccination concept, i.e., small peptides (epitope) can stimulate specific immune response. Advent of bioinformatics proved to be an adjunct in vaccine and drug designing. Genomic study of pathogens aid to identify and analyze the protective epitope. A number of in silico tools have been developed to design immunotherapy as well as peptide-based drugs in the last two decades. These tools proved to be a catalyst in drug and vaccine designing. This review solicits therapeutic peptide databases as well as in silico tools developed for designing peptide-based vaccine and drugs.

1. INTRODUCTION

Evolution of universe and mankind also witnesses evolvement of life-threating diseases. The phylogenetic analysis of extant pathogens moreover suggests that several diseases have been coevolving with humans for millennia. Advances in genetic research have increased our understanding of the antiquity of human–pathogen interactions (Bos et al., 2011; Devault et al., 2014; Rifkin, Potgieter, Ramond, & Cowan, 2017). Molecular level research and advancement of techniques such as combinatorial chemistry, high-throughput screening, and the sequencing of the human genome, proved to be a complement for pharmaceutical industries (Hajduk & Greer, 2007). Scientific studies clinched in the past centuries enhance the development of therapeutics, resulted into increase in life expectancy of human race. In 20th century, the focus was on to discover small molecule-based therapy and antibiotics but last few decades has witnessed a shift in the paradigm toward peptide-based therapy. In 2016, 22 new molecular entities got Food and Drug Administration (FDA) approval among which only 50% are small molecule. Low selectivity, non-specificity, and synthetic nature of small molecules proved to be its disadvantage and their side effects reduced its importance as sole drug candidate (Schneider, 2018).

In contrast, peptides have greater efficacy, selectivity, and specificity. Peptides are naturally occurring biologics and hence safer than synthetic drugs with low or almost no toxicity. Beside this, peptides are nowadays easy to synthesize in low cost, which draws the attention of pharmaceutical industry as well as scientific community (Fosgerau & Hoffmann, 2015). But there are certain drawbacks with peptide- and protein-based therapeutics like low solubility, low circulating half-life, proteolytic degradation, immunogenicity, and physiochemical instability (Otvos & Wade, 2014; Zapadka, Becher, Gomes Dos Santos, & Jackson, 2017). With development of techniques, several strategies like PEGylation, glycosylation, amino acid substitution, and conjugation to serum protein have emerged, which result into the improved pharmacokinetic and pharmacodynamic property of peptides (Di, 2015; Fosgerau & Hoffmann, 2015). These strategies also reduce the immunogenicity and proteolytic cleavage, hence contributing in more potent and advanced peptide-based therapy. Over 60 peptide drugs are already approved by FDA while ~140 are in the clinical trials and around 500–600 in the preclinical trials (Fosgerau & Hoffmann, 2015). The report on global Peptide Therapeutics market predicts the global Peptide Therapeutics market to grow with a CAGR between 9.0% and 9.5% from 2017 to 2023. The global Peptide Therapeutics market was worth USD 21.12 billion in 2016 (New York, 2017).

Rapid increase in peptide-based therapy bring forth peptide informatics. In last two decades, various groups come up with peptide databases and compiled their scattered information from literature, particularly focusing on their therapeutic aspects such as antimicrobial, anticancerous, hemolytic, antihypertensive, etc. These information have been also used to develop computational software and algorithms, which can predict the diverse therapeutic properties of a peptide. Integration of computational and experimental approaches complements each other for designing better therapeutics. Computational virtual screening is more effective in terms of time and cost as compared to conventional hit and trial method of screening thousands of compound experimentally (Bui & Steiner, 2016; Fradera & Babaoglu, 2017; Leelananda & Lindert, 2016). These algorithms help in the development of novel peptide analogs with improved bioactivity (Fig. 1). Several therapeutic peptides have been designed by computational algorithms and proved to be effective through experimental approach (Gautam et al., 2015; Smadbeck et al., 2014).

2. THERAPEUTIC PEPTIDE REPOSITORIES

Peptides once neglected have now gained a wide recognition in the field of therapeutics (Antosova, Mackova, Kral, & Macek, 2009) and have acquired quite an expansion in pharmaceutical market (Peptide Therapeutics market: global industry analysis, size, share, growth, trends, and forecast 2016–24). The strengths like good efficacy, safety, shorter time to market make peptides is one of the favorite therapeutic molecules (Fosgerau &

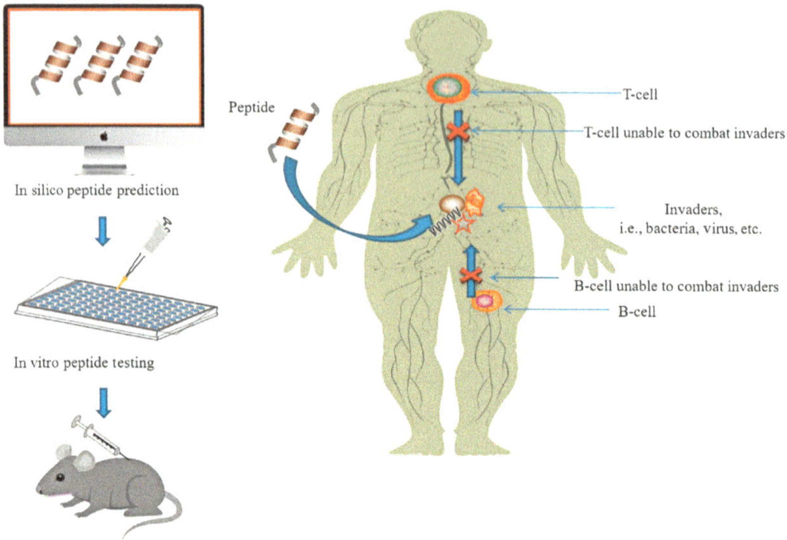

Fig. 1 Schematic representation of peptide-based drug designing.

Hoffmann, 2015). There are some limitations like of chemical and physical instability, prone to hydrolysis and oxidation, shorter half-life and fast elimination, associated with peptide as therapeutics. These barriers can be crossed using various in vivo, in vitro, and in silico tools which are accessible to assess ADME properties of peptides. Some structural modification approaches are also in place to improve the peptide therapeutics (Di, 2015). The resources of therapeutic peptides in this era of the growing peptide market can be the goldmine to screen vaccine candidates. In past the development of peptides databases have grown exponentially (Table 1). These resources can be further categorized into four subtypes which are also explained as follows.

2.1 Immunological Databases

The most multipurpose and vast repository for development of new vaccines, diagnostics, and therapeutics is IEDB (Fleri et al., 2017). This versatile database provides extensive experimentally derived epitope data and primary and secondary prediction tools involved in vaccine design and epitope discovery. Another critical resource of major histocompatibility complex (MHC)-binding peptides is MHCBN, a comprehensive database of MHC-binding and -nonbinding peptides containing 19,777 entries which include 17,129 MHC binders and 2648 MHC nonbinders for more than

Table 1 List of Databases/Repositories Developed for Maintaining Therapeutic Peptide

Name	Description	Weblink	Year
Immunological databases			
EPIMHC (Reche, Zhang, Glutting, & Reinherz, 2005)	MHC-binding peptides	http://imed.med.ucm.es/epimhc/	2005
AntiJen (Toseland et al., 2005)	MHC-binding peptides	http://www.ddg-pharmfac.net/antijen/AntiJen/antijenhomepage.htm	2005
Bcipep (Saha, Bhasin, & Raghava, 2005)	B-cell epitopes	http://webs.iiitd.edu.in/raghava/bcipep/	2005
MHCBN 4.0 (Lata, Bhasin, & Raghava, 2009)	MHC binders and nonbinders	http://webs.iiitd.edu.in/raghava/mhcbn/	2009
IEDB (Vita et al., 2015)	Immune epitope database	http://www.iedb.org/	2015
HLAsupE (Wang, Li, & Wang, 2016)	Database of HLA super type specific epitopes	http://www.immunoinformatics.net/HLAsupE/	2016
PeMtb (Zia et al., 2017)	A database of MHC antigenic peptide of *Mycobacterium tuberculosis*	http://www.pemtb-amu.org/	2017
Disease-specific databases			
Peptaibol (Whitmore & Wallace, 2004)	Antibiotic peptides	http://peptaibol.cryst.bbk.ac.uk/home.shtml	2004
TumorHoPe (Kapoor et al., 2012)	Tumor homing peptides	http://crdd.osdd.net/servers/tumorhope/	2012

Continued

Table 1 List of Databases/Repositories Developed for Maintaining Therapeutic Peptide—cont'd

Name	Description	Weblink	Year
HIPdb (Qureshi, Thakur, & Kumar, 2013)	HIV inhibiting peptides	http://crdd.osdd.net/servers/hipdb/	2013
ParaPep (Mehta et al., 2014)	Antiparasitic peptides	http://webs.iiitd.edu.in/raghava/parapep/	2014
AVPdb (Qureshi, Thakur, Tandon, & Kumar, 2014)	Antiviral peptides	http://crdd.osdd.net/servers/avpdb/	2014
AHTPDB (Kumar, Chaudhary, Sharma, et al., 2015)	Antihypertensive peptides	http://webs.iiitd.edu.in/raghava/ahtpdb/	2015
CancerPPD (Tyagi et al., 2015)	Anticancer peptides and proteins	http://webs.iiitd.edu.in/raghava/cancerppd/	2015
SATPdb (Singh et al., 2016)	Database of structurally annotated therapeutic peptides	http://webs.iiitd.edu.in/raghava/satpdb/	2016
MtbVeb (Dhanda, Vir, et al., 2016)	A web-based platform developed for designing vaccines against existing and emerging strains of mycobacterium tuberculosis	http://webs.iiitd.edu.in/raghava/mtbveb/	2016
PeptiCKDdb (Krochmal et al., 2016)	Peptide- and protein-centric database for the investigation of genesis and progression of chronic kidney disease	http://www.peptickddb.com/Home.cshtml	2016
CancerPDF (Bhalla et al., 2017)	A repository of cancer-associated peptidome found in human biofluids	http://webs.iiitd.edu.in/raghava/cancerpdf/	2017
THPdb (Usmani et al., 2017)	Database of FDA-approved peptide and protein therapeutics	http://webs.iiitd.edu.in/raghava/thpdb/	2017

Antimicrobial peptide databases			
Defensins Knowledgebase (Seebah et al., 2007)	Defensin family of antimicrobial peptides	http://defensins.bii.a-star.edu.sg/	2007
RAPD (Li & Chen, 2008)	Recombinantly produced antimicrobial peptides	http://faculty.ist.unomaha.edu/chen/rapd/index.php	2008
PeptideDB (Liu, Baggerman, Schoofs, & Wets, 2008)	Bioactive peptide database	http://rosettadesigngroup.com/blog/tag/peptidb/	2008
PhytAMP (Hammami, Ben Hamida, Vergoten, & Fliss, 2009)	Antimicrobial plant peptides	http://phytamp.hammamilab.org/	2009
APD2 (Wang, Li, & Wang, 2009)	The antimicrobial peptides (AMPs)	http://aps.unmc.edu/AP/main.php	2009
BACTIBASE (Hammami, Zouhir, Le Lay, Ben Hamida, & Fliss, 2010)	Bacteriocins (natural AMPs)	http://bactibase.hammamilab.org/main.php	2010
YADAMP (Piotto, Sessa, Concilio, & Iannelli, 2012)	Antimicrobial peptides	http://yadamp.unisa.it/	2012
Quorumpeps (Wynendaele et al., 2013)	Quorum sensing peptides	http://quorumpeps.ugent.be/	2013
CAMP (Waghu et al., 2014)	Antimicrobial peptides	http://www.camp.bicnirrh.res.in/	2014

Continued

Table 1 List of Databases/Repositories Developed for Maintaining Therapeutic Peptide—cont'd

Name	Description	Weblink	Year
MilkAMP (Théolier, Fliss, Jean, & Hammami, 2014)	Antimicrobial peptides of dairy origin	http://milkampdb.org/home.php	2014
DBAASP (Gogoladze et al., 2014)	Database of antimicrobial activity and structure of peptides	https://dbaasp.org/home	2014
BaAMPs (Di Luca, Maccari, Maisetta, & Batoni, 2015)	AMPs against microbial biofilms	http://www.baamps.it/	2015
DBAASP v.2 (Pirtskhalava et al., 2016)	Antimicrobial peptides	https://dbaasp.org/home	2016
CAMPR3 (Waghu, Barai, Gurung, & Idicula-Thomas, 2016)	Collection of antimicrobial peptides	http://www.camp.bicnirrh.res.in/	2016
DRAMP (Fan et al., 2016)	A comprehensive data repository of antimicrobial peptides	http://dramp.cpu-bioinfor.org/	2016
APD3 (Wang, Li, et al., 2016)	Antimicrobial peptide database	http://aps.unmc.edu/AP/main.php	2017
InverPep (Gómez, Giraldo, & Orduz, 2017)	A database of invertebrate antimicrobial peptides	http://ciencias.medellin.unal.edu.co/gruposdeinvestigacion/prospeccionydisenobiomoleculas/InverPep/public/home_en	2017

Targeted delivery databases			
SPdb (Choo, Tan, & Ranganathan, 2005)	A signal peptide database	http://proline.bic.nus.edu.sg/spdb/	2008
NeuroPedia (Kim, Bark, Hook, & Bandeira, 2011)	Neuropeptides	http://proteomics.ucsd.edu/Software/NeuroPedia.html	2011
CPPsite (Gautam et al., 2012)	Cell-penetrating peptides	http://webs.iiitd.edu.in/raghava/cppsite/	2012
Brainpeps (Van Dorpe et al., 2012)	Blood–brain barrier peptides	http://brainpeps.ugent.be/	2012
NeuroPep (Wang et al., 2015)	Neuropeptides	http://isyslab.info/NeuroPep/	2015
CPPsite 2.0 (Agrawal et al., 2016)	An updated version of database CPPsite	http://webs.iiitd.edu.in/raghava/cppsite2.0/	2016
Miscellaneous			
Hmrbase (Rashid, Singla, Sharma, Kumar, & Raghava, 2009)	Peptide hormones and their receptors	http://webs.iiitd.edu.in/raghava/hmrbase/	2009
Human urine peptide database (Siwy, Mullen, Golovko, Franke, & Zürbig, 2011)	Naturally occurring human urinary peptides	http://mosaiques-diagnostics.de/mosaiques-diagnostics/human-urinary-proteom-database	2011

Continued

Table 1 List of Databases/Repositories Developed for Maintaining Therapeutic Peptide—cont'd

Name	Description	Weblink	Year
DADP (Novković, Simunić, Bojović, Tossi, & Juretić, 2012)	Anuran defense peptides	http://split4.pmfst.hr/dadp/	2012
ConoServer (Kaas, Westermann, Halai, Wang, & Craik, 2008)	Conopeptides (expressed by snails)	http://www.conoserver.org/	2012
Hemolytik (Gautam et al., 2014)	Hemolytic peptides	http://webs.iiitd.edu.in/raghava/hemolytik/	2014
PEPlife (Mathur et al., 2016)	A repository of the half-life of peptides	http://webs.iiitd.edu.in/raghava/peplife/	2016
PRIDE archive (Vizcaíno et al., 2016)	Database is one of the world-leading data repositories of mass spectrometry (MS)-based proteomics data	https://www.ebi.ac.uk/pride/archive/	2016
CPAD (Thangakani et al., 2016)	Curated protein aggregation database	https://www.iitm.ac.in/bioinfo/CPAD/	2016
MBPDB (Nielsen, Beverly, Qu, & Dallas, 2017)	Milk bioactive peptide database	http://mbpdb.nws.oregonstate.edu/	2017
miPepBase (Garg, Kumari, Kumar, & Kumar, 2017)	Mimicry peptide database	http://proteininformatics.org/mkumar/mipepbase/	2017

400 MHC molecules. In addition to this is another database EPIMHC containing 4867 distinct peptide sequences of MHC-binding peptides and T-cell epitopes from various sources, including 84 tumor-associated antigens (Reche et al., 2005). The additional database published in the same year, AntiJen is a comprehensive resource containing more than 31,000 entries and integrates kinetic, thermodynamic, functional, and cellular data related to immunology and vaccinology (Toseland et al., 2005).

The collection of B-cell epitopes helps in deriving new scales for accurate in silico prediction of linear epitopes. Also it will help the immunologist to understand the complex nature of immunogenic peptides and for the development of vaccines. The Bcipep (Saha et al., 2005) database of experimentally determined linear B-cell epitopes contains 3031 entries from the wide range of pathogens assisting in the development of synthetic peptide vaccines and allergy research.

2.2 Disease-Specific Databases

These databases contain peptides which can be used to design therapeutic peptides targeting specific diseases. Databases such as CancerPPD can be used to develop vaccines for cancer as this database includes 624 unique anticancer peptides (Tyagi et al., 2015). Another repository cancerPDF contains 9692 unique endogenous peptide sequences corresponding to 2230 precursor proteins detected in the biofluids of cancer patients (Bhalla et al., 2017). Another important database called PeptiCKDdb collects peptidomics and proteomics datasets manually extracted from published studies related to chronic kidney disease (CKD) to aid research in the field of CKD through identification of unique biomarkers. Another database specific to different tuberculosis strains is MtbVeb, which is developed for assisting researchers in designing vaccines against emerging Mtb strains using traditional and modern approaches. It contains 103,522 unique peptides (9mers) which have the potential to induce an antibody response and/or are promiscuous binder to MHC alleles and/or have the capability to stimulate T lymphocytes. The ZikaVR database contains the predicted epitopes comprising of B-cell epitopes, T-cell epitopes, and MHC binders (Gupta et al., 2016).

2.3 Antimicrobial Databases

APD2 database is updated version of the diverse set of peptides which has enriched information of peptide families, posttranslational modifications, and binding targets, e.g., anticancer, antiviral, antibacterial, and antifungal. They found that Ala and Gly are frequent in bacterial peptides (Wang et al., 2009). Another resource APD3 contains natural antimicrobial peptides (AMPs) that give information of their sequences and activity (Wang, Guo, Liu, Liu, & Wu, 2016). This database has different categories related to source (phyla/kingdoms), different pathogenic types as well as their activities. InverPep is another repository dedicated to the AMPs with the invertebrate origin (Gómez et al., 2017). It provides information of their structure, physicochemical properties, and source and biological activities. Another significant resource of AMPs is DRAMP containing AMPs with sequence, structure, and biological activities (MIC). The peptides are reported from general AMPs, patents, and drug development (Fan et al., 2016). DBAASP harbors AMPs with diverse information like structure, complexity (monomer/dimer), and synthesis type (ribosomal/synthetic) (Gogoladze et al., 2014). DBAASP v.2 (Pirtskhalava et al., 2016) contains structural information of natural and synthetic peptides with the structure. It contains the complex (monomer/dimer) information apart from the MD simulated structures. CAMP comprises 6756 AMPs with prime focus on the sequence and 3D structures (Waghu et al., 2014). CAMPR3 is a prokaryotic and eukaryotic family specific antimicrobial database (Waghu et al., 2014). Hidden Markov models were used to identify family specific patterns and these models were further used to mine the online resources. This database holds information of family specific patterns/signatures and structure. BaAMPs database is dedicated to the collection of AMPs with the property of disrupting microbial biofilms and not a generic activity against the planktonic microorganisms (Di Luca et al., 2015). HIPdb is dedicated to 981 HIV inhibiting peptides (Qureshi et al., 2013) including their sequence and inhibition (IC50). AVPdb harbors the 2683 antiviral peptides with the sequences, efficacy, modifications, and predicted structures (Qureshi et al., 2014). MilkAMP is database of bioactive peptide of milk origin and contains most of the peptides with antimicrobial properties. It includes sequence, microbiological, and physicochemical properties of the peptides (Théolier et al., 2014). DAPD is a database of 2571 AMPs or other defense peptides originating from 1923 unique bioactive sequences (Novković et al., 2012). YADAMP is dedicated to 2133 antibacterial peptides with activity information of most common strains of bacteria (Piotto et al., 2012). BACTIBASE is antimicrobial database with information of 123 bacteriocins from gram-positive and gram-negative bacteria (Hammami, Zouhir, Ben Hamida, & Fliss, 2007) PhytAMP contains the taxonomic and physicochemical information of AMPs from the plant origin which are usually cysteine-rich peptides (Hammami et al., 2009). RAPD is a resource of recombinant AMPs with information related to expression host, fusion strategy, release, and yield of peptide (Li & Chen, 2008). Defensins Knowledgebase contains sequence, structure, and biological activities profile of 350 defensin family specific peptides (Seebah et al., 2007). Peptaibol database contains the antibiotic and membrane channel-forming inhibitors which exhibit antimicrobial and antifungal properties. This database contains sequence and 3D structure information (NMR/X-ray crystallography) of these types of peptides (Whitmore & Wallace, 2004).

2.4 Targeted Delivery-Related Databases

CPPsite is first database of 843 cell-penetrating peptides (CPPs) with the knowledge of sequence, subcellular localization, physicochemical properties, uptake efficiency, etc. (Gautam et al., 2012). CPPsite 2.0 is updated version of CPPsite with 1850 entries including the chemical modifications, predicted tertiary structure, model system, and cargo information (Agrawal et al., 2016). TumorHoPe is a database of the particular class of peptides which recognize tumor cell or its microenvironment and thus can be used as a delivery vehicle. This database contains 744 peptides having information of sequence, receptor, target cell, etc. (Kapoor et al., 2012).

2.5 Miscellaneous

Along with the above-described peptides, some peptide databases related to food such as MBPDB of bioactive peptides present in milk. Another critical database which has quite a vital utility is PEPlife, a manually curated database of experimentally validated therapeutic peptides and their halflives. Another database Quorumpeps database encompasses the structures and microbial origin as well as functionality responses of the QS-derived signaling peptides. One principal secondary database of therapeutic peptides is SATPdb which is a repository of structurally annotated therapeutic peptides collected and compiled from 20 public domain peptide databases and 2 datasets. One primary resource of raw and processed data identified using mass spectrometry is PRIDE archive. Another vital database is Curated Protein Aggregation Database (CPAD), which has collected results from experimental studies performed by scientific community aimed at understanding protein/peptide aggregation.

3. IN SILICO TOOLS IN PEPTIDE-BASED VACCINE DESIGNING

Vaccination produces a long-lasting effect as it involves both arms (innate as well as adaptive) of immune system. Whole organism-based vaccine has some danger signal and protein associated with it, so, peptide-based vaccines come into picture as they are intrinsically more safer, avoid any nonprotective response, immune evasion, and side effects in terms of autoimmunity (Reche, Flower, Fridkis-Hareli, & Hoshino, 2015). Peptide-based vaccine needs the requirement of adjuvant for their efficacy (Skwarczynski & Toth, 2016). Epitopes are of immense importance for clinical researchers as they offer a huge potential for vaccine development, disease prevention, diagnosis, and treatment. Owing to increase in genomic, proteomics, and epitope mapping techniques, huge amount of data are being generated; and in this context bioinformatics approaches can contributes a lot in the designing of epitope-based vaccine candidates (Potocnakova, Bhide, & Pulzova, 2016). So an ideal vaccine candidate should activate humoral and cell-mediated immunity (CMI).

3.1 In Silico Tools for Humoral Immunity

Humoral immune response protects the body from extracellular foreign invaders such as pathogenic organisms, toxins, etc. Pathogenic organism has antigenic region on its surface, recognized by B-cell antigen receptors (BCR) which in turn induce the B-cell to differentiate and proliferate in to specific antibody-secreting plasma cells. These antibodies protect the host organism from infections by three ways: (1) neutralization, (2) opsonization, and (3) triggering the activation of complement system (Fig. 2).

This antigenic response generated in terms by producing specific antibodies by class switching mechanism of BCR represents the first line of defense against pathogenic bacteria. The antigenic region called the B-cell epitopes (BCE) thus can be used to develop vaccine against pathogenic bacteria. These can be categorized into two functional categories—linear B-cell epitopes which represent the continuous region of amino acid residue; and conformational or distorted epitopes, which constitute 90% of BCE, whose sequences are distantly located and are brought in proximity by physical contacts of three-dimensional protein structure.

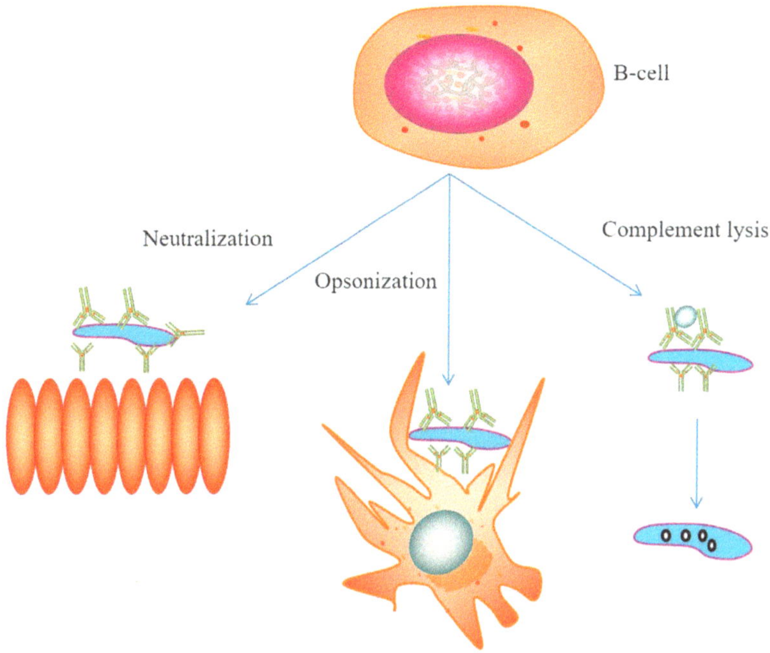

Fig. 2 Mode of action of humoral immunity to protect host from invading pathogens.

3.1.1 Sequence-Based B-Cell Epitope Prediction

Linear B-cell epitope prediction algorithm utilizes the amino acid properties such as hydrophilicity, charge, exposed surface area, and secondary structure (Soria-Guerra, Nieto-Gomez, Govea-Alonso, & Rosales-Mendoza, 2015) and includes BcePred, BepiPred2.0, ABCPred, and BEST (Table 2). BepiPred 2.0 utilizes the random forest algorithm for predicting the epitopes of antigen sequence derived from the antigen–antibody protein structure. APCPred uses the combination of support vector machine and amino acid anchoring pair composition for predicting the linear B-cell epitopes. APCPred have the accuracy of 72.94% and AUC of 0.809.

3.1.2 Structure-Based B-Cell Epitope Prediction

Conformational or discontinuous B-cell prediction methods utilize various algorithm that utilizes the 3D structure and sequence-based information and includes EPCES, Discotope, SEPPA, EpiSearch, MimiPro, MIMEX, etc. DiscoTopepredicts discontinuous B-cell epitopes by combining the surface accessibility and spatial and amino acid statistics to differentiate between

Table 2 List of In Silico Tools for Predicting Epitopes Involved Humoral-Mediated Immunity

Name	Description	Method	Year
Sequence-based B-cell prediction			
ABCPred (Saha & Raghava, 2006)	http://www.imtech.res.in/raghava/abcpred/ Predict B-cell epitope in an antigen sequence	Artificial neural network	2006
BepiPred (Larsen, Lund, & Nielsen, 2006)	http://www.cbs.dtu.dk/services/BepiPred/ Predict the location of linear B-cell epitopes using a combination	Hidden Markov model (HMM) and a propensity scale method	2006
Bcepred (Saha & Raghava, 2007)	http://webs.iiitd.edu.in/raghava/bcepred/ Prediction of B-cell epitopes	Physicochemical properties matrices	2007
BCPREDS (EL-Manzalawy, Dobbs, & Honavar, 2008)	http://ailab.ist.psu.edu/bcpred/ Predict flexible length B-cell epitopes	Support vector machine and string kernels	2008
COBEPro (Sweredoski & Baldi, 2009)	http://scratch.proteomics.ics.uci.edu Predict B-cell epitopes	Support vector machine	2009
SVMTriP (Yao, Zhang, Liang, & Zhang, 2012)	http://sysbio.unl.edu/SVMTriP/ Predict antigenic epitopes 2012	Support vector machine and tri-peptide similarity propensity score	2012
Lbtope (Singh, Ansari, & Raghava, 2013)	http://webs.iiitd.edu.in/raghava/lbtope/ Predict B-cell epitope of fixed and variable length using datasets from IEDB	Support vector machine and K-nearest neighbor algorithm	2013
APCPred (Shen et al., 2015)	http://ccb.bmi.ac.cn/APCpred/ Predict B-cell epitope in an antigen sequence 2015	Support vector machine	2015

BepiPred 2.0 (Jespersen, Peters, Nielsen, & Marcatili, 2017)	http://www.cbs.dtu.dk/services/BepiPred/ Predict B-cell epitope from antigen sequences	Random forest	2017
Structure-based B-cell epitope prediction			
CEP (Kulkarni-Kale, Bhosle, & Kolaskar, 2005)	http://196.1.114.49/cgi-bin/cep.pl Conformational B-cell epitope prediction	CEP algorithm	2005
BEpro (Sweredoski & Baldi, 2008)	http://pepito.proteomics.ics.uci.edu/ Discontinuous B-cell epitope prediction	Based on multiple distance threshold and half-sphere exposure method	2008
ElliPro (Ponomarenko et al., 2008)	http://tools.immuneepitope.org/ellipro/ Identifying discontinuous antibody epitopes in the protein regions of the antigen	Thornton's method residue clustering algorithm	2008
EPCES (Liang, Zheng, Zhang, & Zacharias, 2009)	http://sysbio.unl.edu/EPCES/ Prediction of discontinuous antigenic epitopes on the surface of the protein using	Consensus scoring method	2009
Epitopia (Rubinstein, Mayrose, Martz, & Pupko, 2009)	http://epitopia.tau.ac.il/ Prediction of conformation B-epitope from structure and sequence	Epitopia algorithm	2009
CBTOPE (Ansari & Raghava, 2010)	http://webs.iiitd.edu.in/raghava/cbtope/ Prediction of conformational B-cell from antigenic sequence	Support vector machine	2010
EPSVR (Liang et al., 2010)	http://sysbio.unl.edu/EPSVR/ Prediction of conformational antigenic epitopes	Support vector machine	2010

Continued

Table 2 List of In Silico Tools for Predicting Epitopes Involved Humoral-Mediated Immunity—cont'd

Name	Description	Method	Year
DiscoTope2.0 (Kringelum, Lundegaard, Lund, & Nielsen, 2012)	http://www.cbs.dtu.dk/services/DiscoTope-2.0 Predicts discontinuous B-cell epitope in antigenic structure based on surface accessibility of amino acids	Based on propensity scores and half-sphere exposure	2012
proABC (Olimpieri, Chailyan, Tramontano, & Marcatili, 2013)	http://circe.med.uniroma1.it/proABC/ Predicting the antibody–antigen interactions	Random forest algorithm	2013
SEPPA 2.0 (Spatial epitope prediction server for protein antigens) (Qi et al., 2014)	http://badd.tongji.edu.cn/seppa/ Discontinuous B-epitope prediction in antigen structure	Unit patch of residue triangle SEPPA 2.0 algorithm	2014
LYRA (Lymphocyte receptor automated modeling) (Klausen, Anderson, Jespersen, Nielsen, & Marcatili, 2015)	http://www.cbs.dtu.dk/services/LYRA/ Predicting structures for T-cell receptor, B-cell receptors	Homology modeling and BLOSUM matrix	2015
PEASE (Predicting epitope using antibody sequence) (Sela-Culang, Ashkenazi, Peters, & Ofran, 2015)	http://www.ofranlab.org/PEASE/ Predicting antibody-specific B-cell	PEASE algorithm	2015

epitopes and nonepitope sites. SEPPA uses unit patch of residue triangle which are in 4 A^0 proximity and spatial clustering coefficient to predict the discontinuous B-cell epitopes. This method outperform all other methods with AUC over 0.742 (Table 2). The most complete prediction algorithm for the prediction of continuous and discontinuous epitope prediction includes ElliPro, PepSurf, and Epitopia.

3.2 In Silico Tools for CMI

CMI involves specialized set of lymphocytes, T cells, which recognize foreign antigen represented on to the surface of cell, either through endogenous or exogenous antigen processing. Both the way antigen enters into the cells and antigen processing site determine whether the peptide will be either MHC-I binder or MHC-II binder. Endogenous antigen processing involves recognition and breakage of foreign antigen into smaller fragment or peptide by proteasomes. These peptides are transported to endoplasmic reticulum by transporter-associated proteins (TAP), where peptides get associated with MHC class I molecule. This MHC-I–peptide complex is presented on cell surface via Golgi complex where cytotoxic T-lymphocyte gets into action. While exogenous antigens are presented to cell surface in form of MHC-II–peptide complex through exogenous antigen processing pathway. Antigen-presenting cells (APCs) internalize exogenous antigen by phagocytosis or receptor-mediated endocytosis, and degrade it into smaller fragment. These smaller peptides are routed to endocytic pathway compartment, where it binds with MHC class II and this complex is transported to cell surface, where $CD4^+$ T cells or helper T cells recognize it and act as a bridge to activate CTL response via release of cytokines (Dhanda et al., 2017).

The CMI arm of immune system is always a hot spot to activate the immune system by vaccination. The first step to design a peptide-based vaccine is to discriminate epitope between immunogenic peptide or non-immunogenic as well as assurance of its antigenicity. VaxiJen (Doytchinova & Flower, 2007) is an alignment-independent method for prediction of protective antigens. Epitope interacts with MHC-I or MHC-II grooves by their R group side chains (Falk, Rötzschke, Stevanović, Jung, & Rammensee, 1991; Wieczorek et al., 2017). This fact has been explored to develop various T-cell epitope-mapping algorithm and to identify or predict presumed T-cell epitope (DeLisi & Berzofsky, 1985; Stille, Thomas, Reyes, & Humphreys, 1987).

Initially, direct prediction methods were developed, based on sequential and structural analysis of experimentally verified T-cell epitopes (Stille et al., 1987), such as AMPHI and SOHHA. X-ray crystallography of MHC–peptide complex changed the concept of structure-based prediction of T-cell epitope. Stern et al. (1994) showed in 1994 that, peptide binds to HLA-DR1 in an extended conformation with a pronounced twist. Only 35% of peptide is accessible to solvent and available for interacting with T-cell receptors. This is remarkable discovery and forced development of indirect methods, such as motif-based methods like Epimer and Optimer (Meister, Roberts, Berzofsky, & De Groot, 1995).

3.2.1 Tools to Predict MHC Binders

Prediction of MHC binders draws more attention of scientific community as it provides more specificity. The binding grooves of MHC-I are closed from both side whereas MHC-II binding groove is open. This causes more difficulties in predicting MHC-II binders. ProPred I (Singh & Raghava, 2003), NetMHCcons (Singh & Raghava, 2003), nHLApred (Bhasin & Raghava, 2007), MMBpred (Bhasin & Raghava, 2003), NetCTLpan (Stranzl, Larsen, Lundegaard, & Nielsen, 2010), and RANKPEP (Reche, Glutting, & Reinherz, 2002) are heavily used tools to predict MHC-I binders. Similarly, ProPred (Singh & Raghava, 2001), Consensus, EpiDOCK (Atanasova, Patronov, Dimitrov, Flower, & Doytchinova, 2013), EpiTOP (Nielsen et al., 2003), MHC2Pred (Lata, Bhasin, & Raghava, 2007), HLA-DR4Pred (Bhasin & Raghava, 2004a), and MULTIPRED 2 (Zhang et al., 2011) predict MHC-II binders effectively (Table 3).

3.2.2 Tools Assisting in Antigen Processing

As stated above, the antigen-processing pathway also plays a significant role in deciding the activation of particular immune arm, therefore various prediction tools have been developed to target intermediate step of antigen processing pathways. Pclevage (Bhasin & Raghava, 2005) and NetChop (Keşmir, Nussbaum, Schild, Detours, & Brunak, 2002; Nielsen, Lundegaard, Lund, & Keşmir, 2005) predict proteosomal cleavage site in protein. TAPPred (Bhasin, Lata, & Raghava, 2007) and TAPreg (Diez-Rivero et al., 2010) assist to know the probability of binding a degraded protein or peptide to TAP. EpiJen (Doytchinova, Guan, & Flower, 2006) and WAPP (Dönnes & Kohlbacher, 2005) provide TAP prediction as well as proteosomal cleavage prediction. NetCTLpan (Stranzl et al., 2010), which is an improved version of NetCTL, provides prediction of proteosomal cleavage, transport efficiency by TAP proteins as well as MHC class I binding affinity.

Table 3 List of Significant Computational Methods Available for Predicting MHC Binders and Cell-Mediated Immunity

Name	Description	Method	Year
Tools to predict MHC-I binders			
ProPred 1 (Mustafa & Shaban, 2006)	http://webs.iiitd.edu.in/raghava/propred1/ Promiscuous MHC-I binding prediction server	Matrix based	2003
MmbPred (Bhasin & Raghava, 2003)	http://webs.iiitd.edu.in/raghava/mmbpred/ Mutated MHC binders prediction server	Quantitative matrix based	2003
nHLAPred (Bhasin & Raghava, 2007)	http://webs.iiitd.edu.in/raghava/nhlapred/ MHC-I binding prediction server	Artificial neural network	2006
RANKPEP (Reche, Glutting, Zhang, & Reinherz, 2004)	http://imed.med.ucm.es/Tools/rankpep.html Predict peptide binders to MHC-I and MHC-II molecules from protein sequence/s	Position-specific scoring matrices (PSSMs)	2007
POPI (Tung & Ho, 2007)	http://iclab.life.nctu.edu.tw/POPI/ Server for predicting immunogenicity of MHC class I and II binding peptides	Support vector machine	2007
NetCTLpan (Stranzl et al., 2010)	http://www.cbs.dtu.dk/services/NetCTLpan/ Pan-specific MHC class I pathway epitope prediction	Artificial neural network	2010
NetMHCcons (Karosiene, Lundegaard, Lund, & Nielsen, 2012)	http://www.cbs.dtu.dk/services/NetMHCcons/ Consensus method for MHC-I binding prediction	Combination of ANN-based NetMHC, NetMHCpan, and matrix-based PickPocket	2012

Continued

Table 3 List of Significant Computational Methods Available for Predicting MHC Binders and Cell-Mediated Immunity—cont'd

Name	Description	Method	Year
NetMHCstab (Jørgensen, Rasmussen, Buus, & Nielsen, 2014)	http://www.cbs.dtu.dk/services/NetMHCstab/ Predict stability of peptide–MHC-I complexes	Artificial neural network	2014
NetMHCstabpan (Rasmussen et al., 2016)	http://www.cbs.dtu.dk/services/NetMHCstabpan/ Predicts binding stability of peptides to any known MHC molecule	Artificial neural network	2016
Tools to predict MHC-II binders			
Propred (Singh & Raghava, 2001)	http://webs.iiitd.edu.in/raghava/propred/ Predict MHC class II binding regions in an antigen sequence	Quantitative matrices based	2001
HLA-DR4Pred (Bhasin and Raghava, 2004a)	http://webs.iiitd.edu.in/raghava/hladr4pred/ Prediction of HLA-DRB1*0401 binding peptides in an antigen sequence	Support vector machine and artificial neural network	2004
MHC2Pred (Lata et al., 2007)	http://crdd.osdd.net/raghava/mhc2pred/ Prediction of promiscuous MHC class II binders	Support vector machine	2007
Consensus (Zhao, Zhang, & Cong, 2013)	http://tools.immuneepitope.org/mhcii/ IEDB tool for predicting MHC class II binders	Consensus approach which combines NN-align, SMM-align, and combinatorial library methods	2010
EpiTOP (Dimitrov, Garnev, Flower, & Doytchinova, 2010)	http://www.pharmfac.net/EpiTOP/ Proteochemometric tool for MHC class II binding prediction	Quantitative matrix	2010

MULTIPRED2 (Zhang et al., 2011)	http://cvc.dfci.harvard.edu/multipred2/ Prediction of peptide binding to multiple alleles belonging to HLA class I and class II DR molecules	Combination of MULTPRED, PEPVAC, and NetMHCpan	2011
EpiDOCK (Zhao et al., 2013)	http://epidock.ddg-pharmfac.net/ Docking-based method for predicting MHC-II binders	Docking score-based quantitative matrices (DS-QM)	2013
Important tools for antigen processing			
CTLPred (Bhasin and Raghava, 2004b)	http://webs.iiitd.edu.in/raghava/ctlpred/ Direct method for predicting CTL epitopes	Support vector machine and artificial neural network	2004
TAPPred (Bhasin, Lata, & Raghava, 2007)	http://webs.iiitd.edu.in/raghava/tappred/ Prediction of binding affinity of TAP binders	Cascade support vector machine	2004
Pcleavage (Bhasin & Raghava, 2005)	http://webs.iiitd.edu.in/raghava/pcleavage/ Proteasome/immunoproteasome cleavage sites prediction	Support vector machine	2005
NetChop (Dönnes & Kohlbacher, 2005)	http://www.cbs.dtu.dk/services/NetChop/ Predicting cleavage site for human proteasome	Artificial neural network	2005
NetCTL (Stranzl et al., 2010)	http://www.cbs.dtu.dk/services/NetCTL/ Predict CTL epitopes in protein sequences		2007

Continued

Table 3 List of Significant Computational Methods Available for Predicting MHC Binders and Cell-Mediated Immunity—cont'd

Name	Description	Method	Year
FRED (Feldhahn, Dönnes, Thiel, & Kohlbacher, 2009)	http://www-bs.informatik.uni-tuebingen.de/Software/FRED FRED—a framework for T-cell epitope detection	Integration of multiple tools	2009
TAPreg (Diez-Rivero, Chenlo, Zuluaga, & Reche, 2010)	http://imed.med.ucm.es/Tools/tapreg/ Quantitative modeling of peptide-based peptide binding to TAP	Support vector Machine	2010
NetTepi (Trolle & Nielsen, 2014)	http://www.cbs.dtu.dk/services/NetTepi/ Integrated method for T-cell epitopes prediction	Combination of three prediction types, peptide–MHC-binding affinity, peptide–MHC stability and T-cell propensity	2014
Cytokine-based prediction tools			
IFNepitope (Dhanda, Gupta, et al., 2013)	http://webs.iiitd.edu.in/raghava/ifnepitope/ Prediction of interferon-gamma inducing T-helper (Th1) epitopes	Motif-based and support vector machine	2013
IL4pred (Dhanda, Vir, & Raghava, 2013)	http://webs.iiitd.edu.in/raghava/il4pred/ Prediction of IL4 inducing T-helper (Th2) epitopes	Support vector machine	2013
IL10pred (Nagpal et al., 2017)	http://webs.iiitd.edu.in/raghava/il10pred/ Prediction of interleukin-10 (IL-10) inducing MHC-II binders	Motif-based and support vector machine	2017

3.2.3 T-Cell Epitope and Cytokines Based Prediction

Alongside, some direct methods, like CTLpred (Bhasin & Raghava, 2004b) and NetCTLpan (Stranzl et al., 2010), have been developed to predict CTL epitopes. Recently, Raghava's group focused on interleukin and cytokines based direct methods such as IFNepitope (Dhanda, Gupta, et al., 2013), IL4pred (Dhanda, Vir, & Raghava, 2013), and IL10pred (Nagpal et al., 2017) to assist in designing of peptide-based immunotherapeutic.

4. IN SILICO TOOLS FOR ADJUVANT DESIGNING

Adjuvants have been identified as agents that enhance the induction of protective immunity imparted by vaccination. A limited number of adjuvants like alum and squalene-in-water emulsions have been approved and licensed for clinical use (Pulendran & Ahmed, 2011). A number of recent studies have demonstrated the potential of cationic host defense peptides (HDPs) and their synthetic mimics innate defense regulators (IDRs) as therapeutic adjuvants acting by activation of innate immune system (Hilchie, Wuerth, & Hancock, 2013). Owing to their more recently discovered immunostimulatory and antiinflammatory properties (Hamill, Brown, Jenssen, & Hancock, 2008; Hancock & Sahl, 2006), HDPs are now being studied for therapeutic use especially as vaccine adjuvants (Hancock, Nijnik, & Philpott, 2012; Nicholls, Madera, & Hancock, 2010). Synthetic analogs of HDPs called IDRs have been rationally generated from HDP templates in order to enhance the immunostimulatory potential of the peptide. A few examples of such peptides include IDR-1002, IDR-HH2, and IDR-1018 (Niyonsaba et al., 2013). Thus, a knowledgebase of HDPs is a basic requirement to unveil the cues required for the rational design of immunomodulatory peptides to be used as adjuvants.

Cytokines & Cells Online Pathfinder Encyclopaedia (COPE) is an online resource that compiles the molecules involved in the intercellular communication and defense against foreign invasion across diverse multicellular organisms [COPE: Horst Ibelgaufts' Cytokines & Cells Online Pathfinder Encyclopaedia]. COPE includes MiniCOPE dictionaries that are compilations for specific categories of molecules. One such category is that of "Innate Immunity Defense Peptides" in the MiniCOPE dictionary. This dictionary contains the information for the HDPs including the amino acid sequence that can be used for developing in silico tools for designing IDRs with enhanced immunostimulatory activity than the parent HDP.

Owing to the limited studies of peptides for use as adjuvants, the in silico resources for exploring peptide-based vaccine adjuvants are limited. Thus, the databases that compile the information regarding innate immunity receptors particularly the pattern-recognition receptors (PRRs), their ligands, and their genes become useful to drive the field of adjuvant discovery. PRRdb is a database of PRRs compiled by Lata and Raghava (2008) that provides information of cellular receptors involved in innate immunity. Another valuable resource is InnateDB that has taken a systems approach to compile genes involved in innate immunity and interactions between them (Breuer et al., 2013). Yet, peptides being studied as vaccine adjuvants are very few in numbers and some of these can be found in the database Vaxjo (Sayers, Ulysse, Xiang, & He, 2012), a web-based compilation of adjuvants. Oligonucleotide-based adjuvant designing has been recently explored by Raghva's group and open a new area of tools that can assist scientific community in improved adjuvant designing. VaccineDA (Nagpal et al., 2015), an in silico tool, developed with the aim of designing oligodeoxynucleotide vaccine adjuvants. Similarly, imRNA (Chaudhary, Kumar, et al., 2016) helps in developing RNA-based therapeutics, immunotherapy, and vaccine adjuvants.

5. IN SILICO TOOLS FOR PEPTIDE-BASED DRUG DEVELOPMENT

Protein–protein interaction plays a crucial role in all biological processes. The inhibition of this interaction is considered as an attractive target for the drug development process. Biological entities such as peptides now emerges as an effective tools for the drug development processes as they offers advantages over high-throughput screening of small molecule-based drugs in terms of unique structural property, formulation, synthesis, and delivery. Peptides also offer advantages in terms of absorption per unit mass (Muheem et al., 2016) less systematic toxicity after degradation (Otvos & Wade, 2014), immunogenicity, and small quantity is required to modulate the receptor target (Ratanji, Derrick, Dearman, & Kimber, 2014). Computational approaches have been developed at each step of peptide-based drug development.

5.1 Based on Protein–Peptide Interaction

PeptideMine is a server for designing peptide from protein–protein interactome. EPI-peptide designer is the tool to design the peptide based on epitope–paratope interaction. DockoMatric is a server for generating the cyclic peptide structure to improve the biological activity. PepSite predicts the peptide-binding site in a protein. SPRINT which utilizes the support vector machine algorithm for the prediction of peptide-binding site in a protein.

5.2 Sequence- and Structure-Based Peptide Design

The sequence-based peptide designing involves various physiochemical properties of peptides which optimize the peptide immunogenicity, antibody specificity, multispecies conservation, and robustness in the face of posttranslational modifications, e.g., CellPPD, AVCpred, ABPpred, AVPpred, QSPpred, iACP, to predict the anticancer property of the peptide (Table 4).

Main focus of structure-based peptide design is to mimic the interaction of peptide-binding groove on protein sequence and designing of inhibitors. Various online bioinformatics tools are develop to aid this which identify the proteins predisposed to bind linear epitopes on preselected target proteins, PeptiMap detects the peptide-binding sites on protein surfaces, PEPstrMOD predicts the tertiary structure of peptides. Pepsite 2.0 takes input of PDB file or sequence and predicts the binding of a peptide on to protein structure. Well all these methods have one thing in common that they utilizes the amino acid information with correlation to its spatial proximity in space for to make prediction.

5.3 In Silico Tools for Peptide Antimicrobial Activity Prediction

AMPs represent integral and natural component of innate immune system, considered as "natural antibiotics" and promising candidate to overcome antibiotic resistance. Identification of AMPs through wet lab experiment is time consuming so in silico approaches are required to screen best AMPs. AntiBP predicts the AMPs in a protein sequence, ClassAMP predicts and classify the different AMPs, iAMP-2L predicts as well as classify the functional type of AMPs, iAMPpred predicts the AMPs by utilizing compositional, physiochemical, and structural features into Chou's general pseudo amino acid composition (pseAAC).

5.4 In Silico Tools for Peptide-Based Drug Delivery Vehicles

The need of noninvasive root for drug administration is solely based on peptide-based drug delivery system. The peptide-based delivery system acts

Table 4 List of Computational Methods Available to Assist in Peptide-Based Drug Development

Name	Description	Method	Year
Based on protein–peptide interaction			
PeptideMine (Shameer, Madan, Veeranna, Gopal, & Sowdhamini, 2010)	http://caps.ncbs.res.in/peptidemine/ A webserver for the design of peptides for protein–peptide binding studies derived from protein–protein interactomes	Inteacting sequence space and bioinformatics mashup approach	2010
PepSite (Trabuco, Lise, Petsalaki, & Russell, 2012)	http://pepsite2.russelllab.org/ Prediction of peptide-binding site on protein surface	Protein–peptide interaction based on 3D structure	2012
DockoMatic 2.0 (Bullock et al., 2013)	Standalone	Homology modeling	2013
EFI-Peptide (Viart et al., 2016)	http://www.biocomp.icb.ufmg.br/biocomp/ Tool for designing peptide ligand libraries based on epitope–paratope interactions	I2R interaction	2016
SPRINT (Taherzadeh, Yang, Zhang, Liew, & Zhou, 2016)	http://sparks-lab.org/ Sequence-based prediction of protein–peptide binding sites	Support vector machine	2016
Sequence and structure-based method			
AVPpred (Thakur, Qureshi, & Kumar, 2012)	http://crdd.osdd.net/servers/avppred/ Prediction and designing of antiviral peptides	Support vector machine	2012
CellPPD (Gautam et al., 2013)	http://webs.iiitd.edu.in/raghava/cellppd/ Prediction and designing of highly cell-penetrating peptide	Support vector machine	2013

Peptimap (Lavi et al., 2013)	Standalone Detection of peptide-binding sites on protein surfaces	Ab-initio and FTMap approach	2013
QSPpred (Rajput, Gupta, & Kumar, 2015)	http://crdd.osdd.net/servers/qsppred/ Predicting and designing quorum sensing peptide	Support vector machine	2015
PEPstrMOD (Singh et al., 2015)	http://webs.iiitd.edu.in/raghava/pepstrmod/ Predicts the tertiary structure of small peptides with sequence length varying between 7 and 25 residues	Support vector machine	2015
iACP (Chen, Ding, Feng, Lin, & Chou, 2016)	lin.uestc.edu.cn/server/iACP/ Sequence-based anticancer peptide prediction	Support vector machine	2016
In silico approaches for antimicrobial peptide prediction			
AntiBP2 (Lata, Mishra, & Raghava, 2010)	http://crdd.osdd.net/raghava/antibp2/index.html Antibacterial peptide prediction	Support vector machine	2010
ClassAMP (Joseph, Karnik, Nilawe, Jayaraman, & Idicula-Thomas, 2012)	http://www.bicnirrh.res.in/classamp/ Prediction and classification of antimicrobial peptides	Support vector machine	2012
iAMP-2L (Xiao, Wang, Lin, Jia, & Chou, 2013)	http://www.jci-bioinfo.cn/iAMP-2L Multilabel classification and functional type prediction of antimicrobial peptide	PseAAC and KNN algorithm	2013

Continued

Table 4 List of Computational Methods Available to Assist in Peptide-Based Drug Development—cont'd

Name	Description	Method	Year
APD3 (Wang, Li, et al., 2016)	http://aps.unmc.edu/AP/prediction/prediction_main.php Prediction of antimicrobial peptide	Support vector machine	2016
iAMPpred (Meher, Sahu, Saini, & Rao, 2017)	http://cabgrid.res.in:8080/amppred/ Prediction of antimicrobial peptide	Support vector machine	2017
Peptide-based drug delivery vehicles			
CellPPD (Gautam et al., 2013)	http://webs.iiitd.edu.in/raghava/cellppd/ A webserver for prediction and designing of highly cell-penetrating peptide	Support vector machine and motif based	2013
CPPpred (Holton, Pollastri, Shields, & Mooney, 2013)	http://bioware.ucd.ie/~compass/biowareweb/Server_pages/cpppred.php A webserver for prediction of cell-penetrating peptide	KNN algorithm	2013
C2pred (Tang, Su, Wei, Chen, & Lin, 2016)	http://lin.uestc.edu.cn/server/C2Pred Prediction of cell-penetrating peptide	Support vector machine	2016
Skip-CPP-pred (Wei, Tang, & Zou, 2017)	http://server.malab.cn/SkipCPP-Pred/Index.html Predicting cell-penetrating peptide	K-Skip-2-g feature	2017

Medicine like property of peptide			
Toxinpred (Gupta et al., 2013)	http://webs.iiitd.edu.in/raghava/toxinpred/ A webserver for designing the peptides of desired toxicity	Support vector machine and motif based	2013
AHTpin (Kumar, Chaudhary, Singh Chauhan, et al., 2015)	http://webs.iiitd.edu.in/raghava/ahtpin/ Prediction of antihypertensive peptides	Support vector machines	2015
AntiangioPred (Gupta et al., 2013)	http://webs.iiitd.edu.in/raghava/antiangiopred/ A webserver for predicting the peptide inhibitors of angiogenesis	Support vector machine	2015
HemoPI (Chaudhary, Nagpal, et al., 2016)	http://webs.iiitd.edu.in/raghava/hemopi/ A server for predicting the hemolytic potency of peptide	Support vector and motif based	2016
Signal-3L 2.0 (Zhang & Shen, 2017)	http://www.csbio.sjtu.edu.cn/bioinf/Signal-3L/ Predicting the N-terminal protein signal peptide	Support vector machine	2017

as enhancers (Jitendra, Sharma, Bansal, & Banik, 2011) which includes CPPs and peptide-based nanoparticles. CellPPD, CPPpred, C2pred, and SkipCPP-Pred are webservers which predict the cell-penetrating potential of peptides.

5.5 In Silico Tools for Designing Medicinal Like Property of Peptides

Peptides have some intrinsic properties which make them liable to cleavage, oxidation, isomerization, aggregation, short half-life, less toxicity, etc., which needs to be resolved in order to use them as a therapeutic molecule. These physiochemical properties of peptides can be improved by introduction of stabilizing helices, salt bridges, and disulfide linkages. HemoPI is such server which predicts the hemolytic potential of a peptide, ToxinPred predicts the toxicity of peptides and proteins, PlifePred predicts the half-life of peptide in intestine like environment, AntiAngiopred is a server for the prediction of peptides having antiangiogenic property, and AHTPin is a webserver for the prediction of antihypertensive peptides. The use of such prediction server is expected to be increase in near future as they hold the potential for rational designing of peptide-based drugs.

6. DISCUSSION

Bioinformatics has played a significant role in hastening the rate of drug discovery in the last few decades. Various tools have enabled the aptitude of selecting potential epitopes, without risk of cultivating pathogen as well as screen better possible drug candidate in several other nonpathogenic diseases such as cancer. Conventionally, drug and vaccine designing was labor-intensive and time-consuming process, with toxicity and safety risk. Considerable progress in developing in silico tools not only accelerates drug discovery and vaccine designing process, but also reduces the cost and time by amalgamating computational and experimental approaches. Progression in "omics" also revolutionizes vaccine identification and drug screening. Genomic, proteomic, and transcriptomic study enhances the knowledge of host–pathogen interaction and thus fasten the rate of selecting possible potential epitope.

Last decade witnessed a conquer in "peptidomics" and peptide informatics came as a salvage for pharmaceutical industries. Peptide-based therapeutics, gained an increasing interest, and ~60 drugs reached the market and to help the patients and several hundreds of therapeutics peptides are in

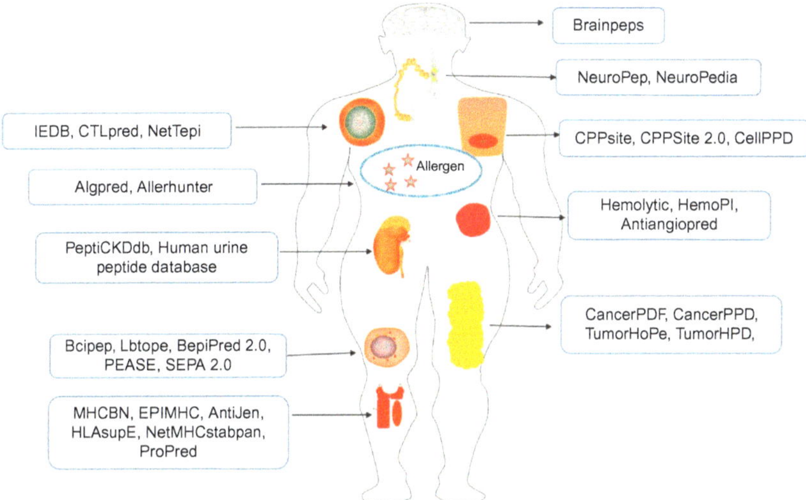

Fig. 3 A brief overview of peptide databases and in silico tools aid to design specific therapeutic peptide.

preclinical and clinical trials developments. Major contributor of peptide-based therapeutics success is potent and yet safe mode of action. Slight cytotoxicity, allergy, or hemolysis might be a challenge in designing peptide therapeutics, but in silico tools such as HemoPI, Algpred, Allerhunter, and ToxinPred provide an option to limit these factors too. Researchers have targeted almost all the fragments of human system which could be a part of battle between health and disease and to sweep the battle thousands of therapeutic peptide has been discovered with significantly positive result (Fig. 3). To retain the effort of these researchers, several groups determined to sustain the everlasting knowledge in form of databases or repositories. This information is also used for developing prediction servers, which proved to be a booster in drug designing process.

Despite the significant growth in peptide-based in silico tools, still designing of vaccine adjuvant is somehow far behind from other aspect of vaccine designing and need to be explored. Self-assembling peptides form nanostructures that provide a scaffold for multiple antigen display. Thus, peptide aggregation leading to increased display of antigens adds a new paradigm to the mechanism of adjuvancy (Azmi, Ahmad Fuaad, Skwarczynski, & Toth, 2014). More recently, peptide mimics of lipopolysaccharides that interact with mammalian Toll-like receptor 4 leading to inflammatory cytokine release have been proposed as a novel class of adjuvants (Shanmugam et al., 2012). This aspect of adjuvancy needs to be explored and computational algorithms are lacking here. Similarly, in case of allergy, autoimmunity, or transplantation, several peptides have been studied and showed significant therapeutic property, but still a repository dedicated to immunosuppressive peptide is lacking, although in silico tools which can predict antiinflammatory peptide has been developed recently (Gupta, Sharma, Shastri, Madhu, & Sharma, 2017).

Despite the tremendous progress in peptide informatics, still further improvement in in silico analysis along with experimental approach is needed to combat endemic diseases, suddenly risen in some part of the world, caused by viruses like Ebola and Zika. Researchers have tried to computationally explore these diseases (Dhanda, Chaudhary, et al., 2016; Gupta et al., 2016) and identified potential vaccine candidates, but still experimental approach is needed to serve the mankind.

7. CONCLUSION

We believe that the field of peptide-based therapeutics continues to grow with the application of traditional rational drug designing approach to improve current weakness of peptides in terms of low membrane permeability, ability of aggregation, prone to hydrolysis, and oxidation. We believe that peptide- and peptide-based drug carrier conjugate will help broaden the applicability of peptide-based therapeutics. We foresee that integration of medicinal chemistry approaches with peptidomimetics, proteomics, and genomic studies will help a lot in designing of peptide-based therapeutics. Taking all factors in to account, we are convinced that peptide therapeutics offers enormous growth potential as therapeutics for the treatment of unmet medical needs.

REFERENCES

Agrawal, P., Bhalla, S., Usmani, S. S., Singh, S., Chaudhary, K., Raghava, G. P. S., et al. (2016). CPPsite 2.0: A repository of experimentally validated cell-penetrating peptides. *Nucleic Acids Research*, *44*, D1098–D1103.

Ansari, H., & Raghava, G. P. (2010). Identification of conformational B-cell epitopes in an antigen from its primary sequence. *Immunome Research*, *6*, 6.

Antosova, Z., Mackova, M., Kral, V., & Macek, T. (2009). Therapeutic application of peptides and proteins: Parenteral forever? *Trends in Biotechnology*, *27*, 628–635.

Atanasova, M., Patronov, A., Dimitrov, I., Flower, D. R., & Doytchinova, I. (2013). EpiDOCK: A molecular docking-based tool for MHC class II binding prediction. *Protein Engineering, Design & Selection*, *26*, 631–634.

Azmi, F., Ahmad Fuaad, A. A. H., Skwarczynski, M., & Toth, I. (2014). Recent progress in adjuvant discovery for peptide-based subunit vaccines. *Human Vaccines & Immunotherapeutics*, *10*, 778–796.

Bhalla, S., Verma, R., Kaur, H., Kumar, R., Usmani, S. S., Sharma, S., et al. (2017). CancerPDF: A repository of cancer-associated peptidome found in human biofluids. *Scientific Reports*, *7*, 1511.

Bhasin, M., Lata, S., & Raghava, G. P. S. (2007). TAPPred prediction of TAP-binding peptides in antigens. *Methods in Molecular Biology*, *409*, 381–386.

Bhasin, M., & Raghava, G. P. S. (2003). Prediction of promiscuous and high-affinity mutated MHC binders. *Hybridoma and Hybridomics*, *22*, 229–234.

Bhasin, M., & Raghava, G. P. S. (2004a). SVM based method for predicting HLA-DRB1*0401 binding peptides in an antigen sequence. *Bioinformatics*, *20*, 421–423.

Bhasin, M., & Raghava, G. P. S. (2004b). Prediction of CTL epitopes using QM, SVM and ANN techniques. *Vaccine*, *22*, 3195–3204.

Bhasin, M., & Raghava, G. P. S. (2005). Pcleavage: An SVM based method for prediction of constitutive proteasome and immunoproteasome cleavage sites in antigenic sequences. *Nucleic Acids Research*, *33*, W202–7.

Bhasin, M., & Raghava, G. P. S. (2007). A hybrid approach for predicting promiscuous MHC class I restricted T cell epitopes. *Journal of Biosciences*, *32*, 31–42.

Bos, K. I., Schuenemann, V. J., Golding, G. B., Burbano, H. A., Waglechner, N., Coombes, B. K., et al. (2011). A draft genome of Yersinia pestis from victims of the black death. *Nature*, *478*, 506–510.

Breuer, K., Foroushani, A. K., Laird, M. R., Chen, C., Sribnaia, A., Lo, R., et al. (2013). InnateDB: Systems biology of innate immunity and beyond—Recent updates and continuing curation. *Nucleic Acids Research*, *41*, D1228–33.

Bui, S., & Steiner, R. A. (2016). New insight into cofactor-free oxygenation from combined experimental and computational approaches. *Current Opinion in Structural Biology*, *41*, 109–118.

Bullock, C., Cornia, N., Jacob, R., Remm, A., Peavey, T., Weekes, K., et al. (2013). DockoMatic 2.0: High throughput inverse virtual screening and homology modeling. *Journal of Chemical Information and Modeling*, *53*, 2161–2170.

Chaudhary, K., Kumar, R., Singh, S., Tuknait, A., Gautam, A., Mathur, D., et al. (2016). A web server and mobile app for computing hemolytic potency of peptides. *Scientific Reports*, *6*, 22843.

Chaudhary, K., Nagpal, G., Dhanda, S. K., & Raghava, G. P. S. (2016). Prediction of immunomodulatory potential of an RNA sequence for designing non-toxic siRNAs and RNA-based vaccine adjuvants. *Scientific Reports*, *6*, 20678.

Chen, W., Ding, H., Feng, P., Lin, H., & Chou, K.-C. (2016). iACP: A sequence-based tool for identifying anticancer peptides. *Oncotarget*, *7*, 16895–16909.

Choo, K. H., Tan, T. W., & Ranganathan, S. (2005). SPdb—A signal peptide database. *BMC Bioinformatics*, *6*, 249.

DeLisi, C., & Berzofsky, J. A. (1985). T-cell antigenic sites tend to be amphipathic structures. *Proceedings of the National Academy of Sciences of the United States of America*, *82*, 7048–7052.

Devault, A. M., McLoughlin, K., Jaing, C., Gardner, S., Porter, T. M., Enk, J. M., et al. (2014). Ancient pathogen DNA in archaeological samples detected with a microbial detection array. *Scientific Reports*, *4*, 4245.

Dhanda, S. K., Chaudhary, K., Gupta, S., Brahmachari, S. K., & Raghava, G. P. S. (2016). A web-based resource for designing therapeutics against Ebola virus. *Scientific Reports*, *6*, 24782.

Dhanda, S. K., Gupta, S., Vir, P., & Raghava, G. P. S. (2013). Prediction of IL4 inducing peptides. *Clinical & Developmental Immunology*, *2013*, 263952.

Dhanda, S. K., Usmani, S. S., Agrawal, P., Nagpal, G., Gautam, A., & Raghava, G. P. S. (2017). Novel in silico tools for designing peptide-based subunit vaccines and immunotherapeutics. *Briefings in Bioinformatics*, *18*, 467–478.

Dhanda, S. K., Vir, P., & Raghava, G. P. S. (2013). Designing of interferon-gamma inducing MHC class-II binders. *Biology Direct*, *8*, 30.
Dhanda, S. K., Vir, P., Singla, D., Gupta, S., Kumar, S., & Raghava, G. P. S. (2016). A web-based platform for designing vaccines against existing and emerging strains of mycobacterium tuberculosis. *PLoS One*, *11*, e0153771.
Di, L. (2015). Strategic approaches to optimizing peptide ADME properties. *The AAPS Journal*, *17*, 134–143.
Di Luca, M., Maccari, G., Maisetta, G., & Batoni, G. (2015). BaAMPs: The database of biofilm-active antimicrobial peptides. *Biofouling*, *31*, 193–199.
Diez-Rivero, C. M., Chenlo, B., Zuluaga, P., & Reche, P. A. (2010). Quantitative modeling of peptide binding to TAP using support vector machine. *Proteins*, *78*, 63–72.
Dimitrov, I., Garnev, P., Flower, D. R., & Doytchinova, I. (2010). EpiTOP—A proteochemometric tool for MHC class II binding prediction. *Bioinformatics*, *26*, 2066–2068.
Dönnes, P., & Kohlbacher, O. (2005). Integrated modeling of the major events in the MHC class I antigen processing pathway. *Protein Science*, *14*, 2132–2140.
Doytchinova, I. A., & Flower, D. R. (2007). VaxiJen: A server for prediction of protective antigens, tumour antigens and subunit vaccines. *BMC Bioinformatics*, *8*, 4.
Doytchinova, I. A., Guan, P., & Flower, D. R. (2006). EpiJen: A server for multistep T cell epitope prediction. *BMC Bioinformatics*, *7*, 131.
EL-Manzalawy, Y., Dobbs, D., & Honavar, V. (2008). Predicting linear B-cell epitopes using string kernels. *Journal of Molecular Recognition*, *21*, 243–255.
Falk, K., Rötzschke, O., Stevanović, S., Jung, G., & Rammensee, H. G. (1991). Allele-specific motifs revealed by sequencing of self-peptides eluted from MHC molecules. *Nature*, *351*, 290–296.
Fan, L., Sun, J., Zhou, M., Zhou, J., Lao, X., Zheng, H., et al. (2016). DRAMP: A comprehensive data repository of antimicrobial peptides. *Scientific Reports*, *6*, 24482.
Feldhahn, M., Dönnes, P., Thiel, P., & Kohlbacher, O. (2009). FRED—A framework for T-cell epitope detection. *Bioinformatics*, *25*, 2758–2759.
Fleri, W., Paul, S., Dhanda, S. K., Mahajan, S., Xu, X., Peters, B., et al. (2017). The immune epitope database and analysis resource in epitope discovery and synthetic vaccine design. *Frontiers in Immunology*, *8*, 278.
Fosgerau, K., & Hoffmann, T. (2015). Peptide therapeutics: Current status and future directions. *Drug Discovery Today*, *20*, 122–128.
Fradera, X., & Babaoglu, K. (2017). Overview of methods and strategies for conducting virtual small molecule screening. *Current Protocols in Chemistry & Biology*, *9*, 196–212.
Garg, A., Kumari, B., Kumar, R., & Kumar, M. (2017). miPepBase: A database of experimentally verified peptides involved in molecular mimicry. *Frontiers in Microbiology*, *8*, 2053.
Gautam, A., Chaudhary, K., Kumar, R., Sharma, A., Kapoor, P., Tyagi, A., et al. (2013). In silico approaches for designing highly effective cell penetrating peptides. *Journal of Translational Medicine*, *11*, 74.
Gautam, A., Chaudhary, K., Singh, S., Joshi, A., Anand, P., Tuknait, A., et al. (2014). Hemolytik: A database of experimentally determined hemolytic and non-hemolytic peptides. *Nucleic Acids Research*, *42*, D444–D449.
Gautam, A., Sharma, M., Vir, P., Chaudhary, K., Kapoor, P., Kumar, R., et al. (2015). Identification and characterization of novel protein-derived arginine-rich cell-penetrating peptides. *European Journal of Pharmaceutics and Biopharmaceutics*, *89*, 93–106.
Gautam, A., Singh, H., Tyagi, A., Chaudhary, K., Kumar, R., Kapoor, P., et al. (2012). CPPsite: A curated database of cell penetrating peptides. *Database (Oxford)*, *2012*, bas015.
Gogoladze, G., Grigolava, M., Vishnepolsky, B., Chubinidze, M., Duroux, P., Lefranc, M.-P., et al. (2014). DBAASP: Database of antimicrobial activity and structure of peptides. *FEMS Microbiology Letters*, *357*, 63–68.

Gómez, E. A., Giraldo, P., & Orduz, S. (2017). InverPep: A database of invertebrate antimicrobial peptides. *Journal of Global Antimicrobial Resistance*, *8*, 13–17.
Gupta, S., Kapoor, P., Chaudhary, K., Gautam, A., Kumar, R., Raghava, G. P. S., et al. (2013). In silico approach for predicting toxicity of peptides and proteins. *PLoS One*, *8*, e73957.
Gupta, A. K., Kaur, K., Rajput, A., Dhanda, S. K., Sehgal, M., Khan, M. S., et al. (2016). ZikaVR: An integrated Zika virus resource for genomics, proteomics, phylogenetic and therapeutic analysis. *Scientific Reports*, *6*, 32713.
Gupta, S., Sharma, A. K., Shastri, V., Madhu, M. K., & Sharma, V. K. (2017). Prediction of anti-inflammatory proteins/peptides: An insilico approach. *Journal of Translational Medicine*, *15*, 7.
Hajduk, P. J., & Greer, J. (2007). A decade of fragment-based drug design: Strategic advances and lessons learned. *Nature Reviews. Drug Discovery*, *6*, 211–219.
Hamill, P., Brown, K., Jenssen, H., & Hancock, R. E. W. (2008). Novel anti-infectives: Is host defence the answer? *Current Opinion in Biotechnology*, *19*, 628–636.
Hammami, R., Ben Hamida, J., Vergoten, G., & Fliss, I. (2009). PhytAMP: A database dedicated to antimicrobial plant peptides. *Nucleic Acids Research*, *37*, D963–8.
Hammami, R., Zouhir, A., Ben Hamida, J., & Fliss, I. (2007). BACTIBASE: A new web-accessible database for bacteriocin characterization. *BMC Microbiology*, *7*, 89.
Hammami, R., Zouhir, A., Le Lay, C., Ben Hamida, J., & Fliss, I. (2010). BACTIBASE second release: A database and tool platform for bacteriocin characterization. *BMC Microbiology*, *10*, 22.
Hancock, R. E. W., Nijnik, A., & Philpott, D. J. (2012). Modulating immunity as a therapy for bacterial infections. *Nature Reviews. Microbiology*, *10*, 243–254.
Hancock, R. E. W., & Sahl, H.-G. (2006). Antimicrobial and host-defense peptides as new anti-infective therapeutic strategies. *Nature Biotechnology*, *24*, 1551–1557.
Hilchie, A. L., Wuerth, K., & Hancock, R. E. W. (2013). Immune modulation by multifaceted cationic host defense (antimicrobial) peptides. *Nature Chemical Biology*, *9*, 761–768.
Holton, T. A., Pollastri, G., Shields, D. C., & Mooney, C. (2013). CPPpred: Prediction of cell penetrating peptides. *Bioinformatics*, *29*, 3094–3096.
Jespersen, M. C., Peters, B., Nielsen, M., & Marcatili, P. (2017). BepiPred-2.0: Improving sequence-based B-cell epitope prediction using conformational epitopes. *Nucleic Acids Research*, *45*, W24–W29.
Jitendra, Sharma, P. K., Bansal, S., & Banik, A. (2011). Noninvasive routes of proteins and peptides drug delivery. *Indian Journal of Pharmaceutical Sciences*, *73*, 367–375.
Jørgensen, K. W., Rasmussen, M., Buus, S., & Nielsen, M. (2014). NetMHCstab—Predicting stability of peptide–MHC-I complexes; impacts for cytotoxic T lymphocyte epitope discovery. *Immunology*, *141*, 18–26.
Joseph, S., Karnik, S., Nilawe, P., Jayaraman, V. K., & Idicula-Thomas, S. (2012). ClassAMP: A prediction tool for classification of antimicrobial peptides. *IEEE/ACM Transactions on Computational Biology and Bioinformatics*, *9*, 1535–1538.
Kaas, Q., Westermann, J.-C., Halai, R., Wang, C. K. L., & Craik, D. J. (2008). ConoServer, a database for conopeptide sequences and structures. *Bioinformatics*, *24*, 445–446.
Kapoor, P., Singh, H., Gautam, A., Chaudhary, K., Kumar, R., & Raghava, G. P. S. (2012). TumorHoPe: A database of tumor homing peptides. *PLoS One*, *7*, e35187.
Karosiene, E., Lundegaard, C., Lund, O., & Nielsen, M. (2012). NetMHCcons: A consensus method for the major histocompatibility complex class I predictions. *Immunogenetics*, *64*, 177–186.
Keşmir, C., Nussbaum, A. K., Schild, H., Detours, V., & Brunak, S. (2002). Prediction of proteasome cleavage motifs by neural networks. *Protein Engineering*, *15*, 287–296.
Kim, Y., Bark, S., Hook, V., & Bandeira, N. (2011). NeuroPedia: Neuropeptide database and spectral library. *Bioinformatics*, *27*, 2772–2773.

Klausen, M. S., Anderson, M. V., Jespersen, M. C., Nielsen, M., & Marcatili, P. (2015). LYRA, a webserver for lymphocyte receptor structural modeling. *Nucleic Acids Research*, *43*, W349–W355.

Kringelum, J. V., Lundegaard, C., Lund, O., & Nielsen, M. (2012). Reliable B cell epitope predictions: Impacts of method development and improved benchmarking. *PLoS Computational Biology*, *8*, e1002829.

Krochmal, M., Fernandes, M., Filip, S., Pontillo, C., Husi, H., Zoidakis, J., et al. (2016). PeptiCKDdb-peptide- and protein-centric database for the investigation of genesis and progression of chronic kidney disease. *Database (Oxford)*, *2016*, baw128.

Kulkarni-Kale, U., Bhosle, S., & Kolaskar, A. S. (2005). CEP: A conformational epitope prediction server. *Nucleic Acids Research*, *33*, W168–71.

Kumar, R., Chaudhary, K., Sharma, M., Nagpal, G., Chauhan, J. S., Singh, S., et al. (2015). AHTPDB: A comprehensive platform for analysis and presentation of antihypertensive peptides. *Nucleic Acids Research*, *43*, D956–62.

Kumar, R., Chaudhary, K., Singh Chauhan, J., Nagpal, G., Kumar, R., Sharma, M., et al. (2015). An in silico platform for predicting, screening and designing of antihypertensive peptides. *Scientific Reports*, *5*, 12512.

Larsen, J. E. P., Lund, O., & Nielsen, M. (2006). Improved method for predicting linear B-cell epitopes. *Immunome Research*, *2*, 2.

Lata, S., Bhasin, M., & Raghava, G. P. S. (2007). Application of machine learning techniques in predicting MHC binders. In *Methods in molecular biology (Clifton, N.J.)* (pp. 201–215).

Lata, S., Bhasin, M., & Raghava, G. P. S. (2009). MHCBN 4.0: A database of MHC/TAP binding peptides and T-cell epitopes. *BMC Research Notes*, *2*, 61.

Lata, S., Mishra, N. K., & Raghava, G. P. (2010). AntiBP2: Improved version of antibacterial peptide prediction. *BMC Bioinformatics*, *11*, S19.

Lata, S., & Raghava, G. P. S. (2008). PRRDB: A comprehensive database of pattern-recognition receptors and their ligands. *BMC Genomics*, *9*, 180.

Lavi, A., Ngan, C. H., Movshovitz-Attias, D., Bohnuud, T., Yueh, C., Beglov, D., et al. (2013). Detection of peptide-binding sites on protein surfaces: The first step toward the modeling and targeting of peptide-mediated interactions. *Proteins: Structure, Function, and Bioinformatics*, *81*, 2096–2105.

Leelananda, S. P., & Lindert, S. (2016). Computational methods in drug discovery. *Beilstein Journal of Organic Chemistry*, *12*, 2694–2718.

Li, Y., & Chen, Z. (2008). RAPD: A database of recombinantly-produced antimicrobial peptides. *FEMS Microbiology Letters*, *289*, 126–129.

Liang, S., Zheng, D., Standley, D. M., Yao, B., Zacharias, M., & Zhang, C. (2010). EPSVR and EPMeta: Prediction of antigenic epitopes using support vector regression and multiple server results. *BMC Bioinformatics*, *11*, 381.

Liang, S., Zheng, D., Zhang, C., & Zacharias, M. (2009). Prediction of antigenic epitopes on protein surfaces by consensus scoring. *BMC Bioinformatics*, *10*, 302.

Liu, F., Baggerman, G., Schoofs, L., & Wets, G. (2008). The construction of a bioactive peptide database in Metazoa. *Journal of Proteome Research*, 7, 4119–4131.

Mathur, D., Prakash, S., Anand, P., Kaur, H., Agrawal, P., Mehta, A., et al. (2016). PEPlife: A repository of the half-life of peptides. *Scientific Reports*, *6*, 36617.

Meher, P. K., Sahu, T. K., Saini, V., & Rao, A. R. (2017). Predicting antimicrobial peptides with improved accuracy by incorporating the compositional, physico-chemical and structural features into Chou's general PseAAC. *Scientific Reports*, 7, 42362.

Mehta, D., Anand, P., Kumar, V., Joshi, A., Mathur, D., Singh, S., et al. (2014). ParaPep: A web resource for experimentally validated antiparasitic peptide sequences and their structures. *Database (Oxford)*, *2014*, bau051.

Meister, G. E., Roberts, C. G., Berzofsky, J. A., & De Groot, A. S. (1995). Two novel T cell epitope prediction algorithms based on MHC-binding motifs; comparison of predicted and published epitopes from Mycobacterium tuberculosis and HIV protein sequences. *Vaccine*, *13*, 581–591.

Muheem, A., Shakeel, F., Jahangir, M. A., Anwar, M., Mallick, N., Jain, G. K., et al. (2016). A review on the strategies for oral delivery of proteins and peptides and their clinical perspectives. *Saudi Pharmaceutical Journal, 24*, 413–428.

Mustafa, A. S., & Shaban, F. A. (2006). ProPred analysis and experimental evaluation of promiscuous T-cell epitopes of three major secreted antigens of mycobacterium tuberculosis. *Tuberculosis (Edinburgh, Scotland), 86*, 115–124.

Nagpal, G., Gupta, S., Chaudhary, K., Dhanda, S. K., Prakash, S., & Raghava, G. P. S. (2015). VaccineDA: Prediction, design and genome-wide screening of oligodeoxynucleotide-based vaccine adjuvants. *Scientific Reports, 5*, 12478.

Nagpal, G., Usmani, S. S., Dhanda, S. K., Kaur, H., Singh, S., Sharma, M., et al. (2017). Computer-aided designing of immunosuppressive peptides based on IL-10 inducing potential. *Scientific Reports, 7*, 42851.

New York, R. (2017). *Peptide therapeutics market: Global industry analysis, trends, market size & forecasts to 2023*. New York: ReportLinker.

Nicholls, E. F., Madera, L., & Hancock, R. E. W. (2010). Immunomodulators as adjuvants for vaccines and antimicrobial therapy. *Annals of the New York Academy of Sciences, 1213*, 46–61.

Nielsen, S. D., Beverly, R. L., Qu, Y., & Dallas, D. C. (2017). Milk bioactive peptide database: A comprehensive database of milk protein-derived bioactive peptides and novel visualization. *Food Chemistry, 232*, 673–682.

Nielsen, M., Lundegaard, C., Lund, O., & Keşmir, C. (2005). The role of the proteasome in generating cytotoxic T-cell epitopes: Insights obtained from improved predictions of proteasomal cleavage. *Immunogenetics, 57*, 33–41.

Nielsen, M., Lundegaard, C., Worning, P., Lauemøller, S. L., Lamberth, K., Buus, S., et al. (2003). Reliable prediction of T-cell epitopes using neural networks with novel sequence representations. *Protein Science, 12*, 1007–1017.

Niyonsaba, F., Madera, L., Afacan, N., Okumura, K., Ogawa, H., & Hancock, R. E. W. (2013). The innate defense regulator peptides IDR-HH2, IDR-1002, and IDR-1018 modulate human neutrophil functions. *Journal of Leukocyte Biology, 94*, 159–170.

Novković, M., Simunić, J., Bojović, V., Tossi, A., & Juretić, D. (2012). DADP: The database of anuran defense peptides. *Bioinformatics, 28*, 1406–1407.

Olimpieri, P. P., Chailyan, A., Tramontano, A., & Marcatili, P. (2013). Prediction of site-specific interactions in antibody–antigen complexes: The proABC method and server. *Bioinformatics, 29*, 2285–2291.

Otvos, L., & Wade, J. D. (2014). Current challenges in peptide-based drug discovery. *Frontiers in Chemistry, 2*, 62.

Piotto, S. P., Sessa, L., Concilio, S., & Iannelli, P. (2012). YADAMP: Yet another database of antimicrobial peptides. *International Journal of Antimicrobial Agents, 39*, 346–351.

Pirtskhalava, M., Gabrielian, A., Cruz, P., Griggs, H. L., Squires, R. B., Hurt, D. E., et al. (2016). DBAASP v.2: An enhanced database of structure and antimicrobial/cytotoxic activity of natural and synthetic peptides. *Nucleic Acids Research, 44*, 6503.

Ponomarenko, J., Bui, H.-H., Li, W., Fusseder, N., Bourne, P. E., Sette, A., et al. (2008). ElliPro: A new structure-based tool for the prediction of antibody epitopes. *BMC Bioinformatics, 9*, 514.

Potocnakova, L., Bhide, M., & Pulzova, L. B. (2016). An introduction to B-cell epitope mapping and in silico epitope prediction. *Journal of Immunology Research, 2016*, 6760830.

Pulendran, B., & Ahmed, R. (2011). Immunological mechanisms of vaccination. *Nature Immunology, 12*, 509–517.

Qi, T., Qiu, T., Zhang, Q., Tang, K., Fan, Y., Qiu, J., et al. (2014). SEPPA 2.0—More refined server to predict spatial epitope considering species of immune host and subcellular localization of protein antigen. *Nucleic Acids Research, 42*, W59–W63.

Qureshi, A., Thakur, N., & Kumar, M. (2013). HIPdb: A database of experimentally validated HIV inhibiting peptides. *PLoS One, 8*, e54908.

Qureshi, A., Thakur, N., Tandon, H., & Kumar, M. (2014). AVPdb: A database of experimentally validated antiviral peptides targeting medically important viruses. *Nucleic Acids Research*, *42*, D1147–D1153.

Rajput, A., Gupta, A. K., & Kumar, M. (2015). Prediction and analysis of quorum sensing peptides based on sequence features. *PLoS One*, *10*, e0120066.

Rashid, M., Singla, D., Sharma, A., Kumar, M., & Raghava, G. P. S. (2009). Hmrbase: A database of hormones and their receptors. *BMC Genomics*, *10*, 307.

Rasmussen, M., Fenoy, E., Harndahl, M., Kristensen, A. B., Nielsen, I. K., Nielsen, M., et al. (2016). Pan-specific prediction of peptide-MHC class I complex stability, a correlate of T cell immunogenicity. *Journal of Immunology*, *197*, 1517–1524.

Ratanji, K. D., Derrick, J. P., Dearman, R. J., & Kimber, I. (2014). Immunogenicity of therapeutic proteins: Influence of aggregation. *Journal of Immunotoxicology*, *11*, 99–109.

Reche, P., Flower, D. R., Fridkis-Hareli, M., & Hoshino, Y. (2015). Peptide-based immunotherapeutics and vaccines 2015. *Journal of Immunology Research*, *2015*, 349049.

Reche, P. A., Glutting, J.-P., & Reinherz, E. L. (2002). Prediction of MHC class I binding peptides using profile motifs. *Human Immunology*, *63*, 701–709.

Reche, P., Glutting, J.-P., Zhang, H., & Reinherz, E. (2004). Enhancement to the RANKPEP resource for the prediction of peptide binding to MHC molecules using profiles. *Immunogenetics*, *56*, 405–419.

Reche, P. A., Zhang, H., Glutting, J.-P., & Reinherz, E. L. (2005). EPIMHC: A curated database of MHC-binding peptides for customized computational vaccinology. *Bioinformatics*, *21*, 2140–2141.

Rifkin, R. F., Potgieter, M., Ramond, J.-B., & Cowan, D. A. (2017). Ancient oncogenesis, infection and human evolution. *Evolutionary Applications*, *10*, 949–964.

Rubinstein, N. D., Mayrose, I., Martz, E., & Pupko, T. (2009). Epitopia: A web-server for predicting B-cell epitopes. *BMC Bioinformatics*, *10*, 287.

Saha, S., Bhasin, M., & Raghava, G. P. S. (2005). Bcipep: A database of B-cell epitopes. *BMC Genomics*, *6*, 79.

Saha, S., & Raghava, G. P. S. (2006). Prediction of continuous B-cell epitopes in an antigen using recurrent neural network. *Proteins*, *65*, 40–48.

Saha, S., & Raghava, G. P. S. (2007). Prediction methods for B-cell epitopes. *Methods in Molecular Biology*, *409*, 387–394.

Sayers, S., Ulysse, G., Xiang, Z., & He, Y. (2012). Vaxjo: A web-based vaccine adjuvant database and its application for analysis of vaccine adjuvants and their uses in vaccine development. *Journal of Biomedicine & Biotechnology*, *2012*, 831486.

Schneider, G. (2018). Automating drug discovery. *Nature Reviews. Drug Discovery*, *17*(2), 97–113. https://doi.org/10.1038/nrd.2017.232. (Epub 2017 Dec 15; Review; PubMed Id: 29242609).

Seebah, S., Suresh, A., Zhuo, S., Choong, Y. H., Chua, H., Chuon, D., et al. (2007). Defensins knowledgebase: A manually curated database and information source focused on the defensins family of antimicrobial peptides. *Nucleic Acids Research*, *35*, D265–8.

Sela-Culang, I., Ashkenazi, S., Peters, B., & Ofran, Y. (2015). PEASE: Predicting B-cell epitopes utilizing antibody sequence. *Bioinformatics*, *31*, 1313–1315.

Shameer, K., Madan, L. L., Veeranna, S., Gopal, B., & Sowdhamini, R. (2010). PeptideMine—A webserver for the design of peptides for protein-peptide binding studies derived from protein–protein interactomes. *BMC Bioinformatics*, *11*, 473.

Shanmugam, A., Rajoria, S., George, A. L., Mittelman, A., Suriano, R., & Tiwari, R. K. (2012). Synthetic toll like receptor-4 (TLR-4) agonist peptides as a novel class of adjuvants. *PLoS One*, *7*, e30839.

Shen, W., Cao, Y., Cha, L., Zhang, X., Ying, X., Zhang, W., et al. (2015). Predicting linear B-cell epitopes using amino acid anchoring pair composition. *BioData Mining*, *8*, 14.

Singh, H., Ansari, H. R., & Raghava, G. P. S. (2013). Improved method for linear B-cell epitope prediction using antigen's primary sequence. *PLoS One*, *8*, e62216.

Singh, S., Chaudhary, K., Dhanda, S. K., Bhalla, S., Usmani, S. S., Gautam, A., et al. (2016). SATPdb: A database of structurally annotated therapeutic peptides. *Nucleic Acids Research*, *44*, D1119–26.

Singh, H., & Raghava, G. P. (2001). ProPred: Prediction of HLA-DR binding sites. *Bioinformatics*, *17*, 1236–1237.

Singh, H., & Raghava, G. P. S. (2003). ProPred1: Prediction of promiscuous MHC class-I binding sites. *Bioinformatics*, *19*, 1009–1014.

Singh, S., Singh, H., Tuknait, A., Chaudhary, K., Singh, B., Kumaran, S., et al. (2015). PEPstrMOD: Structure prediction of peptides containing natural, non-natural and modified residues. *Biology Direct*, *10*, 73.

Siwy, J., Mullen, W., Golovko, I., Franke, J., & Zürbig, P. (2011). Human urinary peptide database for multiple disease biomarker discovery. *Proteomics. Clinical Applications*, *5*, 367–374.

Skwarczynski, M., & Toth, I. (2016). Peptide-based synthetic vaccines. *Chemical Science*, *7*, 842–854.

Smadbeck, J., Peterson, M. B., Zee, B. M., Garapaty, S., Mago, A., Lee, C., et al. (2014). De novo peptide design and experimental validation of histone methyltransferase inhibitors. *PLoS One*, *9*, e95535.

Soria-Guerra, R. E., Nieto-Gomez, R., Govea-Alonso, D. O., & Rosales-Mendoza, S. (2015). An overview of bioinformatics tools for epitope prediction: Implications on vaccine development. *Journal of Biomedical Informatics*, *53*, 405–414.

Stern, L. J., Brown, J. H., Jardetzky, T. S., Gorga, J. C., Urban, R. G., Strominger, J. L., et al. (1994). Crystal structure of the human class II MHC protein HLA-DR1 complexed with an influenza virus peptide. *Nature*, *368*, 215–221.

Stille, C. J., Thomas, L. J., Reyes, V. E., & Humphreys, R. E. (1987). Hydrophobic strip-of-helix algorithm for selection of T cell-presented peptides. *Molecular Immunology*, *24*, 1021–1027.

Stranzl, T., Larsen, M. V., Lundegaard, C., & Nielsen, M. (2010). NetCTLpan: Pan-specific MHC class I pathway epitope predictions. *Immunogenetics*, *62*, 357–368.

Sweredoski, M. J., & Baldi, P. (2008). PEPITO: Improved discontinuous B-cell epitope prediction using multiple distance thresholds and half sphere exposure. *Bioinformatics*, *24*, 1459–1460.

Sweredoski, M. J., & Baldi, P. (2009). COBEpro: A novel system for predicting continuous B-cell epitopes. *Protein Engineering, Design & Selection*, *22*, 113–120.

Taherzadeh, G., Yang, Y., Zhang, T., Liew, A. W.-C., & Zhou, Y. (2016). Sequence-based prediction of protein–peptide binding sites using support vector machine. *Journal of Computational Chemistry*, *37*, 1223–1229.

Tang, H., Su, Z.-D., Wei, H.-H., Chen, W., & Lin, H. (2016). Prediction of cell-penetrating peptides with feature selection techniques. *Biochemical and Biophysical Research Communications*, *477*, 150–154.

Thakur, N., Qureshi, A., & Kumar, M. (2012). AVPpred: Collection and prediction of highly effective antiviral peptides. *Nucleic Acids Research*, *40*, W199–W204.

Thangakani, A. M., Nagarajan, R., Kumar, S., Sakthivel, R., Velmurugan, D., & Gromiha, M. M. (2016). CPAD, curated protein aggregation database: A repository of manually curated experimental data on protein and peptide aggregation. *PLoS One*, *11*, e0152949.

Théolier, J., Fliss, I., Jean, J., & Hammami, R. (2014). MilkAMP: A comprehensive database of antimicrobial peptides of dairy origin. *Dairy Science & Technology*, *94*, 181–193.

Toseland, C. P., Clayton, D. J., McSparron, H., Hemsley, S. L., Blythe, M. J., Paine, K., et al. (2005). AntiJen: A quantitative immunology database integrating functional, thermodynamic, kinetic, biophysical, and cellular data. *Immunome Research*, *1*, 4.

Trabuco, L. G., Lise, S., Petsalaki, E., & Russell, R. B. (2012). PepSite: Prediction of peptide-binding sites from protein surfaces. *Nucleic Acids Research*, *40*, W423–7.

Trolle, T., & Nielsen, M. (2014). NetTepi: An integrated method for the prediction of T cell epitopes. *Immunogenetics*, *66*, 449–456.

Tung, C.-W., & Ho, S.-Y. (2007). POPI: Predicting immunogenicity of MHC class I binding peptides by mining informative physicochemical properties. *Bioinformatics*, *23*, 942–949.

Tyagi, A., Tuknait, A., Anand, P., Gupta, S., Sharma, M., Mathur, D., et al. (2015). CancerPPD: A database of anticancer peptides and proteins. *Nucleic Acids Research*, *43*, D837–D843.

Usmani, S. S., Bedi, G., Samuel, J. S., Singh, S., Kalra, S., Kumar, P., et al. (2017). THPdb: Database of FDA-approved peptide and protein therapeutics. *PLoS One*, *12*, e0181748.

Van Dorpe, S., Bronselaer, A., Nielandt, J., Stalmans, S., Wynendaele, E., Audenaert, K., et al. (2012). Brainpeps: The blood–brain barrier peptide database. *Brain Structure & Function*, *217*, 687–718.

Viart, B., Dias-Lopes, C., Kozlova, E., Oliveira, C. F. B., Nguyen, C., Neshich, G., et al. (2016). EPI-peptide designer: A tool for designing peptide ligand libraries based on epitope–paratope interactions. *Bioinformatics*, *32*, 1462–1470.

Vita, R., Overton, J. A., Greenbaum, J. A., Ponomarenko, J., Clark, J. D., Cantrell, J. R., et al. (2015). The immune epitope database (IEDB) 3.0. *Nucleic Acids Research*, *43*, D405–D412.

Vizcaíno, J. A., Csordas, A., del-Toro, N., Dianes, J. A., Griss, J., Lavidas, I., et al. (2016). 2016 update of the PRIDE database and its related tools. *Nucleic Acids Research*, *44*, D447–56.

Waghu, F. H., Barai, R. S., Gurung, P., & Idicula-Thomas, S. (2016). CAMPR3: A database on sequences, structures and signatures of antimicrobial peptides. *Nucleic Acids Research*, *44*, D1094–7.

Waghu, F. H., Gopi, L., Barai, R. S., Ramteke, P., Nizami, B., & Idicula-Thomas, S. (2014). CAMP: Collection of sequences and structures of antimicrobial peptides. *Nucleic Acids Research*, *42*, D1154–8.

Wang, S., Guo, L., Liu, D., Liu, W., & Wu, Y. (2016). HLAsupE: An integrated database of HLA supertype-specific epitopes to aid in the development of vaccines with broad coverage of the human population. *BMC Immunology*, *17*, 17.

Wang, G., Li, X., & Wang, Z. (2009). APD2: The updated antimicrobial peptide database and its application in peptide design. *Nucleic Acids Research*, *37*, D933–7.

Wang, G., Li, X., & Wang, Z. (2016). APD3: The antimicrobial peptide database as a tool for research and education. *Nucleic Acids Research*, *44*, D1087–D1093.

Wang, Y., Wang, M., Yin, S., Jang, R., Wang, J., Xue, Z., et al. (2015). NeuroPep: A comprehensive resource of neuropeptides. *Database*, *2015*, bav038.

Wei, L., Tang, J., & Zou, Q. (2017). SkipCPP-Pred: An improved and promising sequence-based predictor for predicting cell-penetrating peptides. *BMC Genomics*, *18*, 1–11.

Whitmore, L., & Wallace, B. A. (2004). The Peptaibol database: A database for sequences and structures of naturally occurring peptaibols. *Nucleic Acids Research*, *32*, D593–4.

Wieczorek, M., Abualrous, E. T., Sticht, J., Álvaro-Benito, M., Stolzenberg, S., Noé, F., et al. (2017). Major histocompatibility complex (MHC) class I and MHC class II proteins: Conformational plasticity in antigen presentation. *Frontiers in Immunology*, *8*, 292.

Wynendaele, E., Bronselaer, A., Nielandt, J., D'Hondt, M., Stalmans, S., Bracke, N., et al. (2013). Quorumpeps database: Chemical space, microbial origin and functionality of quorum sensing peptides. *Nucleic Acids Research*, *41*, D655–D659.

Xiao, X., Wang, P., Lin, W.-Z., Jia, J.-H., & Chou, K.-C. (2013). iAMP-2L: A two-level multi-label classifier for identifying antimicrobial peptides and their functional types. *Analytical Biochemistry*, *436*, 168–177.

Yao, B., Zhang, L., Liang, S., & Zhang, C. (2012). SVMTriP: A method to predict antigenic epitopes using support vector machine to integrate tri-peptide similarity and propensity. *PLoS One*, 7, e45152.

Zapadka, K. L., Becher, F. J., Gomes Dos Santos, A. L., & Jackson, S. E. (2017). Factors affecting the physical stability (aggregation) of peptide therapeutics. *Interface Focus*, 7, 20170030.

Zhang, G. L., DeLuca, D. S., Keskin, D. B., Chitkushev, L., Zlateva, T., Lund, O., et al. (2011). MULTIPRED2: A computational system for large-scale identification of peptides predicted to bind to HLA supertypes and alleles. *Journal of Immunological Methods*, *374*, 53–61.

Zhang, Y.-Z., & Shen, H.-B. (2017). Signal-3L 2.0: A hierarchical mixture model for enhancing protein signal peptide prediction by incorporating residue-domain cross-level features. *Journal of Chemical Information and Modeling*, *57*, 988–999.

Zhao, L., Zhang, M., & Cong, H. (2013). Advances in the study of HLA-restricted epitope vaccines. *Human Vaccines & Immunotherapeutics*, *9*, 2566–2577.

Zia, Q., Azhar, A., Ahmad, S., Afsar, M., Hasan, Z., Owais, M., et al. (2017). PeMtb: A database of MHC antigenic peptide of mycobacterium tuberculosis. *Current Pharmaceutical Biotechnology*, *18*, 648–652.

CHAPTER EIGHT

Protein and Peptides for Elderly Health

Nan Shang, Chalamaiah Meram, Nandika Bandara, Jianping Wu[1]
University of Alberta, Edmonton, AB, Canada
[1]Corresponding author: e-mail address: jwu3@ualberta.ca

Contents

Abstract

The world is aging rapidly; thus, the management of elderly health at an advanced age poses a new research challenge. The elderly is vulnerable to not only malnutrition but also hypofunction of all organs and a variety of chronic diseases such as sarcopenia, osteoporosis, gastrointestinal dysfunction, and mental problems. As the major macronutrient, food protein plays an important role in elderly health and well-being. In this chapter, the function of protein and peptide in elderly health as well as their effects on preventing aging-related disease is reviewed.

1. INTRODUCTION

The world is aging at unprecedented rate. According to data from World Population Prospects: the 2015 Revision (United Nations, 2018), between 2015 and 2030, the number of people in the world aged 60 years or over is projected to grow by 56%, from 901 million to 1.4 billion, and by 2050, the global population of older persons is projected to more than

Advances in Protein Chemistry and Structural Biology, Volume 112
ISSN 1876-1623
https://doi.org/10.1016/bs.apcsb.2018.03.003

double its size in 2015, reaching nearly 2.1 billion. These rapid changes in demographics will have profound influence on almost every aspect of our society and life and pose new challenges for elderly nutrition security. Aging is accompanied by many physiological/psychosocial changes (Leslie & Hankey, 2015), and thus, malnutrition is one of the most common concerns in elder population, which is associated with an increased risk of morbidity and mortality (Margetts, Thompson, Elia, & Jackson, 2003). Therefore, meeting the nutrition needs and maintaining a good nutritional status have significant implications for their health and well-being, functional independence, reduced risk of developing disease, and thus promoting continued independent living (Callen, 2010; Leslie & Hankey, 2015).

Protein is one of the major nutrients for human diet, which also plays an important role in elderly health. In addition to providing a source of energy and amino acids for the synthesis of body proteins, food proteins are important biological entities that not only help sustain the overall growth, metabolism, and functioning of cellular metabolisms but also provide health attributes. In this review, aging-related geriatric syndromes and diseases will be discussed together with the effects of protein/peptide on improving elderly health.

2. PROTEIN RECOMMENDATION FOR ELDERLY HEALTH

Proteins are structural nutrients, accounting for 75% of the total body weight on dry basis. They are essential for maintaining a healthy life from the beginning of gestation through the old age. In the human body, proteins are constantly being turned over and maintain in a balance of synthesis and degradation (Phillips, 2017). Adequate intake of dietary protein is essential to maintain the protein content of certain tissues and organs including skin, brain, heart, liver, and kidney for survival (Phillips, 2017).

Protein requirements vary depending on age, body size, gender, physiological states (including pregnancy, illnesses, and fitness), and possibly environment. Current protein requirement for healthy people older than 19 years is 0.80 g/kg/day according to the recommended dietary allowance (RAD) (Phillips, 2017). This requirement has remained unchanged for several years, and a number of reports have challenged that the "recommended intake" of protein is more accurate for "minimal intake" (Phillips, 2017). Some experts suggested to intake 1.2–2.0 g/kg/day protein or higher for elderly adults (Volpi et al., 2013; Wolfe, 2012; Wolfe, Miller, & Miller, 2008). The major health challenge in the aging population is to slow the decline in muscle mass and strength, and maintain a healthy body weight (Phillips, Chevalier, & Leidy, 2016). Thus, a modest increase of protein intake may benefit for elderly health.

When considering protein intake, it is also important to consider for protein source and quality. Dietary animal protein is the primary source of high value protein, iron, vitamin, and other essential nutrients (Chernoff, 2004). It has been found that beef that is rich in the essential amino acid leucine could enhance the rate of postprandial muscle protein synthesis (Nowson & O'Connell, 2015). Milk and egg are also good source of high-quality protein that has potential influence on preventing aging-related disease.

3. PROTEIN/PEPTIDE AND ELDERLY DISEASE

3.1 Sarcopenia

According to the US Census Bureau, ~80 million individuals (20% of the total population) will be above 65 years of age by 2050 (US Census Bureau, 2008). An average of 30%–50% skeletal muscle mass loss has been seen among the people in the age of 40–80 years, while 3% annual decline in muscle functional capacity has been observed in the people above 60 years old (Patel et al., 2013). In 1989, Irwin Rosenberg first used the term "Sarcopenia" to describe the pathophysiological and epidemiological aspects of age-associated skeletal muscle mass deterioration (Marty, Liu, Samuel, Or, & Lane, 2017). With the advancement of the knowledge, sarcopenia is now defined as a multifactorial, age-associated deterioration, and decline of skeletal muscle mass in elderly people (Marty et al., 2017; Walston, 2012).

3.1.1 Pathophysiology of Sarcopenia

Pathophysiology of sarcopenia shows multifactorial contributions in its development. Decreased caloric intake, intracellular oxidative stress, denervation of muscle fibers, decline in hormonal secretion, and increased myostatin signaling are believed to be among the key factors contributing for sarcopenia (Marty et al., 2017; Walston, 2012). Specifically, lack of physical activity, reduced caloric, protein, and vitamin D intake associated with reduced food intake have shown correlation with the diminishing skeletal muscle strength and functionality in older population (Houston et al., 2008; Morley, Anker, & von Haehling, 2014; Nieuwenhuizen, Weenen, Rigby, & Hetherington, 2010). Human hormonal secretions such as testosterone, estrogen, and growth hormone show a rapid decrease with aging in both males and females and are associated with lean muscle loss (Baumgartner, Waters, Gallagher, Morley, & Garry, 1999; Perrini et al., 2010; Samson et al., 2000). Age-related intracellular oxidative stress can cause chronic low-grade inflammation leading to increased production of inflammatory cytokines IL-6 and TNF-α (Wang et al., 2015). In addition, both human clinical and animal trials have shown an increased level of myostatin in aging adults, which can potentially induce muscle atrophy leading to lean muscle loss (Han, Zhou, & Goldberg, 2013; McPherron, Lawler, & Lee, 1997; Yarasheski, Bhasin, Sinha-Hikim, Pak-Loduca, & Gonzalez-Cadavid, 2002). Type II fast-twitch muscle fiber atrophy and muscle fiber denervation will increase the workload on remaining healthy muscle mass, leading to accelerate lean muscle loss in older adults (Nilwik et al., 2013). Recent genome-wide association study showed some evidence on genetical contribution to sarcopenia, where they identified single-nucleotide polymorphisms associated to neural maintenances in muscle fibers (Willems et al., 2017). However, further research into the exact mechanism and factors affecting sarcopenia is essential in developing strategies to prevent and manage sarcopenia in elderly population.

3.1.2 Current Methods in Managing Sarcopenia

Due to the multifactorial nature of sarcopenia development, there were numerous plans developed recently to manage, prevent, and treat the conditions associated with sarcopenia (Naseeb & Volpe, 2017). Medical interventions methods such as reducing catabolic signaling by treating with "Espindolol" (Pötsch et al., 2014), upregulating appetite using peptide hormone "Gherlin" (Porporato et al., 2013), hormone replacement therapy with estrogen (Sipilä et al., 2001), use of androgen/androgen receptor modulators such as testosterone (Bhasin, Woodhouse, & Storer, 2001), myostatin inhibition via antimyostatin antibodies, and faulty myostatin receptors (Attie et al., 2013) were heavily studied in the recent past to reduce symptoms of sarcopenia and increase lean muscle mass in older people (Marty et al., 2017). Whole-body vibration therapy is another method that showed significant improvement in clinical studies, in parameters such as knee dynamic length, leg isometric strength, sit-to-stand performance, and jumping height of adults suffering from sarcopenia (von Stengel, Kemmler, Engelke, & Kalender, 2012). In addition to the above methods, nutritional supplementation and resistance training hold a special place in prevention and management of sarcopenia, mainly due to the easy applicability in home setting, lower cost compared to other medical intervention methods and potential use in preventive care rather than treatment (Bauer et al., 2015; Cramer et al., 2016; Kim et al., 2012). Especially, protein and amino acid supplements with or without coupled with resistance training have been extensively studied in terms of sarcopenia prevention and treatment (Marty et al., 2017; Naseeb & Volpe, 2017). These nutritional and resistance-training interventions were mainly focused on increasing muscle mass and their physical functionality.

3.1.3 Dietary Amino Acids and Proteins in Sarcopenia

Aging is generally associated with a significant decline in food and energy intake, mainly due to the decrease in energy requirement of the body, loss of appetite and hunger (anorexia of aging), changes in digestion, and some disease conditions (Robinson et al., 2017). However, this lower food intake will risk the older individuals due to lack of proper nutrient supplementation. Among the other nutrients, amino acids and protein intake show a significant reduction, directly leading to reduced muscle protein synthesis (Wakimoto & Block, 2001). Therefore, it is essential to understand and use proper nutrient management plan to maintain and increase muscle protein generation.

Institute of Medicine recommends a dietary intake of 10%–35% of energy from protein for a healthy adult; however, wide variability in this range causes some issues in developing effective recommendations (Beasley, Shikany, & Thomson, 2013). A retrospective reassessment on nitrogen balance studies suggested a recommended daily allowance (RDA) of 1.14 g/kg/day for elderly people compared to the RDA value of 0.8 g/kg/day recommended for healthy young individuals (Campbell, Crim, Dallal, Young, & Evans, 1994). However, it is estimated that more than 25% of elderly consumed less than protein amounts of 0.8 g/kg/day, while another 50% of elderly people consumed less than 1.14 g/kg/day protein, which are well below the recommended levels (Kim, Wilson, & Lee, 2010; Rousset, Patureau Mirand, Brandolini, Martin, & Boirie, 2003). The increased RDA values for elderly people are mainly due to the reduced efficiency of protein synthesis in elderly people (Kim et al., 2010).

Emerging evidence on the effect of protein supplementation on sarcopenia provides somewhat controversial evidence on its effect. In some studies, a dietary protein supplementation above the RDA values (1.14 g/kg/day) for adults was effective in prevention and/or mitigation of sarcopenia (Morley et al., 2010; Volpi et al., 2013). In clinical trials, providing 25–30 g of high-quality protein at each meal (with at least 10 g of essential amino acids) showed a maximum increase in muscle protein synthesis (Paddon-Jones & Rasmussen, 2009). In another 24-week study, milk-based protein supplement showed positive effects on physical performance (leg press and leg extension) of frail men and women (65 years and older), when it was combined with strength-training exercise regimen compared to the protein supplement alone (Tieland et al., 2012). Singh et al. (1999) investigated the effect of nutritional supplement with 360 cal, 60% carbohydrate, 17% soy protein, and 23% fat over 10-week period (3 days a week) for frail men and women (72–98 years) with or without resistance-training plan. A 10% increase in type II muscle fibers were observed when the nutritional supplement was combined with a resistance-training regimen (Singh et al., 1999). These variations in muscle protein synthesis in elderly people in different studies can be attributed to the difference in protein source, protein quality, amino acid composition, and experimental conditions (Beasley et al., 2013).

In comparison, the nutritional intervention studies that provided targeted amino acids/protein composition such as essential amino acid supplements or leucine supplements have shown greater improvement in mitigating elderly sarcopenia (Beasley et al., 2013; Robinson et al., 2017). Branched-chain amino acids (BCAAs) such as leucine, isoleucine, and valine show improved muscle protein synthesis, athletic performance, and management of muscle loss in elderly individuals (Borack & Volpi, 2016). Even though several studies have shown different results in the effect of BCAAs on elderly people, a recent meta-analysis concluded the potential of leucine supplements in increasing muscle protein fractional synthetic rate and age-related decline in muscle mass (Xu, Tan, Zhang, Gui, & Yang, 2015). Similar to protein supplementations, BCAAs also showed a greater effect in improving sarcopenia conditions when it coupled with physical exercise (Robinson et al., 2017). In addition, it is extremely important to manage the method of protein/amino acid supplements in dealing with sarcopenia affected elderly individuals. The most practical method for inducing skeletal muscle protein synthesis can be achieved through moderate servings of protein/AA with high biological value in all three meals instead of one dose per day (Paddon-Jones & Rasmussen, 2009).

Even though proteins and amino acids show a beneficial effect in managing and prevention of conditions associated with sarcopenia, it is essential to note that nutritional intervention alone may not be enough in managing the conditions due to multifactorial pathophysiological nature of the condition. Therefore, careful planning with the help of health professional is required in developing a treatment plan for an individual. In addition, further research is required to explore the exact mechanisms and other potential impacts of protein and amino acid supplementation on elderly individual.

3.2 Skeleton Disorder

Bone is a vital and dynamic connective tissue that serves two major functions: provides biomechanical support and protection of soft tissue, and plays a key role in mineral homeostasis (Morgan, Barnes, & Einhorn, 2008; Olsen, Reginato, & Wang, 2000). In addition, bone is the primary site of hematopoiesis and plays important roles in storing minerals, keeping acid–base balance as well as enabling mobility (Morgan et al., 2008). With aging, these functions become impaired, bone becomes more fragile and less able to perform its mechanical functions, and the calcium stores are often depleted (Boskey & Coleman, 2010).

With the expanding of elderly population and the lengthening of life expectancy, bone health for elders will become a big challenge in the future. As a result of the aging process, the bone deteriorates will predispose to osteoporosis first (Demontiero, Vidal, & Duque, 2012). Osteoporosis is the most common bone problem for elderly population affecting 10 million people in the United States over 50 years, and an additional 34 million people are at risk for the disease (Surdykowski, Kenny, Insogna, & Kerstetter, 2010). At the same time, the burden of fracture will also be increased because of the growth of elderly population. Globally, nearly 9 million fragility fractures occur annually (Johnell & Kanis, 2006) and it will increase by ~300% in males and ~200% in females by 2050 based on current trend (Kanis, 2002). Unfortunately, the situation will be even worse if other skeletal disorders are taken into consideration such as osteoarthritis, microcrystal disorders, and Paget's disease. To curb the contribution of the elderly to the total future burden, effective strategies to reduce fracture risk in the elderly are needed. Only relying on drug therapy is not the best option, as the cost of treatment is expensive, and side effects are common (Iuliano & Ayton, 2015). An alternative method such as correcting nutrient deficiencies may be a better approach, especially for protein, calcium, and vitamin D.

Bone is a composite tissue, made up of an inorganic and an organic phase (Morgan et al., 2008). The inorganic phase, also called bone mineral, is an impure form of hydroxyapatite $[Ca_{10}(PO_4)_6(OH)_2]$ that gives the compressive strength to bone tissue (Bonjour, 2011). The organic phase is composed mainly of type I collagen and a variety of noncollagenous proteins (including growth factors such as transforming growth factor-β (TGF-β) and insulin-like growth factor (IGF); cytokines such as the tumor necrosis factors (TNFs) and interleukins (ILs); and extracellular matrix proteins such as osteonectin, bone sialoprotein, osteocalcin, and other phosphoproteins and proteolipids) and cells (Morgan et al., 2008). Thus, sufficient nutrient supply is the essential part for bone health management. In addition to the most well-documented nutrients such as calcium, phosphorus, and vitamin D (Dempster & Raisz, 2015), protein has been considered as a new strategy for osteoporosis prevention as it is the most important and major nutrient for human body. Protein intake is thought to affect bone health in several ways: (1) provides the structure matrix of bone; (2) optimizes IGF-1 levels; (3) increases intestinal calcium absorption (Heaney & Layman, 2008); (4) stimulates the bone cells proliferation, differentiation, and mineralization; (5) inhibits the bone resorption; and (6) inhibits the inflammatory cytokines (Fernandes, Lawrence, & Sun, 2003). Protein had been thought as a double-edged sword for bone health management in the last few decades when some researchers reported that inadequate protein intake could have adverse effects, and considered high protein diet as a risk factor for osteoporosis or bone fracture (Barzel & Massey, 1998; Feskanich, Willett, Stampfer, & Coldltz, 1996; Heaney & Layman, 2008). But now, with more researches and experiments conducted, protein is increasingly recognized as an essential nutrient to bone health. Furthermore, the effect of protein on bone health is likely associated with mineral intake. Milk protein-derived peptides, in particular casein phosphopeptides (CPPs), have been reported to have mineral binding properties (Tsuchita, Suzuki, & Kuwata, 2001). Using Caco-2 and HT-29 cells as models, it was suggested that the phosphorylated "acidic motif" and the preceding N-terminal region of CPP could modulate calcium uptake through the interaction with voltage-operated L-type calcium channel and TRPV6 calcium channel (Ferraretto, Signorile, Gravaghi, Fiorilli, & Tettamanti, 2001; Perego, Cosentino, Fiorilli, Tettamanti, & Ferraretto, 2012). Similarly, egg yolk phosvitin is considered as a bone health benefit bioactive protein deal to its strong affinity to bivalent metal such as calcium and magnesium (Grizzuti & Perlmann, 1973). Jiang and Mine (2000, 2001) reported that the egg yolk phosvitin phosphopeptides could solubilize more calcium than CPP in vitro and increase calcium uptake (Jiang & Mine, 2000, 2001).

Bone is a dynamic organ that undergoes a continual self-regeneration process called remodeling (Demontiero et al., 2012). It is modulated by three bone cells, osteoblasts, osteocytes, and osteoclasts, and has the function to balance new bone formation and local bone resorption (Dempster &

Raisz, 2015). Aging disequilibrates the balance of remodeling and shifts it in a negative direction, favoring greater bone resorption and less bone formation (Demontiero et al., 2012). This accelerated bone loss compromises bone strength and resistance and eventually leads to osteoporosis and fracture (Seeman, 2008). Several reports indicated that bioactive proteins and peptides might have an effect to redress the remodeling balance by promoting bone formation. Growth factors, playing essential roles in bone development and maintenance, could be affected by protein intake. Insulin-like growth factor 1 (IGF-1) has been reported to increase osteoblast activity (Yakar et al., 2002) and the mineralization of bone matrix (Zhang, Wastney, Rosen, Beamer, & Weaver, 2011). A positive association between consumption of protein, red meat, fish, and IGF-1 concentration has been observed in middle-aged and elderly men (Larsson, Wolk, Brismar, & Wolk, 2005). An increased circulating level of IGF-1 was observed in response to the increased dairy products intake as well (Ma et al., 2001). In addition to IGF-1, transforming growth factor-β1 (TGF-β1) and TGF-β2 have been found in human milk that affect several cellular processes, such as cell proliferation and differentiation (Gauthier, Pouliot, & Aubois, 2006; Viñals & Pouysségur, 2011), so may also have beneficial influence for bone health. Aging-associated sarcopenia is another reason of fracture in elders. Sarcopenia, loss of muscle mass, is associated with loss of strength, functional impairment, disability, and mortality contribute to the increased risk of fracture (Roubenoff & Hughes, 2000). Moreover, weak bone condition of elders may also be indirectly affected by other aging-related diseases such as diabetes and obesity.

3.2.1 Aging and Osteoporosis

Osteoporosis, the most popular and serious skeleton disorder, is also called a "silent killer" or "iceberg disease" affecting all over the world. It is characterized by a systemic impairment of bone mass, strength, and microarchitecture, resulting in increased bone fragility and risk of fracture, particularly of the hip, spine wrist, and shoulder (Marcus, Dempster, & Bouxsein, 2016). It is estimated that almost 50% of women older than 50 years old may suffer an osteoporotic fracture, whose incidence is more than the combination of heart attack, stroke, and breast cancer (Marcus et al., 2016).

Due to the limitation of drugs and exist of side effects, bioactive compounds become a new approach to prevent osteoporosis. CPPs have been proved to increase the calcium absorption in growing rats and thus increase calcium accumulation in the femur and tibia (Tsuchita et al., 2001).

The formation of CPPs has been observed during digestion of casein in the gastrointestinal (GI) tract (Boutrou, Henry, & Sanchez-Rivera, 2015; Kasai, Iwasaki, Tanaka, & Kiriyama, 1995; Meisel & Frister, 1989) or during cheese ripening (Nagpal et al., 2011). In addition to milk proteins, egg protein is considered as a calcium absorption enhancer to increase bone mass (Kastenmayer, Barclay, Cavin, & Pridmore-Merten, 2005). Bioactive proteins could also prevent osteoporosis by directly affecting the activity of bone cells. Increasing dietary whey protein showed effects on stimulating osteoblast activity (Takada, Aoe, & Kumegawa, 1996), while suppressing osteoclast-mediated bone resorption (Takada et al., 1997). Among all the milk proteins, lactoferrin is the most extensively studied protein. Lactoferrin potently promotes bone growth by stimulating the proliferation and differentiation of osteoblast and also acts as a survival factor inhibiting the apoptosis (Naot, Palmano, & Cornish, 2012). Meanwhile, Lorget et al. (2002) reported that lactoferrin could reduce in vitro osteoclast differentiation and resorbing activity. In vivo study further confirmed that orally administered lactoferrin could preserve bone mass and microarchitecture in ovariectomized rats (Lorget et al., 2002). Bharadwaj, Naldu, Betageri, Prasadarao, & Naldu (2009) examined the effect of lactoferrin supplement on bone health of postmenopausal women and indicated that the supplement induced a positive effect on bone turnover markers (Bharadwaj et al., 2009). Egg white ovotransferrin (Gentili et al., 1994) and egg yolk soluble protein (Ji, Leem, Kim, & Kim, 2007) could stimulate osteoblast activity and benefit for bone health.

3.2.2 Aging and Arthritis

The prevalence of arthritis increases markedly with age. Although it is not comparable to osteoporosis, osteoarthritis (OA) is also one of the most common aging-related skeleton problems in elder people, as well as the rheumatic arthritis (RA) (Covinsky, 2006). Because arthritis is so ubiquitous in the elderly and is rarely associated with mortality, it is often viewed as a benign condition of aging (Covinsky, 2006). However, severe arthritis could lead to daily living disability and require the assistance of another person (Song, Chang, & Dunlop, 2006), and even mild arthritis could significantly lower the living quality of elders.

Although the primary manifestation of OA and RA seems similar, the underlying pathophysiology of each condition is distinct. OA is characterized by progressive cartilage loss, which is largely mechanical. Thus, maintaining a healthy weight is a key factor in OA prevention (Messier, 2009). High-quality protein consumption can help decrease weight and improve muscle strength (Paddon-Jones et al., 2008), which will benefit elderly OA patients. RA is a chromic, inflammatory, and autoimmune disease, whose etiology still remains unknown. Although the association between diet and RA prevention is still under studying, some researchers claimed that antiinflammatory bioactive food might have a positive effect (Smolen et al., 2018).

Treatments of arthritis pain include both pharmacological agent and nonpharmacological treatment. Because of the health condition of the elders, the use of pharmacological agent should be tempered with caution regarding increased sensitivity of medications, drug–drug interactions, and associated comorbidities (Fitzcharles, Lussier, & Shir, 2010). Self-management strategies and attention to healthy lifestyle including weight control and regular physical activity and exercise are part of the nonpharmacological treatment (Fitzcharles et al., 2010). Although complete pain relief is seldom achieved, modulation of pain and the associated component of sleep disturbance, fatigue, and mood disorder will improve overall life quality for elderly patients (Fitzcharles et al., 2010). Bioactive components with antiinflammatory and immunomodulating effect have been suggested as dietary intervention for arthritis. Thus, protein with antiinflammatory activity from various sources, including milk, meat, egg, and soy (Burris, Ng, & Nagarajan, 2014; Chatterton, Nguyen, Bering, & Sanglld, 2013; McAfee et al., 2010; Wu & Acero-Lopez, 2012), might have the potential for prevention of RA, although there is no direct study.

Bone health is critically important to the overall health, vitality, and quality of life. Nutritional adequacy helps maintain bone strength and quality, as well as to prevent skeleton disease. Proteins appear to be important for bone health management not only because they are building blocks of our bodies but also because they confer a range of bioactivities that are benefit for bone health. Further research is still needed to elucidate the potential benefit of protein intake on the aging-associated skeleton problem as well as the delivery approach.

3.3 Gastrointestinal Disorder

Aging and the accompanying decline of various physiological processes have important effects on the GI tract and are associated with many somatic complaints, ranging from dyspepsia to constipation (D'Souza, 2007). However, without typical manifestations of common conditions in elder people, it is difficult to develop precaution until it becomes serious

(D'Souza, 2007). As the main organ system to get energy and nutrients, GI disorder/disease will lead to not only the development of weight loss, anorexia, and social disadvantages but also increased mortality with serious socioeconomic burden.

The GI tract is an intricate organ system that consists of mouth, esophagus, stomach, small intestine, large intestine, and anus. The liver, pancreas, and gallbladder are also included in the system considered as the solid organs. Aging could cause structural and functional alterations in all of these components in the GI system and finally lead to GI disorders or diseases.

3.3.1 Dysphagia and Esophageal Cancer

Dysphagia, also known as swallowing difficulty, refers to any disruption in the swallowing process from the mouth to the stomach (Aslam & Vaezi, 2013). It is estimated that 15%–40% of individuals over 60 years have dysphagia (Aslam & Vaezi, 2013; Barczi, Sullivan, & Robbins, 2000). Based on the more optimistic prevalence rate of 15% and the 1998 US census data, it is estimated that 6 million adults have dysphagia. These numbers mirror the prevalence in European countries as well (Robbins, Bridges, & Taylor, 2006). The needs of dysphagia management is not only in acute care settings, but also in long-term health protection (Robbins et al., 2006). The consequences of dysphagia vary from social isolation owing to the embarrassment associated with choking or coughing at mealtime, to physical discomfort (e.g., food sticking in the chest), to potentially life-threatening conditions (Robbins et al., 2006), such as dehydration, malnutrition, pneumonia, chronic lung diseases, or even asphyxiation and death (Sura, Madhavan, Carnaby-Mann, & Crary, 2012). Dysphagia may be classified anatomically as either oropharyngeal or esophageal (Aslam & Vaezi, 2013). Oropharyngeal dysphagia is related to the initiation of the swallow, while esophageal dysphagia arises in the body of the esophagus and relates to difficulty in passing food to the stomach (Aslam & Vaezi, 2013). The elderly appear to be more vulnerable for development of dysphagia due to illnesses that affect the swallowing mechanism (Aslam & Vaezi, 2013), as well as additional stressors such as acute illness or certain medications (Robbins et al., 2006). Irrespective of age, dysphagia is commonly associated with certain diseases such as sarcopenia, the age-related loss of skeletal muscle mass, organization, and strength, cerebrovascular accidents, and Parkinson's disease, all of which increase in prevalence with aging (Aslam & Vaezi, 2013; Rosenberg, 1997). Treatment for dysphagia is usually in two major approaches (Robbins et al., 2006). Rehabilitative interventions have the capacity to directly improve the impaired swallow functions through pharmacological actions on the muscle or neural circuitry (Aslam & Vaezi, 2013; Robbins et al., 2006). Compensatory interventions are viewed as the adjustments to the patient, food/liquid, or environment to help reverse disordered physiology and biomechanics on bolus flow (Robbins et al., 2006; Sura et al., 2012).

Dysphagia can result in reduced or altered oral intake of food/liquid which, in turn, can contribute to lowered nutritional status, and malnutrition can further contribute to decreased functional capacity (Sura et al., 2012). Thus, dysphagia may trigger or promote the frailty process among elderly populations (Serra-Prat et al., 2012). Protein intake is indeed important for elderly people. Although dysphagia may change eating habits, the nutritional requirements are still remain same. Elders must find ways to get enough protein in their diet in spite of swallowing difficulties. Since they may eat less food overall as a result of swallow difficulties, it is important to include foods that are especially high in protein.

3.3.2 Gastric Emptying and Gastric Cancer

Many of the various functions of the stomach have been studied in the aging population (Brown, Levine & Zenilman, 2011). The motor activity of the stomach is essential to digestion in both the fasting and fed states (Brown et al., 2011). Many studies have documented that gastric emptying slows with aging (Brown et al., 2011), but the relation between gastric emptying and aging remains controversial with some reports that emptying is accelerated (Kupfer, Heppell, Haggith, & Bateman, 1985) or unchanged (Madsen & Graff, 2004). These disputed results might be due to different techniques (Brogna, Ferrara, Bucceri, Lanteri, & Catalano, 1999), subject populations (Evans, Triggs, Cheung, Broe, & Creasey, 1981; Kao, Lai, Wang, Chen, & Yeh, 1994), or medical comorbidities (Evans et al., 1981). Despite these limitations, the overall weight of evidence supports the notion that healthy aging is associated with a modest slowing of gastric emptying, which will potentially influence postprandial glycemia (O'Donovan et al., 2005). Other observed gastric motor changes with aging include a reduction in postprandial peristalsis and gastric contractile force (Bitar, Greenwood-Van Meerveld, Saad, & Wiley, 2011; Shimamoto et al., 2002). However, whether these motor changes have any role in the increased prevalence of anorexia and dyspeptic symptoms reported in the elderly remains to be determined (Bitar et al., 2011).

Gastric cancer is one of the most common malignancies in the world and is the second leading cause of cancer-related death worldwide (Ferlay et al., 2010). Because the peak incidence of gastric cancer is among patients older than 50 years, it is considered a disease primarily of the middle-aged and elderly (Bitar et al., 2011). Treatments for elderly gastric cancer patients are controversial. Elderly patients have more comorbidities than nonelderly patients and sugary or conventional chemotherapy could be harmful rather than beneficial in some patients (Park et al., 2016). In recent years, the research and development on bioactive peptides are very active, and there are a variety of therapeutic products available. Su, Dong, et al. (2014) reported that anticancer bioactive peptide, purified from goat spleens immunized with human gastric cancer extracts, significantly inhibited gastric cancer cells in vitro and gastric tumors in vivo via repressing cell growth and promoting apoptosis, making it a promising potential biological anticancer drug (Su, Xin, et al., 2014). In addition, an ephrin-B1 C-terminal peptide showed effects on mediating cell migration and invasion, which is important for cancer development. Injection of this peptide into peritoneal cavity of nude mice suppressed gastric cancer development. These results suggested the possible application of ephrin-B1-derived peptide to be a novel protein therapy for gastric cancer, especially in the stage of tumor progression (Tanaka, Kamata, Yanagihara, & Sakai, 2010).

3.3.3 Malabsorption

Over the course of an individual's life, enormous quantities of nutrients are processed through the small intestine (Brown et al., 2011). Thus, the small intestine is extremely important to maintain human health by up taking enough nutrition and energy. Advanced age is associated with a decrease in body weight, reduced caloric intake, and frequent GI complaints. In Canada, 59% of elderly patients admitted to a tertiary care facility were found to be malnourished or at high risk for malnourishment (Woudstra & Thomson, 2002). Besides the factors mentioned earlier, aging may also result in a decline in nutrient absorption and contributes to poor nutritional status. An age-associated decline in glucose absorption has been found in mice (Ferraris & Vinnakota, 1993), as well as in aging humans (Ferraris & Vinnakota, 1993). However, Wallis et al. argued that age alone does not adversely affect absorption of carbohydrate when renal function and bacterial overgrowth of small intestine are taken into consideration (Arora et al., 1989). Similarly, the age-associated lipid absorption also remains controversial. An increased fatty acid and cholesterol uptake associated with aging have been found in rat (Hollander & Dadufalza, 1983) and human study

(Woudstra & Thomson, 2002), while a decline was reported in rabbit model (Thomson, 1980). Majority of the results suggested that the lipid absorption might decrease in the aged population due to a decrease in intestinal brush border membrane phospholipid composition (Brasitus, Yeh, Holt, & Schachter, 1984), membrane fluidity (Brasitus et al., 1984), and reduced gastric lipase and bile acid secretion (Holt & Balint, 1993). Glutamine is the major metabolic fuel of enterocytes (Thomson, 2009). Supplementing the diet of rats with glutamine (20% of diet protein) intermittently prevented the body weight loss with aging, maintained intestinal mass, but did not prevent sarcopenia (Mignon, Beaufrère, Combaret, & Meynial-Denis, 2007). Thus, treatment with glutamine before advanced age may benefit for maintaining intestine integrity and intestinal immune function (Mignon et al., 2007). Calcium absorption is reduced with aging (Bullamore, Wilkinson, Gallagher, Nordin, & Marshall, 1970) without intervention of vitamin D, and iron absorption may be reduced if there is age-associated hypochlorhydria (Thomson, 2009). Vitamin B12 malabsorption may also happen and will be further aggression with small bowel bacterial contamination syndrome (Kassarjian, 1989).

3.3.4 Liver Dysfunction

The liver is responsible for the synthesis of clotting factors and the metabolism of toxins, including anesthetics, analgesics, and sedatives, all of which are important in elderly surgical patients (Brown et al., 2011). Histology reveals that a loss of mitochondria and smooth endoplasmic reticulum in elderly populations suggested the age-related changes on liver cells (Schmucker, 1998). Meanwhile, the volume and blood flow of the liver is also gradually reduced with aging (Kim, Kisseleva, & Brenner, 2015). The liver volume decreases by 20%–40% when one gets older (Kim et al., 2015). But it is still not clear whether a reduction in liver function with age is attributed to physical changes or age-related cellular changes (Mahon & James, 1994). In addition to the decrease of liver function, aging is associated with severity and poor prognosis of various liver diseases including nonalcoholic fatty liver disease, alcoholic liver disease, hepatitis C, and liver transplantation (Kim et al., 2015). In addition to any changes due solely to aging, the liver is particularly susceptible to the toxic effects of environmental agents such as cigarettes and alcohol (Brown et al., 2011). In dietary level, protein deficiency is often associated with liver disease (Mezey, 1982). Thus, proper uptake of protein plays an important role to maintain liver health. Furthermore, studies also investigated the potential of using novel protein/peptides to prevent liver disease. Shi et al. (2011) investigated a novel whey protein isolate rich in lactoperoxidase, lactoferrin, growth factors, and immunoglobulins, in the prevention of fatty liver formation (Shi et al., 2011). However, a study also indicated that usage of whey protein may cause liver damage via inflammatory and apoptotic responses (Gürgen et al., 2015). This conflict might be because of the difference of treatment duration and concentration.

3.3.5 Constipation

Constipation is one of the most common chronic digestive complaints and challenge for elderly population (Schuster, Kosar, & Kamrul, 2015). About 25% of women and 16% of men aged over 65 suffer from constipation compared to 4% of the younger population and 19% of the middle-aged population (Brown et al., 2011). The reasons for constipation in the elderly are multifactorial. Although constipation is not a physiology consequence of normal aging, aging-associated decrease of mobility, medications, underlying disease, and rectal sensory-motor dysfunction may all contribute to it (Schuster et al., 2015). Meanwhile, low-fiber diet, sedentary lifestyle, and side effect of many drugs are also causes of constipation (Brown et al., 2011). From a pathogenetic point of view, chronic constipation can be characterized into (1) slow transit constipation, (2) outlet obstruction, and (3) constipation in irritable bowel syndrome (IBS) (Schuster et al., 2015). Different types of primary constipation may be present alone or coexist in the same patient (Longstreth et al., 2006), which makes it more difficult to prevent.

It is generally assumed that constipation affects the quality of life for aged people (Bharucha, Pemberton, & Locke, 2013) and may produce serious complications, such as acute mental confusion, urinary retention, urinary incontinence, fecal impaction, stercoral ulceration, and bowel perforation (Brown et al., 2011). Diet and lifestyle modification is traditionally considered as the first step of constipation treatment (Tack et al., 2011). A gradually increasing intake of fluids and fibers up to 30 g/day is suggested (Menees, Guentner, Chey, Saad, & Chey, 2015). Other currently available nonpharmacological treatment options are probiotics (De Giorgio et al., 2015). Use of laxatives will be recommended when simple change of lifestyle and diet does not help (Rao & Go, 2010), although it is not the best option for long-term treatment because of the possible side effects.

3.3.6 Colorectal Cancer

Colorectal cancer (CRC) is the third commonest cancer in men, and the second in women (Kordatou, Kountourakis, & Papamichael, 2014), with about 0.6 million deaths worldwide each year (Weitz, Koch, & Debus, 2005). Age is a major risk factor for CRC. Approximately 70% of cases develop over the age of 65 (Kordatou et al., 2014). Surgery is the most common therapy for CRC treatment (Kordatou et al., 2014); however, it may also bring other problems for geriatric patients. Thus, dietary and lifestyle intervention have been used in CRC prevention in addition to pharmacological therapy.

Protein intake has been considered as a double-edged sword of CRC development. Evidences suggested that long-term consumption of red meat or processed meat may increase CRC risk (Chao et al., 2005; Cross et al., 2010; Gonder & Worm, 2005), but not for white meat intake (Bingham, Hughes, & Cross, 2002). This might be because of the carcinogenic heterocyclic amines produced during cooking (Ferrucci et al., 2012; Sinha et al., 1999) and the endogenous formation of nitroso compounds in GI tract by red meat intake (Bingham et al., 2002). Fish is another recommended protein source with potential CRC prevention effect because it is the main dietary source of long-chain n-3 fatty acids (Kordatou et al., 2014). Some observational studies have found an inverse association between fish consumption and CRC risk (Gonder & Worm, 2005; Hall, Chavarro, Lee, Willett, & Ma, 2008; Jedrychowski, Maugeri, Pac, Sochacka-Tatara, & Galas, 2008), while other did not (Kobayashi et al., 2004; Sugawara et al., 2009). Furthermore, dairy products may also benefit for reducing CRC risk and preventing the CRC development by a variety of mechanisms, such as reducing epithelial cell exposure to toxicity, inhibiting proliferation of intestinal mucosa and epithelial cells (Lipkin & Newmark, 1985).

The disproportionate increase of older people leads to a significant increase of functional GI disorders including dysphagia, reflux-related disorders, GI dysfunction, chronic obstipation, etc. (D'Souza, 2007). Gastroenterology in the geriatric patients remains an important aspect of both clinical medicine and nutritional supplementation. The approaches should include consideration of physiological effects of aging on the gut, and the balance of dietary nutrition to protect gut against aging. Protein intake is one of the major nutrients and plays an important role in maintaining gut health. Although there are some debates on the effect of high protein intake and gut health, adequate protein intake is essential for keeping gut function in general.

3.4 Alzheimer's Disease

Alzheimer's disease (AD) is the most common neurodegenerative disorder in the elderly characterized by progressive memory loss and is becoming increasingly widespread throughout the world (Singh, Srivastav, Yadav, Srikrishna, & Perry, 2016). According to the data from the Alzheimer's Association, it is estimated that there will be 13.2 million American AD patients in 2050 (Singh et al., 2016). Although age is the main cause of AD, factors affecting its development are multifactorial including genetics, comorbidities, education level, etc. (Singh et al., 2016). The alarming feature about AD is that, by the time the disease becomes noticeable, the damage is so profound and devastating that makes the diseases incurable (Sadhukhan, Saha, Dutta, Mahalanobish, & Sil, 2018). In the case of AD, the presence of amyloid β peptide accumulation and neurofibrillary tangles (NFTs) induces apoptosis in the neuronal cells (Celik & Sanlier, 2017); but due to the efficient compensatory mechanisms of the brain, there is no behavioral abnormality until the brain can no longer overcome the damage and the disease gets diagnosed (Sadhukhan et al., 2018). With the increase number of AD patients, current estimation of AD health care cost will be greater than $100 billion per year including both medical and social services (Ernst & Hay, 1997; Singh et al., 2016). Although it is still difficult to absolutely diagnose and cure AD, promising researches and development for early detection and treatment are underway (Singh et al., 2016).

The factors that increase the likelihood of developing AD can be divided into two groups: genetic factors and environmental factors. Although the majority of AD cases are not genetically inherited, researchers have shown that those who have a parent, brother, or sister with AD are more likely to develop the disease than individuals who do not have a first degree relative to AD (Singh et al., 2016). Also, some genes are responsible for AD including β-amyloid precursor proteins (APP), presenilin 1 (PSEN1), and presenilin 2 (PSEN2) (Celik & Sanlier, 2017; Singh et al., 2016). In addition, apolipoprotein E (APOE) ε4 allele is effective in the formation of late-onset AD, while APOE ε4 allele has a protective effect (Celik & Sanlier, 2017; Xia & Mo, 2016). In recent years, with the development of epigenetic studies of AD, DNA methylation, histone modification, and problems with microRNAs have been also found to be associated with AD (Celik & Sanlier, 2017). Compared to genetic factors, environmental factors are more complicated, but are modifiable at least to some extent. Aging is the most important factor for AD. Most patients develop AD after 65 years old and the risk of developing AD reaches 50% for individuals above 85 years old (Singh et al., 2016). Other than aging, diet, chemical exposure, smoking, and traumatic brain injury also belongs to environmental factors (Celik & Sanlier, 2017). Studies suggested that the overall healthy life including eating habit, proper socially and physically actives, and abstinent alcohol and tobacco consumption will provide protection against AD (Singh et al., 2016). Furthermore, nutrients and bioactive compounds found to be related to AD-related factors such as epigenetic, amyloid β peptide (Aβ), tau proteins, oxidative stress, and reactive oxygen species (ROS) (Celik & Sanlier, 2017).

AD is characterized pathologically by plaque formation, which is majorly due to Aβ accumulation in the extracellular, and the NFT formation in the intracellular (Celik & Sanlier, 2017). While Aβ plaques are thought to affect neuron–neuron transport in synapses, NFTs block transport of nutrients and other essential molecules to neurons (Celik & Sanlier, 2017). These are mainly confined in the hippocampus region of the brain and the cerebral cortex, the two large forebrain domains related to memory and other higher cognitive functions and lead to the typical clinical symptoms, for example, memory impairment, general cognitive decline, and personality changes associated with AD (Singh et al., 2016). Although the causes of AD are still rather poorly known, the Aβ cascade and tau protein hyperphosphorylation are the theories that have been widely accepted for AD pathogenesis (Hayes, 2014). Thus, compounds that block the formation of Aβ and NFTs may ultimately be clinically useful for treating AD (Cummings, Pike, Shankle, & Cotman, 1996; Hayes, 2014; Wang et al., 2011).

Aβ is a 39–43 amino acid peptide derived from proteolytic processing of APP and forms filaments and senile plaques in the brain (Singh et al., 2016). It is believed that Aβ42 peptide is the most toxic fragments with the capacity to self-aggregate and form clumps of insoluble fibrils in the brain, thereby triggering the formation of senile plaques (Bayer et al., 2006; Singh et al., 2016). It has been postulated that Aβ42 is mainly responsible for initiating a cascade of events leading to neuronal dysfunction, later followed by death (Singh et al., 2016). NFTs are another important characteristic histopathological feature of AD, which are formed by the hyperphosphorylation of microtubule-associated protein tau (MAPT) (Hayes, 2014; Singh et al., 2016). Accumulation of NFTs may cause dysfunction of the normal cytoskeletal architecture of neurons with subsequent death (Singh et al., 2016). Several enzymes including acetylcholinesterase (AChE), prolyl endopeptidase (PEP), or oligopeptidase (POP), and β-site APP-cleaving enzyme (BACE1) are recognized to involve in the development of AD and thought to be suitable targets for drugs and bioactive compounds for potential prevention of Aβ cascade and tau protein hyperphosphorylation (Meakin et al., 2012; Rosenblum & Kozarich, 2003; Yang, Sun, & Chen, 2012).

BACE1 is a critical enzyme for Aβ biosynthesis whose high expression may lead to an increase of Aβ generation and AD development in human (Ghosh & Osswald, 2014; Hayes, 2014; Yang et al., 2012). Inhibition of BACE1 using natural products has provided a positive result in AD therapeutics, especially phenolic compounds such as catechins (Jeon, Bae, Seong, & Song, 2003), lavandulyl flavanones extracted from *Sophora flavescens* (Hwang et al., 2008), and resveratrol (Lee et al., 2007) that have all shown inhibition activity in vitro and in vivo. Recently, several BACE1 inhibition peptides have been designed and synthesized using "in silico conformational structure-based design" and showed inhibition of Aβ production in vivo (Hamada, Ishiura, & Kiso, 2012; Hamada & Kiso, 2013; Raj et al., 2015). However, with the difficulty to cross the blood–brain barrier and to limit the side effects, animal experiments and clinical trials are required before the BACE1 inhibitors are marketed to treat AD (Hayes, 2014). POP or PEP, the proline-specific endopeptidase expressed in the brain, has effects to cleave neuroactive peptides associated to memory, learning, and neurodegeneration (Hayes, 2014). Abnormal POP activity levels were reported in AD patients, suggesting the relation between POP and AD (Maes et al., 2001). Several natural POP inhibitors have been isolated from diary and fish meat. Peptide fractions derived from cod, salmon, and trout hydrolysates showed a strong POP inhibition effect (Sørensen, Kildal, Stepaniak, Pripp, & Sørhaug, 2004). The POP inhibition activities have also been found in water-soluble extracts of cheeses (Sørensen et al., 2004) and human casein peptide (Asano, Nio, & Ariyoshi, 2014). Recently, a tridecapeptide derived from cocoa processing by-products "barquillo" showed POP inhibition as well as the Aβ plaque formation inhibition and Aβ toxicity (Sila et al., 2015). In addition to senile plaques and NFTs, cholinergic abnormality is another major characteristic of AD (Hayes, 2014). AChE is present in the central nervous system regulating the hydrolysis of acetylcholine to choline, which is implicated in the pathogenesis of AD (Hayes, 2014). Thus, AChE inhibitors are used in the treatment of AD. Although most of AChE inhibitors were phytochemicals extracted from plants, protein hydrolysates are also a source of AChE inhibitory peptides. Ahn et al. found that tuna hydrolysates produced by the commercial enzymes showed excellent AChE inhibitory activities, especially the high molecular weight fractions (Ahn, Lee, & Je, 2010). Bioactive peptides denigrated from Chinese fermented soybean food

"Douchi" and "Sufu" also showed inhibition effects of AChE (Chen, Quan, Cheng, Sun, & Li, 2012; Liu et al., 2009). Rice bran is also considered as a source to produce anti-AD bioactive peptides and rich bran-derived peptides were identified with AChE inhibition activity in Hettiarachchy's group (Kannan, Hettiarachchy, & Mahadevan, 2012). The opioid system is associated to cognitive impairment, hyperphosphorylated tau formation, Aβ production, and neuroinflammation, thereby playing an important role in the development of AD (Cai & Ratka, 2012). Dietary proteins such as wheat gluten and casein are rich sources of bioactive and opioid peptides (Zioudrou, Streaty, & Klee, 1979). Exorphins derived from wheat gluten and β-casomorphine-7 derived from casein showed opioid-inducing effects (Zioudrou et al., 1979) and have potential to cross the GI tract and BBB (Trivedi, Zhang, Lopez-Toledano, Clarke, & Deth, 2016). Colostrinin is a proline-rich polypeptide complex isolated from ovine colostrum and is identified as a psycho-immuno enhancer for AD patients (Froud, Wardhaugh, Banks, Saffrey, & Stewart, 2010; Janusz & Zabłocka, 2013; Lemieszewska et al., 2016). Recently, tablets or capsules containing colostrinin are available in many countries as an OTC dietary supplement (Hayes, 2014; Janusz & Zabłocka, 2013). Bioactive peptides precipitated from fermented milk with *Lactobacillus helveticus* IDCC3801 were found to improve APP metabolism and attenuate memory deficit in vivo (Yeon et al., 2010).

In addition to the abnormal accumulation of Aβ and the deposition of NFTs, AD patients are suffering a significant extent of oxidative damage in the brain (Huang, Zhang, & Chen, 2016). Brain membranes are composed of polyunsaturated fatty acid, which makes the brain particularly vulnerable to radical attacks (Pal, Sinha, & Sil, 2013; Sadhukhan et al., 2018). The oxidation of brain proteins is critical to the function of neuron and glial by affecting two enzymes glutamine synthetase and creatine kinase (Pal et al., 2013; Sadhukhan et al., 2018). Both of these two enzymes are markedly reduced in AD brains (Butterfield et al., 1997; Huang et al., 2016). Alteration of glutamate concentration enhanced the excitotoxicity, and oxidative impairment of creatine kinase caused abnormal energy metabolism (Karthick et al., 2015; Sadhukhan et al., 2018; Swomley et al., 2014). Meanwhile, phosphorylation of oxidation can activate the transcription factor NF-κB and stimulate the hyperphosphorylation of tau proteins (Heppner, Ransohoff, & Becher, 2015; Sadhukhan et al., 2018). Furthermore, oxidation of the brain can directly affect DNA, producing strand breaks, sister chromatid exchange, DNA–protein cross-linking, and base modification (Cooke, Evans, Dizdaroglu, & Lunec, 2003; Huang et al., 2016). Based on these, using nutritionally derived antioxidants has become a new approach to delaying the onset and progression of AD (Behl & Moosmann, 2002). Currently, the most well-studied antioxidants for AD prevention are vitamins (e.g., vitamin E, vitamin C, and β-carotene) (Feng & Wang, 2012), polyphenols (e.g., flavonoid, catechins, and resveratrol) (Feng & Wang, 2012; Rezai-Zadeh et al., 2005), and polyunsaturated fatty acids (Nunomura et al., 2006). In addition to these compounds, a lot of antioxidative proteins and peptides from egg, milk, and plant have been identified with validated antioxidative capacity both in vitro and in vivo. Unfortunately, most of these proteins and peptides have not been studied in direct relation to AD. Ahn et al. (2010) reported that the AD attenuation effect of tuna liver protein hydrolysate might be due to the prevention of free radical formation and oxidative damage to DNA (Ahn et al., 2010). Min et al. (2017) also implied that the antioxidative activity of bovine casein-derived peptide might benefit for AD prevention and administration of bovine casein-derived peptide showed prevention of cognitive decline in the AD mice model (Min et al., 2017).

In the past few decades, researches of AD have been largely driven by the amyloid cascade hypothesis and it is still the major pathogenic concept of AD until now. However, new data from preclinical and clinical studies have established that neuroinflammation and immune system-mediated action also contribute to AD pathogenesis. This theory suggests another novel potential therapeutic target for AD, including microglia and several cytokines (Heppner et al., 2015). Based on current knowledge, innate immune cell such as microglia and macrophages are the prime targets for modulating neuroinflammation (Heppner et al., 2015). The nonsteroidal antiinflammatory drugs (NSAIDs) are developed to reduce the risk of AD (Breitner & Zandi, 2001; Heppner et al., 2015; in t' Veld et al., 2001), but there is lack of success in ameliorating AD by the use of broad antiinflammatory drugs (Heppner et al., 2015). Beside of directly intervention the innate immune cells, various immune molecules and related signaling pathway are also considered as AD prevention targets. IL-6, IL-12, and IL-23 all work as part of the Aβ-driven inflammatory response and have been shown to increase in AD patients (Berg et al., 2012; Guerreiro et al., 2007; Heppner et al., 2015). Ablation of IL-6, IL-23, or the IL-12 receptor in an Aβ-overexpressing transgenic mouse model of AD resulted in a substantial reduction in cerebral Aβ burden (Chiang & Koo, 2014; Griffin, 2013; Heppner et al., 2015; Lo, Ho, Cummings, & Kosik, 2014). In addition, CD33 (Bradshaw et al., 2013), CD36 (El Khoury et al., 2003), CD14 (Fassbender et al., 2004), and PPARγ/RXRα (Yamanaka et al., 2012) all play an important role in Aβ formation, phagocytosis, and clearance, which could be considered as a target for AD prevention. Unfortunately, some antiinflammatory drugs have been development, but most of them have not been directly studied for AD prevention. But the use of these bioactive compounds in preventing other diseases suggests that their antiinflammatory ability might work for AD prevention as well. The peptide lactoferricin (residues 17–41 of lactoferrin) suppressed IL-6 production by a human monocyte cell line following stimulation with LPS (Mattsby-Baltzer et al., 1996). A similar effect has been observed for LF11 and lauryl-LF11, peptides derived from human lactoferricin (Andrä et al., 2005). Hen egg lysozyme supplementation significantly reduced intestinal expression of proinflammatory cytokines IL-6 in a porcine model of dextran sodium sulfate-induced colitis (Lee et al., 2009). However, therapy use interference with neuroinflammatory pathways and molecules requires price knowledge about the underlying immune events, which may change during the disease course and most of hypotheses are still under study and validation.

Although researches around the world are focusing on finding treatments, preventive strategies, there is still no cure of AD yet. Drug treatments are available to help relieve symptoms and to counterbalance the neurotransmitter disturbance, but not ultimately therapy (Yiannopoulou & Papageorgiou, 2013). Cholinesterase inhibitors (CIs), *N*-methyl-D-aspartate (NMDA) antagonist, and placebo-controlled trials (RCTs) are currently available and have been approved for the treatment of mild to moderate AD (Yiannopoulou & Papageorgiou, 2013). Based on the Aβ and NTF theory of AD development, "disease-modifying drugs" that capable to stop or at least effectively modify the course of AD have been proposed and under extensive research (Yiannopoulou & Papageorgiou, 2013). Drugs "Tramiprosate," "Colostrinin," "Scyllo-inositol," "Semagacestat," and "Tarenflurbil" targeting to modulate Aβ deposition have been in phase II or III human trial; and drug "Methylene blue," "Lithium," and vaccination aiming to tau deposition modulation are also in clinical or preclinical trials (Yiannopoulou & Papageorgiou, 2013). Meanwhile, treatments modulating inflammation and oxidative damage are also another popular area of "disease-modifying treatments" (Yiannopoulou & Papageorgiou, 2013). Unfortunately, along with Aβ formation, NTF accumulation, oxidative damage, and inflammation, a variety of mechanisms and aging-related disorder can also lead to the generation of AD, such as cholesterol and cardiovascular-related risk. Thus, besides pharmaceutical strategies, nutrition-target treatment became a new trend; especially human nutrition and dietary influences are known to affect brain chemistry, sleeping patterns, mood, and overall mental health

(Hayes, 2014). Several researches have suggested that nutraceutical compounds have the potential to bring changes in brain structure, plasticity, and physiology (Murphy, Dias, & Thuret, 2014), and influence brain health due to the anti-oxidant, antiinflammatory activities, as well as improvement in glycemic control (Selhub, Logan, & Bested, 2014). In the past decades, studies conducted to investigate the potential benefit of nutraceuticals on AD prevention are about polyphenols, vitamins, and unsaturated fatty acids (Sadhukhan et al., 2018). Currently, bioactive protein and peptides are of interest for potential use in the prevention of mental health disorders such as AD. Opioid peptides play an active role in brain health as they are active in the nervous system and are pharmacologically similar to opium (Cai & Ratka, 2012). Opioid peptides are similar to enkephalins as both have affinity for opiate receptor and display opiate-like effects (Cai & Ratka, 2012). Most of these peptides are obtained by the cleavage of milk proteins and others contained in common ailments like cereals (Huebner, Lieberman, Rubino, & Wall, 1984), maize (Zioudrou et al., 1979), rice (Takahashi, Moriguchi, Yoshikawa, & Sasaki, 1994), eggs (Zioudrou et al., 1979), soybean (Ohinata, Agui, & Yoshikawa, 2014), meat (Nyberg, Sanderson, & Glämsta, 1997; Zhao, Garreau, Sannier, & Piot, 1997), and spinach (Stefanucci et al., 2017). Many milk protein hydrolysates, such as β-casein, κ-casein, α-lactalbumin, β-lactoglobulin, and lactoferrin, have been found to possess both opioid agonist and antagonist activities (Zioudrou et al., 1979). Among all, the most extensively studied opioid peptides are fragments of milk β-casein called "β-casomorphins." Trivedi et al. reported that the hydrolytic digestion of casein and gliadin (a wheat-derived protein) releases peptides with opioid activity, and modulated cysteine uptake in cultured human neuronal and GI epithelial cell via activation of opioid receptors (Trivedi et al., 2016). So far, the clinical studies with nutraceuticals are mostly unproductive. Several cognitive performances and behavioral activities in experimental murine indicate that administration of nutraceuticals can significantly attenuate different prevalent biomarkers of AD (Mecocci, Tinarelli, Schulz, & Polidori, 2014). But some animal experiments rejected the hypothesis that administration of nutraceuticals can help with the decline of cognitive and behavioral activities. Actually, the difficulty of AD prevention and treatment is not only because of the poor understanding of AD mechanisms, but also the lack of natural disease animal models (Singh et al., 2016). Although animal models such as dog, rodents, and primates have been used in AD researches, there are still a lot of restrictions that limit the progress of AD program. Thus, although several drugs or nutraceuticals have claimed the effects in animal model, failed in the clinical trails of AD patients (Cummings, Morstorf, & Zhong, 2014; Singh et al., 2016).

With the world going to an aging era, AD has become a serious neurodegenerative disease in the elderly. It not only costs a large amount of health care, and proper therapeutic options are not available yet. Various therapeutic drugs have been studied and investigated, but with limitations. Bioactive components and different nutraceutical ingredients have been taken into consideration to eliminate the detrimental effects of AD at all its stages. Although nutraceuticals might not be enough to cure AD, combination therapy consisting of both drug targeting and diet modulation may be a way to substantially delay the progression and attenuate the development of AD ultimately.

3.5 Cardiovascular Disease

Population aging is associated with the increasing burden of non-communicable diseases such as coronary heart disease (CHD), hypertension, and type 2 diabetes mellitus (T2DM). The prevalence of obesity and diabetes is considerably more among adults aged 65 years and older. Elderly people with obesity and diabetes are at greater risk for developing cardiovascular disease (CVD) and other associated complications. Hyperglycemia, obesity, dyslipidemia, insulin resistance, inflammation, hypertension, and reduced vascular responsiveness are the major known contributors to CVD risk in diabetes (Halter et al., 2014). American Heart Association documented that 68% of elderly people (aged 65 or older) with diabetes die from some types of heart diseases. CVDs cause 17.7 million deaths every year, which accounts for about 31% of all deaths worldwide. CVD is the most important cause of mortality and morbidity in elderly people. The burden of CVDs is huge (30.3%) in older people (aged 60 years and older) followed by cancer (15.1%), pulmonary diseases (9.5%), musculoskeletal diseases (7.5%), and neurological disorders (6.6%) (Prince et al., 2015). The elderly population living with diabetes and CVD has increased significantly in recent years in many countries as the population continues to age and steady enhancement of life expectancy of people living in developing countries.

Hypertension (high blood pressure) has been known as a main risk factor for CVD. Hypertension and insulin resistance are positively associated in elderly population. Elderly people with both hypertension and diabetes have double risk for CVD. Studies reported that high LDL cholesterol, low HDL cholesterol, and high triglycerides levels are often associated with elderly people with diabetes. These abnormal levels of lipids can be risk factors for developing CVD. Diabetic dyslipidemia is a commonly associated disorder with insulin resistance. Studies indicate that cholesterol lowering significantly improves cardiovascular outcomes in diabetes (Schofield, Liu, Rao-Balakrishna, Malik, & Soran, 2016). Another important risk factor for CVD is obesity, which is strongly linked with insulin resistance and high blood pressure. Inflammation and oxidative stress have been reported to play an important role in the underlying mechanisms of aging, diabetes, CVD, and related complications. Increased oxidative stress in elderly is associated with reduced protein synthesis, DNA damage, and degradation that can impair the functional capacity of the body organs.

Overweight older adults with type 2 diabetes have central adiposity and insulin resistance, which enhance dyslipidemia and atherogenesis. Excess fat deposition in the liver stimulates over synthesis of inflammatory proteins that can modify hepatic lipase, cholesterol ester transfer proteins, and lipoprotein lipase. These alternations could result in the formation of functionally damaged LDL and HDL lipoproteins. The damaged LDLs are accumulated into the vessel wall and can enhance the formation of foam cell and atherogenesis. The abnormal HDLs show decreased functional capacities such as cholesterol efflux, antioxidative, antiinflammatory, and antiapoptotic ability, which are important for the stimulation of macrophages and endothelial and smooth muscle cells. The abnormal functions of HDL and LDL may contribute to β-cell failure by enhancing inflammation, amyloid deposition, and increased ROS. These abnormalities finally lead to atherogenesis and plaque formation (Halter et al., 2014). Fig. 1 shows the aging-related changes that are linked to risk factors for diabetes and CVD.

It has been apparent that dietary habits influence CVD risk factors such as obesity, blood pressure (BP), low-density lipoprotein (LDL) cholesterol, glucose–insulin homeostasis, inflammation, oxidative stress, and endothelial function. Protein is an essential macronutrient that is required for the growth and physiological functions of the human body. Even though protein intake is vital in all age groups, depending on the age group its requirement is different to meet the body functions. BCAAs such as leucine, isoleucine, and valine are positively associated with insulin resistance. Higher protein intake may reduce the risk of developing diabetes through the weight loss. However, results of cohort studies indicate that higher protein consumption is associated with an increased risk of type 2 diabetes (T2DM) (Beasley & Wylie-Rosett, 2013). The relative risk (95% CI) of T2DM is 1.22, 1.39, 1.03, 1.03, 0.89, 0.83, and 0.74 for red meat, processed meat, fish, egg, total dairy products, yogurt, and soy, respectively. Studies indicate that total

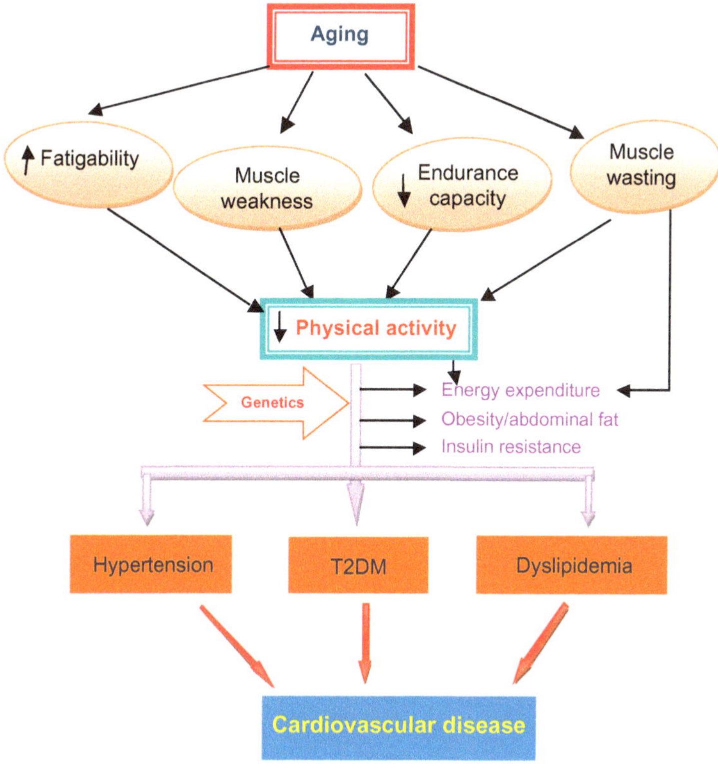

Fig. 1 Aging-related changes linked to muscle, risks for diabetes, and CVD. *T2DM*, type 2 diabetes mellitus. *Modified from Nair, K. S. (2005). Aging muscle. American Journal of Clinical Nutrition, 81, 953–963.*

protein and animal protein consumption could enhance the risk of T2DM in humans. However, plant proteins are inversely associated with the risk of T2DM in females (Tian et al., 2017).

On the other hand, protein-rich diet has a significant impact on CVD through the influence of several risk factors such as elevated blood cholesterol and LDL cholesterol, raised triglycerides, low HDL cholesterol, high blood pressure, and overweight. There has been considerable evidence that consumption of protein can lower the risk of CVD. Several studies have investigated the effects of dietary proteins on CVD in elderly people. Consuming protein-rich food is an important for elderly people with CVD to effectively survive morbidities. Protein nutrition enhances the lean mass gain in older adults with CVD. Table 1 shows the potential mechanisms by which dietary proteins have positive effects on CVD.

Table 1 Potential Mechanisms of Action of Dietary Proteins Through the Modulation of Several Risk Factors for Cardiovascular Disease (Preis, Stampfer, Spiegelman, Willett, & Rimm, 2010)

S. No.	Potential Mechanisms of Action
(1)	Decrease of total cholesterol
(2)	Reduction of total LDL cholesterol
(3)	Decrease of triglycerides
(4)	Increase of HDL cholesterol
(5)	Weight loss
(6)	Decrease in blood pressure

Soy protein is a complete protein as it contains all the essential amino acids that are required for human health and it has been consumed in several countries as a staple food. A population-based longitudinal study in elderly Chinese postmenopausal women revealed that normal consumption of soy protein food was inversely linked with both systolic and diastolic BP. Daily intake of soy protein $\geq$25 g in elderly women aged >60 years linked with a reduction of 4.9 mmHg (−8.0, −1.9 mmHg) in systolic BP and of 2.2 mmHg (−3.8, −0.6 mmHg) in diastolic BP (Yang et al., 2005). The American Heart Association together with the US Food and Drug Administration recommended a daily consumption of $\geq$25 g soy protein to lessen the risk of CVD (Erdman, 2000). The higher ratio of arginine and lysine in soy has been reported to lower insulin and glucagon secretion that would subsequently reduce lipogenesis (Potter, 1995). Additionally, it has been proved in several clinical studies that storage proteins of soy such as globulins 11S and 7S can upregulate LDL receptors in humans (Sirtori et al., 1995).

The consumption of certain dairy products is associated with reduced risk of CVD. Yogurt is a fermented food of milk and has been beneficially associated with CVD. High consumption (>100 g yogurt/day) of yogurt in Australian elderly women (70 years) showed a significantly lower common carotid artery intima-media thickness (CCA-IMT), which is a prognostic of stroke and coronary deaths, and higher HDL cholesterol (Ivey et al., 2011). However, the intake of other dairy products such as milk and cheese has not shown significant benefits on CVD and its risk factors in elderly. A cohort study conducted in older Australians reported that milk consumption (250 mL) had no consistent association with the risk of CHD, stroke, and combined CVD mortality (Louie et al., 2013).

Egg provides several important nutrients for human health including high-quality protein with all essential amino acids. In the past, egg consumption was thought to be negatively associated with CVD and its risk factors. Recently, several intervention studies made it clear that up to seven eggs/week can safely be consumed by healthy adults including elderly without risk of CVD (Fuller, Sainsbury, Caterson, & Markovic, 2015; Geiker, Lytken Larsen, Dyerberg, Stender, & Astrup, 2018). In a study of older adults (>60 years of age), consumption of one egg/day for 5 weeks showed no significant increase in serum LDL cholesterol or HDL cholesterol (Goodrow et al., 2006). A randomized crossover study conducted in 42 healthy elderly population (29 postmenopausal women and 13 men, 60 years old) revealed that three eggs consumption per day for 1 month period did not increase the risk for CVD (Greene et al., 2005).

Fish is a major source of dietary protein for human as it contains high quantity of quality protein with essential amino acids. Consumption of tuna or other broiled or baked fish in elderly is associated with lower risk of fatal CHD. A study conducted in 4775 US elderly adults aged 65–98 years suggested that intake of tuna/other fish five or more times per week has been associated with 28% lower risk of total stroke as compared to consumption of less than once per month. However increasing consumption of fried fish/fish sandwich is also associated with higher risk of total stroke and ischemic stroke in elderly. Intake of fried fish/fish sandwich once per week is associated with higher risk of total stroke (37%) and ischemic stroke (44%), in elderly, as compared with an eating of <1 time/month (Mozaffarian et al., 2005).

In addition to food proteins, amino acid supplements have also showed benefits in elderly people with CVD. A randomized, double-blind, placebo-controlled study performed in a group of 95 outpatients (aged 65–74 years) with chronic heart failure indicated that oral amino acid supplementation twice a day (8 g amino acids per day) for 30 days, in combination with pharmacologic treatment, had improved exercise capacity by improving circulatory function and muscle oxygen consumption (Aquilani et al., 2008; Bauer et al., 2013). Supplementation of a normal diet with essential amino acids and arginine (11 g/twice a day) for 16 weeks in elderly people (67 years mean age) is associated with reduced plasma and liver triglycerides, total cholesterol, and very low density lipoprotein (VLDL)-cholesterol concentrations (Borsheim et al., 2009). However, the exact mechanism of action of amino acid supplementation on CVD and its risk factors is not fully understood.

4. SUMMARY

Aging adversely affects all components of human biological functions and increases the risk for development of aging-related chronic diseases. Protein malnutrition of the elderly is a growing concern, given the fact that their protein requirements are higher than current RAD for healthy people older than 19 years. The role of food protein in preventing skeletomuscular disorders, especially sarcopenia, the loss of muscle for the elderly, has been well established. Current evidence also supports a role of food protein on osteoporosis. In addition, there are bioactive proteins in food proteins that stimulate bone formation in addition to their nutritional function. Furthermore, bioactive peptides derived from food proteins show great potential against CVD, hypertension, and obesity; their prevalence is increasing substantially at the elderly population. Food proteins and peptides are also suggested to play important roles in elder's mental health, such as anorexia and AD. Although it is not well established yet and most of researches are still ongoing, colostrinin extracted from colostrum is available as an OTC dietary supplementation in many countries.

REFERENCES

Ahn, C. B., Lee, K. H., & Je, J. Y. (2010). Enzymatic production of bioactive protein hydrolysates from tuna liver: Effects of enzymes and molecular weight on bioactivity. *International Journal of Food Science and Technology*, *45*(3), 562–568.

Andrä, J., Lohner, K., Blondelle, S. E., Jerala, R., Moriyon, I., Koch, M. H. J., et al. (2005). Enhancement of endotoxin neutralization by coupling of a C12-alkyl chain to a lactoferricin-derived peptide. *The Biochemical Journal*, *385*(1), 135–143.

Aquilani, R., Viglio, S., Iadarola, P., Opasich, C., Testa, A., Dioguardi, F. S., et al. (2008). Oral amino acid supplements improve exercise capacities in elderly patients with chronic heart failure. *American Journal of Cardiology*, *101*, 104E–110E.

Arora, S., Kassarjian, Z., Krasinski, S. D., Croffey, B., Kaplan, M. M., & Russell, R. M. (1989). Effect of age on tests of intestinal and hepatic function in healthy humans. *Gastroenterology*, *96*(6), 1560–1565.

Asano, M., Nio, N., & Ariyoshi, Y. (2014). Inhibition of prolyl endopeptidase by synthetic peptide fragments of human β-casein. *Agricultural and Biological Chemistry*, *55*(3), 825–828.

Aslam, M., & Vaezi, M. F. (2013). Dysphagia in the elderly. *Gastroenterology & Hepatology*, *9*(12), 784–795.

Attie, K. M., Borgstein, N. G., Yang, Y., Condon, C. H., Wilson, D. M., Pearsall, A. E., et al. (2013). A single ascending-dose study of muscle regulator ace-031 in healthy volunteers. *Muscle & Nerve*, *47*(3), 416–423.

Barczi, S. R., Sullivan, P. A., & Robbins, J. (2000). How should dysphagia care of older adults differ? Establishing optimal practice patterns. *Seminars in Speech and Language*, *21*(4), 0347–0364.

Barzel, U. S., & Massey, L. K. (1998). Excess dietary protein can adversely affect bone. *Journal of Nutrition*, *128*(6), 1051–1053.

Bauer, J. I., Biolo, G., Cederholm, T., Cesari, M., Cruz-Jentoft, A. J., Morley, J. E., et al. (2013). Evidence-based recommendations for optimal dietary protein intake in older people: A position paper from the prot-age study group. *Journal of the American Medical Directors Association*, *14*, 542–559.

Bauer, J. M., Verlaan, S., Bautmans, I., Brandt, K., Donini, L. M., Maggio, M., et al. (2015). Effects of a vitamin D and leucine-enriched whey protein nutritional supplement on measures of sarcopenia in older adults, the PROVIDE study: A randomized, double-blind, placebo-controlled trial. *Journal of the American Medical Directors Association*, *16*(9), 740–747.

Baumgartner, R. N., Waters, D. L., Gallagher, D., Morley, J. E., & Garry, P. J. (1999). Predictors of skeletal muscle mass in elderly men and women. *Mechanisms of Ageing and Development*, *107*(2), 123–136.

Bayer, T. A., Wirths, O., Majtényi, K., Hartmann, T., Multhaup, G., Beyreuther, K., et al. (2006). Key factors in Alzheimer's disease: β-Amyloid precursor protein processing, metabolism and intraneuronal transport. *Brain Pathology*, *11*(1), 1–11.

Beasley, J. M., Shikany, J. M., & Thomson, C. A. (2013). The role of dietary protein intake in the prevention of sarcopenia of aging. *Nutrition in Clinical Practice*, *28*(6), 684–690.

Beasley, J. M., & Wylie-Rosett, J. (2013). The role of dietary proteins among persons with diabetes. *Current Atherosclerosis Reports*, *15*(9), 348–359.

Behl, C., & Moosmann, B. (2002). Antioxidant neuroprotection in Alzheimer's disease as preventive and therapeutic approach. *Free Radical Biology & Medicine*, *33*(2), 182–191.

Berg, V. J., Prokop, S., Miller, K. R., Obst, J., Kälin, R. E., Lopategui-Cabezas, I., et al. (2012). Inhibition of IL-12/IL-23 signaling reduces Alzheimer's disease-like pathology and cognitive decline. *Nature Medicine*, *18*(12), 1812–1819.

Bharadwaj, S., Naldu, A. G., Betageri, G. V., Prasadarao, N. V., & Naldu, A. S. (2009). Milk ribonuclease-enriched lactoferrin induces positive effects on bone turnover markers in postmenopausal women. *Osteoporosis International*, *20*(9), 1603–1611.

Bharucha, A. E., Pemberton, J. H., & Locke, G. R. (2013). American gastroenterological association technical review on constipation. *Gastroenterology*, *144*(1), 218–238.

Bhasin, S., Woodhouse, L., & Storer, T. W. (2001). Proof of the effect of testosterone on skeletal muscle. *The Journal of Endocrinology*, *170*(1), 27–38.

Bingham, S. A., Hughes, R., & Cross, A. J. (2002). Effect of white versus red meat on endogenous N-nitrosation in the human colon and further evidence of a dose response. *The Journal of Nutrition*, *132*(11 Suppl), 3522S–3525S.

Bitar, K., Greenwood-Van Meerveld, B., Saad, R., & Wiley, J. W. (2011). Aging and gastrointestinal neuromuscular function: Insights from within and outside the gut. *Neurogastroenterology and Motility: The Official Journal of the European Gastrointestinal Motility Society*, *23*(6), 490–501.

Bonjour, J. P. (2011). Protein intake and bone health. *International Journal for Vitamin and Nutrition Research*, *81*(2–3), 134–142.

Borack, M. S., & Volpi, E. (2016). Efficacy and safety of leucine supplementation in the elderly. *The Journal of Nutrition*, *146*(12), 2625S–2629S.

Borsheim, E., Bui, Q. T., Tissier, S., Cree, M. G., Rønsen, O., Morio, B., et al. (2009). Amino acid supplementation decreases plasma and liver triglycerides in elderly. *Nutrition*, *25*, 281–288.

Boskey, A. L., & Coleman, R. (2010). Aging and bone. *Journal of Dental Research*, *89*(12), 1333–1348.

Boutrou, R., Henry, G., & Sanchez-Rivera, L. (2015). On the trail of milk bioactive peptides in human and animal intestinal tracts during digestion: A review. *Dairy Science & Technology*, *95*(6), 815–829.

Bradshaw, E. M., Chibnik, L. B., Keenan, B. T., Ottoboni, L., Raj, T., Tang, A., et al. (2013). CD33 Alzheimer's disease locus: Altered monocyte function and amyloid biology. *Nature Neuroscience, 16*(7), 848–850.

Brasitus, T. A., Yeh, K. Y., Holt, P. R., & Schachter, D. (1984). Lipid fluidity and composition of intestinal microvillus membranes isolated from rats of different ages. *Biochimica et Biophysica Acta (BBA)—Biomembranes, 778*(2), 341–348.

Breitner, J. C., & Zandi, P. P. (2001). Do nonsteroidal antiinflammatory drugs reduce the risk of Alzheimer's disease? *The New England Journal of Medicine, 345*(21), 1567–1568.

Brogna, A., Ferrara, R., Bucceri, A. M., Lanteri, E., & Catalano, F. (1999). Influence of aging on gastrointestinal transit time. An ultrasonographic and radiologic study. *Investigative Radiology, 34*(5), 357–359.

Brown, N. A., Levine, J. L., & Zenilman, M. E. (2011). Age-related changes in gastrointestinal tract. In R. Rosenthal, M. Zenilman, & M. Katlic (Eds.), *Principles and practice of geriatric surgery*. New York, NY: Springer.

Bullamore, J. R., Wilkinson, R., Gallagher, J. C., Nordin, B. E., & Marshall, D. H. (1970). Effect of age on calcium absorption. *The Lancet, 2*(7672), 535–537.

Burris, R. L., Ng, H. P., & Nagarajan, S. (2014). Soy protein inhibits inflammation-induced VCAM-1 and inflammatory cytokine induction by inhibiting the NF-κB and AKT signaling pathway in apolipoprotein E-deficient mice. *European Journal of Nutrition, 53*(1), 135–148.

Butterfield, D. A., Hensley, K., Cole, P., Subramaniam, R., Aksenov, M., Aksenova, M., et al. (1997). Oxidatively induced structural alteration of glutamine synthetase assessed by analysis of spin label incorporation kinetics: Relevance to Alzheimer's disease. *Journal of Neurochemistry, 68*(6), 2451–2457.

Cai, Z., & Ratka, A. (2012). Opioid system and Alzheimer's disease. *Neuromolecular Medicine, 14*(2), 91–111.

Callen, B. L. (2010). Nutritional screening in community dwelling older adults. *International Journal of Older People Nursing, 6*(4), 272–281.

Campbell, W. W., Crim, M. C., Dallal, G. E., Young, V. R., & Evans, W. J. (1994). Increased protein requirements in elderly people: New data and retrospective reassessments. *The American Journal of Clinical Nutrition, 60*(4), 501–509.

Celik, E., & Sanlier, N. (2017). Effects of nutrient and bioactive food components on Alzheimer's disease and epigenetic. *Critical Reviews in Food Science and Nutrition, 11*, 1–12.

Chao, A., Thun, M. J., Connell, C. J., McCullough, M. L., Jacobs, E. J., Flanders, W. D., et al. (2005). Meat consumption and risk of colorectal cancer. *JAMA, 293*(2), 172–182.

Chatterton, D. E., Nguyen, D. N., Bering, S. B., & Sanglld, P. T. (2013). Anti-inflammatory mechanisms of bioactive milk proteins in the intestine of newborns. *International Journal of Biochemistry and Cell Biology, 45*(8), 1730–1747.

Chen, J., Quan, M. H., Cheng, Y. Q., Sun, J., & Li, L. T. (2012). Acetylcholinesterase inhibitory activity of Chinese sufu (fermented tofu) ethanol-extract. *Food Chemistry, 134*(3), 1263–1266.

Chernoff, R. (2004). Protein and older adults. *Journal of the American College of Nutrition, 23*(6 Suppl), 627S–630S.

Chiang, K., & Koo, E. H. (2014). Emerging therapeutics for Alzheimer's disease. *Annual Review of Pharmacology and Toxicology, 54*(1), 381–405. https://doi.org/10.1146/annurev-pharmtox-011613-135932.

Cooke, M. S., Evans, M. D., Dizdaroglu, M., & Lunec, J. (2003). Oxidative DNA damage: Mechanisms, mutation, and disease. *The FASEB Journal: Official Publication of the Federation of American Societies for Experimental Biology, 17*(10), 1195–1214.

Covinsky, K. (2006). Aging, arthritis, and disability. *Arthritis and Rheumatism, 55*(2), 175–176.

Cramer, J. T., Cruz-Jentoft, A. J., Landi, F., Hickson, M., Zamboni, M., Pereira, S. L., et al. (2016). Impacts of high-protein oral nutritional supplements among malnourished men and women with sarcopenia: A multicenter, randomized, double-blinded, controlled trial. *Journal of the American Medical Directors Association*, *17*(11), 1044–1055.

Cross, A. J., Ferrucci, L. M., Risch, A., Graubard, B. I., Ward, M. H., Park, Y., et al. (2010). A large prospective study of meat consumption and colorectal cancer risk: An investigation of potential mechanisms underlying this association. *Cancer Research*, *70*(6), 2406–2414.

Cummings, J. L., Morstorf, T., & Zhong, K. (2014). Alzheimer's disease drug-development pipeline: Few candidates, frequent failures. *Alzheimer's Research & Therapy*, *6*(4), 37.

Cummings, B. J., Pike, C. J., Shankle, R., & Cotman, C. W. (1996). Beta-amyloid deposition and other measures of neuropathology predict cognitive status in Alzheimer's disease. *Neurobiology of Aging*, *17*(6), 921–933.

De Giorgio, R., Ruggeri, E., Stanghellini, V., Eusebi, L. H., Bazzoli, F., & Chiarioni, G. (2015). Chronic constipation in the elderly: A primer for the gastroenterologist. *BMC Gastroenterology*, *15*(1), 130.

Demontiero, O., Vidal, C., & Duque, G. (2012). Aging and bone loss: New insights for the clinician. *Therapeutic Advances in Musculoskeletal Disease*, *4*(2), 61–76.

Dempster, D. W., & Raisz, L. G. (2015). Bone physiology: Bone cells, modeling, and remodeling. In M. F. Holick & J. W. Nieves (Eds.), *Nutrition and bone health* (2nd ed., pp. 37–56). New York, NY: Humana Press.

D'Souza, A. L. (2007). Ageing and the gut. *Postgraduate Medical Journal*, *83*(975), 44–53.

El Khoury, J. B., Moore, K. J., Means, T. K., Leung, J., Terada, K., Toft, M., et al. (2003). CD36 mediates the innate host response to beta-amyloid. *The Journal of Experimental Medicine*, *197*(12), 1657–1666.

Erdman, J. W. (2000). Soy protein and cardiovascular disease. *Circulation*, *102*, 2555–2559.

Ernst, R. L., & Hay, J. W. (1997). Economic research on Alzheimer disease: A review of the literature. *Alzheimer Disease and Associated Disorders*, *11*(Suppl. 6), 135–145.

Evans, M. A., Triggs, E. J., Cheung, M., Broe, G. A., & Creasey, H. (1981). Gastric emptying rate in the elderly: Implications for drug therapy. *Journal of the American Geriatrics Society*, *29*(5), 201–205.

Fassbender, K., Walter, S., Kühl, S., Landmann, R., Ishii, K., Bertsch, T., et al. (2004). The LPS receptor (CD14) links innate immunity with Alzheimer's disease. *The FASEB Journal: Official Publication of the Federation of American Societies for Experimental Biology*, *18*(1), 203–205.

Feng, Y., & Wang, X. (2012). Antioxidant therapies for Alzheimer's disease. *Oxidative Medicine and Cellular Longevity*, *2012*(3), 472932. 17 pages.

Ferlay, J., Shin, H. R., Bray, F., Forman, D., Mathers, C., & Parkin, D. M. (2010). Estimates of worldwide burden of cancer in 2008: GLOBOCAN 2008. *International Journal of Cancer*, *127*(12), 2893–2917.

Fernandes, G., Lawrence, R., & Sun, D. (2003). Protective role of n-3 lipids and soy protein in osteoporosis. *Prostaglandins, Leukotrienes, and Essential Fatty Acids*, *68*(6), 361–372.

Ferraretto, A., Signorile, A., Gravaghi, C., Fiorilli, A., & Tettamanti, G. (2001). Casein phosphopeptides influence calcium uptake by cultured human intestinal HT-29 tumor cells. *Journal of Nutrition*, *131*(6), 1655–1661.

Ferraris, R. P., & Vinnakota, R. R. (1993). Regulation of intestinal nutrient transport is impaired in aged mice. *The Journal of Nutrition*, *123*(3), 502–511.

Ferrucci, L. M., Sinha, R., Huang, W. Y., Berndt, S. I., Katki, H. A., Schoen, R. E., et al. (2012). Meat consumption and the risk of incident distal colon and rectal adenoma. *British Journal of Cancer*, *106*(3), 608–616.

Feskanich, D., Willett, W. C., Stampfer, M. J., & Coldltz, G. A. (1996). Protein consumption and bone fractures in women. *American Journal of Epidemiology*, *143*(5), 472–479.

Fitzcharles, M. A., Lussier, D., & Shir, Y. (2010). Management of chronic arthritis pain in the elderly. *Drugs & Aging*, *27*(6), 471–490.

Froud, K. E., Wardhaugh, T., Banks, D., Saffrey, M. J., & Stewart, M. G. (2010). Colostrinin alleviates amyloid-beta induced toxicity in rat primary hippocampal cultures. *Journal of Alzheimer's Disease*, *20*(2), 423–426.

Fuller, N. R., Sainsbury, A., Caterson, I. D., & Markovic, T. P. (2015). Egg consumption and human cardio-metabolic health in people with and without diabetes. *Nutrients*, *7*, 7399–7420.

Gauthier, S. F., Pouliot, Y., & Aubois, J. (2006). Growth factors from bovine milk and colostrum: Composition, extraction and biological activities. *Lait*, 99–125.

Geiker, N. R. W., Lytken Larsen, M., Dyerberg, J., Stender, J., & Astrup, A. (2018). Egg consumption, cardiovascular diseases and type 2 diabetes. *European Journal of Clinical Nutrition*, *72*, 44–56.

Gentili, C., Dolina, R., Bet, P., Campanlle, G., Colombatti, A., Cancedda, F. D., et al. (1994). Ovotransferrin and ovotransferrin receptor expression during chondrogenesis and endochondral bone formation in developing chick embryo. *Journal of Cell Biology*, *124*(4), 579–588.

Ghosh, A. K., & Osswald, H. L. (2014). BACE1 (β-secretase) inhibitors for the treatment of Alzheimer's disease. *Chemical Society Reviews*, *43*(19), 6765–6813.

Gonder, U., & Worm, N. (2005). Meat, fish, and colorectal cancer risk: The European prospective investigation into cancer and nutrition. *Journal of the National Cancer Institute*, *97*(23), 1788. author reply 1788–1789.

Goodrow, E. F., Wilson, T. A., Houde, S. C., Vishwanathan, R., Scollin, P. A., Handelman, G., et al. (2006). Consumption of one egg per day increases serum lutein and zeaxanthin concentrations in older adults without altering serum lipid and lipoprotein cholesterol concentrations. *Journal of Nutrition*, *136*, 2519–2524.

Greene, C. M., Zern, T. L., Wood, R. J., Shrestha, S., Aggarwal, D., Sharman, M. J., et al. (2005). Maintenance of the LDL cholesterol:HDL cholesterol ratio in an elderly population given a dietary cholesterol challenge. *The Journal of Nutrition*, *135*, 2793–2798.

Griffin, W. S. T. (2013). Neuroinflammatory cytokine signaling and Alzheimer's disease. *The New England Journal of Medicine*, *368*(8), 770–771.

Grizzuti, K., & Perlmann, G. E. (1973). Binding of magnesium and calcium-ions to phosphoglycoprotein phosvitin. *Biochemistry*, *12*(22), 4399–4403.

Guerreiro, R. J., Santana, I., Brás, J. M., Santiago, B., Paiva, A., & Oliveira, C. (2007). Peripheral inflammatory cytokines as biomarkers in Alzheimer's disease and mild cognitive impairment. *Neuro-Degenerative Diseases*, *4*(6), 406–412.

Gürgen, S. G., Yücel, A. T., Karakuş, A. Ç., Çeçen, D., Özen, G., & Koçtürk, S. (2015). Usage of whey protein may cause liver damage via inflammatory and apoptotic responses. *Human & Experimental Toxicology*, *34*(7), 769–779.

Hall, M. N., Chavarro, J. E., Lee, I.-M., Willett, W. C., & Ma, J. (2008). A 22-year prospective study of fish, n-3 fatty acid intake, and colorectal cancer risk in men. *Cancer Epidemiology, Biomarkers & Prevention: A Publication of the American Association for Cancer Research, cosponsored by the American Society of Preventive Oncology*, *17*(5), 1136–1143.

Halter, J. B., Musi, N., Horne, F. M., Crandall, J. P., Goldberg, A., Harkless, L., et al. (2014). Diabetes and cardiovascular disease in older adults: Current status and future directions. *Diabetes*, *63*, 2578–2589.

Hamada, Y., Ishiura, S., & Kiso, Y. (2012). BACE1 inhibitor peptides: Can an infinitely small Kcat value turn the substrate of an enzyme into its inhibitor? *ACS Medicinal Chemistry Letters*, *3*(3), 193–197.

Hamada, Y., & Kiso, Y. (2013). Advances in the identification of β-secretase inhibitors. *Expert Opinion on Drug Discovery*, *8*(6), 709–731.

Han, H. Q., Zhou, X., & Goldberg, A. L. (2013). Myostatin/activin pathway antagonism: Molecular basis and therapeutic potential. *The International Journal of Biochemistry & Cell Biology*, *45*(10), 2333–2347.

Hayes, M. (2014). Bioactive peptides and their potential use for the prevention of diseases associated with Alzheimer's disease and mental health disorders: Food for thought? *Annals of Psychiatry and Mental Health*, *2*(3), 1017–1025.

Heaney, R. P., & Layman, D. K. (2008). Amount and type of protein influences bone health. *American Journal of Clinical Nutrition*, *87*(5), 1567–1570.

Heppner, F. L., Ransohoff, R. M., & Becher, B. (2015). Immune attack: The role of inflammation in Alzheimer disease. *Nature Reviews. Neuroscience*, *16*(6), 358–372.

Hollander, D., & Dadufalza, V. D. (1983). Increased intestinal absorption of oleic acid with aging in the rat. *Experimental Gerontology*, *18*(4), 287–292.

Holt, P. R., & Balint, J. A. (1993). Effects of aging on intestinal lipid absorption. *American Journal of Physiology. Gastrointestinal and Liver Physiology*, *264*(1), G1–G6.

Houston, D. K., Nicklas, B. J., Ding, J., Harris, T. B., Tylavsky, F. A., Newman, A. B., et al. (2008). Dietary protein intake is associated with lean mass change in older, community-dwelling adults: The health, aging, and body composition (health ABC) study. *The American Journal of Clinical Nutrition*, *87*(1), 150–155.

Huang, W. J., Zhang, X., & Chen, W. W. (2016). Role of oxidative stress in Alzheimer's disease. *Biomedical Reports*, *4*(5), 519–522.

Huebner, F. R., Lieberman, K. W., Rubino, R. P., & Wall, J. S. (1984). Demonstration of high opioid-like activity in isolated peptides from wheat gluten hydrolysates. *Peptides*, *5*(6), 1139–1147.

Hwang, E. M., Ryu, Y. B., Kim, H. Y., Kim, D.-G., Hong, S. G., Lee, J. H., et al. (2008). BACE1 inhibitory effects of lavandulyl flavanones from Sophora flavescens. *Bioorganic & Medicinal Chemistry*, *16*(14), 6669–6674.

in t' Veld, B. A., Ruitenberg, A., Hofman, A., Launer, L. J., van Duijn, C. M., Stijnen, T., et al. (2001). Nonsteroidal antiinflammatory drugs and the risk of Alzheimer's disease. *The New England Journal of Medicine*, *345*(21), 1515–1521.

Iuliano, S., & Ayton, J. (2015). Dietary intakes of expeditions during prolonged sunlight deprivation in polar environment do not support bone health. *International Journal of Circumpolar Health*, *74*(10), 27965.

Ivey, K. L., Lewis, J. R., Hodgson, J. M., Zhu, K., Dhaliwal, S. S., Thompson, P. L., et al. (2011). Association between yogurt, milk, and cheese consumption and common carotid artery intima-media thickness and cardiovascular disease risk factors in elderly women. *American Journal of Clinical Nutrition*, *94*, 234–239.

Janusz, M., & Zabłocka, A. (2013). Colostrinin: A proline-rich polypeptide complex of potential therapeutic interest. *Cellular and Molecular Biology*, *59*(1), 4–11.

Jedrychowski, W., Maugeri, U., Pac, A., Sochacka-Tatara, E., & Galas, A. (2008). Protective effect of fish consumption on colorectal cancer risk. Hospital-based case-control study in Eastern Europe. *Annals of Nutrition & Metabolism*, *53*(3–4), 295–302.

Jeon, S. Y., Bae, K., Seong, Y. H., & Song, K. S. (2003). Green tea catechins as a BACE1 (beta-secretase) inhibitor. *Bioorganic & Medicinal Chemistry Letters*, *13*(22), 3905–3908.

Ji, M., Leem, K., Kim, M., & Kim, H. K. (2007). Egg yolk soluble protein stimulates the proliferation and differentiation of osteoblastic MC3T3-E1 cells. *Bioscience, Biotechnology, and Biochemistry*, *71*(5), 1327–1329.

Jiang, B., & Mine, Y. (2000). Preparation of novel functional oligophosphopeptides from hen egg yolk phosvitin. *Journal of Agricultural and Food Chemistry*, *48*(4), 990–994.

Jiang, B., & Mine, Y. (2001). Phosphopeptides derived from hen egg yolk phosvitin: Effect of molecular size on the calcium-binding properties. *Bioscience, Biotechnology, and Biochemistry*, *65*(5), 1187–1190.

Johnell, O., & Kanis, J. A. (2006). An estimate of the worldwide prevalence and disability associated with osteoporotic fractures. *Osteoporosis International, 17*(12), 1726–1733.
Kanis, J. A. (2002). Diagnosis of osteoporosis and assessment of fracture risk. *The Lancet, 359*(9321), 1929–1936.
Kannan, A., Hettiarachchy, N., & Mahadevan, M. (2012). Peptides derived from rice bran protect cells from obesity and Alzheimer's disease. *International Journal of Biomedical Research, 3*(3), 131–135.
Kao, C. H., Lai, T. L., Wang, S. J., Chen, G. H., & Yeh, S. H. (1994). Influence of age on gastric emptying in healthy Chinese. *Clinical Nuclear Medicine, 19*(5), 401–404.
Karthick, N., Alwin, D., Poornima, K. N., Chitra, V., Saravanan, A., Balakrishnan, D., et al. (2015). Neurobehavioral alterations and brain creatine kinase system changes in chronic renal failure induced male Wistar rats: Impact of erythropoietin supplementation. *Journal of Bioequivalence & Bioavailability*, 7(2), 74–81.
Kasai, T., Iwasaki, R., Tanaka, M., & Kiriyama, S. (1995). Caseinphosphopeptides (CPP) in feces and contents in digestive-tract of rats fed casein and CPP preparations. *Bioscience, Biotechnology, and Biochemistry, 59*, 26–30.
Kassarjian, Z. (1989). Hypochlorhydria: A factor in nutrition. *Annual Review of Nutrition, 9*(1), 271–285.
Kastenmayer, P., Barclay, D., Cavin, E. O., & Pridmore-Merten, S. (2005). In *Calcium absorption enhance*, US Patent No. 20050266140A1. Washington, DC: U.S. Patent and Trademark Office.
Kim, I. H., Kisseleva, T., & Brenner, D. A. (2015). Aging and liver disease. *Current Opinion in Gastroenterology, 31*(3), 184–191.
Kim, H. K., Suzuki, T., Saito, K., Yoshida, H., Kobayashi, H., Kato, H., et al. (2012). Effects of exercise and amino acid supplementation on body composition and physical function in community-dwelling elderly Japanese sarcopenic women: A randomized controlled trial. *Journal of the American Geriatrics Society, 60*(1), 16–23.
Kim, J. S., Wilson, J. M., & Lee, S. R. (2010). Dietary implications on mechanisms of sarcopenia: Roles of protein, amino acids and antioxidants. *The Journal of Nutritional Biochemistry, 21*(1), 1–13.
Kobayashi, M., Tsubono, Y., Otani, T., Hanaoka, T., Sobue, T., Tsugane, S., et al. (2004). Fish, long-chain n-3 polyunsaturated fatty acids, and risk of colorectal cancer in middle-aged Japanese: The JPHC study. *Nutrition and Cancer, 49*(1), 32–40.
Kordatou, Z., Kountourakis, P., & Papamichael, D. (2014). Treatment of older patients with colorectal cancer: A perspective review. *Therapeutic Advances in Medical Oncology, 6*(3), 128–140.
Kupfer, R. M., Heppell, M., Haggith, J. W., & Bateman, D. N. (1985). Gastric emptying and small-bowel transit rate in the elderly. *Journal of the American Geriatrics Society, 33*(5), 340–343.
Larsson, S. C., Wolk, K., Brismar, K., & Wolk, A. (2005). Association of diet with serum insulin-like growth factor I in middle-aged and elderly men. *The American Journal of Clinical Nutrition, 81*(5), 1163–1167.
Lee, S. H., Jun, M., Choi, J. Y., Yang, E. J., Hur, J. M., Bae, K., et al. (2007). Plant phenolics as prolyl endopeptidase inhibitors. *Archives of Pharmacal Research, 30*(7), 827–833.
Lee, M., Kovacs-Nolan, J., Yang, C., Archbold, T., Fan, M. Z., & Mine, Y. (2009). Hen egg lysozyme attenuates inflammation and modulates local gene expression in a porcine model of dextran sodium sulfate (DSS)-induced colitis. *Journal of Agricultural and Food Chemistry, 57*(6), 2233–2240.
Lemieszewska, M., Jakubik-Witkowska, M., Stańczykiewicz, B., Zambrowicz, A., Zabłocka, A., Polanowski, A., et al. (2016). Pro-cognitive properties of the immunomodulatory polypeptide complex, yolkin, from chicken egg yolk and colostrum-derived

substances: Analyses based on animal model of age-related cognitive deficits. *Archivum Immunologiae et Therapiae Experimentalis*, *64*(5), 425–434.
Leslie, W., & Hankey, C. (2015). Aging, nutritional status and health. *Healthcare (Basel, Switzerland)*, *3*(3), 648–658.
Lipkin, M., & Newmark, H. (1985). Effect of added dietary calcium on colonic epithelial-cell proliferation in subjects at high risk for familial colonic cancer. *The New England Journal of Medicine*, *313*(22), 1381–1384.
Liu, Y., Wang, L., Cheng, Y., Saito, M., Yamaki, K., Qiao, Z., et al. (2009). Isoflavone content and anti-acetylcholinesterase activity in commercial douchi (a traditional Chinese salt-fermented soybean food). *Japan Agricultural Research Quarterly*, *43*(4), 301–307.
Lo, A. W., Ho, C., Cummings, J., & Kosik, K. S. (2014). Parallel discovery of Alzheimer's therapeutics. *Science Translational Medicine*, *6*(241), 241cm5.
Longstreth, G. F., Thompson, W. G., Chey, W. D., Houghton, L. A., Mearin, F., & Spiller, R. C. (2006). Functional bowel disorders. *Gastroenterology*, *130*(5), 1480–1491.
Lorget, F., Clough, J., Oliveira, M., Daury, M. C., Sabokbar, A., & Offord, E. (2002). Lactoferrin reduces in vitro osteoclast differentiation and resorbing activity. *Biochemical and Biophysical Research Commuications*, *296*(2), 261–266.
Louie, J. C. Y., Flood, V. M., Burlutsky, G., Rangan, A. M., Gill, T. P., & Mitchel, P. (2013). Dairy consumption and the risk of 15-year cardiovascular disease mortality in a cohort of older Australians. *Nutrients*, *5*, 441–454.
Ma, J., Glovannucci, E., Pollak, M., Chan, J. M., Gazlano, J. M., Willett, W., et al. (2001). Milk intake, circulating levels of insulin-like growth factor-I, and risk of colorectal cancer in men. *Journal of National Cancer Institute*, *93*(17), 1330–1336.
Madsen, J. L., & Graff, J. (2004). Effects of ageing on gastrointestinal motor function. *Age and Ageing*, *33*(2), 154–159.
Maes, M., Monteleone, P., Bencivenga, R., Goossens, F., Maj, M., van West, D., et al. (2001). Lower serum activity of prolyl endopeptidase in anorexia and bulimia nervosa. *Psychoneuroendocrinology*, *26*(1), 17–26.
Mahon, M. M., & James, O. F. W. (1994). Liver disease in the elderly. *Journal of Clinical Gastroenterology*, *18*(4), 330–334.
Marcus, R., Dempster, D. W., & Bouxsein, M. (2016). The nature of osteoporosis. In R. - Marcus, D. Feldman, D. W. Dempster, M. Luckey, & J. A. Cauley (Eds.), *Osteoporosis* (4th ed., pp. 21–30). Burlington, MA: Elsevier, Academic Press.
Margetts, B. M., Thompson, R. L., Elia, M., & Jackson, A. A. (2003). Prevalence of risk of undernutrition is associated with poor health status in older people in the UK. *European Journal of Clinical Nutrition*, *57*(1), 69–74.
Marty, E., Liu, Y., Samuel, A., Or, O., & Lane, J. (2017). A review of sarcopenia: Enhancing awareness of an increasingly prevalent disease. *Bone*, *105*, 276–286.
Mattsby-Baltzer, I., Roseanu, A., Motas, C., Elverfors, J., Engberg, I., & Hanson, L. A. (1996). Lactoferrin or a fragment thereof inhibits the endotoxin-induced interleukin-6 response in human monocytic cells. *Pediatric Research*, *40*(2), 257–262.
McAfee, A. J., McSorley, E. M., Cuskelly, G. J., Moss, B. W., Wallace, J. M., Bonham, M. P., et al. (2010). Red meat consumption: An overview of the risks and benefits. *Meat Science*, *84*(1), 1–13.
McPherron, A. C., Lawler, A. M., & Lee, S. J. (1997). Regulation of skeletal muscle mass in mice by a new TGF-beta superfamily member. *Nature*, *387*(6628), 83–90.
Meakin, P. J., Harper, A. J., Hamilton, D. L., Gallagher, J., McNeilly, A. D., Burgess, L. A., et al. (2012). Reduction in BACE1 decreases body weight, protects against diet-induced obesity and enhances insulin sensitivity in mice. *The Biochemical Journal*, *441*(1), 285–296.
Mecocci, P., Tinarelli, C., Schulz, R. J., & Polidori, M. C. (2014). Nutraceuticals in cognitive impairment and Alzheimer's disease. *Frontiers in Pharmacology*, *5*(E25), 147.

Meisel, H., & Frister, H. (1989). Chemical characterization of bioactive peptides from in vivo digests of casein. *Journal of Dairy Research*, *56*, 343–349.

Menees, S. B., Guentner, A., Chey, S. W., Saad, R., & Chey, W. D. (2015). How do US gastroenterologists use over-the-counter and prescription medications in patients with gastroesophageal reflux and chronic constipation? *The American Journal of Gastroenterology*, *110*(11), 1516–1525.

Messier, S. P. (2009). Obesity and osteoarthritis: Disease genesis and nonpharmacologic weight management. *Medical Clinics of North America*, *93*(1), 145–159.

Mezey, E. (1982). Liver disease and protein needs. *Annual Review of Nutrition*, *2*(1), 21–50.

Mignon, M., Beaufrère, A. M., Combaret, L., & Meynial-Denis, D. (2007). Does long-term intermittent treatment with glutamine improve the well-being of fed and fasted very old rats? *Journal of Parenteral and Enteral Nutrition*, *31*(6), 456–462.

Min, L. J., Kobayashi, Y., Mogi, M., Tsukuda, K., Yamada, A., Yamauchi, K., et al. (2017). Administration of bovine casein-derived peptide prevents cognitive decline in Alzheimer disease model mice. *PLoS One*, *12*(2), e0171515.

Morgan, E. F., Barnes, G. L., & Einhorn, T. A. (2008). The bone organ system: Form and function. In R. Marcus, D. Feldman, D. A. Nelson, & C. J. Rosen (Eds.), *Osteoporosis* (pp. 3–25). Burlington, MA: Elsevier, Academic Press.

Morley, J. E., Anker, S. D., & von Haehling, S. (2014). Prevalence, incidence, and clinical impact of sarcopenia: Facts, numbers, and epidemiology—Update 2014. *Journal of Cachexia, Sarcopenia and Muscle*, *5*(4), 253–259.

Morley, J. E., Argiles, J. M., Evans, W. J., Bhasin, S., Cella, D., Deutz, N. E. P., et al. (2010). Nutritional recommendations for the management of sarcopenia. *Journal of the American Medical Directors Association*, *11*(6), 391–396.

Mozaffarian, D., Longstreth, W. T., Lemaitre, R. N., Manolio, T. A., Kuller, L. H., Burke, G. L., et al. (2005). Fish consumption and stroke risk in elderly individuals: The cardiovascular health study. *Archives of Internal Medicine*, *165*, 200–206.

Murphy, T., Dias, G. P., & Thuret, S. (2014). Effects of diet on brain plasticity in animal and human studies: Mind the gap. *Neural Plasticity*, *2014*(2), 563160. 32 pages.

Nagpal, R. P., Behare, P., Rana, R., Kumar, A., Kumar, M., Arora, S., et al. (2011). Bioactive peptides derived from milk proteins and their health beneficial potentials: An update. *Food and Function*, *2*, 18–27.

Naot, D., Palmano, K., & Cornish, J. (2012). Lactoferrin—A potential anabolic intervention in osteoporosis. In Y. Dionyssiotis (Ed.), *Osteoporosis*. In Tech.

Naseeb, M. A., & Volpe, S. L. (2017). Protein and exercise in the prevention of sarcopenia and aging. *Nutrition Research*, *40*, 1–20.

Nieuwenhuizen, W. F., Weenen, H., Rigby, P., & Hetherington, M. M. (2010). Older adults and patients in need of nutritional support: Review of current treatment options and factors influencing nutritional intake. *Clinical Nutrition*, *29*(2), 160–169.

Nilwik, R., Snijders, T., Leenders, M., Groen, B. B. L., van Kranenburg, J., Verdijk, L. B., et al. (2013). The decline in skeletal muscle mass with aging is mainly attributed to a reduction in type II muscle fiber size. *Experimental Gerontology*, *48*(5), 492–498.

Nowson, C., & O'Connell, S. (2015). Protein requirements and recommendations for older people: A review. *Nutrients*, 7(12), 6874–6899.

Nunomura, A., Castellani, R. J., Zhu, X., Moreira, P. I., Perry, G., & Smith, M. A. (2006). Involvement of oxidative stress in Alzheimer disease. *Journal of Neuropathology and Experimental Neurology*, *65*(7), 631–641.

Nyberg, F., Sanderson, K., & Glämsta, E. L. (1997). The hemorphins: A new class of opioid peptides derived from the blood protein hemoglobin. *Biopolymers*, *43*(2), 147–156.

O'Donovan, D., Hausken, T., Lei, Y., Russo, A., Keogh, J., Horowitz, M., et al. (2005). Effect of aging on transpyloric flow, gastric emptying, and intragastric distribution in healthy humans—Impact on glycemia. *Digestive Diseases and Sciences*, *50*(4), 671–676.

Ohinata, K., Agui, S., & Yoshikawa, M. (2014). Soymorphins, novel μ opioid peptides derived from soy β-conglycinin β-subunit, have anxiolytic activities. *Bioscience, Biotechnology, and Biochemistry*, *71*(10), 2618–2621.

Olsen, B. R., Reginato, A. M., & Wang, W. (2000). Bone development. *Annual Review of Cell and Developmental Biology*, *16*, 191–220.

Paddon-Jones, D., & Rasmussen, B. B. (2009). Dietary protein recommendations and the prevention of sarcopenia. *Current Opinion in Clinical Nutrition and Metabolic Care*, *12*(1), 86–90.

Paddon-Jones, D., Westman, E., Mattes, R. D., Wolfe, R. R., Astrup, A., & Westerterp-Plantenga, M. (2008). Protein, weight management, and satiety. *The American Journal of Clinical Nutrition*, *87*(5), 1558S–1561S.

Pal, P. B., Sinha, K., & Sil, P. C. (2013). Mangiferin, a natural xanthone, protects murine liver in Pb(II) induced hepatic damage and cell death via MAP kinase, NF-κB and mitochondria dependent pathways. *PLoS One*, *8*(2), e56894.

Park, H. J., Ahn, J. Y., Jung, H. Y., Lee, J. H., Jung, K. W., Kim, D. H., et al. (2016). Clinical characteristics and outcomes of gastric cancer patients aged over 80 years: A retrospective case-control study. *PLoS One*, *11*(12), e0167615.

Patel, H. P., Syddall, H. E., Jameson, K., Robinson, S., Denison, H., Roberts, H. C., et al. (2013). Prevalence of sarcopenia in community-dwelling older people in the UK using the European working group on sarcopenia in older people (EWGSOP) definition: Findings from the Hertfordshire cohort study (HCS). *Age and Ageing*, *42*(3), 378–384.

Perego, S., Cosentino, S., Fiorilli, A., Tettamanti, G., & Ferraretto, A. (2012). Casein phosphopeptides modulate proliferation and apoptosis in HT-29 cell line through their interaction with voltage-operated L-type calcium channels. *The Journal of Nutritional Biochemistry*, *23*(7), 808–816.

Perrini, S., Laviola, L., Carreira, M. C., Cignarelli, A., Natalicchio, A., & Giorgino, F. (2010). The GH/IGF1 axis and signaling pathways in the muscle and bone: Mechanisms underlying age-related skeletal muscle wasting and osteoporosis. *The Journal of Endocrinology*, *205*(3), 201–210.

Phillips, S. M. (2017). Current concepts and unresolved questions in dietary protein requirements and supplements in adults. *Frontiers in Nutrition*, *4*, 13.

Phillips, S. M., Chevalier, S., & Leidy, H. J. (2016). Protein "requirements" beyond the RDA: Implications for optimizing health. *Applied Physiology, Nutrition, and Metabolism*, *41*(5), 565–572.

Porporato, P. E., Filigheddu, N., Reano, S., Ferrara, M., Angelino, E., Gnocchi, V. F., et al. (2013). Acylated and unacylated ghrelin impair skeletal muscle atrophy in mice. *The Journal of Clinical Investigation*, *123*(2), 611–622.

Pötsch, M. S., Tschirner, A., Palus, S., von Haehling, S., Doehner, W., Beadle, J., et al. (2014). The anabolic catabolic transforming agent (ACTA) espindolol increases muscle mass and decreases fat mass in old rats. *Journal of Cachexia, Sarcopenia and Muscle*, *5*(2), 149–158.

Potter, S. M. (1995). Overview of proposed mechanisms for the hypocholesterolemic effect of soy. *Journal of Nutrition*, *125*, 606S–611S.

Preis, S. R., Stampfer, M. J., Spiegelman, D., Willett, W. C., & Rimm, E. B. (2010). Dietary protein and risk of ischemic heart disease in middle-aged men. *American Journal of Clinical Nutrition*, *92*, 1265–1272.

Prince, M. J., Wu, F., Guo, Y., Gutierrez Robledo, L. M., O'Donnell, M., Sullivan, R., et al. (2015). The burden of disease in older people and implications for health policy and practice. *The Lancet*, *385*, 549–562.

Raj, N., Helen, A., Manoj, N., Harish, G., Thomas, V., Singh, S., et al. (2015). In silico study of peptide inhibitors against BACE 1. *Systems and Synthetic Biology*, *9*(1–2), 67–72.

Rao, S., & Go (2010). Update on the management of constipation in the elderly: New treatment options. *Clinical Interventions in Aging*, *5*, 163–171.

Rezai-Zadeh, K., Shytle, D., Sun, N., Mori, T., Hou, H., Jeanniton, D., et al. (2005). Green tea epigallocatechin-3-gallate (EGCG) modulates amyloid precursor protein cleavage and reduces cerebral amyloidosis in Alzheimer transgenic mice. *The Journal of Neuroscience*, *25*(38), 8807–8814.

Robbins, J., Bridges, A. D., & Taylor, A. (2006). Oral, pharyngeal and esophageal motor function in aging. *GI Motility Online*. Retrieve at February 08, 2018 from https://www.nature.com/gimo/contents/pt1/full/gimo39.html.

Robinson, S. M., Reginster, J. Y., Rizzoli, R., Shaw, S. C., Kanis, J. A., Bautmans, I., et al. (2017). Does nutrition play a role in the prevention and management of sarcopenia? *Clinical Nutrition*. https://doi.org/10.1016/j.clnu.2017.08.016.

Rosenberg, I. H. (1997). Sarcopenia: Origins and clinical relevance. *The Journal of Nutrition*, *127*(5 Suppl), 990S–991S.

Rosenblum, J. S., & Kozarich, J. W. (2003). Prolyl peptidases: A serine protease subfamily with high potential for drug discovery. *Current Opinion in Chemical Biology*, 7(4), 496–504.

Roubenoff, R., & Hughes, V. A. (2000). Sarcopenia: Current concepts. *The Journals of Gerontology. Series A, Biological Sciences and Medical Sciences*, *55*(12), M716–24.

Rousset, S., Patureau Mirand, P., Brandolini, M., Martin, J. F., & Boirie, Y. (2003). Daily protein intakes and eating patterns in young and elderly French. *British Journal of Nutrition*, *90*(6), 1107–1115.

Sadhukhan, P., Saha, S., Dutta, S., Mahalanobish, S., & Sil, P. C. (2018). Nutraceuticals: An emerging therapeutic approach against the pathogenesis of Alzheimer's disease. *Pharmacological Research*, *129*, 100–114.

Samson, M., Meeuwsen, I. B., Crowe, A., Dessens, J. A., Duursma, S. A., & Verhaar, H. J. (2000). Relationships between physical performance measures, age, height and body weight in healthy adults. *Age and Ageing*, *29*(3), 235–242.

Schmucker, D. L. (1998). Aging and the liver: An update. *The Journals of Gerontology. Series A, Biological Sciences and Medical Sciences*, *53*(5), B315–20.

Schofield, J. D., Liu, Y., Rao-Balakrishna, P., Malik, R. A., & Soran, H. (2016). Diabetes dyslipidemia. *Diabetes Therapy: Research, Treatment and Education of Diabetes and Related Disorders*, 7, 203–219.

Schuster, B. G., Kosar, L., & Kamrul, R. (2015). Constipation in older adults: Stepwise approach to keep things moving. *Canadian Family Physician Médecin de Famille Canadien*, *61*(2), 152–158.

Seeman, E. (2008). Structural basis of growth-related gain and age-related loss of bone strength. *Rheumatology (Oxford, England)*, *47*(Suppl. 4), iv2–8.

Selhub, E. M., Logan, A. C., & Bested, A. C. (2014). Fermented foods, microbiota, and mental health: Ancient practice meets nutritional psychiatry. *Journal of Physiological Anthropology*, *33*(1), 2.

Serra-Prat, M., Palomera, M., Gomez, C., Sar-Shalom, D., Saiz, A., Montoya, J. G., et al. (2012). Oropharyngeal dysphagia as a risk factor for malnutrition and lower respiratory tract infection in independently living older persons: A population-based prospective study. *Age and Ageing*, *41*(3), 376–381.

Shi, J., Tauriainen, E., Martonen, E., Finckenberg, P., Ahlroos-Lehmus, A., Tuomainen, A., et al. (2011). Whey protein isolate protects against diet-induced obesity and fatty liver formation. *International Dairy Journal*, *21*(8), 513–522.

Shimamoto, C., Hirata, I., Hiraike, Y., Takeuchi, N., Nomura, T., & Katsu, K. I. (2002). Evaluation of gastric motor activity in the elderly by electrogastrography and acetate breath test. *Gerontology*, *48*(6), 381–386.

Sila, A., Martinez-Alvarez, O., Haddar, A., Gómez-Guillén, M. C., Nasri, M., Montero, M. P., et al. (2015). Recovery, viscoelastic and functional properties of Barbel skin gelatine: Investigation of anti-DPP-IV and anti-prolyl endopeptidase activities of generated gelatine polypeptides. *Food Chemistry*, *168*, 478–486.

Singh, M. A. F., Ding, W., Manfredi, T. J., Solares, G. S., O'Neill, E. F., Clements, K. M., et al. (1999). Insulin-like growth factor I in skeletal muscle after weight-lifting exercise in frail elders. *American Journal of Physiology. Endocrinology and Metabolism*, *277*(1), 135–143.

Singh, S. K., Srivastav, S., Yadav, A. K., Srikrishna, S., & Perry, G. (2016). Overview of Alzheimer's disease and some therapeutic approaches targeting Aβ by using several synthetic and herbal compounds. *Oxidative Medicine and Cellular Longevity*, *2016*(2), 1–22.

Sinha, R., Chow, W. H., Kulldorff, M., Denobile, J., Butler, J., Garcia-Closas, M., et al. (1999). Well-done, grilled red meat increases the risk of colorectal adenomas. *Cancer Research*, *59*(17), 4320–4324.

Sipilä, S., Taaffe, D. R., Cheng, S., Puolakka, J., Toivanen, J., & Suominen, H. (2001). Effects of hormone replacement therapy and high-impact physical exercise on skeletal muscle in post-menopausal women: A randomized placebo-controlled study. *Clinical Science*, *101*(2), 147–157.

Sirtori, C. R., Lovati, M. R., Manzoni, C., Monetti, M., Pazzucconi, F., & Gatti, E. (1995). Soy and cholesterol reduction: Clinical experience. *Journal of Nutrition*, *125*, 598S–605S.

Smolen, J. S., Aletaha, D., Barton, A., Burmester, G. R., Emery, P., Firestein, G. S., et al. (2018). Rheumatoid arthritis. *Nature Reviews. Disease Primers*, *4*. 18001.

Song, J., Chang, R. W., & Dunlop, D. D. (2006). Population impact of arthritis on disability in older adults. *Arthritis and Rheumatism*, *55*(2), 248–255.

Sørensen, R., Kildal, E., Stepaniak, L., Pripp, A. H., & Sørhaug, T. (2004). Screening for peptides from fish and cheese inhibitory to prolyl endopeptidase. *Nahrung/Food*, *48*(1), 53–56.

Stefanucci, A., Mollica, A., Macedonio, G., Zengin, G., Ahmed, A. A., & Novellino, E. (2017). Exogenous opioid peptides derived from food proteins and their possible uses as dietary supplements: A critical review. *Food Reviews International*, *34*(1), 70–86.

Su, X., Dong, C., Zhang, J., Su, L., Wang, X., Cui, H., et al. (2014). Combination therapy of anti-cancer bioactive peptide with Cisplatin decreases chemotherapy dosing and toxicity to improve the quality of life in xenograft nude mice bearing human gastric cancer. *Cell and Bioscience*, *4*, 7.

Su, L. Y., Xin, H. Y., Liu, Y. L., Zhang, J. L., Xin, H. W., & Su, X. L. (2014). Anticancer bioactive peptide (ACBP) inhibits gastric cancer cells by upregulating growth arrest and DNA damage-inducible gene 45A (GADD45A). *Tumour Biology: The Journal of the International Society for Oncodevelopmental Biology and Medicine*, *35*(10), 10051–10056.

Sugawara, Y., Kuriyama, S., Kakizaki, M., Nagai, M., Ohmori-Matsuda, K., Sone, T., et al. (2009). Fish consumption and the risk of colorectal cancer: The Ohsaki cohort study. *British Journal of Cancer*, *101*(5), 849–854.

Sura, L., Madhavan, A., Carnaby-Mann, G., & Crary, M. (2012). Dysphagia in the elderly: Management and nutritional considerations. *Clinical Interventions in Aging*, *7*, 287–298.

Surdykowski, A. K., Kenny, A. M., Insogna, K. L., & Kerstetter, J. E. (2010). Optimizing bone health in older adults: The importance of dietary protein. *Aging Health*, *6*(3), 345–357.

Swomley, A. M., Förster, S., Keeney, J. T., Triplett, J., Zhang, Z., Sultana, R., et al. (2014). Abeta, oxidative stress in Alzheimer disease: Evidence based on proteomics studies. *Biochimica et Biophysica Acta (BBA)—Molecular Basis of Disease*, *1842*(8), 1248–1257.

Tack, J., Müller-Lissner, S., Stanghellini, V., Boeckxstaens, G., Kamm, M. A., Simren, M., et al. (2011). Diagnosis and treatment of chronic constipation—A European perspective. *Neurogastroenterology and Motility: The Official Journal of the European Gastrointestinal Motility Society*, *23*(8), 697–710.

Takada, Y., Aoe, S., & Kumegawa, M. (1996). Whey protein stimulated the proliferation and differentiation of osteoblastic MC3T3-E1 cells. *Biochemical and Biophysical Research Communications, 223*, 445–449.

Takada, Y., Kobayashi, N., Hiroaki, M., Katoa, K., Yamamuraa, J., Yahiroa, M., et al. (1997). Whey protein suppresses the osteoclast-mediated bone resorption and osteoclast cell formation. *International Dairy Journal*, 7(12), 821–825.

Takahashi, M., Moriguchi, S., Yoshikawa, M., & Sasaki, R. (1994). Isolation and characterization of oryzatensin: A novel bioactive peptide with ileum-contracting and immunomodulating activities derived from rice albumin. *Biochemistry and Molecular Biology International, 33*(6), 1151–1158.

Tanaka, M., Kamata, R., Yanagihara, K., & Sakai, R. (2010). Suppression of gastric cancer dissemination by ephrin-B1-derived peptide. *Cancer Science, 101*(1), 87–93.

Thomson, A. B. (1980). Effect of age on uptake of homologous series of saturated fatty acids into rabbit jejunum. *American Journal of Physiology. Gastrointestinal and Liver Physiology, 239*(5), G363–G371.

Thomson, A. B. R. (2009). Small intestinal disorders in the elderly. *Best Practice & Research. Clinical Gastroenterology, 23*(6), 861–874.

Tian, S., Xu, Q., Jiang, R., Han, T., Sun, C., & Na, L. (2017). Dietary protein consumption and the risk of type 2 diabetes: A systematic review and meta-analysis of cohort studies. *Nutrients, 9*, 982.

Tieland, M., van de Rest, O., Dirks, M. L., van der Zwaluw, N., Mensink, M., van Loon, L. J. C., et al. (2012). Protein supplementation improves physical performance in frail elderly people: A randomized, double-blind, placebo-controlled trial. *Journal of the American Medical Directors Association, 13*(8), 720–726.

Trivedi, M., Zhang, Y., Lopez-Toledano, M., Clarke, A., & Deth, R. (2016). Differential neurogenic effects of casein-derived opioid peptides on neuronal stem cells: Implications for redox-based epigenetic changes. *The Journal of Nutritional Biochemistry, 37*, 39–46.

Tsuchita, H., Suzuki, T., & Kuwata, T. (2001). The effect of casein phosphopeptides on calcium absorption from calcium-fortified milk in growing rats. *British Journal of Nutrition, 85*(1), 5–10.

United Nations. (2018). The world population prospects: 2015 revision, Retrieved January 15, from http://www.un.org/en/development/desa/publications/world-population-prospects-2015-revision.html.

US Census Bureau. (2008). *Table 2, Projections of the population by age groups and sex for the United States: 2010 to 2050 (NP2008-T2)*. Retrieved February 20, 2018, from https://www.census.gov.

Viñals, F., & Pouysségur, J. (2011). Transforming growth factor beta1 (TGF-beta1) promotes endothelia cell survival during in vitro angiogenesis via an autocrine mechanism implicating TGF-alpha signaling. *Molecular and Cellular Biology, 21*(21), 7218–7230.

Volpi, E., Campbell, W. W., Dwyer, J. T., Johnson, M. A., Jensen, G. L., Morley, J. E., et al. (2013). Is the optimal level of protein intake for older adults greater than the recommended dietary allowance? *The Journals of Gerontology Series A: Biological Sciences and Medical Sciences, 68*(6), 677–681.

von Stengel, S., Kemmler, W., Engelke, K., & Kalender, W. A. (2012). Effect of whole-body vibration on neuromuscular performance and body composition for females 65 years and older: A randomized-controlled trial. *Scandinavian Journal of Medicine & Science in Sports, 22*(1), 119–127.

Wakimoto, P., & Block, G. (2001). Dietary intake, dietary patterns, and changes with age: An epidemiological perspective. *The Journals of Gerontology Series A: Biological Sciences and Medical Sciences, 56*, 65–80.

Walston, J. D. (2012). Sarcopenia in older adults. *Current Opinion in Rheumatology, 24*(6), 623–627.

Wang, N., Xu, Q., Xu, S., Qi, Y., Chen, M., Li, H., et al. (2015). High-efficiency exfoliation of layered materials into 2D nanosheets in switchable CO2/surfactant/H2O system. *Scientific Reports*, *5*. 16764.

Wang, F., Zhou, X. L., Yang, Q. G., Xu, W. H., Wang, F., Chen, Y. P., et al. (2011). A peptide that binds specifically to the β-amyloid of Alzheimer's disease: Selection and assessment of anti-β-amyloid neurotoxic effects. *PLoS One*, *6*(11), e27649.

Weitz, J., Koch, M., & Debus, J. (2005). Colorectal cancer: Not an embarrassing problem. *The Lancet*, *365*, 153–165.

Willems, S. M., Wright, D. J., Day, F. R., Trajanoska, K., Joshi, P. K., Morris, J. A., et al. (2017). Large-scale GWAS identifies multiple loci for hand grip strength providing biological insights into muscular fitness. *Nature Communications*, *8*. 16015.

Wolfe, R. R. (2012). The role of dietary protein in optimizing muscle mass, function and health outcomes in older individuals. *The British Journal of Nutrition*, *108*(Suppl. 2), S88–S93.

Wolfe, R. R., Miller, S. L., & Miller, K. B. (2008). Optimal protein intake in the elderly. *Clinical Nutrition*, *27*(5), 675–684.

Woudstra, T., & Thomson, A. B. R. (2002). Nutrient absorption and intestinal adaptation with ageing. *Best Practice & Research. Clinical Gastroenterology*, *16*(1), 1–15.

Wu, J., & Acero-Lopez, A. (2012). Ovotransferrin: Structure, bioactivities, and preparation. *Food Research International*, *46*(2), 480–487.

Xia, W., & Mo, H. (2016). Potential of tocotrienols in the prevention and therapy of Alzheimer's disease. *The Journal of Nutritional Biochemistry*, *31*, 1–9.

Xu, Z., Tan, Z., Zhang, Q., Gui, Q., & Yang, Y. (2015). The effectiveness of leucine on muscle protein synthesis, lean body mass and leg lean mass accretion in older people: A systematic review and meta-analysis. *British Journal of Nutrition*, *113*(1), 25–34.

Yakar, S., Rosen, C. J., Beamer, W. G., Ackert-Bicknell, C. L., Wu, Y., Liu, J. L., et al. (2002). Circulating levels of IGF-1 directly regulate bone growth and density. *Journal of Clinical Investigation*, *110*(6), 771–781.

Yamanaka, M., Ishikawa, T., Griep, A., Axt, D., Kummer, M. P., & Heneka, M. T. (2012). PPARγ/RXRα-induced and CD36-mediated microglial amyloid-β phagocytosis results in cognitive improvement in amyloid precursor protein/presenilin 1 mice. *The Journal of Neuroscience: The Official Journal of the Society for Neuroscience*, *32*(48), 17321–17331.

Yang, G., Shu, X., Jin, F., Zhang, X., Li, H., Li, Q., et al. (2005). Longitudinal study of soy food intake and blood pressure among middle-aged and elderly Chinese women. *The American Journal of Clinical Nutrition*, *81*, 1012–1017.

Yang, H. Q., Sun, Z. K., & Chen, S. D. (2012). Current advances in the treatment of Alzheimer's disease: Focused on considerations targeting Aβ and tau. *Translational Neurodegeneration*, *1*(1), 21.

Yarasheski, K., Bhasin, S., Sinha-Hikim, I., Pak-Loduca, J., & Gonzalez-Cadavid, N. F. (2002). Serum myostatin-immunoreactive protein is increased in 60-92 year old women and men with muscle wasting. *The Journal of Nutrition, Health & Aging*, *6*(5), 343–348.

Yeon, S. W., You, Y. S., Kwon, H. S., Yang, E. H., Ryu, J. S., Kang, B. H., et al. (2010). Fermented milk of *Lactobacillus helveticus* IDCC3801 reduces beta-amyloid and attenuates memory deficit. *Journal of Functional Foods*, *2*(2), 143–152.

Yiannopoulou, K. G., & Papageorgiou, S. G. (2013). Current and future treatments for Alzheimer's disease. *Therapeutic Advances in Neurological Disorders*, *6*(1), 19–33.

Zhang, Q., Wastney, M. E., Rosen, C. J., Beamer, W. G., & Weaver, C. M. (2011). Insulin-like growth factor-1 increases bone calcium accumulation only during rapid growth in female rats. *The Journal of Nutrition*, *141*(11), 2010–2016.

Zhao, Q., Garreau, I., Sannier, F., & Piot, J. M. (1997). Opioid peptides derived from hemoglobin: Hemorphins. *Biopolymers*, *43*(2), 75–98.

Zioudrou, C., Streaty, R. A., & Klee, W. A. (1979). Opioid peptides derived from food proteins. The exorphins. *The Journal of Biological Chemistry*, *254*(7), 2446–2449.

FURTHER READING

European Food Safety Authority (EFSA). (2009). Review of the potential health impact of β-casomorphins and related peptides. *EFSA Journal*, 7(2), 231.

Orimo, H., Ito, H., Suzuki, T., Araki, A., Hosoi, T., & Sawabe, M. (2006). Reviewing the definition of "elderly". *Geriatrics and Gerontology International*, *6*(3), 149–158.

CHAPTER NINE

Peptide Derivatives of Erythropoietin in the Treatment of Neuroinflammation and Neurodegeneration

Ilkcan Ercan*, Kemal Ugur Tufekci*, Ezgi Karaca*, Sermin Genc*,†, Kursad Genc†,1

*International Biomedicine and Genome Institute, Dokuz Eylul University, Balcova, Izmir, Turkey
†Institute of Health Science, Dokuz Eylul University, Balcova, Izmir, Turkey
[1]Corresponding author: e-mail address: kkursadgenc@hotmail.com

Contents

Advances in Protein Chemistry and Structural Biology, Volume 112
ISSN 1876-1623
https://doi.org/10.1016/bs.apcsb.2018.01.007

Abstract

During the past 35 years, recombinant DNA technology has allowed the production of a wide range of hematopoietic and neurotrophic growth factors including erythropoietin (EPO). These have emerged as promising protein drugs in various human diseases. Accumulated evidences have recently demonstrated the neuroprotective effect of EPO in preclinical models of acute and chronic degenerative disorders. Nevertheless, tissue protective effect of EPO could not be translated to the clinical trials because of common lethal thromboembolic events, erythropoiesis and hypertension. Although chemically modified nonerythropoietic analogs of EPO bypass these side effects, high expense, development of antidrug antibodies, and promotion of tumorigenicity are still concern especially in long-term use. As an alternative, nonerythropoietic EPO mimetic peptides can be used as candidate drugs with their high potency and selectivity, easy production, low cost, and immunogenicity properties. Recent experimental studies suggest that these peptides prevent ischemic brain injury and neuroinflammation. The results of clinical trial in patients with neuropathic pain are also promising. Herein, we summarize these studies and review advanced experimental and in silico methods in peptide drug discovery.

ABBREVIATIONS

Aβ amyloid beta
ACTH adrenocorticotropic hormone
ADME absorption, distribution, metabolism, and excretion
BBB blood–brain barrier
CCM corneal confocal microscopy
CEPO carbamoylated EPO
CERA continuous erythropoietin receptor activator
CFA complete Freund's adjuvant
cGMPs current good manufacturing practices
CHBP cyclic helix B peptide
CKD chronic kidney disease
CNTF ciliary neurotrophic factor
CSF cerebrospinal fluid
CSF2Rβ colony-stimulating factor 2 receptor beta
EAE experimental autoimmune encephalomyelitis
EAN experimental allergic neuritis
EPHB4 ephrin receptor B4
EPO erythropoietin
EPOR erythropoietin receptor
ERK1/2 extracellular signal-regulated kinase 1/2
ESAs erythropoiesis-stimulating agents

FDA	the US Food and Drug Administration
fMRI	functional MRI
GEPO	glutaraldehyde EPO
GLP-1	glucagon-like peptide-1
GM-CSF	granulocyte-macrophage colony-stimulating factor
HbA1c	hemoglobin A1c
HBSP	helix B surface peptide
HSA	human serum albumin
IL-3	interleukin 3
IL-5	interleukin 5
IP	intraperitoneal
ITP	immune thrombocytopenic purpura
IV	intravenous
JAK2	janus kinase 2
KA	kainic acid
KDR	kinase domain receptor
LPS	lipopolysaccharide
MOG	myelin oligodendrocyte glycoprotein
NO	nitric oxide
NRF2	nuclear factor (erythroid-derived 2)-like 2
PEG	polyethylene glycol
PI3-K	phosphoinositide-3-kinase
PLP	proteolipid protein
PPIs	protein–protein interactions
PTMs	posttranslational modifications
QST	quantitative sensory testing
rhEPO	recombinant human protein EPO
ROS	reactive oxygen species
RTK	receptor tyrosine kinase
SC	subcutaneous
SHP-1	Src homology region 2 domain-containing phosphatase-1
STAT5	signal transducer and activator of transcription 5
TBI	traumatic brain injury
TNF-α	tumor necrosis factor-alpha
TRPV1	transient receptor potential vanilloid 1
VEGF	vascular endothelial growth factor
VEGFR2	vascular endothelial growth factor receptor 2
βCR	β common receptor (CD131)

1. PEPTIDE DRUGS

Peptides are natural biological or artificially generated synthetic compounds consisting of two or more amino acids linked by peptide (amide) bonds. They are also considered as small proteins because they are less than

50 residues and their molecular weight is less than 10 kDa. So far, around 7000 natural peptides have been identified, which have an important role in the most biological processes such as acting like hormones, growth factors, neurotransmitters, ion channel ligands, and anti-infectives. Formation of peptides is linked with protein degradation and turnover inside cells. Although most of the degraded products are produced as a result of cellular homeostasis, they are hydrolyzed to be recycled in protein synthesis, and regulate protein phosphatases and kinases. The importance of biologically active peptides in normal and pathological cellular events provides the inspiration for pinpointing new therapeutic candidates (Qvit, Rubin, Urban, Mochly-Rosen, & Gross, 2017). Today, there are more than 60 FDA-approved peptide drugs. Nearly 140 clinical trials with peptide drugs are ongoing research, and prec linical studies are carried out with over 150 peptide molecules (Fosgerau & Hoffmann, 2015).

As therapeutic agents, peptides have unique features over proteins or small molecule drugs: They are small in size, have a higher activity, and are highly specific toward their targets. Since peptides are rapidly degraded, they do not accumulate in specific organs (kidney, liver, etc.) (Table 1). So, they are less toxic than small molecule drugs (Marqus, Pirogova, & Piva, 2017). An added benefit of using peptides in treatment is that they are less immunogenic when compared to antibodies or proteins and also have minimal drug–drug interaction (McGregor, 2008). In addition to these advantages, therapeutic peptides have also some drawbacks. They have a short half-life and low oral bioavailability, since proteolytic enzymes of the digestive system can break them down quickly. They also have difficulties in crossing the physiological

Table 1 Advantages and Disadvantages of Peptide Drugs

Advantages	Disadvantages
High activity	Low oral bioavailability
Low toxicity	Short half-life
High potency and selectivity	Low stability
Low accumulation in tissues	Low membrane permeability
Broad range of targets	Tendency for aggregation
High chemical and biological diversity	Risk of immunogenic effects
Minimal drug–drug interactions	High production cost

barriers due to their hydrophilic character. Their high conformational flexibility appears to be an advantage, though sometimes this results in a lack of selectivity and leads to their interactions with different targets (Vlieghe, Lisowski, Martinez, & Khrestchatisky, 2010).

2. STRATEGIES TO IMPROVE THE PHARMACOKINETIC PROPERTIES OF PEPTIDE DRUGS

Different strategies can be applied to improve the biological effects of peptide drugs (Table 2). Modifications of peptides during formulation such as cyclization, peptide stapling, imposing N-methylation, or using a transporter system, increasing permeability using surfactants, phospholipids, fatty acids, and nanoparticles are available approaches that can improve the absorption of peptide drugs (Di, 2015; Furman, Chiu, & Hunter, 2015). Both diffusion and convection mechanisms provide biodistribution of peptide drugs, which can be enhanced via conjugation to endogenous proteins or structural modification of peptides (Di, 2015).

The metabolism of the peptide drugs occurs in blood, liver, kidneys, and small intestine by proteolytic degradation. Degraded peptide metabolites and fragments may continue their biological actions (Otvos & Wade, 2014). Half-lives of peptide drugs may be extended as demonstrated for adrenocorticotropic hormone (ACTH), parathormone, and relaxin (Otvos & Wade, 2014). Structure modification of peptides, such as inclusion of nonnatural amino acids; conjugation to macromolecules (polyethylene glycol (PEG), XTEN or antibody Fc portion) and using protease inhibitors are effective strategies to improve stability of the peptide (Blaskovich, 2016; Furman et al., 2015; Goldflam & Ullman, 2015; Vhora, Patil, Bhatt, & Misra, 2015). Macromolecules for conjugation should be selected carefully. For example, a PEGylated erythropoietic peptide, peginesatide, was quickly withdrawn from the market due to unexplained serious adverse drug reactions (Bennett, Jacob, Hymes, Usvyat, & Maddux, 2014; Hermanson, Bennett, & Macdougall, 2016). Moreover, degradable polymers such as XTEN are replaced with PEG to enhance peptide stability. Peptides are rapidly filtered through glomeruli du to their small size and their limited tubular uptake. Renal clearance of peptide drugs could be modified by their conjugation to polymers, albumin-binding small molecules, or increasing their plasma protein binding (Diao & Meibohm, 2013).

Physicochemical properties of peptides are critical for their absorption, distribution, metabolism, and excretion (ADME) properties. Prediction of

Table 2 Strategies to Improve Poor Pharmacokinetic Properties of Peptide Drugs

Properties	Solutions	Mechanisms
Low absorption	N-methylation, cyclization, introducing intramolecular hydrogen bonds, incorporation of positively charged amino acids, repetitive arginine-containing modules, stapled peptides	Reduce hydrogen bonding potential, increase rigidity, reduce intermolecular hydrogen bonds and flexibility, enhance absorption
	Using the vitamin B12 multivitamin uptake system	Use transporter system
	Bile salts, phospholipids, FAs, glycerides, surfactants, EDTA, heparin	Enhance absorption, increase solubility and paracellular permeability
	Mildly perturbing mucosal surfaces, oral permeability enhancers, cell-penetrating peptides	Enhance transcellular permeation
	Incorporation of various sugars, binding to endogenous proteins, conjugation to nanoparticles, biotin conjugates	Enhance absorption
Insufficient biodistribution	Cyclization, C-terminal amide, N-terminal acetylation, N-terminal ring formation	Protect C and N terminus
	Conjugation	Increase membrane transport
Rapid metabolism	Replacing L-amino acids with D-amino acids, modification of amino acids (αMet-AAs, N-me-AAs, β-AAs), side chain modification, disulfide-rich peptides, asparagine deamidation/aspartic acid isomerization, acylation	Protect from proteolysis
	Conjugation to macromolecules, antibody Fc portion	Protect from proteolysis, prolong half-life
	Protease inhibitors	Protect from proteolysis
Rapid excretion	Conjugation to large polymers (PEG, PSA, XTEN, HES, PVP)	Reduce glomerular filtration, prolong half-life
	Covalent linkage to albumin-binding small molecules	
	Fusion to long-live plasma proteins	

ADME properties with in silico tools can improve drug potential of peptides. Besides, transporters and permeability enhancers could take essential role in peptide pharmacokinetic features.

3. PEPTIDE DRUG DISCOVERY

3.1 The Empirical Approach

The empirical approach is the most commonly used method to discover bioactive peptides that are obtained from food proteins. Protein extract from foods is hydrolyzed by using proteolytic enzymes or by fermentation to obtain peptide fragments (Darewicz, Borawska, Vegarud, Minkiewicz, & Iwaniak, 2014; Ha et al., 2015). Those fragments can then be fractionated based on their structural properties and can further be tested for their benefits in vitro. Furthermore, the active fractions of the peptides can be defined using liquid chromatography–tandem mass spectrometry (Garcia-Tejedor et al., 2014). Using this approach, Bidasolo et al. discovered functional peptides after digesting donkey's milk casein with pepsin and corolase enzymes (Bidasolo, Ramos, & Gomez-Ruiz, 2012). Even though this approach eases the process of novel bioactive peptide discovery in protein samples, it usually neglects the less concentrated peptides present in the media. Also, this method is costly and time consuming for screening process. Therefore, computational protein design emerges as an economical alternative to this approach.

3.2 Computational Design of Peptide Ligands for Protein Receptors

The majority of the computational design methods utilize a rational strategy while constructing the peptide of interest. In the rational strategy, availability of the receptor structure obtained either by structural biology techniques (i.e., X-ray crystallography, nuclear magnetic resonance spectroscopy, electron microscopy) or homology modeling is assumed. When the receptor structure is at hand, binding of the desired peptide can be modeled by using different algorithms. Such designs ultimately aim at blocking or enhancing specific protein–protein interactions (PPIs) (Modell, Blosser, & Arora, 2016). There is a growing interest in designing peptides with such a structure-based approach, not only because it is cost effective but also because it can provide a physics-based model on predicting the design's affinity toward its receptor (Kilburg & Gallicchio, 2016; Sable & Jois, 2015).

3.2.1 General Rules for Receptor–Peptide Binding

Peptides are flexible by nature, which allows them to refold upon binding their receptors. To compensate this entropic cost, they form weak interactions with their partners. Their flexibility and weak binding nature make the description of their complex formation very difficult (London, Raveh, & Schueler-Furman, 2013; Trellet, Melquiond, & Bonvin, 2013). To that end, a noteworthy effort has been invested in unveiling the receptor–peptide binding patterns. This effort has been embodied in gathering databases of available receptor–peptide structures [PepBind (Das, Sharma, Kumar, Krishna, & Mathur, 2013), DOMMINO (Kuang et al., 2012), peptiDB (London, Movshovitz-Attias, & Schueler-Furman, 2010), PixelDB (Frappier, Duran, & Keating, 2017), THPdb (Usmani et al., 2017)]. Statistical analyses performed on these databases have delivered a number of peptide characteristics:

- Peptides are more complex and flexible than standard medicinal compounds. They tend to establish extended interactions with their protein partners (Kilburg & Gallicchio, 2016).
- Compared to PPIs, they form a more packed interface enriched in main chain hydrogen bonding (London et al., 2013).
- Receptor–peptide interactions are driven mainly by hotspots[a] positioned on the receptor (London et al., 2013; Sable & Jois, 2015).
- Core-binding sites on peptides are mostly made of hydrophobic residues, i.e., *Leu*, *Met*, and *Phe* (Frappier et al., 2017).
- Peptide binding generally does not induce conformational change in the receptor (Vanhee et al., 2011).
- Protein–peptide interactions follow structural folds that are observed within monomeric proteins (Vanhee et al., 2009).

Current knowledge on receptor–peptide binding has shaped the content of the receptor–peptide structure prediction algorithms. In Table 3, you will find a brief collection of these prediction tools and also how they make use of the current knowledge in predicting different types of receptor–peptide binding.

3.2.2 Available Methods

Prediction of a receptor–peptide complex involves two layers of complexity: prediction of (i) the peptide's binding site on the receptor and (ii) the structure of the bound peptide on the receptor (Peterson et al., 2017). As addressing

[a] If replacement of an amino acid residue by alanine lowers the free energy of binding at least by 2 kcal/mol, then this residue is classified as a "hotspot." Frequently observed hotspot residues are Tyr, Trp, and Arg.

Table 3 Structure-Based Methods for the Prediction of Receptor–Peptide Binding

Method and Availability	Category	Input Peptide	Reference
Data-driven tools			
Rosetta FlexPepDock ab initio http://flexpepdock.furmanlab.cs.huji.ac.il	Docking	Extended backbone conformation	Raveh, London, Zimmerman, and Schueler-Furman (2011)
HADDOCK http://haddock.science.uu.nl/services/HADDOCK2.2/haddock.php	Docking	Ensemble of three canonical peptide conformations: an a-helix, extended, or polyproline II	Trellet et al. (2013)
GalaxyPepDock http://galaxy.seoklab.org/pepdock	Knowledge-based modeling (uses similar protein–peptide interactions as templates)	Peptide sequence	Lee, Heo, Lee, and Seok (2015)
PepComposer	Knowledge-based modeling (uses motifs found in monomeric proteins as backbone scaffolds)	No input is required, plausible peptide sequences that can bind to a given pocket are listed	Obarska-Kosinska, Iacoangeli, Lepore, and Tramontano (2016)
AnchorDock	Molecular dynamics (uses ANCHORSmap's output)	Peptide sequence	Ben-Shimon and Niv (2015)
Ab initio tools			
MedusaDock	Adapted molecular dynamics (with highly weighted electrostatics)	Peptide sequence	Dagliyan, Proctor, D'Auria, Ding, and Dokholyan (2011)

Continued

Table 3 Structure-Based Methods for the Prediction of Receptor–Peptide Binding—cont'd

Method and Availability	Category	Input Peptide	Reference
FoldX Suite http://foldxsuite.crg.eu	Knowledge-based modeling (uses motifs found in monomeric proteins)	Peptide sequence	Verschueren, Vanhee, Rousseau, Schymkowitz, and Serrano (2013)
PiPreD http://www.bioinsilico.org/PIPRED/	Knowledge-based modeling (uses precompiled and bespoken library of structural motifs, extracted from protein complexes)	Peptide sequence	Oliva and Fernandez-Fuentes (2015)
pepATTRACT http://bioserv.rpbs.univ-paris-diderot.fr/services/pepATTRACT	Docking	Peptide sequence	de Vries, Rey, Schindler, Zacharias, and Tuffery (2017)
IDP-LZerD http://kiharalab.org/proteindocking/idplzerd.php	Docking	Peptide sequence (can also account peptides having a sequence length >15 amino acids)	Peterson, Roy, Christoffer, Terashi, and Kihara (2017)
CABS-dock http://biocomp.chem.uw.edu.pl/CABSdock/	Docking	Peptide sequence	Kurcinski, Jamroz, Blaszczyk, Kolinski, and Kmiecik (2015)

both problems is a challenging task, the majority of the structure prediction methods necessitate knowledge of the peptide-binding site on the receptor (data-driven modeling). Binding site information can be obtained via different sorts of binding experiments (e.g., mutagenesis) or bioinformatic predictions (e.g., Pepsite (Trabuco, Lise, Petsalaki, & Russell, 2012), Peptimap (Lavi et al., 2013), ANCHORSmap (Ben-Shimon & Eisenstein, 2010)). Some other methods, on the other hand, perform a global search on the receptor (ab initio modeling) for finding the optimum binding pocket. Available data-driven and ab initio algorithms are constructed either as docking,[b] knowledge-based,[c] or molecular dynamics[d] modeling packages (Kilburg & Gallicchio, 2016; London et al., 2013; Sable & Jois, 2015). Based on the available information on the peptide under study, any of the mentioned packages can be utilized.

3.3 Phage Display

Phage display method was first defined by G. P. Smith et al. in 1985 to express cloned antigens on viral surface (Smith, 1985). It is a combinatorial technology which attracts a great deal of attention for the future of drug discovery. Other than phage display, different molecular display techniques can be also employed in peptide discovery, namely, yeast and bacterial display, ribosome display, mRNA display, CIS, and covalent antibody display (Wu, Liu, Lu, & Wu, 2016). In phage display method, coat protein of the phage is engineered to express random sequence peptides or antibodies on the surface. Afterward, the phage library is tested on desired target in terms of target-ligand binding (Hamzeh-Mivehroud, Alizadeh, Morris, Church, & Dastmalchi, 2013). This method is a robust tool in drug discovery principally for peptide drug identification (Rothe, Hosse, & Power, 2006). Phage display method enables researchers to construct libraries and rapidly isolate and identify specific protein interactions of molecular targets (Noren & Noren, 2001). Furthermore, this technique facilitates ligand identification for receptors, enzyme blocker discovery, protein/DNA–protein interactions, cDNA expression screening, antibody epitope mapping, antibody engineering, and vaccine design (Hamzeh-Mivehroud et al., 2013). Phage display has several advantages over traditional screening techniques, considering its ease of use, cost effectiveness, agility, and more importantly high-throughput screening

[b] Docking depicts plausible binding modes of the peptide on the receptor.

[c] Knowledge-based modeling builds receptor–peptide interactions by following structurally related interaction patterns deposited in structural databases.

[d] Whereas molecular dynamics look for stable contacts within a time course of the simulation.

ability of peptide and peptide variants. In phage display, after binding of phage to target, selected phages are eluted and phage DNA is sequenced for identification (Omidfar & Daneshpour, 2015). Afterward, selected ligands can be structurally analyzed to produce specifics of the receptor–ligand interaction (Omidfar & Daneshpour, 2015). Bioactive peptide screen using phage display is a powerful approach for also food-derived peptides. The newly discovered peptides using phage display technique can then be cloned and overexpressed to increase yield in hybrid technology combining solid and liquid synthesis.

3.4 Integrated Approach

As an alternative to the isolated use of empirical or computational methods, integrated approaches can be employed in bioactive peptide discovery. These approaches combine empirical and computational tools, in order to maximize the advantages of both methods in designing and testing therapeutic peptides in the functional food and health applications (Daliri, Lee, & Oh, 2017). As an example, in silico determined peptides derived from food proteins can be chemically synthesized, which can lead to the discovery of novel peptides (Udenigwe & Aluko, 2011).

3.5 Peptide Production and Manufacturing

Apart from erythropoietin (EPO)-derived peptides, there are many other peptide types that are studied preclinically and clinically. For the production of such peptides, there are alternative methods and definite determinants, like size and their nature. So, peptides can be produced by several methods, namely, chemical synthesis, recombinant expression, and in vitro translation.

Chemical peptide synthesis can be used to produce peptides less than 100 aa length. This method allows researchers to synthesize peptides composed of nonnatural or D-amino acids, pseudo-peptide bonds, and directed disulfide bonds to allow broader diversity than other peptide synthesis methods (Furman et al., 2015). Substitution of natural amino acids to unnatural ones also results in higher plasma stability as they can escape from proteases (Vlieghe et al., 2010). Furthermore, constructing synthetic combinations of peptide library that is composed of random peptides can enhance drug discovery rate and candidate selection.

Chemical peptide synthesis has been first performed by Vigneaud et al. in 1953 (Vigneaud et al., 1953), and since then, novel techniques have been developed for synthesis of the peptides in terms of coupling efficiency, diversity of amino acids, and addition of functional groups. Current advancements in the synthetic peptide synthesis by automated systems have accelerated this process, while the yield of peptide synthesis is greatly affected by coupling efficiencies. As the length of peptide increases, poorer coupling reactions are encountered in the final yield of pure peptide (Furman et al., 2015).

In the case that pseudo-peptide bond and unnatural amino acids are not required, recombinant protein expression methods may be the best option for manufacturing. In this method a fusion protein, human serum albumin (HSA), Fc, and XTEN polymer, is attached to N- or T-termini of the protein to be manufactured (Doknic, Stojanovic, & Popovic, 2014; Kenanova et al., 2010). This system also requires a host organism to express protein, such as transgenic plants and animals, bacteria, yeast, insect cells, and mammalian cells. Furthermore, there are also some commercially available recombinant peptide therapeutics available including nesiritide, teriparatide, and salmon calcitonin (Chandrudu, Simerska, & Toth, 2013; Vlieghe et al., 2010).

Another method for protein manufacturing includes cell-free (in vitro) translation technologies. Those technologies include necessary components for translation machinery obtained from cell extracts of *Escherichia coli* S30, rabbit reticulocytes, wheat germ, *Leishmania*, *Thermus*, or human (Furman et al., 2015). This technique has gained some advantages over traditional cell-based methods (Murray & Baliga, 2013). First, this method eliminated the limitations of cell wall or membranes by adding chaperones, isomerases, or posttranslational modifier enzymes to achieve proper translation and folding. Second, it allows more efficient use of reactor volumes and eliminates cell viability limitations (Harris & Jewett, 2012).

Nevertheless, cell-free systems have been generally utilized in small-scale peptide production, resulting in batch-to-batch variability and improper folding in the case of scale-ups. Recent advancements could allow manufacturers to linearly scale up under cGMP conditions, which is transferrable to industrial scale (Uhlig et al., 2014). As an example, 100 L batch of biologically active granulocyte-macrophage colony-stimulating factor (GM-CSF), a 127-amino acid protein can give yield of 700 mg/L in 10-h period (Hipolito & Suga, 2012). So, this method can be a prominent alternative to traditional peptide generation methods. FDA's guideline specific for quality control in peptide manufacturing includes guidelines for the specifications of GMP peptides (Commission Directive, 2003).

Peptide drug preparations must be stable, sterile, and pyrogen- and particulate-free. Especially during long-term storage microbial contamination of peptide preparations results in loss of biological activity by proteolysis. Thus, preservative formulations in ingredients are edited (Bak et al., 2015;

Furman et al., 2015). Peptides must be stored in suitable containers to avoid their adsorption on surfaces. Appropriate container retains stability of the peptides during processing and storage (Bak et al., 2015).

Currently, most peptide drugs are administered by parenteral route. However, orally bioavailable peptides are more suitable for patient's convenience, compliance, and adherence. There are only a few oral peptide drugs in the market, for example, cyclosporine, desmopressin, and calcitonin (Bak et al., 2015). These drugs have also spray formulations for intranasal delivery. The disadvantage of peroral and intranasal administration is intra- and inter-individual variability of bioavailability. Oral formulations often interact with food. As an alternative, peptide drug delivery through intestinal patch system can be used instead of oral usage (Banerjee & Mitragotri, 2017). This innovative approach facilitates paracellular uptake of peptides from intestinal wall to systemic circulation.

Because unmodified peptides with short half-life are eliminated before they act, they should be provided in frequent high doses through parenteral delivery (Diao & Meibohm, 2013). For example, long-acting analog of glucagon-like peptide-1 (GLP-1) is used for the normalization of blood glucose (Deacon, 2004). Another example is depot injection form of leuprorelin that provides slow, continuous, and controlled release of the drug into the systemic circulation. Monthly intramuscular injection of leuprorelin stimulates the release of follicle-stimulating hormone and luteinizing hormone (Periti, Mazzei, & Mini, 2002).

4. ERYTHROPOIETIN

4.1 Erythropoietin and Erythropoietin Receptor

EPO is a 34-kDa glycoprotein hormone, cytokine, and hematopoietic growth factor, belonging to the type I cytokine superfamily (Broxmeyer, 2013; Genc, Koroglu, & Genc, 2004a). The primary production of EPO is in the liver during fetal development. After birth, production subsequently changes to the kidney (Souma, Suzuki, & Yamamoto, 2015). EPO binds to its receptor (EPOR) on erythroid precursor cells in bone marrow and sustains their viability, proliferation, and differentiation in normoblast (Genc, Koroglu, & Genc, 2004b). A few amount of EPO was first isolated from 2500 L of urine of aplastic anemia patients by Goldwasser and Kung (Miyake, Kung, & Goldwasser, 1977). Obtaining pure human protein was a pioneering approach for the partial identification of amino acid sequence, construction of isolated DNA probes according to tryptic fragments, and finally gene cloning

(Jacobs et al., 1985; Lin et al., 1985). Right after cloning, DNA-derived recombinant human protein EPO (rhEPO) was successfully used to treat anemia in end-stage renal disease in clinical trials, and the US Food and Drug Administration (FDA) approved to use EPO for the treatment of anemia related with renal failure, prematurity, human immunodeficiency virus infection, cancer, and chronic inflammatory disease (Fisher, 2010; Genc et al., 2004a).

EPO gene encodes a 193-amino acid sequence protein at the beginning, and after posttranslational modifications (PTMs), the remainder sequence leads to a mature protein containing 165 amino acids to which four glycans are attached (Rahbek-Nielsen et al., 1997; Ribatti, 2008). Three N-glycans and one O-glycan are making up EPO's carbohydrate part (~40% of the protein). N-glycans provide protection to EPO from proteases and prevent disruption of its receptor-binding affinity (Jelkmann, 2013).

EPOR is a 59-kDa glycoprotein and member of cytokine receptor superfamily (Jelkmann, 2011). It is encoded by the human EPOR gene, located in the short (p) arm of chromosome 19 at position 13.2 (Winkelmann, Penny, Deaven, Forget, & Jenkins, 1990), containing 8 exons that synthesize 508 residues. The first 24 amino acids code a signal peptide, which is cut for maturation. After glycosylation, phosphorylation, and ubiquitination processes, mature human EPOR protein is produced (Farrell & Lee, 2004). EPOR has three major domains: an extracellular domain, including two pairs of cysteine residues to provide binding of the ligand and also conserved motif (WSXWS); a single transmembrane domain; and an intracellular domain lacking catalytic action (Farrell & Lee, 2004). EPOR preexists on the surface of erythroid progenitors as homodimers, which experiences a conformational change upon binding of its ligand EPO. This change brings its intracellular domains into close position (Livnah et al., 1999; Remy, Wilson, & Michnick, 1999). Dimerization of EPOR subunits assembles with janus kinase 2 (JAK2) and other members of molecular signaling machinery. This results in autophosphorylation and activation of JAK2 kinases (Youssoufian, Longmore, Neumann, Yoshimura, & Lodish, 1993). Phosphorylation of tyrosine residues leads to several downstream signaling pathways. One of them is induction of several antiapoptotic proteins (Bcl-2 and Bcl-XL) by active phosphoinositide-3-kinase (PI3-K) (Broxmeyer, 2013). One of the other pathways is dimerization and translocation of signal transducer and activator of transcription 5 (STAT5) to the nucleus where proliferation and cell survival genes are transcribed with the induced STAT5. STAT5 has a role in stimulation of mitochondrial antiapoptotic protein BCL-XL as well (Sargin, Friedrichs, El-Kordi, & Ehrenreich, 2010). Activation of extracellular signal-regulated kinase 1/2 (ERK1/2) by EPOR signaling stimulates nuclear translocation of nuclear factor (erythroid-derived 2)-like 2 (Nrf2) and plays pivotal role in neuroprotective effects of EPO in SHSY-5Y cell line and primary rat cortical neurons (Genc, Egrilmez, & Genc, 2010; Zhang, Zhang, Zhao, & Sun, 2017). Inactivation of the receptor is mediated by dephosphorylation of Src homology region 2 domain-containing phosphatase-1 (SHP-1). Overall, proliferation, inhibition of apoptosis, and differentiation are the net effect of EPOR stimulation (Sargin et al., 2010).

4.2 EPOR-Associated Receptors

Recently, it has been shown that EPO binds to various receptors that mediate diverse effects of EPO in different tissues. The diversity of biological roles of same molecule is achieved by the expression of receptor type and affinity of molecules.

The β common receptor (βCR; CD131) is a member of type-1 cytokine receptor family. βCR, also known as a hemopoietin receptor, shares a common motif (WSXWS) in the extracellular part with EPO and has a long intracellular tail as well (Watowich, 2011). βCR is encoded by the human colony-stimulating factor 2 receptor beta (CSF2Rβ) gene which is located at the chromosome 22 (position 22q13.1) and involves 14 exons (Debeljak, Solar, & Sytkowski, 2014). βCR has a common beta chain subunit of the high-affinity receptor for GM-CSF, interleukin 3 (IL-3), and interleukin 5 (IL-5) (Hermanson et al., 2016). Binding of specific cytokine α receptor subunits to the βCR by disulfide bonds indicates ligand specified as observed for EPOR. The interaction of EPOR and βCR receptor was first hypothesized to play a role in hematopoiesis based on the phosphorylation of βCR in hematopoietic cells with EPO signaling. Moreover, the protection effect of EPO observed in nonhematopoietic tissues (kidney, vessels, heart, and brain) can be appear through EPOR–βCR heterodimer or EPOR–β CR–EPOR heterotrimer (Brines & Cerami, 2006). The βCR and EPOR expression has been examined in a variety of tissues including central and peripheral nervous system, heart, kidney, retina, endothelium, and muscle (Brines & Cerami, 2012). Its signaling pathway is still not well understood; however, it is anticipated that the signaling pathways activated by EPOR–β CR heterodimer are different from the signaling pathway occurred by EPOR dimerization. Consequently, the only clear fact is that βCR–EPOR heterocomplex has a role in tissue protective response (Brines & Cerami, 2012; Ghezzi & Brines, 2004).

Ephrin receptors which form the largest family of receptor tyrosine kinase (RTK) family and their ligands, the ephrins, have a role in several developmental processes, mainly in the nervous system (Debeljak et al., 2014). The protein encoded by this gene was specified as survival factor in several cancers and also binds to ephrin-B2 ligand to play an important role in vascular development (Debeljak et al., 2014). Identification of the EphB4 as an alternative EPOR has many consequences including opportunities such as anti-EphB4 approaches to clear away the stimulatory effects of EPO on tumor growth (Pradeep et al., 2015). Glioma cells exhibits higher EphB4 expression and overexpression of EphB4 leads to growth of glioma cells (Chen et al., 2013). In 2013, there was a new European patent application made for NEPOR, a novel tissue protective EPO-binding receptor complex. NEPOR is composed of EPOR heteroreceptor dimer EPOR–EPHB4, homoreceptor EPOR–EPOR, and EPHB4–EPHB and other components stemming from ephrin biology (Debeljak et al., 2014).

EPO and vascular endothelial growth factor (VEGF) have effects on the endothelium and endothelial progenitor cells to increase the number and performance properties that are dependent on nitric oxide (NO) stimulated by both EPO and VEGF (Sautina et al., 2010). VEGF has many receptors and among them, vascular endothelial growth factor receptor 2 (VEGFR2) has a serious role in EPO function. VEGFR2 is a kinase domain receptor (KDR) and its gene is located on chromosome 4 (position: 4q12), encoding 30 exons (KDR Kinase Insert Domain Receptor, 2017). VEGFR2 is stimulated to become active in neoangiogenesis in hypoxic endothelial cells and be a member of the EPOR–βCR complex which is essential for production of nitric oxide with the addition of VEGFR2 (Brines & Cerami, 2012; Kimakova, Solar, Solarova, Komel, & Debeljak, 2017). In addition, βCR also colloborates with VEGFR2 in other cellular systems and forms homo or heterotrimers (Brines & Cerami, 2012).

Unfortunately, the expression pattern of the different EPOR complexes and their activation in different cell types are not well understood yet.

4.3 Adverse Effects of Recombinant Human Erythropoietin

The erythropoietic activity of EPO has proven very beneficial for the treatment of anemia. Both biosynthetic and synthetic analogs of EPO are available in the market (Jelkmann, 2013). However, several side effects can emerge according to desired erythropoietic activity, such as pure red cell aplasia, tumor growth, and thromboembolic and hypertension complications. According to

FDA statement, when compared with the patients that did not treat with EPO, treatment with EPO analogs resulted in earlier progression or sooner death (Uversky & Redwan, 2016).

Some clarifications for the adverse effect of rhEPO in anemic cancer patients include tumor promotion via signaling involving EPO or erythropoiesis-stimulating agents (ESAs) and EPOR. According to this hypothesis, several preclinical studies indicate the existence of EPO-induced proliferation of cancer cells, migration and invasion of cancer cells, stimulation of chemotaxis, increase in drug resistance, interference of apoptosis, and protection from ischemic injury. In addition, the presence of EPOR mRNA or EPOR protein has been observed in several tumor types among different organs (Bennett et al., 2016; Uversky & Redwan, 2016). EPO induces proliferation, migration, and invasion of several cancer cells including glioma cells (Alural, Ayyildiz, Tufekci, Genc, & Genc, 2017; Peres et al., 2011). Besides, EPO has suppressed sensitivity in glioma cells toward chemotherapeutic agents (Alural et al., 2017; Peres et al., 2011). The patients with tumors which have EPORs might not be affected from the harmful proliferative effects of ESAs (Erslev, 1993; Uversky & Redwan, 2016). Although it is not determined how ESAs affect the tumor control, two potential mechanisms were propounded by McKinney and Arcasoy for tumor progression by rhEPO therapy (McKinney & Arcasoy, 2011). rhEPO may have an effect locally in tumor cells or other cell types found in tumor microenvironment, and rhEPO may alter tumor biology by raising the systemic effects or increasing specific systemic toxicities (Debeljak et al., 2014).

Another unwanted effect of ESAs is an increase in arterial blood pressure, possibly caused by hypertension in the treatment of patients with chronic kidney disease (CKD) (Provatopoulou & Ziroyiannis, 2011). Rise in blood pressure can be connected to increase in blood viscosity (Jelkmann, 2013; Provatopoulou & Ziroyiannis, 2011).

Nearly all biopharmaceuticals induce an immune response, but their effects can be remissible or serious. Antibodies to therapeutic proteins can also react with endogenous molecules to increase toxicity, while they can be efficient for the treatment of the target tissue (Uversky & Redwan, 2016). Similarly, pure red cell aplasia was emerged from patients with CDK who receive recombinant erythropoietin in late 1990s (Casadevall, Eckardt, & Rossert, 2005). Immunogenicity studies have shown that pure red cell aplasia is caused by the arising of autoantibodies against rhEPO (Hermanson et al., 2016; Uversky & Redwan, 2016).

The studies provide an evidence that thromboembolic complications and death risk were increased with the ESAs treatment. Endothelial cells are activated by EPO to express adhesion molecules which induce plug accumulation in vessels with platelets and leucocytes by thrombus formation (Jelkmann, 2013). ESA treatment increased stroke risk in a clinical study involving 4038 patients with type 2 diabetes and CKD (Pfeffer et al., 2009).

4.4 Neuroprotective Effects of Recombinant Human Erythropoietin

EPO and EPOR are expressed by other tissues, including heart, kidney, and central and peripheral nervous system. Several in vitro and in vivo studies demonstrated neuroprotective, cardioprotective, and renoprotective effects of EPO (Nekoui & Blaise, 2017).

EPOR is expressed in neuronal cell lines, primary cultured hippocampal and cortical neurons, and murine hippocampal and cerebral cortical areas (Masuda et al., 1993; Morishita, Masuda, Nagao, Yasuda, & Sasaki, 1997). Western blotting experiment by using protein samples obtained from PC12 cells revealed that the size of EPOR was smaller than in erythroid cells (Masuda et al., 1993). Recent studies with new antibodies against EPOR developed using novel genetic strategies also confirmed EPOR expression in nervous system (Maxwell et al., 2015; Ott et al., 2015). Conditional knockdown of EPOR leads to embryonic and adult neurogenesis deficits, suggesting that EPO and EPOR are essential for embryonic neural development and regeneration of neurons in ischemic injury (Tsai et al., 2006). EPO expression was detected in primary astrocyte cell culture and culture supernatant, supporting that EPO is produced and secreted by astrocyte (Masuda et al., 1994). EPO and EPOR are expressed by Schwann cells in rat sciatic nerve and their expression increased in crush injury (Li, Gonias, & Campana, 2005). In vitro neuroprotective effects were first demonstrated by Sakanaka et al. against glutamate toxicity in 1997 (Morishita et al., 1997). After 1 year, in vivo neuroprotective effects of EPO against ischemic damage in Mongolian gerbils have been reported (Sakanaka et al., 1998). Then, various in vitro and in vivo preclinical studies showed a wide range of neuroprotective effects of EPO in acute and chronic degenerative disorders (Sargin et al., 2010). These promising preclinical results encouraged the EPO clinical studies in neurodegenerative disorders.

4.5 Clinical Failure of Recombinant Human Erythropoietin in Neurological Disorders

EPO and EPOR have been shown to exist and function in several nonerythroid tissues including the brain, heart, and vascular endothelium. In vitro and in vivo studies demonstrated that EPO exerts tissue protective effects in a number of conditions (Solling, 2012). Thus, it was considered that the tissue protective effects of EPO may translate to humans. But, only a few multicentered and large clinical trials with EPO have been published until today. Clinical trials with EPO have been carried out to reduce tissue damage especially in cardiac ischemia, stroke, and kidney diseases (Lund, Lundby, & Olsen, 2014). A large-scale multicentered study with 522 stroke patients determined that EPO increases mortality without any positive effect on neurological outcome (Ehrenreich et al., 2009). EPO also did not exhibit cardioprotective and nephroprotective effect in most of clinical studies (Solling, 2012). Tissue protective effects of EPO in preclinical studies may not have been transferred to the clinic due to various reasons. Major reason for this discrepancy could be the differences in anatomy and physiology between rodents and humans (Reagan-Shaw, Nihal, & Ahmad, 2008; Shanks, Greek, & Greek, 2009). Differences in the dosage and timing of treatment may also explain clinical translation failure (Lund et al., 2014). rhEPO has been used in animal studies at higher doses than human studies. Lower dose of EPO than used in animal studies does not exhibit therapeutic effect in clinical studies. Lack of replication and reproducibility in experimental animal studies with EPO due to inappropriate methodology is common (Steward, Popovich, Dietrich, & Kleitman, 2012). The absence of ideal animal models that represent all the features of the diseases is another factor limiting the translation of knowledge from preclinical studies to the clinic (Mak, Evaniew, & Ghert, 2014). Another limitation in preclinical evaluation is the inability to consider comorbid diseases and possible interactions with other drugs.

5. CHEMICALLY MODIFIED AND NONERYTHROPOIETIC ERYTHROPOIETIN ANALOGS

The emergence of side effects related to EPO therapy has led to the development of modified forms of EPO with no erythropoietic effect but with retained tissue protective effect. EPO molecule could be altered by several ways such as carbamoylation, amidation, and acylation (Chen, Yang, & Zhang, 2015; de Lange et al., 2017; Torup & Leist, 2006). Carbamoylated EPO (CEPO) exhibits neuroprotective effect in wide range of animal models of neurological disorders including ischemic stroke, spinal cord injury, and diabetic neuropathy binding to CD131/EPO heteroreceptor (Chen et al., 2015). The results of completed clinical safety studies with CEPO are not published yet. Glutaraldehyde EPO (GEPO) was obtained by chemical modification of the lysine residues in rhEPO (Chattong et al., 2013). GEPO interacts with heterotrimeric receptor of EPO and increases BCL2 expression (Chattong et al., 2013). GEPO protects kidney cells against silver nanoparticle exposure and ischemia reperfusion injury (Chattong et al., 2013; Sooklert et al., 2016). The effects of GEPO in the nervous system have not been evaluated to date. Neuro-EPO is a low sialic acid-containing EPO that exhibits neuroprotective effect in experimental focal cerebral ischemia and Alzheimer's disease (Rodriguez Cruz, Strehaiano, Rodriguez Obaya, Garcia Rodriguez, & Maurice, 2017; Teste et al., 2012). Randomized open-label safety study with intranasal Neuro-EPO in healthy volunteers was reported with no safety issues (Lagarto et al., 2012). Phase I/II clinical trials have just started in the treatment of stroke and spinocerebellar ataxia (World Health Organization, 2017a, 2017b). Less acidic modified EPO glycoform called neuropoietin does not exhibit significant in vivo hematopoietic activity and retains neuroprotective effect in vitro (Mattio et al., 2011). The ability of the neuropoietin to pass blood–brain barrier (BBB) faster than rhEPO gives advantages to neuropoietin for faster penetration into damaged tissue (Etcheverrigaray et al., 2011). In vivo studies evaluating neuroprotective effect of neuropoietin in animal model of neurodegenerative disorders have not performed yet.

Although chemically modified nonerythropoietic analogs of EPO circumvent its adverse effects, high expense, and require parenteral administration, potential development of antidrug antibodies and promotion of tumorigenicity should be considered especially in chronic applications (Collino, Thiemermann, Cerami, & Brines, 2015).

6. GROWTH FACTOR MIMETIC PEPTIDE DRUGS

Growth factor mimetic peptides are promising candidates in the treatment of various diseases with low or no side effects of growth factors. They have advantages as a therapeutic molecule in terms of oral usage, longer half-life, crossing BBB, and less immunogenicity. One of the growth hormone mimetic peptides, P021, is an 11-mer ciliary neurotrophic factor (CNTF) mimetic peptide generated by Kazim and Iqbal (2016). Its half-life is longer than 3 h, and it has higher stability in artificial intestinal fluid and the capacity to cross the BBB. Perorally treatment of 3 × Tg Alzheimer's disease mice with P021 for 12 months decreased tau pathology (Kazim & Iqbal, 2016). Thrombopoietin mimetic peptides are one of first-generation growth factor mimetics entering clinical use. They are used for the treatment of immune thrombocytopenic purpura (ITP), hepatitis C-associated thrombocytopenia, and myelodysplastic syndrome (Gutti et al., 2016).

Shortly after recombinant EPO, peptide and nonpeptide erythropoiesis-inducing agents were developed. EMP-1 is a first developed synthetic peptide with ESA activity (Johnson et al., 1998). EMP-1 did not exhibit potent erythropoietic activity. But, it is very useful to study the activation mechanisms of EPOR (Guarnieri, 2017).

Continuous erythropoietin receptor activator (CERA) is a 60-kDA molecule; its half-life has been extended to 135 h by modification in formulation (Gutti et al., 2016). Erythropoietic activity of CERA is not superior to rhEPO and darbepoetin (Saglimbene et al., 2017).

Peginesatide is another peptide ESA that has potent erythropoietic activities. It is synthetic PEGylated dimeric peptide in which the primary sequence is unrelated to EPO (Gutti et al., 2016). Peginesatide binds and activates EPOR and stimulates erythropoiesis in vitro and in vivo. Phase III studies reported higher cardiovascular events and mortality in patients treated with peginesatide (Hermanson et al., 2016; Locatelli, Choukroun, Truman, Wiggenhauser, & Fliser, 2016). Peginesatide was withdrawn from the market due to unexplained serious adverse drug reactions (anaphylactic reactions) in February 2013 (Hermanson et al., 2016).

7. ERYTHROPOIETIN MIMETIC PEPTIDES

10 EPO mimetic peptides have been recently discovered. They were demonstrated to be beneficial in in vitro and in vivo models of neurological disorders (Tables 4 and 5).

7.1 EMP-1

EMP-1 is a member of cyclic peptide family with low erythropoietic activity. It was isolated and characterized by Wrighton et al. in 1996 (Wrighton et al., 1996). EMP-1 is a 2-kDa agonistic peptide whose sequence does not have any homology with EPO's primary sequence (Middleton et al., 1999; Wrighton et al., 1996). EMP-1 can bind EPOR and lead to tyrosine phosphorylation in several erythroid and neuronal cell lines.

The effects of EMP-1 on neuronal cells were evaluated in vitro studies too (Kawakami et al., 2000, 2001), where EMP-1 was shown to inhibit Ca^{2+}-induced dopamine release in a concentration-dependent manner (Kawakami et al., 2000). The maximum effect was seen at 10 μM concentration of EMP-1. Western blotting showed that EMP-1 increased JAK2 phosphorylation and dephosphorylation of GAP-43 which has an important role in the regulation of neurotransmitter release from presynaptic neurons. EMP-1

Table 4 In Vitro Neuronal Effects of EPO Mimetic Peptides

Peptide	Cell Type	Insult Model	Dose	Time	Effects	Signaling Mechanism	Reference
EMP-1	PC12	Ca^{2+} induction	0–10 μM	10 min	Inhibits DA release, dephosphorylates GAP-43	JAK2	Kawakami, Iwasaki, Sato, and Takahashi (2000)
	Rat CGNs Rat hippocampal neurons	Ca^{2+} induction, chemical ischemia (glucose-free 1 mM KCN)	0–25 μM	10 min	Suppresses Glu release and neuronal death	NA	Kawakami, Sekiguchi, Sato, Kozaki, and Takahashi (2001)
Epopeptide AB	NS20Y SK-N-MC PC 12	Serum starvation	0–5 ng/mL	0–48 h	Stimulates differentiation, inhibits neuronal death	ERK1/2	Campana, Misasi, and O'Brien (1998)
ARA 290	Rat primary motoneurons	Kainic acid	1.8 nM	48 h	Protects against kainic acid-induced excitotoxic death	NA	Brines et al. (2008)
Epobis	Rat CGNs Rat hippocampal neurons	Low K^{+} and kainic acid	0.1–10 μM	24 h	Stimulates neurite outgrowth, inhibits neuronal death	JAK2/STAT5	Pankratova et al. (2012)
	Rat motor neurons	None	0.04–0.098 μM	24 h	Stimulates neurite outgrowth	NA	Dmytriyeva, Pankratova, Korshunova, and Walmod (2016)
	Rat microglia	LPS	0.9 μM	24 h	Suppresses TNF-α secretion	NA	

Continued

Table 4 In Vitro Neuronal Effects of EPO Mimetic Peptides—cont'd

Peptide	Cell Type	Insult Model	Dose	Time	Effects	Signaling Mechanism	Reference
Epotris	Rat CGNs Rat hippocampal neurons	Low K^+ and kainic acid	0.1–10 μM	24 h	Stimulates neurite outgrowth, inhibits of neuronal death	JAK2/STAT5	Pankratova et al. (2012)
JM-4	MOG peptide enriched T cells PC12		1 ng/mL	72 h	Suppresses proinflammatory cytokine production	NA	Yuan, Wang, Lu, Maeda, and Dowling (2015)
	PC12	$A\beta_{25-35}$ peptide	1 ng/mL	2 h pretreatment (26 h)	Inhibits neuronal death	NA	Yuan et al. (2015)
MK-X	PC12	H_2O_2	0–5 ng/mL	0–7 h	Attenuates neuronal death	ERK1/2, AKT, P38	Yoo, Cho, Moon, Yu, and Moon (2016)
	Rat cortical neuron	Glutamate-induced excitotoxicity	1 pM	15 min to 24 h	Ameliorates mitochondrial damage, inhibits caspase-dependent and -independent cell death pathways	JAK2/STAT5, ERK1/2, AKT	Yoo et al. (2017)

Table 5 In Vivo Neuronal Effects of EPO Mimetic Peptides

Peptide	Model Organism	Disease Model	Dose	Time	Administration Route	Outcome	Reference
Epopeptide AB	Mouse	None	20 μg/day	NA	Right superior gluteus muscle (7 days)	Induces motor endplate sprouting	Campana et al. (1998)
	Mongolian gerbil	3-min forebrain ischemia	4/20 ng/day	Just after ischemia/reperfusion	Lateral ventricle with osmotic minipump (7 days)	Protects against neuron death	Nagao et al. (2005)
ARA 290	Rat	Middle cerebral artery occlusion	1.5 nmol/kg	Immediately after a 1-h arterial occlusion	IV	Reduces infarct volume	Brines et al. (2008)
	Rat	Sciatic nerve compression injury	0.3 nmol/kg	Immediately after compression of sciatic nerve	IV (single injection)	Reduces sciatic nerve compression-induced damage	
	Rat	Diabetic retinal edema	1.5 nmol/kg	After confirming diabetes	IP (5 days per week, for 3 weeks)	Prevents development of retinal edema	

Continued

Table 5 In Vivo Neuronal Effects of EPO Mimetic Peptides—cont'd

Peptide	Model Organism	Disease Model	Dose	Time	Administration Route	Outcome	Reference
	Rat	Status epilepticus	1.5 nmol/kg, IV; 10 mg/kg, SC	2–5 min following status epilepticus	IV, SC	Promotes neuronal differentiation and survival, attenuated cognitive deficits in epileptic animals	Seeger et al. (2011)
	Akita mouse	Diabetic autonomic neuropathy	36.1 μg/kg	After 15–16 weeks	SC (5 days per week, for 7 weeks)	Decreases neuritic dystrophy in celiac sympathetic ganglia	Schmidt et al. (2011)
	Rat	Sciatic nerve dissection-induced neuritis	30 or 120 μg/kg	1, 2, 3, and 4 days following surgery	IP	Prevents development of mechanical allodynia	Pulman, Smith, Mengozzi, Ghezzi, and Dilley (2013)
ARA 290	Rat	Spared nerve injury-induced chronic neuropathic pain	3, 10, 30, and 60 μg/kg	1, 3, 6, 8, and 10 days postsurgery	IP	Reduces allodynia, suppresses microglial reactivity	Swartjes et al. (2014)
	Rat	Experimental autoimmune neuritis	30 mg/kg	14 days	IP	Reduces paraparesis, inflammation, enhances nerve repair, promotes phagocytic activity	Liu et al. (2014)

	Rat	EAE	35 or 70 μg/kg	Daily from Day 7 to Day 18	IP	Reduces disease severity, decreased number of inflammatory cells in spinal cord, and helper T cells in lymph nodes	Chen et al. (2014)
	Mouse	Capsaicin-induced pain in DRG and TG dissection	1 and 10 nM	15 min before capsaicin	SC	Inhibits TRPV1 channel activity, and relieves the mechanical hypersensitivity induced by capsaicin	Zhang, Yu, and Zhang (2016)
	Rat	Traumatic brain injury	60 μg/kg	3- and 12-h following injury	IP	Decreases TBI-induced pulmonary injury	Liu et al. (2017)
Epobis	Rat	EAE (MBP and CFA)	10 mg/kg	10 days after EAE induction	SC (5 days)	Delays the clinical signs of EAE, improves social memory	Dmytriyeva et al. (2016)
Epotris	Rat	Kainic acid-induced seizures	10 mg/kg	2–48 h pre	SC (single dose)	Attenuates seizures, decreases neurodegeneration and mortality	Pankratova et al. (2010)
	Rat	Status epilepticus	10 mg/kg	2–48 h pre	SC (single dose)	Attenuates microglial activation in the thalamus and modulates hippocampal neurogenesis	Zellinger et al. (2011)

Continued

Table 5 In Vivo Neuronal Effects of EPO Mimetic Peptides—cont'd

Peptide	Model Organism	Disease Model	Dose	Time	Administration Route	Outcome	Reference
JM-4	Mouse (SJL/J, C57BL/6)	EAE (PLP peptide/MOG peptide 35–55)	250 μg/kg/day	24–48 h after the onset of neurologic signs	IV	Reduces clinical sign of EAE, demyelination, axonal damage, and immune cells in peripheral lymphatics	Yuan et al. (2015)
	Mouse	TBI (cortical impact injury)	10 μg/animal	15 min, 3, 9, or 24 h after TBI	IP (repeated)	Suppresses neuron death, improves neurological deficits	Wang et al. (2016)
MK-X	Mouse	Cerebral ischemia	3.6 μg/kg	2 h after	IP (single)	Decreases the infarction area, crosses the BBB	Yoo et al. (2017)

also inhibits Ca^{2+}- and chemical ischemia-induced Glu release from cerebellar granule neurons (CGNs) via EPOR–JAK2 activation (Kawakami et al., 2000). EMP-1 suppressed neuronal death induced by the chemical ischemia for 20 min in hippocampal slices culture.

7.2 Epopeptide AB

Epopeptide AB was identified by Campana et al. in 1998 (Campana et al., 1998). Epopeptide AB has 17 aa residues in the predicted A–B loop of EPO. It dose-dependently induces neurite outgrowth and choline acetyltransferase activity of neuroblastoma cells with the maximal dose of 250 pg/mL. Functional study with EPOR antibody confirmed that differentiation induction with Epopeptide AB is EPOR mediated. This study also exhibited neuroprotective effect against serum starvation-induced neuronal cell death. Epopeptide AB had no effect on cell proliferation in EPO responsive mouse spleen cells and human erythroleukemic TF-1 cells, suggesting lack of erythropoietic activity of this peptide. 20 μg/mL Epopeptide AB injection into gluteus muscle induced motor endplate sprouting in vivo (Campana et al., 1998).

The continuous infusion of Epopeptide AB with minipump for 7 days into the lateral ventricle of mongolian gerbils just after ischemia/reperfusion protected neurons from ischemic damage (Nagao et al., 2005). Epopeptide AB increased the response latency in the step-down passive avoidance task and hippocampal CA1 neuron density in ischemic injury.

7.3 ARA 290

ARA 290 [also called as helix B surface peptide (HBSP)] is an 11-mer peptide, computationally modeled based on the three-dimensional structure of EPO's helix B, responsible for EPO's interaction with the EPOR/βCR heteromer (Brines et al., 2008). It was shown that ARA 290 lacks hematopoietic properties of EPO (Brines et al., 2008). To gain favorable storage conditions, HBSP was modified by the spontaneous cyclization of N-terminal glutamine. The pharmaceutical product pyroglutamate HBSP (p-HBSP) is stable at least 2 years at 4°C or up to 12 months at 25°C (Collino et al., 2015).

ARA 290 has been extensively investigated by in vitro, in vivo, and clinical studies. The first in vitro study which shows neuroprotective effects was examined by Brines et al. in motor neurons. ARA 290 at 1.8 nM concentration inhibited kainic acid (KA)-induced excitotoxic cell death (Brines et al., 2008). In the same study, Brines et al. also reported the protective roles of ARA 290 in middle cerebral artery occlusion injury model. In this study, administration of peptide 1 hour after occlusion adequately reduced infarct volume. Regarding traumatic brain injury (TBI), a recent study reported that ARA 290 also prevented from acute lung injury following TBI (Liu et al., 2017). In this study, ARA 290 was administered 3 and 12 h after weight drop-induced TBI model in rats and the author found decreased total bronchoalveolar lavage fluid protein, attenuated pulmonary injury, reduced CD68+ macrophages, and inflammatory markers.

Brines et al. studied the neuroprotective roles of ARA 290 in sciatic nerve compression injury in sciatic nerve compression model. Single IV injection of ARA 290 immediately after compression successfully reduced compression-induced damage. In addition, it has been reported that ARA 290 was able to prevent mechanical allodynia in both the spared nerve injury model, based on sciatic nerve surgical transection, and neuritis model in rats (Pulman et al., 2013). In this study, ARA 290 was administered intraperitoneally for 4 days post-op. In peripheral nervous system injuries, local delivery methods such as slow releasing with implantable pumps may be an alternative for a better efficiency.

In a model of status epilepticus in rats, ARA 290 notably promoted neuronal differentiation and survival, and attenuated cognitive deficits in epileptic animals (Seeger et al., 2011). In this model ARA 290 was given 2–5 min following status epilepticus by both IV and SC ways. Subcutaneous administration was repeated 24 and 72 h following status epilepticus.

ARA 290 has also been investigated in animal models of neuropathy. The first study in neuropathy models showed the reduction in the microglial activation and therefore the suppression of the neuroinflammatory response and chronic neuropathic pain (Swartjes et al., 2014). In this study, ARA 290 was tested in spared nerve injury-induced chronic neuropathic pain and administered intraperitoneally on 1, 3, 6, 8, and 10 days post-op. In another neuropathic pain model induced by capsaicin, ARA 290 was administered 15 min before capsaicin and inhibited transient receptor potential vanilloid 1 (TRPV1) channel activity, which serves as the receptor for capsaicin (Zhang et al., 2016). Moreover, ARA 290 caused the relief of capsaicin-induced mechanical hypersensitivity. The effect of ARA 290 was also investigated in neuritic dystrophy and neuropathy in Akita diabetic mouse (Schmidt et al., 2011). Akita mice, which has spontaneous point mutation of insulin 2 gene, is a genetic model of type 1 diabetes, which became diabetic at 3–4 weeks of age. ARA 290 treatment of these mice started at age 18–19 weeks via SC route. After 7 weeks, the effect of ARA 290 on metabolic parameters and sympathetic ganglia was evaluated with measurement of body weight, blood glucose, hemoglobin A1c (HbA1c), and histological analysis of sympathetic ganglia.

ARA 290 reduced neuritic dystrophy without affecting metabolic parameters and neurodegeneration (Liu et al., 2014).

There are also solid data on the neuroprotective effects of ARA 290 in experimental autoimmune encephalomyelitis (EAE) model, mimicking the human multiple sclerosis. Therapeutic administration of ARA 290 to rats reduced severity and prevented demyelination in the EAE, presumably by suppressing the expression of inflammatory cytokines in the spinal cord, repressing lymphocyte proliferation, and modifying T helper cell differentiation outcomes (Chen et al., 2014). ARA 290 also suppressed clinical findings of experimental allergic neuritis (EAN), enhanced nerve regeneration and remyelination, and inhibited nerve inflammation (Liu et al., 2014).

There is some evidence showing the beneficial effects of ARA 290 in animal models of diabetes mellitus and its complications, such as diabetic retinal edema models in rats. Regarding diabetic retinal edema, after confirming diabetes in rats, intraperitoneal ARA 290 treatment has been applied for 3 weeks and 5 days per week (Brines et al., 2008). In this model, ARA 290 treatment effectively prevented development of retinal edema in diabetes. Reengineered ARA 290 into elastin biopolymers accelerated wound healing via inhibiting apoptosis of keratinocytes and increasing angiogenesis in the wound bed in diabetic mice (Devalliere et al., 2017).

The promising data obtained from preclinical studies encouraged researchers to launch some clinical studies for neurological disorders (Table 6). The initial clinical studies were commenced in patients with sarcoidosis or type 1 or 2 diabetes mellitus. The first clinical study on ARA 290 was published by Heij et al. in 2012 (Heij et al., 2012). This double-blind, randomized, placebo-controlled trial was performed with 22 sarcoidosis patients. 2 mg intravenous ARA 290 was administered on Monday, Wednesday, and Friday for four consecutive weeks. After 1 week of treatment, a significant reduction in pain scores was observed. At the end of treatment period, small fiber neuropathy screening list scores were improved, while brief pain inventory score and fatigue assessment scale scores were reduced. No safety issues or side effects were reported by clinical or laboratory assessments. In another study on sarcoidosis patients, 26 subjects were recruited and asked to perform injection of 4 mg ARA 290 for 28 days (Dahan et al., 2013). As opposed to first trial, subcutaneous formulation of ARA 290 was utilized to allow daily self-injection of drug. This study resulted in increased density of both dermal and retinal nerve fibers in the ARA 290-treated group. Furthermore, a significant improvement in neuropathic pain and autonomic symptoms, with no effect on hemoglobin concentrations, was reported.

Table 6 Clinical Studies With ARA 290

Patient Group	Study Type	*N* (D/P)	Dose	Route	Time to Treatment	Follow-Up	Methods	Outcome	Side Effects	Reference
Small fiber neuropathy in sarcoidosis	Single-site, double-blind, placebo-controlled	22 (12/10)	2 mg	IV	3 days per week for 4 weeks		Self-report, QST	Improves symptoms and quality of life, recovery in QST	None reported	Heij et al. (2012)
Small nerve fiber loss and damage in sarcoidosis	Single-site, double-blind, placebo-controlled	38 (21/17)	4 mg	SC	Daily for 28 days	12 weeks	Self-report, QST, skin biopsy, 6-min walk test, CCM	Increases median nerve fiber area in retina and skin, and small fiber neuropathy screening list score, elevation in thermal pain threshold, and exercise capacity	Weight loss (1 patient)	Dahan et al. (2013)
Type 2 diabetes with neuropathic symptoms	Single-site, double-blind, placebo-controlled	48 (24/24)	4 mg	SC	Daily for 28 days	56 days	Self-report, QST, CCM	Increases mean corneal fiber density, improves HbA1c and lipid profiles, decreases neuropathic symptoms and QST	No difference between ARA 290 and placebo	Brines et al. (2015)

Healthy volunteers	Double-blind, randomized, parallel group	36	2 mg	IV	Single	After 6–7 days	Affective symptoms and mood states evaluation and fMRI	Suppresses neural responses to happy faces in fusiform gyrus and recognition of happy and disgust facial expressions, enhances categorization of positive and negative words and attention to positive emotional pictures	None reported	Cerit et al. (2015)
Sarcoidosis-associated small nerve fiber loss and neuropathic pain	Double-blind, randomized, placebo-controlled, two centers	64 (48/16)	1, 4, and 8 mg	SC	Daily for 28 days	Week 2 and 4	Self-report, QST, CCM, retinal tomography, skin biopsy	Increases small nerve fiber abundance in the cornea and skin, fiber length in 4 mg group, improves functional capacity	Suicidal ideation (1 patient), injection site pain, diarrhea, fatigue, nausea, headache	Culver et al. (2017)

CCM, corneal confocal microscopy; *QST*, quantitative sensory testing.

Furthermore, a double-blind, randomized, and controlled trial of subcutaneous ARA 290 was performed in type 2 diabetes with neuropathic symptoms (Brines et al., 2015). In this study, 42 patients were recruited and 4 mg subcutaneous ARA 290 was injected daily for 28 days. Neuropathic symptoms, HbA1c levels, and lipid profiles were monitored throughout the 56-day observation period. ARA 290 administration resulted in decrease in PainDetect questionnaire scores and improvement of HbA1c and lipid profiles. Their observations suggest that ARA 290 may improve metabolic control and diminish neuropathic symptoms in type 2 diabetes patients. Similar to these findings, parent protein EPO protects pancreatic β cells, increases angiogenesis in the islet, and reduces inflammation (Choi et al., 2010). EPO is also involved in the regulation of glucose and energy metabolism through the hypothalamic–pituitary axis (Dey & Noguchi, 2017).

The another published clinical study on ARA 290 was performed to test antidepressant properties. In this trial, 36 healthy volunteers were admitted and single dose of 2 mg intravenous ARA 290 was injected to ARA 290-receiving group (Cerit et al., 2015). Patients were examined by categorization, free recall, recognition, visual probe tasks, and functional magnetic resonance imaging or functional MRI (fMRI) analysis. The results of fMRI analysis suggest that ARA 290 caused lowered neural responses to happy faces in fusiform gyrus and recognition of happy and disgust facial expressions. Furthermore, ARA 290 enhanced categorization of positive and negative words and increased attention to positive emotional pictures. Also, in this study no potential safety issues and side effects were reported.

The effect of ARA 290 (cibinetide as marketing name) was evaluated in sarcoidosis-associated small nerve fiber loss which quantified with corneal confocal microscopy (Culver et al., 2017; Oudejans, Niesters, Brines, Dahan, & van Velzen, 2017; van Velzen et al., 2014). There, ARA 290 enhanced small nerve fiber in the cornea and skin (Culver et al., 2017).

The results of a phase II clinical trial (EudraCT number 2015-001940-12) in diabetic macular edema with ARA290 have not been published yet (EU Clinical Trials Register, 2017).

7.4 Epobis

Epobis is composed of an 18-mer sequence motif that mimics the C-terminal part of the AB loop of EPO. Epobis was synthesized as a tetramer by Pankratova et al. (2012). Epobis induced neurite outgrowth in CGNs and hippocampal neurons via EPOR with an optimal concentration of 1 μM.

It stimulates neurite outgrowth in primary cultures of rat motor neurons at a lower dose (maximal stimulation at 0.33 μM) (Dmytriyeva et al., 2016). STAT5 signaling pathway is activated in hippocampal neurons by Epobis (Pankratova et al., 2012). It also protected primary neurons against low potassium- and KA-induced apoptotic cell death (Pankratova et al., 2012). Apart from neurons, Epobis alters microglial responses. Treatment of primary rat microglia with 0.9 μM Epobis decreased lipopolysaccharide (LPS)-induced tumor necrosis factor-alpha (TNF-α) secretion (Dmytriyeva et al., 2016). Peptide was detectable in cerebrospinal fluid (CSF) 2 h after a single SC injection of Epobis, suggesting that it can cross BBB. Epobis treatment at 10-mg/kg dose did not decrease clinical signs and microglial activation in EAE induced by complete Freund's adjuvant (CFA), but it delayed onset of EAE clinical findings. Additionally, Epobis improved social memory in healthy adult rats at long-term period.

7.5 Epotris

Epotris is a tetrameric 20-amino acid-containing peptide which corresponds to helix C region of EPO (Pankratova et al., 2010). Epotris did not demonstrate hematopoietic effect both in vitro and in vivo. It induced neurite outgrowth both in hippocampal neurons and CGNs with an optimal concentration of 0.3 mM. Functional studies by using EPOR short-hairpin RNA and JAK2 inhibitor showed that neurotrophic effect of Epotris is mediated by EPOR and the downstream JAK2–STAT5 signaling pathway. Epotris also protected primary neurons against low potassium- and KA-induced neuronal cell death.

Systemically administered biotinylated Epotris was also detected in CSF 2 h after SC injection, suggesting it can cross the BBB (Pankratova et al., 2010). Epotris exhibited a significant delay in seizure onset, decreased frequency of high-grade seizures, and mortality in a mouse model of KA-induced seizure (Pankratova et al., 2010). Peptide protected neurons against KA-induced degeneration and attenuated status epilepticus-associated expansion of the hippocampal neuronal progenitor cells (Pankratova et al., 2010; Zellinger et al., 2011). Epotris also diminished the microglial activation in the thalamus (Zellinger et al., 2011).

7.6 JM-4

JM-4 is a 19-mer peptide derived from the AB loop of EPO (Yuan et al., 2015). JM-4 blocked amyloid beta 25–35 ($A\beta_{25-35}$)-induced apoptotic cell death in PC12 cell line in vitro (Yuan et al., 2015). Peptide passed BBB after

IV injection of radiolabeled 3H-JM4 in normal mouse (Wang et al., 2016). JM-4 peptide demonstrated neuroprotective effect in EAE and TBI model (Wang et al., 2016; Yuan et al., 2015). JM-4 therapy for 7 days delayed the onset of disease and reduced clinical signs of both proteolipid protein (PLP)- and myelin oligodendrocyte glycoprotein (MOG)-induced EAE (Yuan et al., 2015). Histopathological analysis showed that JM-4 prevented demyelination and axonal damage. JM-4 treatment suppressed inflammatory cells in peripheral lymphatics and proinflammatory cytokine production by MOG peptide-specific T cells. JM-4 therapy decreased neuronal cell death and lesion size, and improved functional recovery in TBI (Wang et al., 2016).

7.7 MK-X

MK-X was derived from α-helix C in the low-affinity binding site of EPO by Yoo et al in 2016 (Yoo et al., 2016). MK-X protected PC12 cells from hydrogen peroxide-induced cell death with an optimal concentration of 1.2 ng/mL. MK-X peptide also decreased reactive oxygen species (ROS) production of PC12 cells. It activated ERK1/2 and Akt signaling pathways, inhibited p38 phosphorylation, but did not lead to JAK2–STAT5 activation under the oxidative stress conditions. It also inhibited mitochondrial dysfunction and caspase-dependent and -independent cell death caused by glutamate-induced oxidative stress (Yoo et al., 2017). Both JAK2–STAT5 and ERK1/2 signaling pathways contribute to its neuroprotective effects. Single injection of MK-X (3.6 μg/kg) reduced infarction area in mouse model of ischemic stroke induced by middle cerebral artery occlusion (Yoo et al., 2017).

7.8 NP1 and NP2

NP1 is a 28-mer peptide originated from helix B region of EPO. NP2 is a shorter peptide (18 mer) originated from helix A region (Colella et al., 2011). Gene delivery of this EPO mimetic peptide into subretina exhibits protective effect against light-induced and genetic retinitis pigmentosa degeneration. Subretinal delivery of NP1 and NP2 gene encoding vectors did not alter hematocrit levels in rats and mice, suggesting that these peptides do not have erythropoietic activity in vivo.

7.9 AF41676 and AF43136

AF41676 and AF43136 are dimeric nonerythropoietic peptides, which have structural relationship to the peptide portion of peginesatide (Kiss et al., 2016).

Intravenous administration of AF41676 and AF43136 5 min before the onset of reperfusion decreased infarct size in acute myocardial infarction in rats. The effect of these peptides in the CNS has not been studied yet.

7.10 Cyclic Helix B Peptide

Cyclic helix B peptide (CHBP) is a proteolysis-resistant novel peptide which derived from HBSP (Yang et al., 2014). Its metabolic stability and half-life are higher than HBSP. CHBP protected kidney tissue against ischemia reperfusion and prevented renal allograft rejection (Yang et al., 2014, 2015). It also exhibits protective effects against mitochondrial dysfunction and apoptosis in mesenchymal stem cells via Nrf2/Sirt3/FoxO3a pathway activation (Wang et al., 2017). Neuroprotective effects of CHBP have not been evaluated.

8. FUTURE PERSPECTIVES AND CONCLUSION

Peptide-based drugs continue to attract increasing interest due to their easy synthetic production, controlled metabolism, and specific and high-affinity binding capacities. Moreover, the costs and immunogenic features of peptides are lower than protein drugs. Despite these favorable properties, their poor pharmacokinetic features (low permeability, bioavailability, stability, and oral bioavailability) and lack of appropriate routes of administration are disadvantages of the therapeutic use of peptides. Recent scientific and technological advancements in peptide drug discovery contribute to the recovery of sub-optimal properties of peptide-based drugs. New tools available for the computational design of peptide drugs, cell-free high-throughput screening and production methods, peptidomics approaches, novel formulation technologies, and innovative drug delivery strategies such as nanocarriers and cell-penetrating peptides will be helpful in improving the drug-like properties of the peptides, while maintaining their target potency. Another strategy for enhancing the effects of peptide drugs could be the design and development of multifunctional peptides. Multifunctional peptides that target multiple EPO receptors may exhibit strong beneficial effects. Combined therapy with EPO mimetic peptides and nonpeptide selective activator of EPO receptor, STS-E412, could also enhance their therapeutic effects (Miller et al., 2015). Accurate identification of dose, dosing frequency, duration of therapy, therapeutic window, and administration timing in preclinical studies improve the success of peptide drugs in clinical trials. Pharmacogenomic evaluation, personalized drug dosage, and monitorization of pharmacokinetics and pharmacodynamics of peptides are helpful approaches for succeeding with peptide clinical trials. As an important basic research area, EPO mimetic peptide drugs may be used to clarify the activation mechanisms of EPOR and EPOR-associated receptors.

All in all, peptide-based drugs are promising means in the treatment of neurodegenerative and neuroinflammatory disorders with unknown cure.

REFERENCES

Alural, B., Ayyildiz, Z. O., Tufekci, K. U., Genc, S., & Genc, K. (2017). Erythropoietin promotes glioblastoma via miR-451 suppression. *Vitamins and Hormones*, *105*, 249–271. https://doi.org/10.1016/bs.vh.2017.03.002.

Bak, A., Leung, D., Barrett, S. E., Forster, S., Minnihan, E. C., Leithead, A. W., et al. (2015). Physicochemical and formulation developability assessment for therapeutic peptide delivery—A primer. *The AAPS Journal*, *17*(1), 144–155. https://doi.org/10.1208/s12248-014-9688-2.

Banerjee, A., & Mitragotri, S. (2017). Intestinal patch systems for oral drug delivery. *Current Opinion in Pharmacology*, *36*, 58–65. https://doi.org/10.1016/j.coph.2017.08.005.

Bennett, C. L., Jacob, S., Hymes, J., Usvyat, L. A., & Maddux, F. W. (2014). Anaphylaxis and hypotension after administration of peginesatide. *The New England Journal of Medicine*, *370*(21), 2055–2056. https://doi.org/10.1056/NEJMc1400883.

Bennett, C. L., Lai, S. Y., Sartor, O., Georgantopoulos, P., Hrushesky, W. J., Henke, M., et al. (2016). Consensus on the existence of functional erythropoietin receptors on cancer cells. *JAMA Oncology*, *2*(1), 134–136. https://doi.org/10.1001/jamaoncol.2015.3940.

Ben-Shimon, A., & Eisenstein, M. (2010). Computational mapping of anchoring spots on protein surfaces. *Journal of Molecular Biology*, *402*(1), 259–277. https://doi.org/10.1016/j.jmb.2010.07.021.

Ben-Shimon, A., & Niv, M. Y. (2015). AnchorDock: Blind and flexible anchor-driven peptide docking. *Structure*, *23*(5), 929–940. https://doi.org/10.1016/j.str.2015.03.010.

Bidasolo, I. B., Ramos, M., & Gomez-Ruiz, J. A. (2012). In vitro simulated gastrointestinal digestion of donkeys' milk. Peptide characterization by high performance liquid chromatography–tandem mass spectrometry. *International Dairy Journal*, *24*(2), 146–152. https://doi.org/10.1016/j.idairyj.2011.04.014.

Blaskovich, M. A. (2016). Unusual amino acids in medicinal chemistry. *Journal of Medicinal Chemistry*, *59*(24), 10807–10836. https://doi.org/10.1021/acs.jmedchem.6b00319.

Brines, M., & Cerami, A. (2006). Discovering erythropoietin's extra-hematopoietic functions: Biology and clinical promise. *Kidney International*, *70*(2), 246–250. https://doi.org/10.1038/sj.ki.5001546.

Brines, M., & Cerami, A. (2012). The receptor that tames the innate immune response. *Molecular Medicine*, *18*, 486–496. https://doi.org/10.2119/molmed.2011.00414.

Brines, M., Dunne, A. N., van Velzen, M., Proto, P. L., Ostenson, C. G., Kirk, R. I., et al. (2015). ARA 290, a nonerythropoietic peptide engineered from erythropoietin, improves metabolic control and neuropathic symptoms in patients with type 2 diabetes. *Molecular Medicine*, *20*, 658–666. https://doi.org/10.2119/molmed.2014.00215.

Brines, M., Patel, N. S., Villa, P., Brines, C., Mennini, T., De Paola, M., et al. (2008). Non-erythropoietic, tissue-protective peptides derived from the tertiary structure of erythropoietin. *Proceedings of the National Academy of Sciences of the United States of America*, *105*(31), 10925–10930. https://doi.org/10.1073/pnas.0805594105.

Broxmeyer, H. E. (2013). Erythropoietin: Multiple targets, actions, and modifying influences for biological and clinical consideration. *The Journal of Experimental Medicine*, *210*(2), 205–208. https://doi.org/10.1084/jem.20122760.

Campana, W. M., Misasi, R., & O'Brien, J. S. (1998). Identification of a neurotrophic sequence in erythropoietin. *International Journal of Molecular Medicine*, *1*(1), 235–241.

Casadevall, N., Eckardt, K. U., & Rossert, J. (2005). Epoetin-induced autoimmune pure red cell aplasia. *Journal of the American Society of Nephrology*, *16*(Suppl. 1), S67–69.
Cerit, H., Veer, I. M., Dahan, A., Niesters, M., Harmer, C. J., Miskowiak, K. W., et al. (2015). Testing the antidepressant properties of the peptide ARA290 in a human neuropsychological model of drug action. *European Neuropsychopharmacology*, *25*(12), 2289–2299. https://doi.org/10.1016/j.euroneuro.2015.09.005.
Chandrudu, S., Simerska, P., & Toth, I. (2013). Chemical methods for peptide and protein production. *Molecules*, *18*(4), 4373–4388. https://doi.org/10.3390/molecules18044373.
Chattong, S., Tanamai, J., Kiatsomchai, P., Nakatsu, M., Sereemaspun, A., Pimpha, N., et al. (2013). Glutaraldehyde erythropoietin protects kidney in ischaemia/reperfusion injury without increasing red blood cell production. *British Journal of Pharmacology*, *168*(1), 189–199. https://doi.org/10.1111/j.1476-5381.2012.02123.x.
Chen, T., Liu, X., Yi, S., Zhang, J., Ge, J., & Liu, Z. (2013). EphB4 is overexpressed in gliomas and promotes the growth of glioma cells. *Tumour Biology*, *34*(1), 379–385. https://doi.org/10.1007/s13277-012-0560-7.
Chen, H., Luo, B., Yang, X., Xiong, J., Liu, Z., Jiang, M., et al. (2014). Therapeutic effects of nonerythropoietic erythropoietin analog ARA290 in experimental autoimmune encephalomyelitis rat. *Journal of Neuroimmunology*, *268*(1–2), 64–70. https://doi.org/10.1016/j.jneuroim.2014.01.006.
Chen, J., Yang, Z., & Zhang, X. (2015). Carbamylated erythropoietin: A prospective drug candidate for neuroprotection. *Biochemistry Insights*, *8*(Suppl. 1), 25–29. https://doi.org/10.4137/BCI.S30753.
Choi, D., Schroer, S. A., Lu, S. Y., Wang, L., Wu, X., Liu, Y., et al. (2010). Erythropoietin protects against diabetes through direct effects on pancreatic beta cells. *The Journal of Experimental Medicine*, *207*(13), 2831–2842. https://doi.org/10.1084/jem.20100665.
Colella, P., Iodice, C., Di Vicino, U., Annunziata, I., Surace, E. M., & Auricchio, A. (2011). Non-erythropoietic erythropoietin derivatives protect from light-induced and genetic photoreceptor degeneration. *Human Molecular Genetics*, *20*(11), 2251–2262. https://doi.org/10.1093/hmg/ddr115.
Collino, M., Thiemermann, C., Cerami, A., & Brines, M. (2015). Flipping the molecular switch for innate protection and repair of tissues: Long-lasting effects of a non-erythropoietic small peptide engineered from erythropoietin. *Pharmacology & Therapeutics*, *151*, 32–40. https://doi.org/10.1016/j.pharmthera.2015.02.005.
Commission Directive 2003/63/EC of 25 June 2003 amending Directive 2001/83/EC of the European Parliament and of the Council on the Community code relating to medicinal products for human use. H. a. Consumers (2003).
Culver, D. A., Dahan, A., Bajorunas, D., Jeziorska, M., van Velzen, M., Aarts, L., et al. (2017). Cibinetide improves corneal nerve fiber abundance in patients with sarcoidosis-associated small nerve fiber loss and neuropathic pain. *Investigative Ophthalmology & Visual Science*, *58*(6), BIO52–BIO60. https://doi.org/10.1167/iovs.16-21291.
Dagliyan, O., Proctor, E. A., D'Auria, K. M., Ding, F., & Dokholyan, N. V. (2011). Structural and dynamic determinants of protein-peptide recognition. *Structure*, *19*(12), 1837–1845. https://doi.org/10.1016/j.str.2011.09.014.
Dahan, A., Dunne, A., Swartjes, M., Proto, P. L., Heij, L., Vogels, O., et al. (2013). ARA 290 improves symptoms in patients with sarcoidosis-associated small nerve fiber loss and increases corneal nerve fiber density. *Molecular Medicine*, *19*, 334–345. https://doi.org/10.2119/molmed.2013.00122.
Daliri, E. B., Lee, B. H., & Oh, D. H. (2017). Current trends and perspectives of bioactive peptides. *Critical Reviews in Food Science and Nutrition*. https://doi.org/10.1080/10408398.2017.1319795. (Epub ahead of print).
Darewicz, M., Borawska, J., Vegarud, G. E., Minkiewicz, P., & Iwaniak, A. (2014). Angiotensin I-converting enzyme (ACE) inhibitory activity and ACE inhibitory peptides of

salmon (Salmo salar) protein hydrolysates obtained by human and porcine gastrointestinal enzymes. *International Journal of Molecular Sciences*, *15*(8), 14077–14101. https://doi.org/10.3390/ijms150814077.

Das, A. A., Sharma, O. P., Kumar, M. S., Krishna, R., & Mathur, P. P. (2013). PepBind: A comprehensive database and computational tool for analysis of protein-peptide interactions. *Genomics, Proteomics & Bioinformatics*, *11*(4), 241–246. https://doi.org/10.1016/j.gpb.2013.03.002.

Deacon, C. F. (2004). Therapeutic strategies based on glucagon-like peptide 1. *Diabetes*, *53*(9), 2181–2189.

Debeljak, N., Solar, P., & Sytkowski, A. J. (2014). Erythropoietin and cancer: The unintended consequences of anemia correction. *Frontiers in Immunology*, *5*, 563. https://doi.org/10.3389/fimmu.2014.00563.

de Lange, I. M., Rump, P., Neuteboom, R. F., Augustijn, P. B., Hodges, K., Kistemaker, A. I., et al. (2017). Male patients affected by mosaic PCDH19 mutations: Five new cases. *Neurogenetics*, *18*(3), 147–153. https://doi.org/10.1007/s10048-017-0517-5.

Devalliere, J., Dooley, K., Hu, Y., Kelangi, S. S., Uygun, B. E., & Yarmush, M. L. (2017). Co-delivery of a growth factor and a tissue-protective molecule using elastin biopolymers accelerates wound healing in diabetic mice. *Biomaterials*, *141*, 149–160. https://doi.org/10.1016/j.biomaterials.2017.06.043.

de Vries, S. J., Rey, J., Schindler, C. E. M., Zacharias, M., & Tuffery, P. (2017). The pepATTRACT web server for blind, large-scale peptide-protein docking. *Nucleic Acids Research*, *45*, W361–W364. https://doi.org/10.1093/nar/gkx335.

Dey, S., & Noguchi, C. T. (2017). Erythropoietin and hypothalamic-pituitary axis. *Vitamins and Hormones*, *105*, 101–120. https://doi.org/10.1016/bs.vh.2017.02.007.

Di, L. (2015). Strategic approaches to optimizing peptide ADME properties. *The AAPS Journal*, *17*(1), 134–143. https://doi.org/10.1208/s12248-014-9687-3.

Diao, L., & Meibohm, B. (2013). Pharmacokinetics and pharmacokinetic-pharmacodynamic correlations of therapeutic peptides. *Clinical Pharmacokinetics*, *52*(10), 855–868. https://doi.org/10.1007/s40262-013-0079-0.

Dmytriyeva, O., Pankratova, S., Korshunova, I., & Walmod, P. S. (2016). Epobis is a nonerythropoietic and neuroprotective agonist of the erythropoietin receptor with anti-inflammatory and memory enhancing effects. *Mediators of Inflammation*, *2016*, 1346390. https://doi.org/10.1155/2016/1346390.

Doknic, M., Stojanovic, M., & Popovic, V. (2014). Novel long-acting GH preparations. *Pediatric Endocrinology Reviews*, *12*(2), 206–212.

Ehrenreich, H., Weissenborn, K., Prange, H., Schneider, D., Weimar, C., Wartenberg, K., et al. (2009). Recombinant human erythropoietin in the treatment of acute ischemic stroke. *Stroke*, *40*(12), e647–656. https://doi.org/10.1161/STROKEAHA.109.564872.

Erslev, A. J. (1993). The discovery of erythropoietin. *ASAIO Journal*, *39*(2), 89–92.

Etcheverrigaray, M., Ceaglio, N., Mattio, M., Oggero, M., Amadeo, I., Forno, G., et al. (2011). Neuroprotective activity of a new erythropoietin formulation with increased penetration in the central nervous system. *BMC Proceedings*, *5*(Suppl. 8), P3. https://doi.org/10.1186/1753-6561-5-S8-P3.

EU Clinical Trials Register. (2017). *EU Clinical Trials Register*. Retrieved from https://www.clinicaltrialsregister.eu/.

Farrell, F., & Lee, A. (2004). The erythropoietin receptor and its expression in tumor cells and other tissues. *The Oncologist*, *9*(Suppl. 5), 18–30. https://doi.org/10.1634/theoncologist.9-90005-18.

Fisher, J. W. (2010). Landmark advances in the development of erythropoietin. *Experimental Biology and Medicine (Maywood, N.J.)*, *235*(12), 1398–1411. https://doi.org/10.1258/ebm.2010.010137.

Fosgerau, K., & Hoffmann, T. (2015). Peptide therapeutics: Current status and future directions. *Drug Discovery Today*, *20*(1), 122–128. https://doi.org/10.1016/j.drudis.2014.10.003.

Frappier, V. I., Duran, M., & Keating, A. E. (2017). PixelDB: Protein-peptide complexes annotated with structural conservation of the peptide binding mode. *Protein Science*, *27*, 276–285. https://doi.org/10.1002/pro.3320.

Furman, J. L., Chiu, M., & Hunter, M. J. (2015). Early engineering approaches to improve peptide developability and manufacturability. *The AAPS Journal*, *17*(1), 111–120. https://doi.org/10.1208/s12248-014-9681-9.

Garcia-Tejedor, A., Sanchez-Rivera, L., Castello-Ruiz, M., Recio, I., Salom, J. B., & Manzanares, P. (2014). Novel antihypertensive lactoferrin-derived peptides produced by Kluyveromyces marxianus: Gastrointestinal stability profile and in vivo angiotensin I-converting enzyme (ACE) inhibition. *Journal of Agricultural and Food Chemistry*, *62*(7), 1609–1616. https://doi.org/10.1021/jf4053868.

Genc, K., Egrilmez, M. Y., & Genc, S. (2010). Erythropoietin induces nuclear translocation of Nrf2 and heme oxygenase-1 expression in SH-SY5Y cells. *Cell Biochemistry and Function*, *28*(3), 197–201. https://doi.org/10.1002/cbf.1639.

Genc, S., Koroglu, T. F., & Genc, K. (2004a). Erythropoietin and the nervous system. *Brain Research*, *1000*(1–2), 19–31. https://doi.org/10.1016/j.brainres.2003.12.037.

Genc, S., Koroglu, T. F., & Genc, K. (2004b). Erythropoietin as a novel neuroprotectant. *Restorative Neurology and Neuroscience*, *22*(2), 105–119.

Ghezzi, P., & Brines, M. (2004). Erythropoietin as an antiapoptotic, tissue-protective cytokine. *Cell Death and Differentiation*, *11*(Suppl. 1), S37–44. https://doi.org/10.1038/sj.cdd.4401450.

Goldflam, M., & Ullman, C. G. (2015). Recent advances toward the discovery of drug-like peptides de novo. *Frontiers in Chemistry*, *3*, 69. https://doi.org/10.3389/fchem.2015.00069.

Guarnieri, F. (2017). Analysis of the asymmetry of activated EPO receptor enables designing small molecule agonists. *Vitamins and Hormones*, *105*, 19–37. https://doi.org/10.1016/bs.vh.2017.03.004.

Gutti, U., Pasupuleti, S. R., Sahu, I., Kotipalli, A., Undi, R. B., Kandi, R., et al. (2016). Erythropoietin and thrombopoietin mimetics: Natural alternatives to erythrocyte and platelet disorders. *Critical Reviews in Oncology/Hematology*, *108*, 175–186. https://doi.org/10.1016/j.critrevonc.2016.11.002.

Ha, G. E., Chang, O. K., Jo, S. M., Han, G. S., Park, B. Y., Ham, J. S., et al. (2015). Identification of antihypertensive peptides derived from low molecular weight casein hydrolysates generated during fermentation by Bifidobacterium longum KACC 91563. *Korean Journal for Food Science of Animal Resources*, *35*(6), 738–747. https://doi.org/10.5851/kosfa.2015.35.6.738.

Hamzeh-Mivehroud, M., Alizadeh, A. A., Morris, M. B., Church, W. B., & Dastmalchi, S. (2013). Phage display as a technology delivering on the promise of peptide drug discovery. *Drug Discovery Today*, *18*(23–24), 1144–1157. https://doi.org/10.1016/j.drudis.2013.09.001.

Harris, D. C., & Jewett, M. C. (2012). Cell-free biology: Exploiting the interface between synthetic biology and synthetic chemistry. *Current Opinion in Biotechnology*, *23*(5), 672–678. https://doi.org/10.1016/j.copbio.2012.02.002.

Heij, L., Niesters, M., Swartjes, M., Hoitsma, E., Drent, M., Dunne, A., et al. (2012). Safety and efficacy of ARA 290 in sarcoidosis patients with symptoms of small fiber neuropathy: A randomized, double-blind pilot study. *Molecular Medicine*, *18*, 1430–1436. https://doi.org/10.2119/molmed.2012.00332.

Hermanson, T., Bennett, C. L., & Macdougall, I. C. (2016). Peginesatide for the treatment of anemia due to chronic kidney disease—An unfulfilled promise. *Expert Opinion on Drug Safety*, *15*(10), 1421–1426. https://doi.org/10.1080/14740338.2016.1218467.

Hipolito, C. J., & Suga, H. (2012). Ribosomal production and in vitro selection of natural product-like peptidomimetics: The FIT and RaPID systems. *Current Opinion in Chemical Biology*, *16*(1–2), 196–203. https://doi.org/10.1016/j.cbpa.2012.02.014.

Jacobs, P., Cravador, A., Loriau, R., Brockly, F., Colau, B., Chuchana, P., et al. (1985). Molecular cloning, sequencing, and expression in Escherichia coli of human preprourokinase cDNA. *DNA*, *4*(2), 139–146. https://doi.org/10.1089/dna.1985.4.139.

Jelkmann, W. (2011). Regulation of erythropoietin production. *The Journal of Physiology*, *589*(Pt. 6), 1251–1258. https://doi.org/10.1113/jphysiol.2010.195057.

Jelkmann, W. (2013). Physiology and pharmacology of erythropoietin. *Transfusion Medicine and Hemotherapy*, *40*(5), 302–309. https://doi.org/10.1159/000356193.

Johnson, D. L., Farrell, F. X., Barbone, F. P., McMahon, F. J., Tullai, J., Hoey, K., et al. (1998). Identification of a 13 amino acid peptide mimetic of erythropoietin and description of amino acids critical for the mimetic activity of EMP1. *Biochemistry*, *37*(11), 3699–3710. https://doi.org/10.1021/bi971956y.

Kawakami, M., Iwasaki, S., Sato, K., & Takahashi, M. (2000). Erythropoietin inhibits calcium-induced neurotransmitter release from clonal neuronal cells. *Biochemical and Biophysical Research Communications*, *279*(1), 293–297. https://doi.org/10.1006/bbrc.2000.3926.

Kawakami, M., Sekiguchi, M., Sato, K., Kozaki, S., & Takahashi, M. (2001). Erythropoietin receptor-mediated inhibition of exocytotic glutamate release confers neuroprotection during chemical ischemia. *Journal of Biological Chemistry*, *276*(42), 39469–39475. https://doi.org/10.1074/jbc.M105832200.

Kazim, S. F., & Iqbal, K. (2016). Neurotrophic factor small-molecule mimetics mediated neuroregeneration and synaptic repair: Emerging therapeutic modality for Alzheimer's disease. *Molecular Neurodegeneration*, *11*(1), 50. https://doi.org/10.1186/s13024-016-0119-y.

KDR Kinase Insert Domain Receptor. (2017). Retrieved from https://www.ncbi.nlm.nih.gov/gene/3791.

Kenanova, V. E., Olafsen, T., Salazar, F. B., Williams, L. E., Knowles, S., & Wu, A. M. (2010). Tuning the serum persistence of human serum albumin domain III: Diabody fusion proteins. *Protein Engineering, Design & Selection*, *23*(10), 789–798. https://doi.org/10.1093/protein/gzq054.

Kilburg, D., & Gallicchio, E. (2016). Recent advances in computational models for the study of protein–peptide interactions (1st ed. Vol. 105): Elsevier Inc.

Kimakova, P., Solar, P., Solarova, Z., Komel, R., & Debeljak, N. (2017). Erythropoietin and its angiogenic activity. *International Journal of Molecular Sciences*, *18*(7), 1519. https://doi.org/10.3390/ijms18071519.

Kiss, K., Csonka, C., Paloczi, J., Pipis, J., Gorbe, A., Kocsis, G. F., et al. (2016). Novel, selective EPO receptor ligands lacking erythropoietic activity reduce infarct size in acute myocardial infarction in rats. *Pharmacological Research*, *113*(Pt. A), 62–70. https://doi.org/10.1016/j.phrs.2016.08.013.

Kuang, X., Han, J. G., Zhao, N., Pang, B., Shyu, C. R., & Korkin, D. (2012). DOMMINO: A database of macromolecular interactions. *Nucleic Acids Research*, *40*(D1), D501–D506. https://doi.org/10.1093/nar/gkr1128.

Kurcinski, M., Jamroz, M., Blaszczyk, M., Kolinski, A., & Kmiecik, S. (2015). CABS-dock web server for the flexible docking of peptides to proteins without prior knowledge of the binding site. *Nucleic Acids Research*, *43*(W1), W419–W424. https://doi.org/10.1093/nar/gkv456.

Lagarto, A., Bueno, V., Sanchez, J. A., Couret, M., Valdes, O., Barzaga, P., et al. (2012). Short-term intra-nasal erythropoietin administration with low sialic acid content is without toxicity or erythropoietic effects. *Current Neurovascular Research*, *9*(4), 233–238.

Lavi, A., Ngan, C. H., Movshovitz-Attias, D., Bohnuud, T., Yueh, C., Beglov, D., et al. (2013). Detection of peptide-binding sites on protein surfaces: The first step toward the modeling and targeting of peptide-mediated interactions. *Proteins: Structure, Function and Bioinformatics*, *81*(12), 2096–2105. https://doi.org/10.1002/prot.24422.

Lee, H., Heo, L., Lee, M. S., & Seok, C. (2015). GalaxyPepDock: A protein-peptide docking tool based on interaction similarity and energy optimization. *Nucleic Acids Research*, *43*(W1), W431–435. https://doi.org/10.1093/nar/gkv495.

Li, X., Gonias, S. L., & Campana, W. M. (2005). Schwann cells express erythropoietin receptor and represent a major target for Epo in peripheral nerve injury. *Glia*, *51*(4), 254–265. https://doi.org/10.1002/glia.20202.

Lin, F. K., Suggs, S., Lin, C. H., Browne, J. K., Smalling, R., Egrie, J. C., et al. (1985). Cloning and expression of the human erythropoietin gene. *Proceedings of the National Academy of Sciences of the United States of America*, *82*(22), 7580–7584.

Liu, Y., Lu, J., Wang, X., Chen, L., Liu, S., Zhang, Z., et al. (2017). Erythropoietin-derived peptide protects against acute lung injury after rat traumatic brain injury. *Cellular Physiology and Biochemistry*, *41*(5), 2037–2044. https://doi.org/10.1159/000475434.

Liu, Y., Luo, B., Han, F., Li, X., Xiong, J., Jiang, M., et al. (2014). Erythropoietin-derived nonerythropoietic peptide ameliorates experimental autoimmune neuritis by inflammation suppression and tissue protection. *PLoS One*, *9*(3), e90942. https://doi.org/10.1371/journal.pone.0090942.

Livnah, O., Stura, E. A., Middleton, S. A., Johnson, D. L., Jolliffe, L. K., & Wilson, I. A. (1999). Crystallographic evidence for preformed dimers of erythropoietin receptor before ligand activation. *Science*, *283*(5404), 987–990.

Locatelli, F., Choukroun, G., Truman, M., Wiggenhauser, A., & Fliser, D. (2016). Once-monthly continuous erythropoietin receptor activator (C.E.R.A.) in patients with hemodialysis-dependent chronic kidney disease: Pooled data from phase III trials. *Advances in Therapy*, *33*(4), 610–625. https://doi.org/10.1007/s12325-016-0309-6.

London, N., Movshovitz-Attias, D., & Schueler-Furman, O. (2010). The structural basis of peptide-protein binding strategies. *Structure*, *18*(2), 188–199. https://doi.org/10.1016/j.str.2009.11.012.

London, N., Raveh, B., & Schueler-Furman, O. (2013). Peptide docking and structure-based characterization of peptide binding: From knowledge to know-how. *Current Opinion in Structural Biology*, *23*(6), 894–902. https://doi.org/10.1016/j.sbi.2013.07.006.

Lund, A., Lundby, C., & Olsen, N. V. (2014). High-dose erythropoietin for tissue protection. *European Journal of Clinical Investigation*, *44*(12), 1230–1238. https://doi.org/10.1111/eci.12357.

Mak, I. W., Evaniew, N., & Ghert, M. (2014). Lost in translation: Animal models and clinical trials in cancer treatment. *American Journal of Translational Research*, *6*(2), 114–118.

Marqus, S., Pirogova, E., & Piva, T. J. (2017). Evaluation of the use of therapeutic peptides for cancer treatment. *Journal of Biomedical Science*, *24*(1), 21. https://doi.org/10.1186/s12929-017-0328-x.

Masuda, S., Nagao, M., Takahata, K., Konishi, Y., Gallyas, F., Jr., Tabira, T., et al. (1993). Functional erythropoietin receptor of the cells with neural characteristics. Comparison with receptor properties of erythroid cells. *The Journal of Biological Chemistry*, *268*(15), 11208–11216.

Masuda, S., Okano, M., Yamagishi, K., Nagao, M., Ueda, M., & Sasaki, R. (1994). A novel site of erythropoietin production. Oxygen-dependent production in cultured rat astrocytes. *The Journal of Biological Chemistry*, *269*(30), 19488–19493.

Mattio, M., Ceaglio, N., Oggero, M., Perotti, N., Amadeo, I., Orozco, G., et al. (2011). Isolation and characterization of a subset of erythropoietin glycoforms with cytoprotective but minimal erythropoietic activity. *Biotechnology Progress*, *27*(4), 1018–1028. https://doi.org/10.1002/btpr.633.

Maxwell, P., Melendez-Rodriguez, F., Matchett, K. B., Aragones, J., Ben-Califa, N., Jaekel, H., et al. (2015). Novel antibodies directed against the human erythropoietin receptor: Creating a basis for clinical implementation. *British Journal of Haematology, 168*(3), 429–442. https://doi.org/10.1111/bjh.13133.
McGregor, D. P. (2008). Discovering and improving novel peptide therapeutics. *Current Opinion in Pharmacology, 8*(5), 616–619. https://doi.org/10.1016/j.coph.2008.06.002.
McKinney, M., & Arcasoy, M. O. (2011). Erythropoietin for oncology supportive care. *Experimental Cell Research, 317*(9), 1246–1254. https://doi.org/10.1016/j.yexcr.2011.03.003.
Middleton, S. A., Barbone, F. P., Johnson, D. L., Thurmond, R. L., You, Y., McMahon, F. J., et al. (1999). Shared and unique determinants of the erythropoietin (EPO) receptor are important for binding EPO and EPO mimetic peptide. *The Journal of Biological Chemistry, 274*(20), 14163–14169.
Miller, J. L., Church, T. J., Leonoudakis, D., Lariosa-Willingham, K., Frigon, N. L., Tettenborn, C. S., et al. (2015). Discovery and characterization of nonpeptidyl agonists of the tissue-protective erythropoietin receptor. *Molecular Pharmacology, 88*(2), 357–367. https://doi.org/10.1124/mol.115.098400.
Miyake, T., Kung, C. K., & Goldwasser, E. (1977). Purification of human erythropoietin. *Journal of Biological Chemistry, 252*(15), 5558–5564.
Modell, A. E., Blosser, S. L., & Arora, P. S. (2016). Systematic targeting of protein–protein interactions. *Trends in Pharmacological Sciences, 37*(8), 702–713. https://doi.org/10.1016/j.tips.2016.05.008.
Morishita, E., Masuda, S., Nagao, M., Yasuda, Y., & Sasaki, R. (1997). Erythropoietin receptor is expressed in rat hippocampal and cerebral cortical neurons, and erythropoietin prevents in vitro glutamate-induced neuronal death. *Neuroscience, 76*(1), 105–116.
Murray, C. J., & Baliga, R. (2013). Cell-free translation of peptides and proteins: From high throughput screening to clinical production. *Current Opinion in Chemical Biology, 17*(3), 420–426. https://doi.org/10.1016/j.cbpa.2013.02.014.
Nagao, M., Wen, T. C., Okamoto, M., Irie, K., Takaku, T., & Sakanaka, M. (2005). In vivo neuroprotective activity of epopeptide AB against ischemic damage. *Cytotechnology, 47*(1–3), 139–144. https://doi.org/10.1007/s10616-005-3758-3.
Nekoui, A., & Blaise, G. (2017). Erythropoietin and nonhematopoietic effects. *The American Journal of the Medical Sciences, 353*(1), 76–81. https://doi.org/10.1016/j.amjms.2016.10.009.
Noren, K. A., & Noren, C. J. (2001). Construction of high-complexity combinatorial phage display peptide libraries. *Methods, 23*(2), 169–178. https://doi.org/10.1006/meth.2000.1118.
Obarska-Kosinska, A., Iacoangeli, A., Lepore, R., & Tramontano, A. (2016). PepComposer: Computational design of peptides binding to a given protein surface. *Nucleic Acids Research, 44*(W1), W522–528. https://doi.org/10.1093/nar/gkw366.
Oliva, B., & Fernandez-Fuentes, N. (2015). Knowledge-based modeling of peptides at protein interfaces: PiPreD. *Bioinformatics, 31*(9), 1405–1410. https://doi.org/10.1093/bioinformatics/btu838.
Omidfar, K., & Daneshpour, M. (2015). Advances in phage display technology for drug discovery. *Expert Opinion on Drug Discovery, 10*(6), 651–669. https://doi.org/10.1517/17460441.2015.1037738.
Ott, C., Martens, H., Hassouna, I., Oliveira, B., Erck, C., Zafeiriou, M. P., et al. (2015). Widespread expression of erythropoietin receptor in brain and its induction by injury. *Molecular Medicine, 21*, 803–815. https://doi.org/10.2119/molmed.2015.00192.
Otvos, L., Jr., & Wade, J. D. (2014). Current challenges in peptide-based drug discovery. *Frontiers in Chemistry, 2*, 62. https://doi.org/10.3389/fchem.2014.00062.

Oudejans, L. C., Niesters, M., Brines, M., Dahan, A., & van Velzen, M. (2017). Quantification of small fiber pathology in patients with sarcoidosis and chronic pain using cornea confocal microscopy and skin biopsies. *Journal of Pain Research, 10*, 2057–2065. https://doi.org/10.2147/JPR.S142683.

Pankratova, S., Gu, B., Kiryushko, D., Korshunova, I., Kohler, L. B., Rathje, M., et al. (2012). A new agonist of the erythropoietin receptor, Epobis, induces neurite outgrowth and promotes neuronal survival. *Journal of Neurochemistry, 121*(6), 915–923. https://doi.org/10.1111/j.1471-4159.2012.07751.x.

Pankratova, S., Kiryushko, D., Sonn, K., Soroka, V., Kohler, L. B., Rathje, M., et al. (2010). Neuroprotective properties of a novel, non-haematopoietic agonist of the erythropoietin receptor. *Brain, 133*(Pt. 8), 2281–2294. https://doi.org/10.1093/brain/awq101.

Peres, E. A., Valable, S., Guillamo, J. S., Marteau, L., Bernaudin, J. F., Roussel, S., et al. (2011). Targeting the erythropoietin receptor on glioma cells reduces tumour growth. *Experimental Cell Research, 317*(16), 2321–2332. https://doi.org/10.1016/j.yexcr.2011.06.011.

Periti, P., Mazzei, T., & Mini, E. (2002). Clinical pharmacokinetics of depot leuprorelin. *Clinical Pharmacokinetics, 41*(7), 485–504. https://doi.org/10.2165/00003088-200241070-00003.

Peterson, L. X., Roy, A., Christoffer, C., Terashi, G., & Kihara, D. (2017). Modeling disordered protein interactions from biophysical principles. *PLoS Computational Biology, 13*(4), e1005485. https://doi.org/10.1371/journal.pcbi.1005485.

Pfeffer, M. A., Burdmann, E. A., Chen, C. Y., Cooper, M. E., de Zeeuw, D., Eckardt, K. U., et al. (2009). A trial of darbepoetin alfa in type 2 diabetes and chronic kidney disease. *The New England Journal of Medicine, 361*(21), 2019–2032. https://doi.org/10.1056/NEJMoa0907845.

Pradeep, S., Huang, J., Mora, E. M., Nick, A. M., Cho, M. S., Wu, S. Y., et al. (2015). Erythropoietin stimulates tumor growth via EphB4. *Cancer Cell, 28*(5), 610–622. https://doi.org/10.1016/j.ccell.2015.09.008.

Provatopoulou, S. T., & Ziroyiannis, P. N. (2011). Clinical use of erythropoietin in chronic kidney disease: Outcomes and future prospects. *Hippokratia, 15*(2), 109–115.

Pulman, K. G., Smith, M., Mengozzi, M., Ghezzi, P., & Dilley, A. (2013). The erythropoietin-derived peptide ARA290 reverses mechanical allodynia in the neuritis model. *Neuroscience, 233*, 174–183. https://doi.org/10.1016/j.neuroscience.2012.12.022.

Qvit, N., Rubin, S. J., Urban, T. J., Mochly-Rosen, D., & Gross, E. R. (2017). Peptidomimetic therapeutics: Scientific approaches and opportunities. *Drug Discovery Today, 22*(2), 454–462. https://doi.org/10.1016/j.drudis.2016.11.003.

Rahbek-Nielsen, H., Roepstorff, P., Reischl, H., Wozny, M., Koll, H., & Haselbeck, A. (1997). Glycopeptide profiling of human urinary erythropoietin by matrix-assisted laser desorption/ionization mass spectrometry. *Journal of Mass Spectrometry, 32*(9), 948–958. https://doi.org/10.1002/(SICI)1096-9888(199709)32:9<948::AID-JMS551>3.0.CO;2-C.

Raveh, B., London, N., Zimmerman, L., & Schueler-Furman, O. (2011). Rosetta FlexPepDock ab-initio: Simultaneous folding, docking and refinement of peptides onto their receptors. *PLoS One, 6*(4), e18934. https://doi.org/10.1371/journal.pone.0018934.

Reagan-Shaw, S., Nihal, M., & Ahmad, N. (2008). Dose translation from animal to human studies revisited. *The FASEB Journal, 22*(3), 659–661. https://doi.org/10.1096/fj.07-9574LSF.

Remy, I., Wilson, I. A., & Michnick, S. W. (1999). Erythropoietin receptor activation by a ligand-induced conformation change. *Science, 283*(5404), 990–993.

Ribatti, D. (2008). Erythropoietin, the first century. *Leukemia Research, 32*(8), 1169–1172. https://doi.org/10.1016/j.leukres.2008.01.018.

Rodriguez Cruz, Y., Strehaiano, M., Rodriguez Obaya, T., Garcia Rodriguez, J. C., & Maurice, T. (2017). An intranasal formulation of erythropoietin (Neuro-EPO) prevents memory deficits and amyloid toxicity in the APPSwe transgenic mouse model of Alzheimer's disease. *Journal of Alzheimer's Disease, 55*(1), 231–248. https://doi.org/10.3233/JAD-160500.

Rothe, A., Hosse, R. J., & Power, B. E. (2006). In vitro display technologies reveal novel biopharmaceutics. *The FASEB Journal, 20*(10), 1599–1610. https://doi.org/10.1096/fj.05-5650rev.

Sable, R., & Jois, S. (2015). Surfing the protein-protein interaction surface using docking methods: Application to the design of PPI inhibitors. *Molecules, 20*(6), 11569–11603. https://doi.org/10.3390/molecules200611569.

Saglimbene, V. M., Palmer, S. C., Ruospo, M., Natale, P., Craig, J. C., & Strippoli, G. F. (2017). Continuous erythropoiesis receptor activator (CERA) for the anaemia of chronic kidney disease. *Cochrane Database of Systematic Reviews 8*, CD009904. https://doi.org/10.1002/14651858.CD009904.pub2.

Sakanaka, M., Wen, T. C., Matsuda, S., Masuda, S., Morishita, E., Nagao, M., et al. (1998). In vivo evidence that erythropoietin protects neurons from ischemic damage. *Proceedings of the National Academy of Sciences of the United States of America, 95*(8), 4635–4640.

Sargin, D., Friedrichs, H., El-Kordi, A., & Ehrenreich, H. (2010). Erythropoietin as neuroprotective and neuroregenerative treatment strategy: Comprehensive overview of 12 years of preclinical and clinical research. *Best Practice & Research. Clinical Anaesthesiology, 24*(4), 573–594. https://doi.org/10.1016/j.bpa.2010.10.005.

Sautina, L., Sautin, Y., Beem, E., Zhou, Z., Schuler, A., Brennan, J., et al. (2010). Induction of nitric oxide by erythropoietin is mediated by the {beta} common receptor and requires interaction with VEGF receptor 2. *Blood, 115*(4), 896–905. https://doi.org/10.1182/blood-2009-04-216432.

Schmidt, R. E., Feng, D., Wang, Q., Green, K. G., Snipes, L. L., Yamin, M., et al. (2011). Effect of insulin and an erythropoietin-derived peptide (ARA290) on established neuritic dystrophy and neuronopathy in Akita (Ins2 Akita) diabetic mouse sympathetic ganglia. *Experimental Neurology, 232*(2), 126–135. https://doi.org/10.1016/j.expneurol.2011.05.025.

Seeger, N., Zellinger, C., Rode, A., Roloff, F., Bicker, G., Russmann, V., et al. (2011). The erythropoietin-derived peptide mimetic pHBSP affects cellular and cognitive consequences in a rat post-status epilepticus model. *Epilepsia, 52*(12), 2333–2343. https://doi.org/10.1111/j.1528-1167.2011.03302.x.

Shanks, N., Greek, R., & Greek, J. (2009). Are animal models predictive for humans? *Philosophy, Ethics, and Humanities in Medicine, 4*, 2. https://doi.org/10.1186/1747-5341-4-2.

Smith, G. P. (1985). Filamentous fusion phage: Novel expression vectors that display cloned antigens on the virion surface. *Science, 228*(4705), 1315–1317.

Solling, C. (2012). Organ-protective and immunomodulatory effects of erythropoietin—An update on recent clinical trials. *Basic & Clinical Pharmacology & Toxicology, 110*(2), 113–121. https://doi.org/10.1111/j.1742-7843.2011.00820.x.

Sooklert, K., Chattong, S., Manotham, K., Boonwong, C., Klaharn, I. Y., Jindatip, D., et al. (2016). Cytoprotective effect of glutaraldehyde erythropoietin on HEK293 kidney cells after silver nanoparticle exposure. *International Journal of Nanomedicine, 11*, 597–605. https://doi.org/10.2147/IJN.S95654.

Souma, T., Suzuki, N., & Yamamoto, M. (2015). Renal erythropoietin-producing cells in health and disease. *Frontiers in Physiology 6*, 167. https://doi.org/10.3389/fphys.2015.00167.

Steward, O., Popovich, P. G., Dietrich, W. D., & Kleitman, N. (2012). Replication and reproducibility in spinal cord injury research. *Experimental Neurology, 233*(2), 597–605. https://doi.org/10.1016/j.expneurol.2011.06.017.

Swartjes, M., van Velzen, M., Niesters, M., Aarts, L., Brines, M., Dunne, A., et al. (2014). ARA 290, a peptide derived from the tertiary structure of erythropoietin, produces long-term relief of neuropathic pain coupled with suppression of the spinal microglia response. *Molecular Pain, 10,* 13. https://doi.org/10.1186/1744-8069-10-13.

Teste, I. S., Tamos, Y. M., Cruz, Y. R., Cernada, A. M., Rodriguez, J. C., Martinez, N. S., et al. (2012). Dose effect evaluation and therapeutic window of the neuro-EPO nasal application for the treatment of the focal ischemia model in the Mongolian gerbil. *ScientificWorldJournal 2012,* 607498. https://doi.org/10.1100/2012/607498.

Torup, L., & Leist, M. (2006). Development of non-erythropoietic erythropoietin variants for neuroprotection. In A. Höke (Ed.), *Erythropoietin and the nervous system: Novel therapeutic options for neuroprotection* (pp. 211–219). Boston, MA: Springer US.

Trabuco, L. G., Lise, S., Petsalaki, E., & Russell, R. B. (2012). PepSite: Prediction of peptide-binding sites from protein surfaces. *Nucleic Acids Research, 40*(W1), 423–427. https://doi.org/10.1093/nar/gks398.

Trellet, M., Melquiond, A. S., & Bonvin, A. M. (2013). A unified conformational selection and induced fit approach to protein-peptide docking. *PLoS One, 8*(3), e58769. https://doi.org/10.1371/journal.pone.0058769.

Tsai, P. T., Ohab, J. J., Kertesz, N., Groszer, M., Matter, C., Gao, J., et al. (2006). A critical role of erythropoietin receptor in neurogenesis and post-stroke recovery. *The Journal of Neuroscience, 26*(4), 1269–1274. https://doi.org/10.1523/JNEUROSCI.4480-05.2006.

Udenigwe, C. C., & Aluko, R. E. (2011). Chemometric analysis of the amino acid requirements of antioxidant food protein hydrolysates. *International Journal of Molecular Sciences, 12*(5), 3148–3161. https://doi.org/10.3390/ijms12053148.

Uhlig, T., Kyprianou, T., Martinelli, F. G., Oppici, C. A., Heiligers, D., Hills, D., et al. (2014). The emergence of peptides in the pharmaceutical business: From exploration to exploitation. *EuPA Open Proteomics, 4*(Suppl. C), 58–69. https://doi.org/10.1016/j.euprot.2014.05.003.

Usmani, S. S., Bedi, G., Samuel, J. S., Singh, S., Kalra, S., Kumar, P., et al. (2017). THPdb: Database of FDA-approved peptide and protein therapeutics. *PLoS One, 12*(7), 1–12. https://doi.org/10.1371/journal.pone.0181748.

Uversky, V. N., & Redwan, E. M. (2016). Erythropoietin and co.: Intrinsic structure and functional disorder. *Molecular BioSystems, 13*(1), 56–72. https://doi.org/10.1039/c6mb00657d.

Vanhee, P., Stricher, F., Baeten, L., Verschueren, E., Lenaerts, T., Serrano, L., et al. (2009). Protein-peptide interactions adopt the same structural motifs as monomeric protein folds. *Structure, 17*(8), 1128–1136. https://doi.org/10.1016/j.str.2009.06.013.

Vanhee, P., van der Sloot, A. M., Verschueren, E., Serrano, L., Rousseau, F., & Schymkowitz, J. (2011). Computational design of peptide ligands. *Trends in Biotechnology, 29*(5), 231–239. https://doi.org/10.1016/j.tibtech.2011.01.004.

van Velzen, M., Heij, L., Niesters, M., Cerami, A., Dunne, A., Dahan, A., et al. (2014). ARA 290 for treatment of small fiber neuropathy in sarcoidosis. *Expert Opinion on Investigational Drugs, 23*(4), 541–550. https://doi.org/10.1517/13543784.2014.892072.

Verschueren, E., Vanhee, P., Rousseau, F., Schymkowitz, J., & Serrano, L. (2013). Protein-peptide complex prediction through fragment interaction patterns. *Structure, 21*(5), 789–797. https://doi.org/10.1016/j.str.2013.02.023.

Vhora, I., Patil, S., Bhatt, P., & Misra, A. (2015). Protein- and peptide-drug conjugates: An emerging drug delivery technology. *Advances in Protein Chemistry and Structural Biology, 98,* 1–55. https://doi.org/10.1016/bs.apcsb.2014.11.001.

Vigneaud, V. d., Ressler, C., Swan, C. J. M., Roberts, C. W., Katsoyannis, P. G., & Gordon, S. (1953). The synthesis of an octapeptide amide with the hormonal activity of oxytocin. *Journal of the American Chemical Society, 75*(19), 4879–4880. https://doi.org/10.1021/ja01115a553.

Vlieghe, P., Lisowski, V., Martinez, J., & Khrestchatisky, M. (2010). Synthetic therapeutic peptides: Science and market. *Drug Discovery Today, 15*(1–2), 40–56. https://doi.org/10.1016/j.drudis.2009.10.009.

Wang, B., Kang, M., Marchese, M., Rodriguez, E., Lu, W., Li, X., et al. (2016). Beneficial effect of erythropoietin short peptide on acute traumatic brain injury. *Neurotherapeutics, 13*(2), 418–427. https://doi.org/10.1007/s13311-015-0418-y.

Wang, S., Zhang, C., Niyazi, S., Zheng, L., Li, J., Zhang, W., et al. (2017). A novel cytoprotective peptide protects mesenchymal stem cells against mitochondrial dysfunction and apoptosis induced by starvation via Nrf2/Sirt3/FoxO3a pathway. *Journal of Translational Medicine, 15*(1), 33. https://doi.org/10.1186/s12967-017-1144-5.

Watowich, S. S. (2011). The erythropoietin receptor: Molecular structure and hematopoietic signaling pathways. *Journal of Investigative Medicine, 59*(7), 1067–1072. https://doi.org/10.2310/JIM.0b013e31820fb28c.

Winkelmann, J. C., Penny, L. A., Deaven, L. L., Forget, B. G., & Jenkins, R. B. (1990). The gene for the human erythropoietin receptor: Analysis of the coding sequence and assignment to chromosome 19p. *Blood, 76*(1), 24–30.

World Health Organization. (2017a). *NeuroEPO– Type 2 spinocerebellar ataxia phase I-II.* Retrieved from http://apps.who.int/trialsearch/Trial2.aspx?TrialID=RPCEC00000187.

World Health Organization. (2017b). *NeuroEPO-stroke. Phase I-II.* Retrieved from http://apps.who.int/trialsearch/Trial2.aspx?TrialID=RPCEC00000185.

Wrighton, N. C., Farrell, F. X., Chang, R., Kashyap, A. K., Barbone, F. P., Mulcahy, L. S., et al. (1996). Small peptides as potent mimetics of the protein hormone erythropoietin. *Science, 273*(5274), 458–464.

Wu, C. H., Liu, I. J., Lu, R. M., & Wu, H. C. (2016). Advancement and applications of peptide phage display technology in biomedical science. *Journal of Biomedical Science 23*, 8. https://doi.org/10.1186/s12929-016-0223-x.

Yang, C., Xu, Z., Zhao, Z., Li, L., Zhao, T., Peng, D., et al. (2014). A novel proteolysis-resistant cyclic helix B peptide ameliorates kidney ischemia reperfusion injury. *Biochimica et Biophysica Acta, 1842*(11), 2306–2317. https://doi.org/10.1016/j.bbadis.2014.09.001.

Yang, C., Zhang, Y., Wang, J., Li, L., Wang, L., Hu, M., et al. (2015). A novel cyclic helix B peptide inhibits dendritic cell maturation during amelioration of acute kidney graft rejection through Jak-2/STAT3/SOCS1. *Cell Death & Disease 6*, e1993. https://doi.org/10.1038/cddis.2015.338.

Yoo, S. J., Cho, B., Lee, D., Son, G., Lee, Y. B., Soo Han, H., et al. (2017). The erythropoietin-derived peptide MK-X and erythropoietin have neuroprotective effects against ischemic brain damage. *Cell Death & Disease, 8*(8), e3003. https://doi.org/10.1038/cddis.2017.381.

Yoo, S. J., Cho, B., Moon, C., Yu, S. W., & Moon, C. (2016). Neuroprotective effects of an erythropoietin-derived peptide in PC12 cells under oxidative stress. *CNS & Neurological Disorders Drug Targets, 15*(8), 927–934.

Youssoufian, H., Longmore, G., Neumann, D., Yoshimura, A., & Lodish, H. F. (1993). Structure, function, and activation of the erythropoietin receptor. *Blood, 81*(9), 2223–2236.

Yuan, R., Wang, B., Lu, W., Maeda, Y., & Dowling, P. (2015). A distinct region in erythropoietin that induces immuno/inflammatory modulation and tissue protection. *Neurotherapeutics, 12*(4), 850–861. https://doi.org/10.1007/s13311-015-0379-1.

Zellinger, C., Seeger, N., Hadamitzky, M., Fischborn, S., Russmann, V., Wendt, H., et al. (2011). Impact of the erythropoietin-derived peptide mimetic Epotris on the histopathological consequences of status epilepticus. *Epilepsy Research, 96*(3), 241–249. https://doi.org/10.1016/j.eplepsyres.2011.06.009.

Zhang, W., Yu, G., & Zhang, M. (2016). ARA 290 relieves pathophysiological pain by targeting TRPV1 channel: Integration between immune system and nociception. *Peptides*, *76*, 73–79. https://doi.org/10.1016/j.peptides.2016.01.003.

Zhang, D. X., Zhang, L. M., Zhao, X. C., & Sun, W. (2017). Neuroprotective effects of erythropoietin against sevoflurane-induced neuronal apoptosis in primary rat cortical neurons involving the EPOR-Erk1/2-Nrf2/Bach1 signal pathway. *Biomedicine & Pharmacotherapy*, *87*, 332–341. https://doi.org/10.1016/j.biopha.2016.12.115.

CHAPTER TEN

The Structure/Function Relationship in Antimicrobial Peptides: What Can we Obtain From Structural Data?

Marlon H. Cardoso[*,†,‡,2], **Karen G.N. Oshiro**[‡,2], **Samilla B. Rezende**[‡], **Elizabete S. Cândido**[†,‡], **Octávio L. Franco**[*,†,‡,1]

[*]Programa de Pós-Graduação em Patologia Molecular, Faculdade de Medicina, Universidade de Brasília, Brasília, Brazil
[†]Centro de Análises Proteômicas e Bioquímicas, Programa de Pós-Graduação em Ciências Genômicas e Biotecnologia, Universidade Católica de Brasília, Brasília, Brazil
[‡]S-inova Biotech, Programa de Pós-Graduação em Biotecnologia, Universidade Católica Dom Bosco, Campo Grande, Brazil
[1]Corresponding author: e-mail address: ocfranco@gmail.com

Contents

Abstract

Antimicrobial peptides (AMPs) have been widely isolated from most organisms in nature. This class of antimicrobials may undergo changes in their sequence for improved physicochemical properties, including charge, hydrophobicity, and hydrophobic moment. It is known that such properties may be directly associated with AMPs' structural arrangements and, consequently, could interfere in their modes of action against microorganisms. In this scenario, biophysical methodologies, such as nuclear magnetic resonance spectroscopy, X-ray crystallography, and cryo-electron

[2] These authors contributed equally.

Advances in Protein Chemistry and Structural Biology, Volume 112
ISSN 1876-1623
https://doi.org/10.1016/bs.apcsb.2018.01.008

microscopy, allied to in silico approaches, including molecular modeling, docking, and dynamics nowadays represent an enormous first step for the structural elucidation of AMPs, leading to further structure–function annotation. In this context, this chapter will focus on the main atomic-level experimental and computational tools used for the structural elucidation of AMPs that have assisted in the investigation of their functions.

1. INTRODUCTION

The indiscriminate use of antibiotics by the population has become a public concern, mainly because it has led to multidrug-resistance (Cao et al., 2017). Particularly in hospitals and communities, where the arsenal of antibiotics is no longer effective to counter bacterial infections, this situation is heading toward a crisis regarding antibiotic resistance (Xie et al., 2017). Thus, efforts have been in progress by the research community to develop favorable alternatives to restrain such infections. In this context, antimicrobial peptides (AMPs) appear as promising candidates to overcome this phenomenon, since they belong to a class of multifunctional molecules produced by most organisms in nature as their first line of defense (Nissen-Meyer & Nes, 1997). Over the years, reports have shown that AMP functions are not straightforward, but are rather a complex map involving structure–function relationships, allowing AMP to adapt and develop new functions (Fjell, Hiss, Hancock, & Schneider, 2011). For example, overviews of AMPs structural diversity have been underway by clustering AMPs according to their backbone conformations and fold (Fjell, Hiss, et al., 2011). This promiscuous behavior ensures a highly effective defense in living organisms since AMPs production is faster in response to microbial threats than those proteins expressed by the adaptive immune system (Dennison, Wallace, Harris, & Phoenix, 2005).

It is important to understand how AMPs are arranged at the structural level in order to further elucidate their mechanisms of action toward specific targets. This correlation has represented a focal point in AMP studies and has been reported as the "AMP structure–function relationship." Studies focusing on the development of more efficient AMPs with improved antimicrobial activity have proposed a series of rational design strategies, which mainly involve amino acid substitutions and modifications of physicochemical properties, thus shedding light on how AMPs are arranged structurally to act against different bacterial strains. In this context, diverse methodologies have helped researchers to understand AMP structure based on experimental techniques and computational simulations, as well as the combination of both. Among the experimental approaches, we can cite nuclear magnetic resonance (NMR), X-ray crystallography, and cryo-electron microscopy (cryo-EM). In addition, along with experimental methodologies, in silico strategies, including molecular modeling, docking, and dynamics, nowadays represent essential tools for the deep understanding of AMPs' interactions with their targets. Thus, this chapter will focus on experimental and computational strategies that have been applied in the elucidation and prediction of AMP 3D structure and on how these methodologies have helped AMP structure–function relationship studies in macromolecular complexes, including AMPs/membranes, AMPs/proteins, AMPs/ribosomes, among others.

2. EXPERIMENTAL APPROACHES FOR THE STRUCTURAL ELUCIDATION OF AMPs

2.1 Solution NMR

One of the main characteristics of AMPs is their ability to bind to bacterial surfaces, which are constituted of diverse phospholipids, carbohydrates, and proteins (Su, Li, & Hong, 2013). In this context, AMP structures have been widely investigated in membrane-like environments, generating peptide–membrane complexes suitable for NMR studies (Hong, Zhang, & Hu, 2012). In addition, even when AMPs are not bound to bacterial model membranes, their small size and promiscuous structural arrangements (usually environment dependent) make them attractive objects for 1D, 2D, and 3D NMR analysis.

Among the NMR strategies employed for AMP structural studies, we can cite solution NMR and solid-state NMR. The first approach has been largely used to determine the 3D structure of AMPs (Table 1) in diverse solutions, including water (Rosengren et al., 2003), mixtures of water with fluorinated organic solvents (Resende et al., 2008), and micelles (e.g., sodium dodecyl sulfate (SDS) and dodecylphosphocholine (DPC) micelles) (Campagna, Saint, Molle, & Aumelas, 2007). Solution NMR of AMPs in water is particularly unusual and mainly applied to the study of highly stable peptides (Rosengren et al., 2003). However, even though a few AMPs have had their structures determined in water, an increasing number of works with solution NMR have highlighted the importance of studying AMPs in water/fluorinated organic solvent mixtures (e.g., water/2,2,2-trifluoroethanol (TFE)) and membrane mimicking systems, which are expected to better simulate biological conditions such as bacterial surfaces.

Table 1 Summary of the AMPs Cited in the Text That Were Studied Either by Solution NMR or Solid-State NMR

AMPs	Structural Arrangement	Experimental Condition	Suggested Mode of Action	References
Solution NMR				
Kalata B1	Cyclic; antiparallel β-sheet	H_2O; D_2O	Membrane disruption	Rosengren, Daly, Plan, Waine, and Craik (2003)
Magainin 2	α-Helix	H_2O; 2.3% TFE-d_3; D_2O	Pore formation	Marion, Zasloff, and Bax (1988)
Phylloseptin-1, -2, and -3	α-Helix	H_2O; D_2O; 60% TFE-d_2	Membrane permeabilization	Resende et al. (2008)
Cm-p5	α-Helix	D_2O; 40% TFE-d_2	Membrane active—under investigation	Lopez-Abarrategui et al. (2015)
Aurein 2.2 and 2.3	α-Helix	H_2O; D_2O; 25% TFE-d_3	Membrane disruption	Pan et al. (2007)
Citropin 1.1	α-Helix	H_2O; 50% TFE-d_3	Detergent-like	Wegener et al. (1999)
Uperin 3.6	α-Helix	H_2O; 50% TFE-d_3	Carpet-like	Chia, Carver, Mulhern, and Bowie (1999)
Protegrin-2, -3, and -5	Antiparallel β-sheet	H_2O; DPC-d_{28}	Pore formation	Usachev, Efimov, Kolosova, Filippov, and Klochkov (2015), Usachev, Efimov, Kolosova, Klochkova, et al. (2015), and Usachev et al. (2017)
Gad-1 and -2	α-Helix	D_2O; 150 mM SDS-d_{25}	Membrane active—under investigation	McDonald et al. (2015)
Cn-AMP1	α-Helix	H_2O; D_2O; 100 mM SDS-d_{25}	Under investigation	Santana et al. (2015)
SET-M33	α-Helix	H_2O; D_2O; 100 mM SDS-d_{25}	Membrane permeability/Intracellular	van der Weide et al. (2017)

Polymyxin B and E	Cyclic; Envelop-like fold	H_2O; D_2O; 10 μg.mL^{-1} LPS	Membrane destabilization	Pristovsek and Kidric (1999)
Melittin	α-Helix	D_2O; 10 μg.mL^{-1} LPS	LPS recognition/Pore formation	Bhunia, Domadia, and Bhattacharjya (2007)
Temporin-1TL	α-Helix	D_2O; 9–18 μM LPS	Outer membrane permeabilization	Bhunia, Saravanan, Mohanram, Mangoni, and Bhattacharjya (2011)
S1-Nal-Nal	α-Helix	D_2O; 20 mM NH_2PO_4; 50 μM LPS	LPS neutralization	Yu et al. (2017)
VG13P	β-Turn	D_2O; H_2O; 0–70 μM LPS	LPS neutralization	Datta et al. (2017)
Solid-state NMR				
Protegrin-1	Antiparallel β-sheet	POPC-d_{31}; POPG-d_{31}; POPC-d_{31}/POPG (3:1)	Pore formation	Kwon, Waring, and Hong (2013)
Magainin 2 and aurein 3.3	α-Helix	POPC-d_{31}; POPC-d_{31}/POPG; POPC-d_{31}/chol	Pore formation	Chen, Lee, and Huang (2003)
MSI-89 and MSI-594	α-Helix	POPC-d_{31}; POPC-d_{31}/POPG (3:1)	Carpet-like/Toroidal pores	Ramamoorthy, Thennarasu, Lee, Tan, and Maloy (2006)
Magainin 2 and PGLa (synergy)	α-Helix	POPC/POPG	Pore formation	Strandberg et al. (2016)
Polymyxin B	Cyclic; Envelop-like fold	Deuterated membrane phospholipids in *E. coli*	Membrane destabilization	Tardy-Laporte et al. (2013)

D_2O, deuterium oxide; *DPC*, dodecylphosphocholine; *H_2O*, water; *LPS*, lipopolysaccharide; *POPC*, 1-palmitoyl-2-oleoyl-*sn*-glycero-3-phosphatidylcholine; *POPG*, 1-palmitoyl-2-oleoyl-*sn*-glycero-3-phosphatidylglycerol; *SDS*, sodium dodecyl sulfate; *TFE*, 2,2,2-trifluoroethanol.

The most commonly used organic solvent to determine the structure of AMPs consists of mixtures of water and TFE in different proportions (Haney, Hunter, Matsuzaki, & Vogel, 2009). TFE is known to act as a cosolvent on the stability of secondary structure in peptides, including the formation of α-helix, β-sheet, and β-hairpins (Roccatano, Colombo, Fioroni, & Mark, 2002). Moreover, small-angle X-ray-scattering, NMR, and molecular dynamics (MD) studies have shown that TFE displaces water molecules around the peptides, thus favoring intrapeptide hydrogen bonds, without disrupting hydrophobic interactions within the peptides (Fioroni, Diaz, Burger, & Berger, 2002; Jasanoff & Fersht, 1994; Roccatano et al., 2002). Marion et al. (1988) performed one of the first studies regarding water/TFE mixtures for reliable solution magainin 2 NMR analyses. In this work, the authors showed that magainin 2 structural conformations are solvent dependent, transiting from random coil conformation in water to α-helix in a small amount of TFE (2.3%) (Marion et al., 1988). Ever since, several works have applied water/TFE mixtures to determine the structure of water-unstable AMPs (Haney et al., 2009). Resende et al. (2008), for example, have reported the solution NMR structure of three phylloseptins derived from the frog *Phyllomedusa hypochondriallis*. In this study, different water/TFE proportions (10%, 30%, and 60% v:v) were evaluated by circular dichroism, and 60% TFE (where higher helical contents were obtained) was selected for NMR spectroscopy. Solution NMR in 60% TFE resulted in the structural determination of all the phylloseptins studied, which adopted well-defined α-helical conformations that were suggested to play a crucial role in the antimicrobial activities of these AMPs by membrane permeabilization (Resende et al., 2008). In addition, an antifungal hydrophilic peptide derived from the coastal mollusk *Cenchritis muricatus*, named Cm-p5, was also studied by solution NMR in 40% TFE resulting in the determination of a α-helix arrangement, which was further used in membrane studies through molecular modeling and dynamics simulations (Lopez-Abarrategui et al., 2015). Similar works using water/TFE mixtures have been performed for other AMPs, including citropin 1.1 (Wegener et al., 1999), hylaseptin P1 (Prates et al., 2004), and uperin 3.6 (Chia et al., 1999).

Despite the advantages of using TFE as a cosolvent to determine 3D structures of AMPs, it is still far from mimicking the complexity of bacterial membrane environments. In this situation, the use of deuterated SDS and DPC micelles has gained great attention in the antimicrobial field. These micelles simulate better membrane interfaces due to the presence of the hydrophobic acyl chain and hydrophilic head groups (Haney et al., 2009). Currently, numerous AMP structures deposited in the Protein Data Bank (PDB, https://www.rcsb.org) have been determined by solution NMR in the presence of SDS and DPC micelles (Table 1). As an example, a series of high-resolution NMR studies have been in progress by Klochkov's research group regarding the influence of micelles for the characterization of protegrins (Usachev, Efimov, Kolosova, Filippov, et al., 2015; Usachev, Efimov, Kolosova, Klochkova, et al., 2015; Usachev et al., 2017). The authors have shown that this class of AMPs adopt either parallel or antiparallel β-sheets in the presence of perdeuterated DPC, also indicating protegrins' dimerization and oligomerization as an initial step for membrane pore formation (Usachev, Efimov, Kolosova, Filippov, et al., 2015; Usachev, Efimov, Kolosova, Klochkova, et al., 2015; Usachev et al., 2017), as previously proposed (Capone et al., 2010). The structure–function relationship of histidine-rich AMPs has also been studied by high-resolution NMR (McDonald et al., 2015). For example, McDonald et al. (2015) have reported the structural characterization of two histidine-rich AMPs, called Gad-1 and Gad-2, derived from Atlantic cod. In this work, the authors described the great structural plasticity of these AMPs under different physiological conditions. Furthermore, solution NMR in the presence of SDS micelles helped the authors to conclude that maximum antimicrobial properties for both AMPs rely on the appropriated balance between amphipathic central/C-terminal regions and mixed hydrophobic/hydrophilic N-terminus (McDonald et al., 2015). Solution NMR analysis in the presence of SDS micelles was also performed by Santana et al. (2015) with a short disulfide-free multifunctional peptide from green coconut water (*Cn*-AMP1). It was found that *Cn*-AMP1 might adopt α-helical conformations from residues 1 to 6 (2 turns of helix) in SDS. Moreover, a deuterium exchange experiment indicates that Gly^4, Arg^5, and Met^9 are less accessible to solvent, and *Cn*-AMP1 present a flexible C-terminal region. All these findings have helped authors to elucidate the multiple bioactivities of *Cn*-AMP1, including antifungal and anticancer activities (Santana et al., 2015). More recently, van der Weide et al. (2017) have also shown the structural plasticity of a multimeric AMP (SET-M33) from an unstructured state to α-helix in the presence of SDS micelles, where SET-M33 arranged itself on the surface of micelles. In the same work, the authors also highlight the role of Gram-negative lipopolysaccharides (LPS) as an initial target for further AMP interactions with bacterial membranes (van der Weide et al., 2017).

LPS (bacterial endotoxin) has been investigated as a promising target for AMP function, and NMR strategies are underway to investigate AMP–LPS structures and atomic interactions (Table 1). In this context, several well-known AMPs, including polymyxins (Pristovsek & Kidric, 1999), melittins (Bhunia et al., 2007), and temporins (Bhunia et al., 2011), have been submitted to solution NMR experiments in the presence of LPS micelles. Yu et al. (2017) have shown the role of LPS-induced inflammations in patients, as well as LPS neutralization by an AMP (S1-Nal-Nal) through diverse solution NMR analysis in LPS micelles. LPS is also known to mediate sepsis and septic shocks. With that in mind, Datta et al. (2017) investigated the antiendotoxin activity of a short mutated AMP (VG13P) toward LPS micelles, showing that VG13P adopts two consecutive β-turns at both N-terminus and central region. This seems to improve antiendotoxin activities compared to its parent AMP. Interestingly, in contrast, reports have also shown that in some cases LPS can induce AMP self-association rendering them inactive (Mangoni et al., 2008). For instance, Saravanan et al. (2013) performed solution NMR analysis of temporin-1TA (AMP derived from skin secretion of frogs) in the presence of LPS micelles, which induced helical oligomeric structures in temporin-1Ta that may make the translocation of this AMP across bacterial outer membrane difficult (Fig. 1B). The

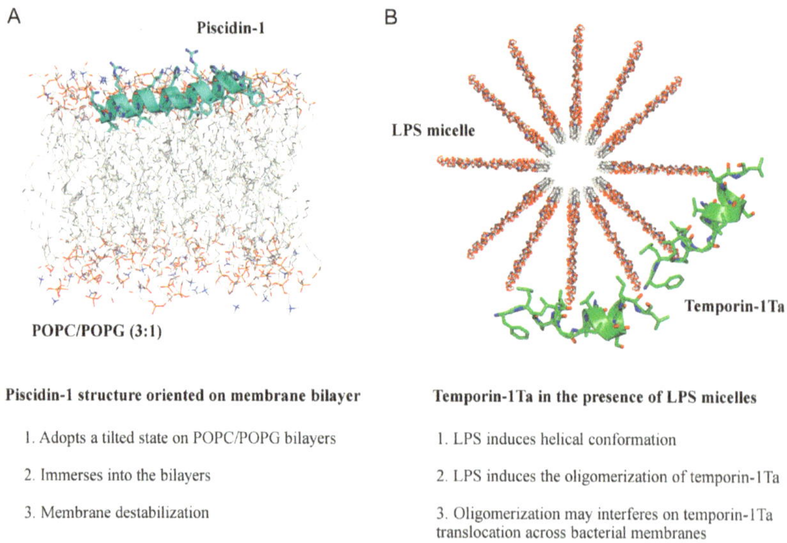

Fig. 1 Schematic representation of a membrane-active AMP (piscidin-1) oriented in a tilted (T-state) on POPC/POPG bilayers (A). Structural visualization of temporin-1Ta oligomerization in contact with LPS micelles (B). Structural visualization and image processing were performed from the solid-state NMR and solution NMR structures available in Protein Data Bank (PDB entries: 2mcu and 2maa, respectively).

authors also highlight the importance of solution NMR studies in micelle-environment as useful information for the future development of improved AMPs (Saravanan et al., 2013).

Apart from these methodologies, an increasing number of works have focused on the NMR studies of AMPs in membrane bilayers and intact bacterial cells, thus greatly and/or fully understanding how AMP structures define their selectivity for specific bacterial strains. These approaches are mainly evaluated by solid-state NMR (SSNMR) and will be discussed with more detail in the following paragraphs.

2.2 Solid-State NMR

SSNMR represents the most suitable technique by which to study AMP structure, dynamics, and topology in liquid-disordered lipid bilayers (Fillion & Auger, 2015). Moreover, depending on which nucleus is investigated, useful information may be acquired for both lipids and peptides (Fillion & Auger, 2015). Otherwise, some disadvantages in the usage of SSNMR must be highlighted, as well as possible alternatives to overcome these obstacles. It is known that SSNMR lacks resolution compared to solution NMR. This happens due to anisotropic interactions, including chemical shift, dipolar coupling, and quadrupolar coupling, which cause an overlap in spectral resonances from individual sites (Bechinger & Salnikov, 2012). Thus, two main strategies have been applied to increase SSNMR resolution, including fast spinning around the magic angle, which averages the anisotropic interactions, resulting in residual isotropic contributions; and the uniaxial alignment of the samples relative to the magnetic field direction (experiments with oriented samples).

NMR experiments performed in oriented samples, for example, have allowed the study of peptide–membrane interactions involving the AMPs protegrin-1 (Kwon et al., 2013), magainin 2, and aurein 3.3 (Kim, Spano, Park, & Wi, 2009). Among the most common types of oriented samples described in the literature are the use of bicelles and phospholipids mechanically oriented between glass plates (Fillion & Auger, 2015). This approach allied to ^{31}P and deuterium NMR, as well as SSNMR strategies that required AMP labeling, including ^{15}N and ^{19}F SSNMR, have provided useful information on membrane perturbation caused by AMPs (Fillion & Auger, 2015). Ramamoorthy et al. (2006), for example, have thoroughly studied the structure and mechanism of action of magainin 2 and melittin-derived peptides, known as MSI-78 and MSI-594. In this study, both ^{15}N-labeled peptides were evaluated in 1-palmitoyl-2-oleoyl-*sn*-glycero-3-phosphatidylcholine (POPC) and 3:1 POPC/1-palmitoyl-2-oleoyl-*sn*-glycero-3-phosphatidylglycerol (POPG) bilayers. Those authors observed that MSI-78 adopts an α-helical conformation, which is oriented perpendicularly to the bilayers normal. Interestingly, however, the presence of anionic POPG in the 3:1 POPC/POPG membranes seems to enhance the interactions between MSI-78 and lipids, thus reducing the peptide heterogeneity in these bilayers compared to the ones containing only neutral POPC (Ramamoorthy et al., 2006). Unlike MSI-78, MSI-594 presented a single chemical shift in both bilayers, suggesting an in-plane orientation (Ramamoorthy et al., 2006). Regarding the peptide interactions with POPC lipid headgroups, both MSI-78 and MSI-594 could affect the phospholipids' distribution. However, when in contact with POPG membrane models, both peptides seem to induce the formation of nonbilayer structures (Ramamoorthy et al., 2006). These findings were discussed in the light of AMP–bilayer interactions, allowing the authors to propose membrane disruption mechanisms of action for MSI-78 and MSI-594 via carpet and toroidal-type models.

Interestingly, works have also explored how the synergistic activities of two different AMPs can be molecularly studied by SSNMR in membrane bilayers (Strandberg et al., 2016). It has been reported that magainin 2 and PGLa alone in contact with POPC/POPG bilayers adopt a surface state (S-state), lying flat on the membrane surface (Fig. 2) (Strandberg et al., 2016). However, when synergy conditions are established, for example, 1:1 magainin 2/PGLa, both AMPs reorientate on POPC/POPG, where a transmembrane inserted state (I-state) is observed for PGLa, whereas magainin 2 turns to a tilted state (T-state) (Fig. 2) (Strandberg et al., 2016). Based on that, the authors hypothesized that PGLa and magainin 2

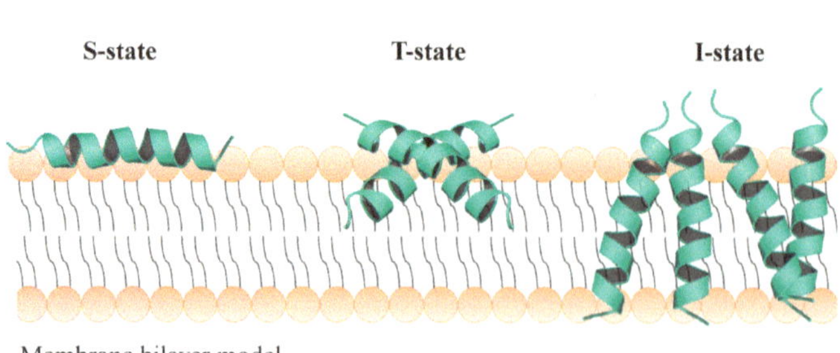

Fig. 2 Schematic representation of magainin 2 in membrane models on its surface state (S-state), tilted state (T-state), and transmembrane inserted state (I-state), respectively.

may form a complex in membranes that in turn favors the formation of stable water-filled pores, thus assisting in a better understanding of the synergistic modes of action of these two AMPs against bacterial strains (Strandberg, Zerweck, Wadhwani, & Ulrich, 2013; Tremouilhac, Strandberg, Wadhwani, & Ulrich, 2006).

NMR studies using model membrane represent a huge step in the investigation of AMP mechanisms of action. However, more recently, increasing reports have introduced the evaluation of AMP structures in intact bacterial strains by ^{2}H SSNMR (Booth, Warschawski, Santisteban, Laadhari, & Marcotte, 2017). Tardy-Laporte et al. (2013) developed a protocol to deuterate *Escherichia coli* membrane phospholipids, further evaluating the structure and effect of polymyxin B on the bacterial surface. More recently, the same research group has also shown the interactions of two AMPs, carein 1.1 and aurein 1.2, with intact *E. coli* and *Bacillus subtilis* using deuterium SSNMR (Laadhari, Arnold, Gravel, Separovic, & Marcotte, 2016). The authors observed that carein 1.1 and aurein 1.2 decreased the order of the phospholipids' acyl chains in both bacterial strains tested. However, higher concentrations of both peptides were required against *E. coli*, which might be explained due to its more complex cell wall characteristic of Gram-negative bacteria compared to Gram-positive ones, including the presence of anionic teichoic and lipoteichoic acids, as well as LPS (Laadhari et al., 2016).

2.3 X-Ray Crystallography

A large number of AMPs have had their structures determined by X-ray crystallography. X-ray studies with human beta-defensins, for example, have shown evidence of high-order oligomerization in this class of AMPs (Hoover et al., 2000). Similarly, plant defensins have been structurally characterized, revealing a conserved dimeric cationic grip conformation to mediate microbial cell lysis (Baxter et al., 2015). Interestingly, disulfide-rich AMPs can be notoriously difficult to crystallize. In this context, researchers have explored the use of racemic crystallization procedures to determine the structure of the cyclotide kalata-B1 (Wang, King, Northfield, Ojeda, & Craik, 2014). Apart from these structural characterization studies, AMP structure–function relationships have also been explored by X-ray crystallography through the evaluation of AMP atomic interactions with both bacterial membrane models and intracellular targets (Yang, Weiss, Lehrer, & Huang, 2000).

Similarly to what is observed in SSNMR, X-ray crystallography studies have been underway to understand the influence of AMP on mimetic membrane models. Yang et al. (2000) have reported, for the first time, the crystallization of membrane pores spontaneously formed by magainin 2 and protegrin-1 in 1,2-dimyristoyl-*sn*-glycero-3-phosphocholine (DMPC)/ 1,2-dimyristoyl-*sn*-glycero-3-phosphorylglycerol (DMPG) membrane multilayers, providing accurate information on toroidal pore formation by these AMPs. In addition, it has been shown that increasing concentrations of magainin 2 cause a decrease in bilayer thickness, which has been largely correlated with magainin 2 pore formation mechanisms (Chen et al., 2003). In addition, Qian, Wang, Yang, and Huang (2008) performed the pore reconstruction of the AMP alamethicin in multilayers, where eight alamethicin helices formed barrel-stave pore types. Moreover, in a robust work, X-ray crystallography, NMR spectroscopy, and MD simulations were put together to elucidate the membrane-disruptive mechanism of the human dermcidin AMP (Song et al., 2013). In this study, the authors concluded that dermcidin adopts zinc-connected trimmers of antiparallel helical pairs to architect high-conductance transmembrane channels responsible for its antimicrobial activities (Song et al., 2013). Furthermore, apart from structural studies using membrane models, some works have also shown the reconstruction of bacterial membrane lipid extracts into bilayers for the characterization of AMPs' modes of action. Staudegger et al. (2000), for example, reported the evidence for cubic phase formation by the AMP gramicidin S in bilayers formed from the total membrane lipid of *Acholeplasma laidlawii* B and *E. coli*, also observing an increase in the negative curvature of these membranes, mediating bacterial cell lysis.

Apart from membrane-related mechanism of actions, AMPs are also known to act on specific intracellular bacterial targets (Fjell, Hiss, et al., 2011). These features have been well investigated in proline-rich AMPs. The insect-sourced proline-rich adepcin and oncocin, for example, have shown to bind to 70 kDa bacterial heat shock protein DnaK (Fig. 3A) (Krizsan et al., 2014). In addition, Roy, Lomakin, Gagnon, and Steitz (2015) have demonstrated that an oncocin derivative, called Onc112, binds *Thermus thermophilus* 70S ribosomes, thus blocking both the peptidyl tranferase center and the peptide-exit tunnel of these ribosomal subunits, directly interfering in bacterial protein synthesis (Fig. 3B). Further studies have shown similar findings through crystallographic studies of bactenecin, pyrrhocoricin, and metalnikowin bonds to *T. thermophilus* 70S ribosomes (Gagnon et al., 2016).

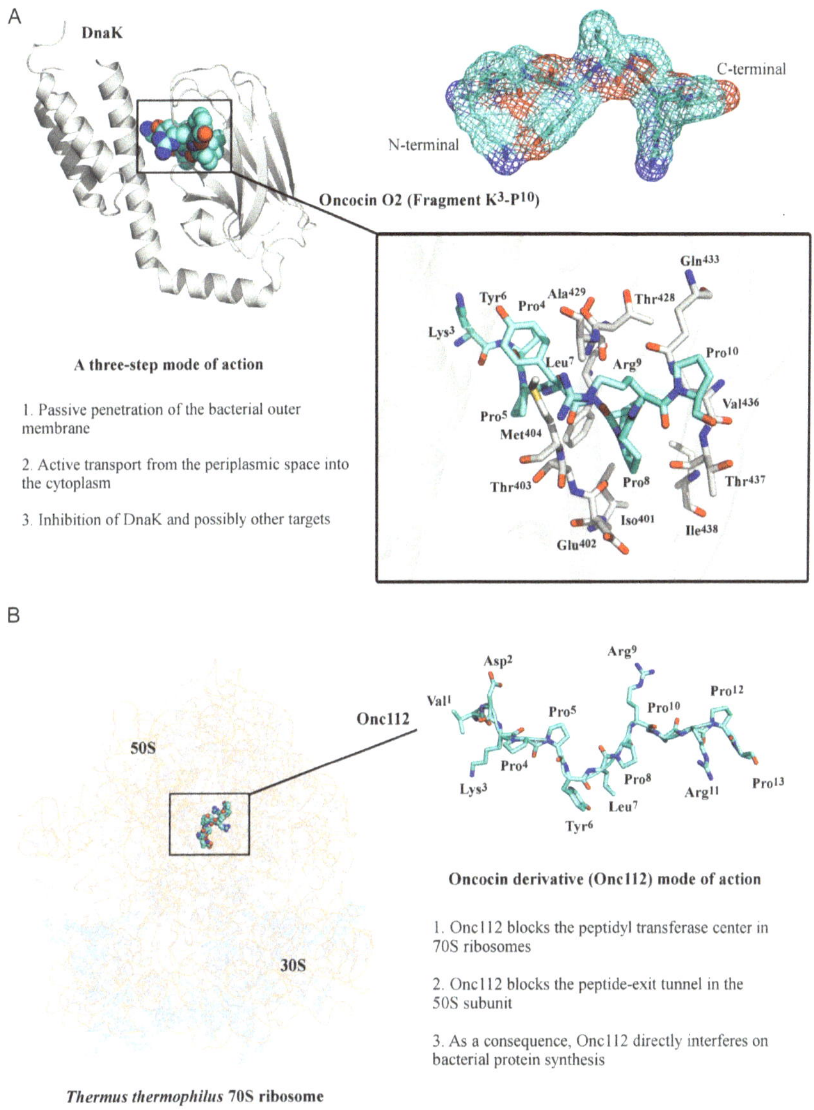

Fig. 3 A three-step mode of action proposed for the AMP oncocin O2, highlighting its interactions with DnaK (A). Amino acid residues from DnaK that are mainly involved in the stabilization of oncocin O2 in the binding site are represented as *white sticks*. Structural visualization and image processing were performed from the crystal structure of oncocin O2/DnaK available in Protein Data bank (PDB entry: 3qnj). Structural visualization of the AMP Onc112 bond to the 50S subunit of a *Thermus thermophilus* 70S ribosome (B). The proposed mechanisms of action for this AMP are also highlighted. Structural visualization and image processing were performed from the crystal structure of Onc112 bond to *T. thermophilus* 70S ribosome available in Protein Data Bank (PDB entry: 4z8c).

2.4 Cryo-Electron Microscopy

The effective use of cryo-EM for the structural resolution of biomolecules became a reality in the past decade, with the advances provided by the pioneering discoveries of the recent Nobel Prize winners Jacques Dubochet, Joachim Frank, and Richard Henderson (Cressey & Callaway, 2017). When it comes to AMP structural analysis, cryo-EM has been used for supporting studies of complexes' stability, as reported by Boge et al. (2016). These authors demonstrated that AMPs could successfully be loaded into liquid cubic/hexagonal crystalline gels and nanoparticles. They examined lyotropic liquid crystalline (LC) structures, formed by cubic glycerol monooleate/water (GMO/water gel) and hexagonal glycerol monooleate/oleic acid/water (GMO/OA water gel) as carriers for AMPs. Moreover, the authors demonstrated the formation of vesicles derived from high-pressure homogenizer used to make dispersions, resulting in ordered internal gel structures. However, the GMO/water gel was more sensitive to AMP addition than the GMO/AO water gel complex. It was shown that AMPs' nature interferes in the curvature of cubic gels (peptide net charge and hydrophobicity levels are important factors for the curvature variation), where GMO/AO displays a fuzzier structure than GMO/water gel. The hexagonal GMO/OA water complex was stronger for the addition of any of the tested AMPs and did not disturb the LC structure (Boge et al., 2016).

The study of complex systems such as human cells/bacterial biofilms/ AMPs has also been explored. Haisma et al. (2016) reported the use of cryo-scanning electron microscopy to visualize the activity of the peptide P60.4Ac (derived from LL-37) on *Staphylococcus aureus* methicillin-resistant strain LUH14616 biofilms, on human epidermal models. The 24 h biofilms were exposed to P60.4Ac in PBS for 1 h and then analyzed. These experiments revealed that mature biofilm on the stratum corneum of the epidermal models 24 h after colonization was completely eradicated by the peptide during a 1 h treatment (Haisma et al., 2016).

Natural complex biological systems, including biofilm/human cell arrangement, still constitute a challenge for researchers, so approaches using synthetic mimetic membranes are interesting alternatives for structural and biological studies. In this frontage, the well-known magainin 2 was used as the seed sequence for studies of AMP optimization, aiming at Gram-negative bacteria antimicrobial activity. Han, Mei, Khant, and Ludtke (2009) used synthetic vesicles composed of different proportions of POPC/POPG and DMPC/POPG in contact with this peptide. Their results revealed that POPC/POPG responds differently from DMPC/POPG to magainin 2 regarding its behavior, demonstrating that the magainin–lipid interactions are highly stochastic, and it should not be assumed that all vesicles would be impacted equally by the presence of magainin, which was further observed with the differentiation of vesicles into two classes (rough and smooth). The authors demonstrated that magainin induced surface perturbations in synthetic lipid vesicles, and these results can be attributed to a random distribution of pores with ~80 Å of diameter. Also working with magainin 2, Lichtenecker et al. (2013) used cryo-EM to characterize the membrane-associated mechanism of action of a new D-/L-peptide with antibiotic potential, thus confirming the disturbance of bacterial membrane by these AMPs. In this work, magainin 2 was also explored regarding its interactions with lipid membranes using a combination of cryo-EM and neutron/X-ray scattering to improve the quality and accuracy of single particle resolution (Lichtenecker et al., 2013). However, cryo-EM of bacterial cell membranes did not show any evidence for pore formation, as already reported for magainin (Han et al., 2009), but the formation of small vesicles was seen in the presence of active D-/L-peptides (Lichtenecker et al., 2013). Under these circumstances, cryo-EM exhibited in a nondestructive way that the compounds have disturbed the bacterial membrane integrity (Lichtenecker et al., 2013). Similarly, Wessman, Morin, Reijmar, and Edwards (2010) reported the use of cryogenic transmission electron microscopy (cryo-TEM) in combination with turbidity and leakage assays to evaluate and compare the membrane modifications resulting in effects from magainin 2, mellitin, and alamethicin on POPC-based liposomes in distinct compositions. Mellitin and alamethicin were able to induce similar structural effects on liposomes, whereas magainin 2 seems to have a preference for curved lipid surfaces (Wessman et al., 2010). In the field of lipopeptides, Hamley et al. (2013) evaluated the self-assembly of surfactin, plipastin, and mycosubtilin from *B. subtilis* in aqueous solution using cryo-TEM and X-ray scattering, revealing that surfactin and plipastin aggregate into 2 nm-radius spherical micelles, whereas mycosubtilin self-assembles into extended nanotapes (with a variable width 15–80 nm) based on bilayer ordering of the lipopeptides. Usually, the length and isomeric nature of the fatty acid chain contribute to the activity of the surfactant molecule and the intermolecular hydrophobic interaction, which leads to micelle formation (Hamley et al., 2013).

In summary, cryo-EM provides direct imaging in systems closely related to in vivo conditions. This technique has shown to be a useful tool for studying peptide–lipid interactions, even when the resulting structures cannot be determined individually. Moreover, cryo-EM provides fine control over the image included in the power spectrum calculation. Still, despite the power of this technique, it lacks the flexibility provided by some approaches such as solution NMR or X-ray crystallography, but this has been overcome by combining cryo-EM with neutron/X-ray scattering experiments. Altogether, the experimental approaches discussed in the last topics have helped researchers to better understand AMP mechanisms of actions under different conditions and targets. Furthermore, complementary strategies have been applied, including the usage of reliable computational approaches in peptide drug design and AMP structural characterization. In this context, here we will be dedicated to in silico approaches currently used in AMP structure–function studies.

3. COMPUTATIONAL APPROACHES FOR AMPs STRUCTURE PREDICTION

3.1 Molecular Modeling, Dynamics, and Docking

One of the technical features used in computational biology is the description of molecular systems at atomic level (Vriend, 1990), which has become increasingly important in the field of protein engineering and drug design (Fjell, Jenssen, Cheung, Hancock, & Cherkasov, 2011; Vriend, 1990). In relation to experimental techniques, molecular modeling (MM) allows the user a "fast" methodology to reliably predict protein structures for diverse purposes, including drug rational design, protein function analysis, molecular interactions, antigenic behavior, and redesign of proteins and/or peptides with greater stability and specificity (Schlick, 2010).

Some authors have shown a correlation between the prediction of AMPs' 3D structure and function. Irazazabal et al. (2016), for example, performed the comparative modeling of a mastoparan-L analogue (I_5R_8), showing that I_5R_8 was able to form an amphipathic α-helix that favored membrane perturbation. Similarly, an AMP isolated from *Anoplius samariensis* (anoplin) had its sequence submitted to comparative modeling according to mastoparan-X atomic coordinates (Konno et al., 2001). Both peptides presented similar helical conformation that was further related to similar biological activities between mastoparan-X and anoplin, as potent mast cell degranulation inductors, and broad spectrum antimicrobial activity against Gram-positive and -negative bacteria (Konno et al., 2001). In addition, Tavares et al. (2012) have reported the comparison of the structure of a glycine-rich natural peptide (Pg-AMP1) isolated from *Psidium guajava* with its analogue, showing that both present α-helical conformations with regions rich in glycine and serine residues, which confer flexibility on these molecules. According to the authors, both Pg-AMP1 and its recombinant can assume different conformations, given the variations observed in their flexible regions, and that might be a key point for their bactericidal activities (Tavares et al., 2012). Ab initio modeling has also played a crucial role in AMP structural investigation. In this context, Mandal et al. (2009) have reported the ab initio modeling of AMPs isolated from green coconut water, revealing that the arrangement of amino acids along the sequence of the peptides could favor the formation of helices, and thus interfere in their action.

Structural data of proteins and peptides such as AMPs are of clinical interest, since physicochemical characteristics and structure may interfere in cellular recognition functions and play a fundamental role in optimizing antiinfection therapies (Fjell, Hiss, et al., 2011). Dynamics simulations are important tools for understanding the physical basis of the structure and function of biological macromolecules (Karplus & McCammon, 2002). MD methods are used, for example, for the study of protein folding routes (Kubitzki, de Groot, & Seeliger, 2009). These simulations have become a standard technique in protein studies and are routinely applied to a wide range of conformational challenges (Schlick, 2010).

As described in Section 2.1, the usage of TFE as cosolvent has been widely explored in AMPs structural studies, including in MD simulations. Mehrnejad, Naderi-Manesh, and Ranjbar (2007) performed comparative MD simulation analyses of magainin in water, TFE/water in 2, 4, and 8 M urea solutions, reporting higher helical content for this AMP in water/TFE compared to water alone. In the field of membrane bilayers, Perrin et al. (2014) have used high-resolution structures from the homologues of piscidin 1 (P1) and piscidin 3 (P3), associated with MD to provide better structural information and guidance (Fig. 1A). In this study, MD simulations of the peptides under the same experimental conditions could provide an independent determination of conformation and orientation for these AMPs. Together, NMR and MD provided a reference for the structural determination of interfaced helices, which may be suitable for the study of other amphipathic AMPs in lipid bilayers (Perrin et al., 2014). Moreover, MD simulations have been performed in magainin MG-H2-phospholipid model membrane systems, thus investigating the mechanism by which this AMP may act. In this study, several copies of the peptide were randomly placed in solution near the membrane, and it was observed that magainin

MG-H2 induces the formation of toroidal pores in the bilayer (Leontiadou, Mark, & Marrink, 2006). MD simulations have increasingly been an integral part of the refinement of protein structures solved both by NMR and X-ray diffraction (Karplus & McCammon, 2002). In recent years, advanced computational techniques based on MD have been offered. These techniques include thermodynamic integrations, free energy disturbance, linear interaction energy, molecular mechanics, and methods related to surface areas (Åqvist, Luzhkov, & Brandsdal, 2002). In addition, MD is now being used to directly estimate the effect of water molecules in the system, as well as providing more accurate estimates of binding free energy values in a receptor–ligand complex (Okimoto et al., 2009).

MD techniques associated with molecular coupling techniques appear as an attractive tool for the study of the molecular interactions between AMPs and biological membranes, bacterial DNA, proteins of interest, among others. Combining all the in silico techniques described earlier, Mandal et al. (2013) performed structural analyses, as well as ligand–receptor interaction studies to characterize a new AMP, called Ps-AFP1, isolated from *Pisum sativum*. Molecular docking followed by MD simulations was performed to predict the interactions of this peptide with *N*-acetylglucosamine trimers. As a result, both couplings and MD simulations clearly indicated the stability generated by the Ps-AFP1/*N*-acetylglucosamine complex, suggesting that this peptide may interact with the chitin molecules constituent of the fungi membranes, thus explaining, in part, its activities against *Colletotrichum gloeosporioides*, *Aspergillus terreus*, *Candida albicans*, *Fusarium oxysporum*, and *Fusarium solani* (Mandal et al., 2013). Another work reporting the structure–function of AMPs through in vitro/in silico methodologies was carried out by Cardoso et al. (2016). In this work, the authors correlated the action of a polar fish-derived peptide (*Pa*-MAP 1.9) against Gram-negative bacteria (*Enterococcus faecalis*, *E. coli*, and *Klebsiella pneumoniae*) with computational simulations to evaluate peptide–membrane interactions. They demonstrated that *Pa*-MAP1.9 was able to interact with mimetic bacterial membranes and vesicles by adopting a well-defined α-helical conformation, which favored the orientation and insertion of this peptide in lipid bilayers to carry out its activities (Cardoso et al., 2016). MM associated with molecular dynamics and docking provides a way to explore structural/dynamic patterns to test and develop hypotheses, interpret and extend experimental data, as well as extend basic laws governing molecular structure, flexibility, and function (Lindahl, 2008; Okimoto et al., 2009; Schlick, 2010).

4. OUTLOOK

AMPs have been explored for the past century, providing enormous changes in our ability to treat infectious diseases worldwide. Even so, we still have a lack of effective new antibiotics for the treatment of Gram-negative and -positive bacterial infections and, together with emerging multidrug resistance, this creates an urgent challenge for global health. In this scenario, the rapid and efficient development of new antimicrobial strategies represents a critical need. The development of AMPs for clinical use is full of challenges, which are now being overcome due to greater understanding of how peptides' structure influences their mechanisms of action. The elucidation of 3D structures of AMPs has been fundamental to the determination of their functions. Whereas membrane interaction aspects remain controversial, it is now possible to facilitate the design of peptide analogs with greater antimicrobial activity and minimal toxicity. Continued investigation of these aspects should deliver improvements, allowing AMPs to become a feasible complement to conventional antibiotics. In this context, the precise understanding of AMP–target interactions through diverse and rapidly evolving techniques will be of great advantage in AMPs development.

The ways by which AMPs bind to their targets can be investigated, as previously discussed here, either in solid-state interaction (due to cocrystallization or soaking methods) through X-ray crystallography or in a more dynamic interaction with their targets, as allowed by NMR spectroscopy. NMR approaches provide an amazing range of details, not only for static structures, but also by allowing us to visualize mobile biological systems throughout their individuality, revealing the chemical and molecular mechanisms, which is unique for magnetic resonance imaging. Despite the impressive amount of information delivered, this technique still has shortcomings in terms of sensitivity and resolution. With the improvement of these two aspects, lower amounts of samples would be required, which is quite important, since many biological systems cannot be made in large quantities. NMR is an inherently low-throughput method, compared to X-ray crystallography, for the deliver of information; the development of a fast and efficient data collection flow that allows high efficiency in NMR spectroscopy assignments would be extremely helpful for further AMP studies.

It is known that crystallography has been widely used for the structural analysis of AMPs and also macromolecular complexes. However, like

NMR, X-ray crystallography also presents its own challenges for AMP elucidation. The use of sub-mm^3 crystals is now commonplace for data collection, allowing the atomic resolution of a few small proteins and single biological systems. However, sample damage due to radiation is still a big issue, as are the needs for reduction in crystal volume or data collection time. Besides, X-ray crystallography still has sample solubility as a bottleneck, mainly in macromolecular studies, considering that the targets of many AMPs are cell membranes or other insoluble molecules. The fields of data validation, data mining, and data management still require user-friendly approaches, which demand great human and computational efforts. Regarding future steps, it is believed that the screening of fragments based on crystallography analysis will become more widely used as libraries of compounds that will be better optimized, allowing laboratories and databases to become even more helpful in handling experiments and structural data.

In the field of cryo-EM-based determination of structures at very high resolution, the quality of the sample remains as the major bottleneck. In contrast, this approach allows the determination of structures with near-atomic resolution, appearing as an alternative for the determination of macromolecular structures without crystallization, also being less affected by solubility problems, even though it requires a homogeneous sample. A successful cryo-TEM experiment is certainly the future in developing cryo-EM as a powerful method to study AMP-membrane systems (such as vesicle fusion or the mechanics of membrane-penetrating peptides). Thereafter, crystallography and NMR will remain in coexistence with cryo-EM, since both techniques can be integrated to enhance details of biological systems; this represents an integrated arrangement of methods to help biologists and chemists with functional biology questions, and with fully exploring these methods.

The structural analysis of AMPs through laboratory experiments is often an expensive approach. At this point, NMR spectroscopy, cryo-EM, and X-ray crystallography should be integrated with computational techniques to gain in-depth understanding into AMP structure and function (Perilla et al., 2017). What is expected in the next few years is automation in the collection of data, and more accurate analysis of the data deposited in databases, avoiding the accumulation of errors for future researchers who depend on this information. In addition, it is extremely interesting to integrate automated structural results from different sources, generating a complementarity in the available information. At this point, the algorithms developed by bioinformatics will be of great value for integrating the structural biochemical information generated, making it possible to integrate the knowledge engendered by different techniques and to apply it to the rational design of new AMPs. Hence, the use of multifunctional and efficient computational tools has become essential to select the best molecule prior to in vitro tests. Computational approaches (including MM, MD, and molecular docking) have been demonstrating great advantages in engendering knowledge for the prediction of AMP structure, and their mechanism of action. The improvement of algorithms for AMP predictions has been used as a powerful weapon in the design and redesign of molecules that are more effective and selective for specific targets. In silico methodologies associated with structural biophysics approaches have enabled the characterization of an arsenal of AMPs that can be used as future prototypes for antiinfectious drugs. While great challenges remain in this field, the discovery and clinical development of ideal therapeutic peptides and their mimetics are likely to be enhanced by the latest advances in understanding their mechanism of action, stability, and improved formulation strategies. These findings have reinforced the clinical relevance of AMP as promising candidates to the elaboration of new peptide-based therapies.

REFERENCES

Åqvist, J., Luzhkov, V. B., & Brandsdal, B. O. (2002). Ligand binding affinities from MD simulations. *Accounts of Chemical Research*, *35*, 358–365.

Baxter, A. A., Richter, V., Lay, F. T., Poon, I. K., Adda, C. G., Veneer, P. K., et al. (2015). The tomato defensin TPP3 binds phosphatidylinositol (4,5)-Bisphosphate via a conserved dimeric cationic grip conformation to mediate cell lysis. *Molecular and Cellular Biology*, *35*, 1964–1978.

Bechinger, B., & Salnikov, E. S. (2012). The membrane interactions of antimicrobial peptides revealed by solid-state NMR spectroscopy. *Chemistry and Physics of Lipids*, *165*, 282–301.

Bhunia, A., Domadia, P. N., & Bhattacharjya, S. (2007). Structural and thermodynamic analyses of the interaction between melittin and lipopolysaccharide. *Biochimica et Biophysica Acta*, *1768*, 3282–3291.

Bhunia, A., Saravanan, R., Mohanram, H., Mangoni, M. L., & Bhattacharjya, S. (2011). NMR structures and interactions of temporin-1Tl and temporin-1Tb with lipopolysaccharide micelles: Mechanistic insights into outer membrane permeabilization and synergistic activity. *The Journal of Biological Chemistry*, *286*, 24394–24406.

Boge, L., Bysell, H., Ringstad, L., Wennman, D., Umerska, A., Cassisa, V., et al. (2016). Lipid-based liquid crystals as carriers for antimicrobial peptides: Phase behavior and antimicrobial effect. *Langmuir*, *32*, 4217–4228.

Booth, V., Warschawski, D. E., Santisteban, N. P., Laadhari, M., & Marcotte, I. (2017). Recent progress on the application of 2H solid-state NMR to probe the interaction of antimicrobial peptides with intact bacteria. *Biochimica et Biophysica Acta*, *1865*, 1500–1511.

Campagna, S., Saint, N., Molle, G., & Aumelas, A. (2007). Structure and mechanism of action of the antimicrobial peptide piscidin. *Biochemistry*, *46*, 1771–1778.

Cao, H., Pradhan, A., Karns, J., Wolfgang, D., Hovingh, E., Vinyard, B., et al. (2017). Prevalence and risk factors for antimicrobial resistance on US dairy operations. *Journal of Animal Science*, *95*, 131–132.
Capone, R., Mustata, M., Jang, H., Arce, F. T., Nussinov, R., & Lal, R. (2010). Antimicrobial protegrin-1 forms ion channels: Molecular dynamic simulation, atomic force microscopy, and electrical conductance studies. *Biophysical Journal*, *98*, 2644–2652.
Cardoso, M. H., Ribeiro, S. M., Nolasco, D. O., de la Fuente-Núñez, C., Felício, M. R., Gonçalves, S., et al. (2016). A polyalanine peptide derived from polar fish with anti-infectious activities. *Scientific Reports*, *6*, 21385.
Chen, F. Y., Lee, M. T., & Huang, H. W. (2003). Evidence for membrane thinning effect as the mechanism for peptide-induced pore formation. *Biophysical Journal*, *84*, 3751–3758.
Chia, B. C., Carver, J. A., Mulhern, T. D., & Bowie, J. H. (1999). The solution structure of uperin 3.6, an antibiotic peptide from the granular dorsal glands of the Australian toadlet, *Uperoleia mjobergii*. *The Journal of Peptide Research*, *54*, 137–145.
Cressey, D., & Callaway, E. (2017). Cryo-electron microscopy wins chemistry Nobel. *Nature*, *550*, 167.
Datta, A., Jaiswal, N., Ilyas, H., Debnath, S., Biswas, K., Kumar, D., et al. (2017). Structural and dynamic insights into a glycine-mediated short analogue of a designed peptide in lipopolysaccharide micelles: Correlation between compact structure and anti-endotoxin activity. *Biochemistry*, *56*, 1348–1362.
Dennison, S. R., Wallace, J., Harris, F., & Phoenix, D. A. (2005). Amphiphilic α-helical antimicrobial peptides and their structure/function relationships. *Protein and Peptide Letters*, *12*, 31–39.
Fillion, M., & Auger, M. (2015). Oriented samples: A tool for determining the membrane topology and the mechanism of action of cationic antimicrobial peptides by solid-state NMR. *Biophysical Reviews*, 7, 311–320.
Fioroni, M., Diaz, M. D., Burger, K., & Berger, S. (2002). Solvation phenomena of a tetrapeptide in water/trifluoroethanol and water/ethanol mixtures: A diffusion NMR, intermolecular NOE, and molecular dynamics study. *Journal of the American Chemical Society*, *124*, 7737–7744.
Fjell, C. D., Hiss, J. A., Hancock, R. E., & Schneider, G. (2011). Designing antimicrobial peptides: Form follows function. *Nature Reviews Drug Discovery*, *11*, 37–51.
Fjell, C. D., Jenssen, H., Cheung, W. A., Hancock, R. E., & Cherkasov, A. (2011). Optimization of antibacterial peptides by genetic algorithms and cheminformatics. *Chemical Biology & Drug Design*, 77, 48–56.
Gagnon, M. G., Roy, R. N., Lomakin, I. B., Florin, T., Mankin, A. S., & Steitz, T. A. (2016). Structures of proline-rich peptides bound to the ribosome reveal a common mechanism of protein synthesis inhibition. *Nucleic Acids Research*, *44*, 2439–2450.
Haisma, E. M., Goblyos, A., Ravensbergen, B., Adriaans, A. E., Cordfunke, R. A., Schrumpf, J., et al. (2016). Antimicrobial peptide P60.4Ac-containing creams and gel for eradication of methicillin-resistant *Staphylococcus aureus* from cultured skin and airway epithelial surfaces. *Antimicrobial Agents and Chemotherapy*, *60*, 4063–4072.
Hamley, I. W., Dehsorkhi, A., Jauregi, P., Seitsonen, J., Ruokolainen, J., Coutte, F., et al. (2013). Self-assembly of three bacterially-derived bioactive lipopeptides. *Soft Matter*, *9*, 9572–9578.
Han, M., Mei, Y., Khant, H., & Ludtke, S. J. (2009). Characterization of antibiotic peptide pores using cryo-EM and comparison to neutron scattering. *Biophysical Journal*, *97*, 164–172.
Haney, E. F., Hunter, H. N., Matsuzaki, K., & Vogel, H. J. (2009). Solution NMR studies of amphibian antimicrobial peptides: Linking structure to function? *Biochimica et Biophysica Acta*, *1788*, 1639–1655.

Hong, M., Zhang, Y., & Hu, F. (2012). Membrane protein structure and dynamics from NMR spectroscopy. *Annual Review of Physical Chemistry*, *63*, 1–24.

Hoover, D. M., Rajashankar, K. R., Blumenthal, R., Puri, A., Oppenheim, J. J., Chertov, O., et al. (2000). The structure of human beta-defensin-2 shows evidence of higher order oligomerization. *The Journal of Biological Chemistry*, *275*, 32911–32918.

Irazazabal, L. N., Porto, W. F., Ribeiro, S. M., Casale, S., Humblot, V., Ladram, A., et al. (2016). Selective amino acid substitution reduces cytotoxicity of the antimicrobial peptide mastoparan. *Biochimica et Biophysica Acta*, *1858*, 2699–2708.

Jasanoff, A., & Fersht, A. R. (1994). Quantitative determination of helical propensities from trifluoroethanol titration curves. *Biochemistry*, *33*, 2129–2135.

Karplus, M., & McCammon, J. A. (2002). Molecular dynamics simulations of biomolecules. *Nature Structural & Molecular Biology*, *9*, 646–652.

Kim, C., Spano, J., Park, E. K., & Wi, S. (2009). Evidence of pores and thinned lipid bilayers induced in oriented lipid membranes interacting with the antimicrobial peptides, magainin-2 and aurein-3.3. *Biochimica et Biophysica Acta*, *1788*, 1482–1496.

Konno, K., Hisada, M., Fontana, R., Lorenzi, C. C., Naoki, H., Itagaki, Y., et al. (2001). Anoplin, a novel antimicrobial peptide from the venom of the solitary wasp *Anoplius samariensis*. *Biochimica et Biophysica Acta*, *1550*, 70–80.

Krizsan, A., Volke, D., Weinert, S., Strater, N., Knappe, D., & Hoffmann, R. (2014). Insect-derived proline-rich antimicrobial peptides kill bacteria by inhibiting bacterial protein translation at the 70S ribosome. *Angewandte Chemie International Edition*, *53*, 12236–12239.

Kubitzki, M. B., de Groot, B. L., & Seeliger, D. (2009). Protein dynamics: From structure to function. In D. J. Rigden (Ed.), *From protein structure to function with bioinformatics* (pp. 217–249). Netherlands: Springer.

Kwon, B., Waring, A. J., & Hong, M. (2013). A 2H solid-state NMR study of lipid clustering by cationic antimicrobial and cell-penetrating peptides in model bacterial membranes. *Biophysical Journal*, *105*, 2333–2342.

Laadhari, M., Arnold, A. A., Gravel, A. E., Separovic, F., & Marcotte, I. (2016). Interaction of the antimicrobial peptides caerin 1.1 and aurein 1.2 with intact bacteria by 2H solid-state NMR. *Biochimica et Biophysica Acta*, *1858*, 2959–2964.

Leontiadou, H., Mark, A. E., & Marrink, S. J. (2006). Antimicrobial peptides in action. *Journal of the American Chemical Society*, *128*, 12156–12161.

Lichtenecker, R. J., Ellinger, B., Han, H. M., Jadhav, K. B., Baumann, S., Makarewicz, O., et al. (2013). Iterative antimicrobial candidate selection from informed D-/L-peptide dimer libraries. *Chembiochem*, *14*, 2492–2499.

Lindahl, E. R. (2008). Molecular dynamics simulations. In: A. Kukol (Ed.), *Molecular modeling of proteins Methods in Molecular Biology*, *443*, 3–23.

Lopez-Abarrategui, C., McBeth, C., Mandal, S. M., Sun, Z. J., Heffron, G., Alba-Menendez, A., et al. (2015). Cm-p5: An antifungal hydrophilic peptide derived from the coastal mollusk *Cenchritis muricatus* (Gastropoda: Littorinidae). *FASEB Journal*, *29*, 3315–3325.

Mandal, S. M., Dey, S., Mandal, M., Sarkar, S., Maria-Neto, S., & Franco, O. L. (2009). Identification and structural insights of three novel antimicrobial peptides isolated from green coconut water. *Peptides*, *30*, 633–637.

Mandal, S. M., Porto, W. F., Dey, P., Maiti, M. K., Ghosh, A. K., & Franco, O. L. (2013). The attack of the phytopathogens and the trumpet solo: Identification of a novel plant antifungal peptide with distinct fold and disulfide bond pattern. *Biochimie*, *95*, 1939–1948.

Mangoni, M. L., Epand, R. F., Rosenfeld, Y., Peleg, A., Barra, D., Epand, R. M., et al. (2008). Lipopolysaccharide, a key molecule involved in the synergism between temporins in inhibiting bacterial growth and in endotoxin neutralization. *The Journal of Biological Chemistry*, *283*, 22907–22917.

Marion, D., Zasloff, M., & Bax, A. (1988). A two-dimensional NMR study of the antimicrobial peptide magainin 2. *FEBS Letters*, *227*, 21–26.
McDonald, M., Mannion, M., Pike, D., Lewis, K., Flynn, A., Brannan, A. M., et al. (2015). Structure-function relationships in histidine-rich antimicrobial peptides from Atlantic cod. *Biochimica et Biophysica Acta*, *1848*, 1451–1461.
Mehrnejad, F., Naderi-Manesh, H., & Ranjbar, B. (2007). The structural properties of magainin in water, TFE/water, and aqueous urea solutions: Molecular dynamics simulations. *Proteins: Structure, Function, and Bioinformatics*, *67*, 931–940.
Nissen-Meyer, J., & Nes, I. F. (1997). Ribosomally synthesized antimicrobial peptides: Their function, structure, biogenesis, and mechanism of action. *Archives of Microbiology*, *167*, 67–77.
Okimoto, N., Futatsugi, N., Fuji, H., Suenaga, A., Morimoto, G., Yanai, R., et al. (2009). High-performance drug discovery: Computational screening by combining docking and molecular dynamics simulations. *PLoS Computational Biology*, *5*, e1000528.
Pan, Y. L., Cheng, J. T., Hale, J., Pan, J., Hancock, R. E., & Straus, S. K. (2007). Characterization of the structure and membrane interaction of the antimicrobial peptides aurein 2.2 and 2.3 from Australian southern bell frogs. *Biophysical Journal*, *92*, 2854–2864.
Perilla, J. R., Zhao, G., Lu, M., Ning, J., Hou, G., Byeon, I. L., et al. (2017). CryoEM structure refinement by integrating NMR chemical shifts with molecular dynamics simulations. *The Journal of Physical Chemistry B*, *121*, 3853–3863.
Perrin, B. S., Tian, Y., Fu, R., Grant, C. V., Chekmenev, E. Y., Wieczorek, W. E., et al. (2014). High-resolution structures and orientations of antimicrobial peptides piscidin 1 and piscidin 3 in fluid bilayers reveal tilting, kinking, and bilayer immersion. *Journal of the American Chemical Society*, *136*, 3491–3504.
Prates, M. V., Sforca, M. L., Regis, W. C., Leite, J. R., Silva, L. P., Pertinhez, T. A., et al. (2004). The NMR-derived solution structure of a new cationic antimicrobial peptide from the skin secretion of the anuran Hyla punctata. *The Journal of Biological Chemistry*, *279*, 13018–13026.
Pristovsek, P., & Kidric, J. (1999). Solution structure of polymyxins B and E and effect of binding to lipopolysaccharide: An NMR and molecular modeling study. *Journal of Medicinal Chemistry*, *42*, 4604–4613.
Qian, S., Wang, W., Yang, L., & Huang, H. W. (2008). Structure of the alamethicin pore reconstructed by X-ray diffraction analysis. *Biophysical Journal*, *94*, 3512–3522.
Ramamoorthy, A., Thennarasu, S., Lee, D. K., Tan, A., & Maloy, L. (2006). Solid-state NMR investigation of the membrane-disrupting mechanism of antimicrobial peptides MSI-78 and MSI-594 derived from magainin 2 and melittin. *Biophysical Journal*, *91*, 206–216.
Resende, J. M., Moraes, C. M., Prates, M. V., Cesar, A., Almeida, F. C., Mundim, N. C., et al. (2008). Solution NMR structures of the antimicrobial peptides phylloseptin-1, -2, and -3 and biological activity: The role of charges and hydrogen bonding interactions in stabilizing helix conformations. *Peptides*, *29*, 1633–1644.
Roccatano, D., Colombo, G., Fioroni, M., & Mark, A. E. (2002). Mechanism by which 2,2,2-trifluoroethanol/water mixtures stabilize secondary-structure formation in peptides: A molecular dynamics study. *Proceedings of the National Academy of Sciences of the United States of America*, *99*, 12179–12184.
Rosengren, K. J., Daly, N. L., Plan, M. R., Waine, C., & Craik, D. J. (2003). Twists, knots, and rings in proteins. Structural definition of the cyclotide framework. *The Journal of Biological Chemistry*, *278*, 8606–8616.
Roy, R. N., Lomakin, I. B., Gagnon, M. G., & Steitz, T. A. (2015). The mechanism of inhibition of protein synthesis by the proline-rich peptide oncocin. *Nature Structural & Molecular Biology*, *22*, 466–469.

Santana, M. J., de Oliveira, A. L., Queiroz Junior, L. H., Mandal, S. M., Matos, C. O., Dias Rde, O., et al. (2015). Structural insights into Cn-AMP1, a short disulfide-free multifunctional peptide from green coconut water. *FEBS Letters*, *589*, 639–644.

Saravanan, R., Joshi, M., Mohanram, H., Bhunia, A., Mangoni, M. L., & Bhattacharjya, S. (2013). NMR structure of temporin-1 ta in lipopolysaccharide micelles: Mechanistic insight into inactivation by outer membrane. *PLoS One*, *8*, e72718.

Schlick, T. (2010). *Molecular modeling and simulation: An interdisciplinary guide: An interdisciplinary guide* (2nd ed.). New York: Springer Science & Business Media.

Song, C., Weichbrodt, C., Salnikov, E. S., Dynowski, M., Forsberg, B. O., Bechinger, B., et al. (2013). Crystal structure and functional mechanism of a human antimicrobial membrane channel. *Proceedings of the National Academy of Sciences of the United States of America*, *110*, 4586–4591.

Staudegger, E., Prenner, E. J., Kriechbaum, M., Degovics, G., Lewis, R. N., McElhaney, R. N., et al. (2000). X-ray studies on the interaction of the antimicrobial peptide gramicidin S with microbial lipid extracts: Evidence for cubic phase formation. *Biochimica et Biophysica Acta*, *1468*, 213–230.

Strandberg, E., Horn, D., Reisser, S., Zerweck, J., Wadhwani, P., & Ulrich, A. S. (2016). 2H-NMR and MD simulations reveal membrane-bound conformation of magainin 2 and its synergy with PGLa. *Biophysical Journal*, *111*, 2149–2161.

Strandberg, E., Zerweck, J., Wadhwani, P., & Ulrich, A. S. (2013). Synergistic insertion of antimicrobial magainin-family peptides in membranes depends on the lipid spontaneous curvature. *Biophysical Journal*, *104*, L9–11.

Su, Y., Li, S., & Hong, M. (2013). Cationic membrane peptides: Atomic-level insight of structure-activity relationships from solid-state NMR. *Amino Acids*, *44*, 821–833.

Tardy-Laporte, C., Arnold, A. A., Genard, B., Gastineau, R., Morancais, M., Mouget, J. L., et al. (2013). A (2)H solid-state NMR study of the effect of antimicrobial agents on intact *Escherichia coli* without mutating. *Biochimica et Biophysica Acta*, *1828*, 614–622.

Tavares, L. S., Rettore, J. V., Freitas, R. M., Porto, W. F., do Nascimento Duque, A. P., de Lacorte Singulani, J., et al. (2012). Antimicrobial activity of recombinant Pg-AMP1, a glycine-rich peptide from guava seeds. *Peptides*, *37*, 294–300.

Tremouilhac, P., Strandberg, E., Wadhwani, P., & Ulrich, A. S. (2006). Synergistic transmembrane alignment of the antimicrobial heterodimer PGLa/magainin. *The Journal of Biological Chemistry*, *281*, 32089–32094.

Usachev, K. S., Efimov, S. V., Kolosova, O. A., Filippov, A. V., & Klochkov, V. V. (2015). High-resolution NMR structure of the antimicrobial peptide protegrin-2 in the presence of DPC micelles. *Journal of Biomolecular NMR*, *61*, 227–234.

Usachev, K. S., Efimov, S. V., Kolosova, O. A., Klochkova, E. A., Aganov, A. V., & Klochkov, V. V. (2015). Antimicrobial peptide protegrin-3 adopt an antiparallel dimer in the presence of DPC micelles: A high-resolution NMR study. *Journal of Biomolecular NMR*, *62*, 71–79.

Usachev, K. S., Kolosova, O. A., Klochkova, E. A., Yulmetov, A. R., Aganov, A. V., & Klochkov, V. V. (2017). Oligomerization of the antimicrobial peptide Protegrin-5 in a membrane-mimicking environment. Structural studies by high-resolution NMR spectroscopy. *European Biophysics Journal*, *46*, 293–300.

van der Weide, H., Brunetti, J., Pini, A., Bracci, L., Ambrosini, C., Lupetti, P., et al. (2017). Investigations into the killing activity of an antimicrobial peptide active against extensively antibiotic-resistant *K. pneumoniae* and *P. aeruginosa*. *Biochimica et Biophysica Acta*, *1859*, 1796–1804.

Vriend, G. (1990). WHAT IF: A molecular modeling and drug design program. *Journal of Molecular Graphics*, *8*, 52–56.

Wang, C. K., King, G. J., Northfield, S. E., Ojeda, P. G., & Craik, D. J. (2014). Racemic and quasi-racemic X-ray structures of cyclic disulfide-rich peptide drug scaffolds. *Angewandte Chemie International Edition*, *53*, 11236–11241.

Wegener, K. L., Wabnitz, P. A., Carver, J. A., Bowie, J. H., Chia, B. C., Wallace, J. C., et al. (1999). Host defence peptides from the skin glands of the Australian blue mountains tree-frog *Litoria citropa*. Solution structure of the antibacterial peptide citropin 1.1. *European Journal of Biochemistry*, *265*, 627–637.

Wessman, P., Morin, M., Reijmar, K., & Edwards, K. (2010). Effect of alpha-helical peptides on liposome structure: A comparative study of melittin and alamethicin. *Journal of Colloid and Interface Science*, *346*, 127–135.

Xie, J., Peters, B. M., Li, B., Li, L., Yu, G., Xu, Z., et al. (2017). Clinical features and antimicrobial resistance profiles of important Enterobacteriaceae pathogens in Guangzhou representative of Southern China, 2001–2015. *Microbial Pathogenesis*, *107*, 206–211.

Yang, L., Weiss, T. M., Lehrer, R. I., & Huang, H. W. (2000). Crystallization of antimicrobial pores in membranes: Magainin and protegrin. *Biophysical Journal*, *79*, 2002–2009.

Yu, H. Y., Chen, Y. A., Yip, B. S., Wang, S. Y., Wei, H. J., Chih, Y. H., et al. (2017). Role of beta-naphthylalanine end-tags in the enhancement of antiendotoxin activities: Solution structure of the antimicrobial peptide S1-Nal-Nal in complex with lipopolysaccharide. *Biochimica et Biophysica Acta*, *1859*, 1114–1123.

CHAPTER ELEVEN

Linear Analogues of the Lipopeptide Battacin With Potent In Vitro Activity Against *S. aureus*

Hugh D. Glossop, Esperanza Pearl, Gayan H. De Zoysa, Vijayalekshmi Sarojini[1]
School of Chemical Sciences, The University of Auckland, Auckland, New Zealand
[1]Corresponding author: e-mail address: v.sarojini@auckland.ac.nz

Contents

Abstract

Eight linear analogues of the lipopeptide battacin were evaluated for their antibacterial activity against the Gram-positive *Staphylococcus aureus*. Of this library, the enantiomeric lipopeptide analogue **9.4** exhibited nanomolar inhibitory activity (MIC = 200 nmol) against *S. aureus*. Furthermore, this lipopeptide was resilient toward degradation conditions when exposed to rat serum proteases for up to 8 h.

1. INTRODUCTION

Staphylococcus aureus is one of the most frequently isolated pathogens in the hospital environment and is implicated in potentially life-threatening infections such as hospital-acquired pneumonia, toxic shock syndrome, sepsis, and acute or right-sided endocarditis (Archer, 1998; Emori & Gaynes, 1993). Ever since the first reported outbreak of methicillin-resistant *S. aureus*

Advances in Protein Chemistry and Structural Biology, Volume 112
ISSN 1876-1623
https://doi.org/10.1016/bs.apcsb.2018.03.004

(MRSA) in the 1960s, this multidrug-resistant pathogen has remained the most commonly recognized antibiotic-resistant pathogen affecting many parts of the world (Barrett, McGehee, & Finland, 1968). MRSA is typically resistant to most β-lactam drugs, and has developed resistance against other classical antibiotic classes such as aminoglycosides and fluoroquinolones (Chambers & Deleo, 2009; Dalhoff, 2012; Fuda, Suvorov, Vakulenko, & Mobashery, 2004; Schmitz et al., 1999). Vancomycin-intermediate and vancomycin-resistant *S. aureus* strains have also been isolated (Fridkin, 2001). The multitude of virulence factors and the intrinsic evasion of immune system processes (Foster, 2005; Liu, 2009) combined with the emergence of pandrug-resistant strains make *S. aureus* an important target for novel non-classical antimicrobial agents such as lipopeptides.

Battacin is a natural cyclic lipopeptide produced by the soil bacterium *Paenibacillus tianmuensis* with broad-spectrum antibacterial activity against MDR Gram-negative bacteria (Qian et al., 2012). Battacin lipopeptides are structurally similar to the polymyxins but are considered to have a safer toxicity profile (De Zoysa, Cameron, Hegde, Raghothama, & Sarojini, 2015). Our previous in-depth structure–activity relationship (SAR) studies identified a linear analogue of battacin with improved antibacterial activity toward both Gram-positive and Gram-negative pathogens (De Zoysa et al., 2015). This linear lipopeptide amide, acylated by 4-methyl hexanoic acid at the N-terminus (herein known as compound **9**), exhibited potent (2.5–5 μM) broad-spectrum antibacterial activity against both Gram-positive and Gram-negative bacteria. Modifications to the chirality of key residues of compound **9** as well as the N-terminal acyl chain length were found to separately improve or reduce the antimicrobial activity by more than threefold in either direction. This chapter reports the subsequent SAR studies on compound **9** that yielded a library of lipopeptide antibiotics (Table 1) with potent activity against *S. aureus* that are discussed herein.

2. RESULTS AND DISCUSSION

2.1 Rationale Behind the Selection of Analogues

The parent compound **9** is amidated at the C-terminus. C-terminal amidation of peptides is a technique employed to increase the net cationic charge by one unit compared to the C-terminal acid version. Additionally, C-terminal amidation has been shown to improve the stability of peptides against exopeptidases (Adessi & Soto, 2002). However, no clear correlation exists between

Table 1 Amino Acid Sequences of the Peptide Analogues[a]

Peptide	Sequence
9	R¹-D-Dab-L-Dab-L-Dab-L-Leu-D-Phe-L-Dab-L-Dab-L-Leu-NH_2
9.1	R¹-D-Dab-L-Dab-L-Dab-L-Leu-D-Phe-L-Dab-L-Dab-L-Leu-**OH**
9.2	R¹-D-Dab-**D-Dab-D-Dab-D-Leu**-D-Phe-**D-Dab-D-Dab-D-Leu**-NH_2
9.3	R¹**-L-Dab**-L-Dab-L-Dab-L-Leu-**L-Phe**-L-Dab-L-Dab-L-Leu-NH_2
9.4	R¹**-L-Dab-D-Dab-D-Dab-D-Leu-L-Phe-D-Dab-D-Dab-D-Leu**-NH_2
9.5	**R²**-D-Dab-L-Dab-L-Dab-L-Leu-D-Phe-L-Dab-L-Dab-L-Leu-NH_2
9.6	**R³**-D-Dab-L-Dab-L-Dab-L-Leu-D-Phe-L-Dab-L-Dab-L-Leu-NH_2
9.7	**R⁴**-D-Dab-L-Dab-L-Dab-L-Leu-D-Phe-L-Dab-L-Dab-L-Leu-NH_2
9.8	**R⁵**-D-Dab-L-Dab-L-Dab-L-Leu-D-Phe-L-Dab-L-Dab-L-Leu-NH_2
9.9	**R⁶**-D-Dab-L-Dab-L-Dab-L-Leu-D-Phe-L-Dab-L-Dab-L-Leu-NH_2

[a]R^1, 4-methyl-hexanoyl; R^2, hexanoyl; R^3, pentanoyl; R^4, butanoyl; R^5, propanoyl; R^6, acetyl. Changes to the original sequence **9** are highlighted in bold.

the C-terminal modification and antibacterial activity (Gallis et al., 1988; Mor & Nicolas, 1994; Nakajima, Qu, & Natori, 1987). Lipopeptide analogue **9.1** was synthesized as the C-terminal acid version of **9** to probe the importance of C-terminal modification on antibacterial activity. Analogues **9.2** and **9.3** were designed with solely D-amino acids (**9.2**) and vice versa with all L-amino acids (**9.3**) to probe the importance of D-amino acid substitution on antibacterial activity and stability because the parent compound featured two D-amino acids in the same positions as the natural battacin lipopeptide. An increase in the proportion of D-amino acids was expected to increase the protease resistance of the lipopeptide analogues, as peptide bonds made up of D-amino acids are seldom recognized by serum proteases (Adessi & Soto, 2002). Even though this method is commonly used in the literature to increase serum stability of peptides, some groups have reported that D-amino acid substitutions may not lead to significant changes in the observed antibacterial activity (Lee et al., 2004; Oren & Shai, 1997; Shai & Oren, 1996; Wade et al., 1990). Analogue **9.4** is the enantiomeric version of the parent peptide, essentially a mirror of the analogue **9** chirality, and was synthesized to identify if the enantiomer has greater activity. Even though our previous study showed that deletion of the fatty acid component led to complete loss of antibacterial activity, the influence of fatty acid chain length, particularly fatty acids shorter than six carbon atoms long, has not been thoroughly investigated (De Zoysa et al., 2015). Therefore, analogues **9.5–9.9** were generated with systematically shortened acyl chain lengths each, from C6 (hexanoyl) to C2 (acetyl). This brief library was included in this study to identify the optimum fatty acid chain length required to maintain a high antibacterial activity in these lipopeptides.

2.2 In Vitro Antibacterial Activity

The antibacterial activity of the peptides was investigated through broth dilution method: minimal inhibitory concentration (MIC) assays against *S. aureus*. As evident from Table 2, majority of the modified analogues showed more potent antibacterial activity than the parent peptide **9**. The all D- and all L-lipopeptides, **9.2** and **9.3**, were active at 780 nmol. Shared activity between all D- and L-analogues is also consistent with literature reports (Wade et al., 1990). The C-terminal acid peptide **9.1** showed a slight reduction in activity. The diminished activity was likely due to an overall reduced net cationic charge of **9.1** which could be implicated in weaker electrostatic interactions with negatively charged phospholipids in the *S. aureus* membrane. The enantiomeric peptide **9.4** was the most potent analogue (0.2 μM) identified from the screening assay. This peptide is 12-fold more active than the parent peptide **9**. Peptide **9** exhibits its antibacterial action via membrane lysis as previously reported by us (De Zoysa et al., 2015). Given its structural similarity to the

Table 2 Antibacterial Activity of the Peptides

Peptide	*S. aureus*
9	2.5
9.1	3.25
9.2	0.78
9.3	0.78
9.4	0.2
9.5	1.56
9.6	1.56
9.7	6.25
9.8	12.5
9.9	12.5

Each compound was tested as triplicates and the assay was repeated on 3 independent days.

parent peptide, **9.4** is highly likely to have retained the membrane lytic mechanism of action of **9**. The hexanoyl (**9.5**) and pentanoyl (**9.6**) fatty acid analogues showed about 1.6 times improved activity against *S. aureus*, while the remaining fatty acid-conjugated analogues are 2–4 times less active than the parent. These results indicate that a minimum chain length of five to seven carbons can conserve low micromolar activity in this peptide family against *S. aureus*.

2.3 Serum Stability Assay of Battacin Peptides

In vitro stability studies in serum or plasma are considered to be indicative of stability under physiological conditions and provide useful information on the potential bioavailability of the drug under development. The stability of analogues **9–9.4** was analyzed toward enzymatic degradation in rat serum. A serum-only sample served as the control for stability studies using analytical reversed-phase high-performance liquid chromatography (RP-HPLC). The serum-only sample showed several peaks between 11 and 18 min (Fig. 1).

In order to obtain a quantitative comparison of the degradation of the analogues upon exposure to serum, the decrease in peak intensity of the analogues upon exposure to serum through various time intervals was plotted over time (Fig. 2). This was achieved by calculating the peak area at each serum exposure time point and the percentage decrease in peak area worked out assuming the peak area at $t=0$ to be 100%. The serum peak at 18 min retention time showed no net change over time and was used as an internal standard to normalize the peak area observed for the target analogue.

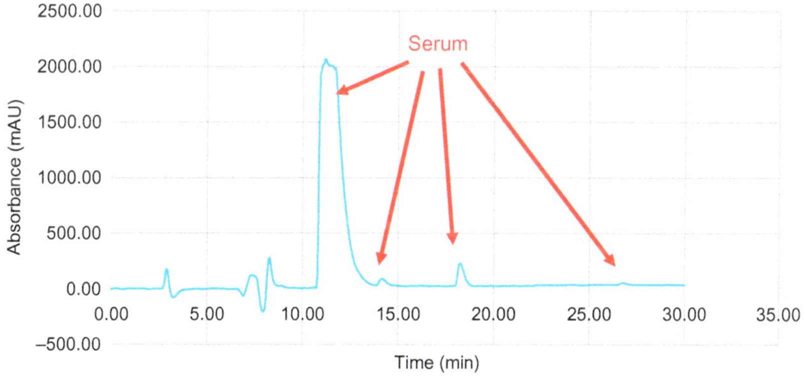

Fig. 1 RP-HPLC chromatogram of 25% rat serum in PBS ("serum-only" sample). Proteins have been precipitated by addition of 15% TCA and centrifugation.

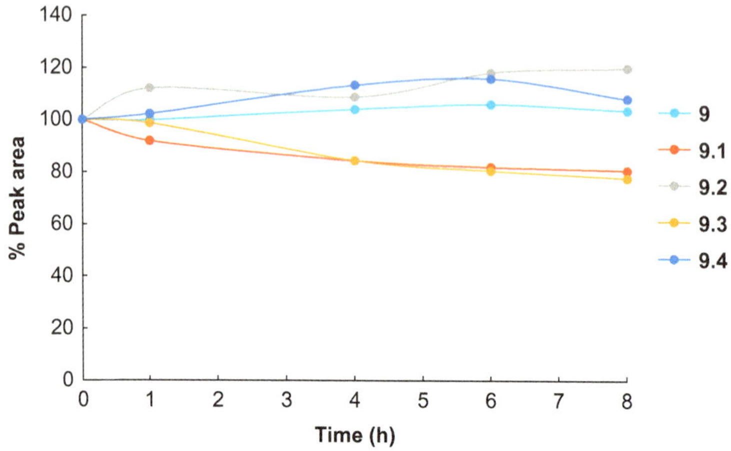

Fig. 2 Serum stability profiles of analogues **9–9.4**. Percentage peak area was determined from the integration of the target analogue peaks of the RP-HPLC chromatograms and normalized against the serum peak at 18 min as an internal standard. The target analogue peak area of the $t = 0$ h aliquot was taken as 100%.

Previous literature studies have shown that 1 h exposure to serum is sufficient to detect peptide degradation, with a highly unstable control peptide showing almost full degradation and comparable levels of the degradation products (Nguyen et al., 2010).

The all L-analogue, **9.3**, was the least stable in serum. The peak area of **9.3** chromatogram had decreased by over 40% after the 8-h exposure period. This rapid degradation indicates that the substitution of the two D-amino acids (D-Dab and D-Phe) to their L counterparts causes the peptides to be sensitive to proteases.

The C-terminal acid analogue, **9.1**, also showed significant degradation after 8 h in serum, unlike the amidated parent peptide **9** (Fig. 2). This agrees with previous findings that C-terminal amidation confers greater resistance to enzymatic degradation (Stromstedt, Pasupuleti, Schmidtchen, & Malmsten, 2009). Unsurprisingly, the parent compound **9** did not show significant degradation after 8 h in serum. The enantiomeric analogue **9.4** showed a similar stability profile to **9**. This high level of stability of compound **9** and its enantiomer **9.4** can be attributed to the presence of D-amino acids as well as C-terminal amidation. The "all-D" analogue **9.2** showed the greatest stability with no degradation observed after 8 h in serum. This is consistent with the enhanced protease resistance of D-amino acid-containing peptides reported in the literature (Adessi & Soto, 2002).

3. CONCLUSIONS

Inhibitory assays of a library of battacin analogues against *S. aureus* revealed a number of structural modifications that improved the anti-staphylococcal activity of these compounds. In particular, a five-carbon-atom-long N-terminal fatty acyl chain was found to be the minimum length required to retain low micromolar activity against *S. aureus*. Additionally, the C-terminal acid lipopeptide **9.1** showed a slightly higher MIC than the parent compound **9**. The reduced antibacterial activity of the C-terminal acid is likely a result of the reduced overall charge due to the ionizable carboxylic acid group which could result in weaker electrostatic interactions with the bacterial membrane or increased protease susceptibility. Most interestingly, this SAR study identified the enantiomeric lipopeptide **9.4** which exhibited potent activity with a MIC of 200 nmol, over 12-fold higher than its enantiomer, analogue **9**. This is further evidence toward the reliance upon L- and D-amino acid substitution and even L/D patterning for activity in antimicrobial lipopeptides since **9.4** was more active than the full D-amino acid analogue **9.2**. These battacin analogues were not only highly active but also had excellent stability in rat serum for up to 8 h which may suggest stability in vivo too. Unsurprisingly, both all L peptide **9.3** and C-terminal acid **9.1** showed the most degradation at the 8-h mark. Both the parent peptide **9** and the all D-amino acid peptide **9.2** showed no peptide degradation up to 8 h. Overall, the results presented in this chapter have revealed one candidate lipopeptide, **9.4**, to have excellent activity compared to the original compound **9** and could be further developed as a promising antimicrobial treatment against *S. aureus* infections.

4. EXPERIMENTAL SECTION

4.1 Chemicals, Reagents, and Bacterial Cultures

All reagents were used as obtained from the supplier. Amino acids and coupling reagents were purchased from either Alabiochem (SuZhou, China) or GL Biochem (Shanghai, China). Tentagel resin was purchased from Peptides International (Louisville, USA). Piperidine, DIPEA, and the fatty acids were purchased from Sigma Aldrich (New Zealand).

S. aureus was obtained from the microbial culture collection of the School of Biological Sciences at the University of Auckland. The strain was stored in 50% glycerol at −20°C and −80°C. For routine use, bacteria were plated on Mueller–Hinton broth agar plates. Single colonies were replated every 2 weeks. All media for bacterial growth and glassware were autoclaved at 120°C for 1 h using a Tomy SX 500E high-pressure steam sterilizer.

4.2 Peptide Synthesis

4.2.1 C-Terminal Amide *(De Zoysa et al., 2015)*

The C-terminal lipopeptide analogues investigated in this chapter have been synthesized following standard solid-phase peptide synthesis protocols using Fmoc chemistry. Experimental details of this synthesis are reported in our previous publication on this family of peptides (De Zoysa et al., 2015). The crude peptides were purified to homogeneity using RP-HPLC on a GE Pharmacia ÄKTA purifier 10 system on a Phenomenex Luna 5 μm C_{18} 100 Å (250 × 10 mm) column using 0.1% TFA in water as solvent A and 0.1% TFA and 0.9% water in 99% acetonitrile as solvent B at a flow rate of 4 mL/min. Homogeneity (>95% purity) of the purified peptides was determined by analytical RP-HPLC. Analytical reversed-phase HPLC was performed using a Phenomenex Luna 5 μm C_{18} 100 Å (250 × 4.6 mm) column using the same solvent system as above at a flow rate of 1 mL/min. The identity of the peptides was established using electrospray ionization mass spectrometry (ESIMS) recorded on a Bruker micrOTOF-Q mass spectrometer.

4.2.2 C-Terminal Acid *(Barlos, Chatzi, Gatos, & Stavropoulos, 1991)*

The peptide **9.1** was synthesized using the following protocol. 2-Chlorotrityl chloride resin (1.58 meq/g, 0.1 mmol, 63.29 mg) was swollen in DCM (10 min) and drained. Fmoc-Leu-OH (4 equiv., 0.4 mmol, 141.7 mg) in DCM (2 mL) and DIPEA (5 equiv., 175 μL) was added to the resin and agitated on a shaker (10 min). A further 175 μL of DIPEA was added and the mixture agitated on a shaker (50 min). The reaction mixture was washed with DCM (3 × 3 mL) and a solution of DCM, MeOH, and DIPEA (8:1.5:0.5, ~3 mL) added. This was agitated on a shaker (10 min) and washed sequentially with DCM and DMF (3 × 3 mL of each). Subsequent cleavage of the Fmoc-protecting group was carried out immediately which was followed by general protocol for elongation of the peptide chain (C-terminal amide).

4.2.3 Minimal Inhibitory Concentration Assay *(Wiegand, Hilpert, & Hancock, 2008)*

Briefly, bacteria were grown in Mueller–Hinton broth (1.25%) prepared in sterilized Milli-Q water which was autoclaved at 120°C for 1 h. A single colony of the bacterial strain was transferred to 20 mL of the MHB and grown at 37°C overnight. The optical density (OD) of the overnight culture was measured and adjusted to a McFarland standard 0.5 (1×10^8 cfu/mL) followed by 1 in 100 dilution to obtain 1×10^6 CFU/mL. Stock solution of the peptides and the antibiotic controls were prepared in MHB, and diluted as necessary. For the assay 50 μL each of the peptide solutions at various concentrations and the diluted bacterial culture were incubated together in 96-well microtiter plates, with three replicates of each peptide concentration tested against each bacterial strain. MIC was determined as the minimum concentration where no growth was observed using optical density (OD_{600}) measurements which were conducted using an EnSpire Multimode plate reader at the Faculty of Medical and Health Sciences, University of Auckland. The experiment was then repeated on 3 independent days.

4.2.4 Serum Stability Assay *(Nguyen et al., 2010)*

Fresh rat serum was obtained from the Vernon Jansen Unit (VJU). The lipopeptides (**9** and **9.6–9.9**) solutions used at 1 mg/mL in 1 mL of MQ H_2O. Rat serum (225 μL) was mixed with 675 μL of PBS (0.01 mM, pH 7.4) in a 1.5-mL Eppendorf tube to give 900 μL of a 25% serum solution to which 100 μL of 1 mg/mL peptide solution was added and the serum–peptide incubation mixture incubated. 100 μL aliquots of the incubation mixture were taken at set time intervals, mixed with 20 μL of 15% TCA, and centrifuged (4500 rpm, 7 min, 4°C) to precipitate out the serum proteins. Time interval chosen were 0, 1, 2, 4, 6, 8, and 24 h. The 0-h aliquot was taken directly after the peptide had been added, and before the incubation mixture had been returned to the incubator.

REFERENCES

Adessi, C., & Soto, C. (2002). Converting a peptide into a drug: Strategies to improve stability and bioavailability. *Current Medicinal Chemistry*, *9*(9), 963–978.

Archer, G. L. (1998). Staphylococcus aureus: A well-armed pathogen. *Clinical Infectious Diseases*, *26*(5), 1179–1181.

Barlos, K., Chatzi, O., Gatos, D., & Stavropoulos, G. (1991). 2-Chlorotrityl chloride resin. *International Journal of Peptide and Protein Research*, *37*, 513–520.

Barrett, F. F., McGehee, R. F., Jr., & Finland, M. (1968). Methicillin-resistant Staphylococcus aureus at Boston City Hospital: Bacteriologic and epidemiologic observations. *New England Journal of Medicine*, *279*(9), 441–448.

Chambers, H. F., & Deleo, F. R. (2009). Waves of resistance: Staphylococcus aureus in the antibiotic era. *Nature Reviews. Microbiology*, 7(9), 629–641. https://doi.org/10.1038/nrmicro2200.

Dalhoff, A. (2012). Global fluoroquinolone resistance epidemiology and implications for clinical use. *Interdisciplinary Perspectives on Infectious Diseases*, *2012*, 37. https://doi.org/10.1155/2012/976273.

De Zoysa, G. H., Cameron, A. J., Hegde, V. V., Raghothama, S., & Sarojini, V. (2015). Antimicrobial peptides with potential for biofilm eradication: Synthesis and structure activity relationship studies of battacin peptides. *Journal of Medicinal Chemistry*, *58*(2), 625–639.

Emori, T. G., & Gaynes, R. P. (1993). An overview of nosocomial infections, including the role of the microbiology laboratory. *Clinical Microbiology Reviews*, *6*(4), 428–442.

Foster, T. J. (2005). Immune evasion by Staphylococci. *Nature Reviews. Microbiology*, *3*(12), 948–958. https://doi.org/10.1038/nrmicro1289.

Fridkin, S. K. (2001). Vancomycin-intermediate and -resistant Staphylococcus aureus: What the infectious disease specialist needs to know. *Clinical Infectious Diseases*, *32*(1), 108–115.

Fuda, C., Suvorov, M., Vakulenko, S. B., & Mobashery, S. (2004). The basis for resistance to beta-lactam antibiotics by penicillin-binding protein 2a of methicillin-resistant Staphylococcus aureus. *The Journal of Biological Chemistry*, *279*(39), 40802–40806. https://doi.org/10.1074/jbc.M403589200.

Gallis, B., Mehl, J., Prickett, K., Martin, J., Merriam, J., March, C., et al. (1988). Antimicrobial activity of synthetic bactenecin. *Biotechnology Therapeutics*, *1*(4), 335–346.

Lee, D., Powers, J., Pflegerl, K., Vasil, M., Hancock, R., & Hodges, R. (2004). Effects of single D-amino acid substitutions on disruption of β-sheet structure and hydrophobicity in cyclic 14-residue antimicrobial peptide analogs related to gramicidin S. *The Journal of Peptide Research*, *63*(2), 69–84.

Liu, G. Y. (2009). Molecular pathogenesis of Staphylococcus aureus infection. *Pediatric Research*, *65*(5 Pt. 2), 71R–77R. https://doi.org/10.1203/PDR.0b013e31819dc44d.

Mor, A., & Nicolas, P. (1994). The NH2-terminal alpha-helical domain 1-18 of dermaseptin is responsible for antimicrobial activity. *Journal of Biological Chemistry*, *269*(3), 1934–1939.

Nakajima, Y., Qu, X., & Natori, S. (1987). Interaction between liposomes and sarcotoxin IA, a potent antibacterial protein of Sarcophaga peregrina (flesh fly). *Journal of Biological Chemistry*, *262*(4), 1665–1669.

Nguyen, L. T., Chau, J. K., Perry, N. A., De Boer, L., Zaat, S. A., & Vogel, H. J. (2010). Serum stabilities of short tryptophan- and arginine-rich antimicrobial peptide analogs. *PLoS One*, *5*(9), e12684.

Oren, Z., & Shai, Y. (1997). Selective lysis of bacteria but not mammalian cells by diastereomers of melittin: Structure–function study. *Biochemistry*, *36*(7), 1826–1835.

Qian, C.-D., Wu, X.-C., Teng, Y., Zhao, W.-P., Li, O., Fang, S.-G., et al. (2012). Battacin (octapeptin B5), a new cyclic lipopeptide antibiotic from Paenibacillus tianmuensis active against multidrug-resistant Gram-negative bacteria. *Antimicrobial Agents and Chemotherapy*, *56*(3), 1458–1465.

Schmitz, F. J., Fluit, A. C., Gondolf, M., Beyrau, R., Lindenlauf, E., Verhoef, J., et al. (1999). The prevalence of aminoglycoside resistance and corresponding resistance genes in clinical isolates of Staphylococci from 19 European hospitals. *The Journal of Antimicrobial Chemotherapy*, *43*(2), 253–259.

Shai, Y., & Oren, Z. (1996). Diastereomers of cytolysins, a novel class of potent antibacterial peptides. *Journal of Biological Chemistry*, *271*(13), 7305–7308.

Stromstedt, A. A., Pasupuleti, M., Schmidtchen, A., & Malmsten, M. (2009). Evaluation of strategies for improving proteolytic resistance of antimicrobial peptides by using variants of EFK17, an internal segment of LL-37. *Antimicrobial Agents and Chemotherapy*, *53*(2), 593–602. https://doi.org/10.1128/AAC.00477-08.

Wade, D., Boman, A., Wåhlin, B., Drain, C. M., Andreu, D., Boman, H. G., et al. (1990). All-D amino acid-containing channel-forming antibiotic peptides. *Proceedings of the National Academy of Sciences of the United States of America*, *87*(12), 4761–4765. https://doi.org/10.1073/pnas.87.12.4761.

Wiegand, I., Hilpert, K., & Hancock, R. E. (2008). Agar and broth dilution methods to determine the minimal inhibitory concentration (MIC) of antimicrobial substances. *Nature Protocols*, *3*(2), 163–175.

FSC
www.fsc.org
MIX
Papier | Fördert
gute Waldnutzung
FSC® C083411

Kolibri 360 GmbH
Ferdinand-Jühlke-Straße 7
99095 Erfurt, Deutschland
produktsicherheit@kolibri360.de